Forensic Microbiology

American Academy of Forensic Sciences

Published and forthcoming titles in the Forensic Science in Focus series

Published

The Global Practice of Forensic Science
Douglas H. Ubelaker (Editor)

Forensic Chemistry: Fundamentals and Applications
Jay A. Siegel

Forensic Microbiology
David O. Carter, Jeffery K. Tomberlin, M. Eric Benbow and Jessica L. Metcalf

Forthcoming

The Future of Forensic Science
Daniel A. Martell

Humanitarian Forensics and Human Identification
Paul Emanovsky and Shuala M. Drawdy

Forensic Anthropology: Theoretical Framework and Scientific Basis
Clifford Boyd and Donna Boyd

Forensic Microbiology

EDITED BY

David O. Carter
Chaminade University of Honolulu, USA

Jeffery K. Tomberlin
Texas A&M University, USA

M. Eric Benbow
Michigan State University, USA

Jessica L. Metcalf
Colorado State University, USA

WILEY

Registered Office
John Wiley & Sons Ltd, The Atrium, Southern Gate, Chichester, West Sussex, PO19 8SQ, UK

Editorial Offices
111 River Street, Hoboken, NJ 07030, USA
9600 Garsington Road, Oxford, OX4 2DQ, UK
The Atrium, Southern Gate, Chichester, West Sussex, PO19 8SQ, UK

For details of our global editorial offices, customer services, and more information about Wiley products visit us at www.wiley.com.

Wiley also publishes its books in a variety of electronic formats and by print-on-demand. Some content that appears in standard print versions of this book may not be available in other formats.

Library of Congress Cataloging-in-Publication data applied for

ISBN: 9781119062554

Cover Design: Wiley
Cover Image: Courtesy of David O. Carter

Set in 10.5/13.5pt Meridien by SPi Global, Pondicherry, India

10 9 8 7 6 5 4 3 2 1

This work is dedicated to all donors (human and nonhuman) that allowed for much of this research to be completed. You are not forgotten and your contribution to science is timeless.

David O. Carter would like to thank his wife, Charlotte, for her never-ending support and encouragement.

Jeffery K. Tomberlin would like to thank his family (Laura, Celeste, and Jonah) for putting up with his late nights with his computer, constant texting about work, and simply allowing him to enjoy his pursuit of knowledge. He would like to extend a special thanks to Tawni L. Crippen for being patient with him as he explored the world of microbiology under her tutelage and his constant leaning on her sterile bench tops. She has truly opened his eyes to the wonderful world of bacteria and their importance in nature. You are a great friend and colleague.

M. Eric Benbow would like to thank Melissa, Arielle, and Alia for their tremendous support, encouragement, and understanding while he travels to continue his passion that leads to efforts like this project. He would also like to thank Jen Pechal for her extra attention to reviewing several of the chapters of this book and for "maintaining the ship" of writing and productivity while he explores new avenues of science like that related to many topics of the book.

Jessica L. Metcalf would like to thank her forensic collaborators for being excellent and patient colleagues, coauthors, and coeditors. She would like to thank Rob Knight for his support of pursuing microbial tools for forensic science. She would also like to thank her family for their support and contributions to her scientific endeavors.

Contents

About the editors

Dr. David O. Carter is Director and Associate Professor of Forensic Sciences at Chaminade University of Honolulu. He also serves as Principal Investigator of the Laboratory of Forensic Taphonomy. His primary research interest is the decomposition of human remains, particularly in tropical environments. Current research projects focus on the structure and function of antemortem and postmortem microbial communities: using microbiomes as spatial and temporal evidence. He is interested in understanding the relationships between decomposing remains, microbial communities, and the environment. Dr. Carter's ultimate goal is to get quality science and technology in the hands of first responders and investigators.

Dr. Carter is an active member of the forensic science community with a significant interest in undergraduate education. He is a Fellow in the Pathology/Biology Section of the American Academy of Forensic Sciences and has recently served as Program Chair for the Pathology/Biology Section. Dr. Carter also serves on the Medicolegal Death Investigation Subcommittee in the Organization of Scientific Area Committees, a joint endeavor between the US Department of Justice and US Department of Commerce. He incorporates this experience into undergraduate education where he plays an active role in curriculum development, assessment, academic advising, and recruiting.

Dr. Jeffery K. Tomberlin is an Associate Professor and Co-Director of the Forensic & Investigative Sciences Program and Principle Investigator of the Forensic Laboratory for Investigative Entomological Sciences (FLIES) Facility (forensicentomology.tamu.edu) in the Department of Entomology at Texas A&M University. Research in the FLIES Facility examines species interactions on ephemeral resources such as vertebrate carrion, decomposing plant material and animal wastes in order to better understand the mechanisms regulating arthropod behavior as related to arrival, colonization and succession patterns. The goals of his program are to refine current methods used by entomologists in forensic investigations. His research also is focused on waste management in confined animal facilities and the production of alternate protein sources for use as livestock, poultry, and aquaculture feed. Since arriving on campus at Texas A&M University in 2007, eight Ph.D. and fourteen M.S. students have completed their degrees under his supervision.

Dr. Tomberlin has been very active within the forensic science community. He, along with a colleague, initiated the first forensic entomology conference in North America as well as the formation of the North America Forensic Entomology Association of which he served as its first president. He is also a Fellow in the American Academy of Forensic Sciences and has served as the Chair of the Pathology/Biology Section. Dr. Tomberlin is also one of 18 entomologists board certified by the American Board of Forensic Entomology (ABFE). He has served a number of roles with the ABFE including Secretary and Chair.

Dr. M. Eric Benbow is an Associate Professor in the Department of Entomology and the Department of Osteopathic Medical Specialties at Michigan State University. The research in his lab focuses on microbial-invertebrate community interactions in aquatic ecosystems, disease systems and carrion decomposition. These research foci use basic science to inform applications in areas such as human health, natural resources management and forensics. Dr. Benbow has authored or co-authored a collection of over 120 peer-reviewed papers, book chapters, and proceedings, many of which relate to carrion decomposition ecology, forensic entomology and forensic microbiology. He has served on two National Research Council committees related to aquatic ecology, and is regularly invited as a speaker at international and national academic meetings related to decomposition ecology and forensics. Dr. Benbow has led workshops at the international level discussing experimental design, statistical analyses and the importance of novel basic ecological concepts in advancing the field of carrion ecology and applications in forensics. Dr. Benbow was part of the inaugural executive committee for the North American Forensic Entomology Association (NAFEA) where he served as the Editor-in- Chief of the annual NAFEA Newsletter and NAFEA Webmaster (www.nafea.net) for eight years. He was the president of NAFEA from 2012–2013 and has served as an expert witness and worked on several cases that involved insects as evidence during investigations or water resource litigation. He continues a recognized and collaborative research program in microbe-insect interactions that supports undergraduate and graduate students and postdoctoral associates. Dr. Benbow continues to mentor and co-mentor students and postdoctoral associates through research and teaching. He sees the future of ecology, evolution and applications in forensics to be fundamentally in the hands of students and early career scientists worldwide.

Dr. Jessica L. Metcalf is an Assistant Professor in the Department of Animal Sciences at Colorado State University. Her research uses high-throughput sequencing and bioinformatic tools in an ecological and evolutionary framework to understand changes of the human microbiome during life and after death. Her microbiome research projects span the fields of forensic science, medicine, agriculture, and vertebrate ecology. A main focus of her research is developing a microbial stopwatch for estimating the postmortem interval. Dr. Metcalf strives to provide foundational microbiome science that can be developed into tools for the justice system.

Dr. Metcalf has a background in evolutionary biology, ancient DNA, experimental microbial ecology, and microbiome science. Her training includes expertise in generating genomic data from a broad range of sample types, including fossil remains and other low-biomass samples, and analyzing large genomic data sets. Dr. Metcalf is currently building her microbiome science program as part of the microbiome cluster at Colorado State University.

List of contributors

Kate M. Barnes
Department of Natural Sciences
University of Derby
Derby, UK

M. Eric Benbow
Department of Entomology and
Department of Osteopathic Medical Specialties
Michigan State University
East Lansing, MI, USA

Sibyl R. Bucheli
Department of Biological Sciences
Sam Houston State University
Huntsville, TX, USA

Michael Bunce
Trace and Environmental DNA (TrEnD)
Laboratory, Department of Environment
and Agriculture
Curtin University
Perth, Western Australia, Australia

Zachary M. Burcham
Department of Biological Sciences
Mississippi State University
Starkville, MS, USA

Danielle M. Butzbach
Toxicology Section
Forensic Science SA
Adelaide, South Australia, Australia

David O. Carter
Laboratory of Forensic Taphonomy, Forensic
Sciences Unit, Division of Natural Sciences
and Mathematics
Chaminade University of Honolulu
Honolulu, HI, USA

Jared W. Castle
School of Chemical and Physical Sciences
Flinders University
Adelaide, South Australia, Australia

Tawni L. Crippen
Southern Plains Agricultural Research Center
Agricultural Research Service
United States Department of Agriculture
College Station, TX, USA

Franklin E. Damann
Central Identification Laboratory—Offutt
Air Force Base
Defense POW/MIA Accounting Agency
Bellevue, NE, USA

David R. Foran
Forensic Science Program,
School of Criminal Justice
and
Department of Integrative Biology
Michigan State University
East Lansing, MI, USA

Jack A. Gilbert
Department of Ecology and Evolution and
Department of Surgery
University of Chicago
Chicago, IL
and
Institute for Genomics and Systems
Biology
Argonne National Laboratory
Lemont, IL
and
Marine Biological Laboratory
Woods Hole, MA, USA

Valery Hedouin
Lille University School of Medicine, CHU Lille
Forensic Taphonomy Unit, UTML, Lille, France

James M. Hopkins
Forensic Science Program,
School of Criminal Justice
Michigan State University
East Lansing, MI, USA
Current address:
St. Jude Children's Research Hospital
Memphis, TN, USA

Embriette R. Hyde
Knight Laboratory, Department of Pediatrics
University of California, San Diego
San Diego, CA, USA

Miranda M.E. Jans
Central Identification Laboratory—Hickam
Air Force Base
Defense POW/MIA Accounting Agency
Honolulu, HI, USA

Ellen M. Jesmok
Forensic Science Program,
School of Criminal Justice
Michigan State University
East Lansing, MI, USA
Current address:
Minnesota Bureau of Criminal Apprehension
Laboratory
St. Paul, MN, USA

Heather R. Jordan
Department of Biological Sciences
Mississippi State University
Starkville, MS, USA

Emily N. Junkins
Laboratory of Forensic Taphonomy, Forensic
Sciences Unit, Division of Natural Sciences
and Mathematics
Chaminade University of Honolulu
Honolulu, HI, USA
and
Department of Microbiology
and Plant Biology, University of Oklahoma
Norman, OK, USA

K. Paul Kirkbride
School of Chemical and Physical Sciences
Flinders University
Adelaide, South Australia, Australia

Rob Knight
Department of Pediatrics
and
Department of Computer Science and
Engineering
University of California, San Diego
San Diego, CA, USA

Whitney A. Kodama
Laboratory of Forensic Taphonomy, Forensic
Sciences Unit, Division of Natural Sciences
and Mathematics
Chaminade University of Honolulu
Honolulu, HI, USA

Simon Lax
Department of Ecology and Evolution
University of Chicago, Chicago, IL
and
Institute for Genomics and Systems Biology
Argonne National Laboratory
Lemont, IL, USA

Claire E. Lenehan
School of Chemical and Physical Sciences
Flinders University
Adelaide, South Australia, Australia

Aaron M. Lynne
Department of Biological Sciences
Sam Houston State University
Huntsville, TX, USA

Vadim Mesli
Lille University School of Medicine, CHU Lille
Forensic Taphonomy Unit, UTML
Lille, France

Jessica L. Metcalf
Department of Animal Sciences
Colorado State University
Fort Collins, CO, USA

Dáithí C. Murray
Trace and Environmental DNA (TrEnD)
Laboratory, Department of Environment
and Agriculture
Curtin University
Perth, Western Australia, Australia

Christel Neut
Clinical Bacteriology Laboratory
INSERM U995 LIRIC, University of Lille
Lille, France

Jennifer L. Pechal
Department of Entomology
Michigan State University
East Lansing, MI, USA

Mohammad Raza
Department of Medical Microbiology and
Virology
Sheffield Teaching Hospitals NHS Foundation
Trust
Sheffield, UK

Frank Reith
School of Biological Sciences
Sprigg Geobiology Centre
University of Adelaide
and
CSIRO Land and Water
Environmental Contaminant Mitigation
and Technologies, PMB2
Adelaide, South Australia, Australia

Elisabeth J. Ridgway
Department of Medical Microbiology and
Virology
Sheffield Teaching Hospitals NHS
Foundation Trust
Sheffield, UK

Baneshwar Singh
Department of Forensic Science
Virginia Commonwealth University
Richmond, VA, USA

Bala M. Subramanian
Department of Medical Microbiology and
Virology
Sheffield Teaching Hospitals NHS
Foundation Trust
Sheffield, UK

Jeffery K. Tomberlin
Department of Entomology
Texas A&M University
College Station, TX, USA

Silvana R. Tridico
Forensic Science & Wildlife Matters
Adelaide, South Australia, Australia

G. Stewart Walker
School of Chemical and Physical Sciences
Flinders University
Adelaide, South Australia, Australia

Foreword

In recent years, technology improvements have enabled forensic scientists to analyze smaller and smaller samples. However, with new and expanded capabilities often comes the challenge of providing appropriate meaning to that evidence. Use of information from microbial communities in forensic investigations may open new avenues for answering important questions in crime-solving efforts. This book provides a look into current applications, and future possibilities, with microbial analysis aiding forensic investigations.

The Human Microbiome effort over the past decade or so has built on information gained from the earlier Human Genome Project. Likewise, forensic microbiology will in many cases build on previous work with human forensic DNA testing. Forensic DNA analysis with human DNA began in 1984 with the pioneering work of Alec Jeffreys at Leicester University in the United Kingdom and a *Nature* publication the following year (Gill et al., 1985). Forensic microbiology is poised to have a similar trajectory, with several research groups having their work disseminated in several high impact publications. It remains to be seen whether recent advances in forensic microbiology will impact forensic science as human DNA testing, but it nevertheless has the potential to be equally revolutionary in some cases (Fierer et al., 2010). Bioterrorism and forensic pathology are two important areas where microbial analysis plays an important role, but these applications are being expanded as we understand the temporal and spatial value of microorganisms.

Perhaps in a future day the microbial environment around crime scenes and persons of interest will be rapidly assessed to answer important investigative questions. Massively parallel (next-generation) sequencing can provide microbial and human DNA sequences on an unprecedented scale.

This book is part of a larger effort by the American Academy of Forensic Sciences and Wiley to bring forensic science issues and capabilities into focus. It is hoped that the completed product is satisfying to both the authors and their future readers.

John M. Butler, Ph.D.
NIST Fellow and Special Assistant to the Director for Forensic Science
National Institute of Standards and Technology
Gaithersburg, MD, USA

References

Fierer, N., Lauber, C.L., Zhou, N. *et al.* (2010) Forensic identification using skin bacterial communities. *Proceedings of the National Academy of Sciences of the United States of America*, **107**, 6477–6481.

Gill, P., Jeffreys, A.J. and Werrett, D.J. (1985) Forensic application of DNA "fingerprints". *Nature*, **318**, 577–579.

Series preface

The forensic sciences represent diverse, dynamic fields that seek to utilize the very best techniques available to address legal issues. Fueled by advances in technology, research and methodology, as well as new case applications, the forensic sciences continue to evolve. Forensic scientists strive to improve their analyses and interpretations of evidence and to remain cognizant of the latest advancements. This series results from a collaborative effort between the American Academy of Forensic Sciences (AAFS) and Wiley to publish a select number of books that relate closely to the activities and objectives of the AAFS. The book series reflects the goals of the AAFS to encourage quality scholarship and publication in the forensic sciences. Proposals for publication in the series are reviewed by a committee established for that purpose by the AAFS and also reviewed by Wiley.

The AAFS was founded in 1948 and represents a multidisciplinary professional organization that provides leadership to advance science and its application to the legal system. The 11 sections of the AAFS consist of Criminalistics, Digital and Multimedia Sciences, Engineering Sciences, General, Pathology/Biology, Questioned Documents, Jurisprudence, Anthropology, Toxicology, Odontology, and Psychiatry and Behavioral Science. There are over 7000 members of the AAFS, originating from all 50 States of the United States and many countries beyond. This series reflects global AAFS membership interest in new research, scholarship, and publication in the forensic sciences.

Douglas H. Ubelaker
Senior Scientist
Smithsonian Institution
Washington, DC, USA
Series Editor

Preface

Microorganisms have been used as physical evidence in "forensic" investigations for nearly as long as their existence has been known. These organisms have been used to reconstruct events as diverse as cause of death, food spoilage, and bioterrorism. Such diverse applications represent our definition of modern forensic microbiology. Ultimately, this text will serve as a resource for those interested in the relationship between microbiology and forensic science, particularly criminal and medicolegal death investigations.

We have approached the text with the perspective that forensic microbiology primarily draws from three fundamental disciplines: medical microbiology, environmental microbiology, and food microbiology (Figure P.1). The current text was inspired by a workshop presented at the 2014 American Academy of Forensic Sciences Conference in

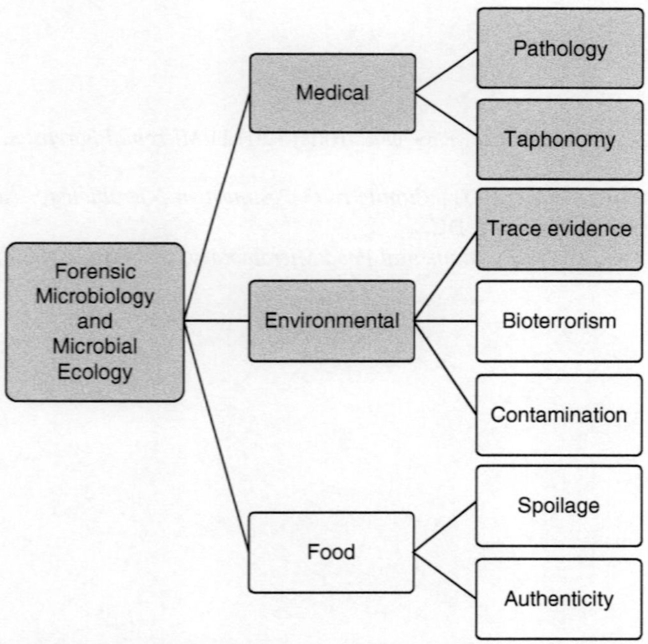

Figure P.1 Forensic microbiology and microbial ecology is a broad field that contributes to investigations ranging from pathology and bioterrorism to the quality of food and the environment. This text will focus on newly emerging areas of forensic microbiology related to postmortem microbiology and trace evidence (indicated in gray). The text will also serve to direct readers to established texts on other aspects of forensic microbiology, particularly bioterrorism and food microbiology

Seattle, Washington. From this workshop germinated the topics covered in the text. Topics reviewed include primarily newly emerging applications of microbiology to criminal and medicolegal death investigations. To do this, the contents focus on the characterization and interpretation of microbial communities through culturing and sequencing techniques. Our goal is to position forensic microbiology in a modern context of improved microbiome sequencing and analytical techniques beyond those that have been traditionally covered in microbial forensics that include bioterrorism and pathology (Caplan and Koontz, 2001; Budowle et al., 2011; Ray and Bhunia, 2013). Recent developments in sequencing technology and corresponding discoveries have allowed scientists to examine microbial communities at unprecedented resolution and in multidisciplinary contexts. The current text will highlight the potential new applications that these technologies can provide.

February 2017

David O. Carter
Jeffery K. Tomberlin
M. Eric Benbow
Jessica L. Metcalf

References

Budowle, B., Schutzer, S.E., Breeze, R.G. *et al.* (eds) (2011) *Microbial Forensics*, 2nd edn, Academic Press, Burlington.

Caplan, M.J. and Koontz, F.P. (2001) *Cumitech 35 Postmortem Microbiology*, American Society for Microbiology Press, Washington, DC.

Ray, B. and Bhunia, A. (2013) *Fundamental Food Microbiology*, 5th edn, CRC Press, Boca Raton.

CHAPTER 1
A primer on microbiology

David O. Carter[1], Emily N. Junkins[1,2] and Whitney A. Kodama[1]

[1] *Laboratory of Forensic Taphonomy, Forensic Sciences Unit, Division of Natural Sciences and Mathematics, Chaminade University of Honolulu, Honolulu, HI, USA*

[2] *Department of Microbiology and Plant Biology, University of Oklahoma, Norman, OK, USA*

1.1 Introduction

In many ways microorganisms are an ideal form of physical evidence. They can be found virtually everywhere and are certainly present in every habitat occupied by humans. Thus, microbes could be collected from every scene under a forensic investigation, yet not all microorganisms are everywhere; like many forms of trace evidence, some microbes are found only in certain locations due to having a preferred habitat, much like how insects, birds, and reptiles have a preferred habitat range. Another valuable characteristic of microorganisms is that many of them can transform themselves into a highly durable structure that is designed to survive harsh conditions, which increases the likelihood of their survival and discovery. Considering all of these attributes, it is probably not surprising that microorganisms have been used as physical evidence since the early days of forensic science, particularly to establish the cause of death (e.g., MacCallum and Hastings, 1899). Forensic microbiology has since grown into an exciting discipline relevant to several areas of forensic science including medicolegal death investigation (Caplan and Koontz, 2001; Forbes *et al.*, 2016), bioterrorism (Budowle *et al.*, 2011), and product authenticity (Brzezinski and Craft, 2012). It will be absolutely fascinating to learn of the new discoveries in forensic microbiology over the next few decades.

Historically microbes have been used almost exclusively as spatial evidence—physical evidence that is used to associate people with diseases, objects, and/or locations (Locard, 1930a, b, c; Caplan and Koontz, 2001; Tridico *et al.*, 2014; Wiltshire *et al.*, 2014; Young *et al.*, 2015). This application is similar to the use of any other form of trace evidence, such as soil (Bisbing, 2016), paint (Kirkbride, 2016), glass (Almirall and Trejos, 2016), and fibers (Houck, 2016). However, recent research has shown that microorganisms

Forensic Microbiology, First Edition. Edited by David O. Carter, Jeffery K. Tomberlin, M. Eric Benbow and Jessica L. Metcalf.

represent a relatively unique form of physical evidence that can also serve as temporal evidence, evidence that is used to establish a timeline. This application uses the ability of microorganisms to respond rapidly to changes in their environment (e.g., Carter and Tibbett, 2006), and these changes are temporally predictable (Metcalf *et al.*, 2013; Pechal *et al.*, 2014; Guo *et al.*, 2016; Metcalf *et al.*, 2016), with an apparent ability to serve as an estimate of the postmortem interval (Chapter 2) and human habitation (Chapter 13) interval.

We are currently in an exciting time when multiple research groups around the world are leading advances in postmortem microbiology and trace microbiology (Fierer *et al.*, 2010; Benbow *et al.*, 2015; Lax *et al.*, 2015; Metcalf *et al.*, 2016). These advances are occurring rapidly and have great potential to significantly change how microorganisms are used as physical evidence. Microorganisms will likely play a greater role as physical evidence in the future, so the purpose of the current chapter is to provide an introduction to some fundamental aspects of microbiology and microbial ecology to help the reader develop an appreciation for the vast diversity of microorganisms and how they can be used to identify a location or time period of investigative interest. It is not possible for this chapter to review all known microorganisms, so the contents hereinafter will place an emphasis on bacteria that are of interest to the most recent research relevant to the scope of this book, postmortem microbiology and trace microbiology (e.g., Benbow *et al.*, 2015; Iancu *et al.*, 2015; Finley *et al.*, 2016; Metcalf *et al.*, 2016). However, domains Archaea and Eukarya are also highly relevant, and the current chapter will reference relevant work, when possible, that focuses on these very important taxa within a forensic context.

1.2 Microbial characteristics

Microorganisms can differ in many ways including their morphology, method of movement (motility), metabolic strategy, environmental requirements, and several other characteristics (Brown, 2015). The current chapter will address this issue with relative simplicity by describing how microorganisms survive with a brief description of some relevant taxa.

1.2.1 Microbial taxonomy and function

Classification of life has proven to be a challenge. Presently, there are a number of opinions on how we should organize organisms in terms of their relationship to one another. Although not the focus of this chapter, this topic is of great importance as it impacts our ability to assess microbial communities in general. Thus, we suggest individuals with an interest in forensic microbiology remain cognizant of the ever-shifting landscape of microbial taxonomy.

For this text, we focus our discussion on three major groups of microorganisms organized as domains: Archaea, Bacteria, and Eukarya (Woese *et al.*, 1990) although a new

Table 1.1 Phyla of domain Bacteria included in the list of prokaryotic names with standing in nomenclature, http://www.bacterio.net

Acidobacteria	Cyanobacteria	Nitrospira
Actinobacteria	Deferribacteres	Planctomycetes
Aquificae	Deinococcus–Thermus	Proteobacteria
Armatimonadetes	Dictyoglomi	Spirochaetes
Bacteroidetes	Elusimicrobia	Synergistetes
Caldiserica	Fibrobacteres	Tenericutes
Chlamydiae	Firmicutes	Thermodesulfobacteria
Chlorobi	Fusobacteria	Thermomicrobia
Chloroflexi	Gemmatimonadetes	Thermotogae
Chrysiogenetes	Lentisphaerae	Verrucomicrobia

perspective on this classification was presented recently (Hug *et al.*, 2016). The List of Prokaryotic Names with Standing in Nomenclature (http://www.bacterio.net) currently divides Bacteria into 30 phyla (Table 1.1) and Archaea into five phyla (Crenarchaeota, Euryarchaeota, Korarchaeota, Nanoarchaeota, Thaumarchaeota). These microbes can vary morphologically, with spherical (cocci; Figure 1.1a) and rod-shaped (historically termed bacilli; Figure 1.1b) being the most common. Variations on these general morphologies exist (Figure 1.1c), as do other morphologies such as club-shaped cells (coryneform; Figure 1.1d) and curved rods (e.g., vibrio). Thus, referring to a bacterium as a rod or a coccus is a helpful way to begin the identification process.

As the term microbiology indicates, microbes are small. A bacterial cell will likely have a diameter of 1–5 µm (see Figure 1.1), which means that microscopy is necessary to view individual microbial cells. Thus, the shape and size of microbial cells can be used for a general identification, usually to exclude possible identities. Other commonly used characteristics to identify microbes include the reaction to the Gram stain and the ratio of nucleotides in a cell, which is presented as guanine–cytosine (GC) content.

Stains play a significant role in the identification of microorganisms. For example, the Gram stain was developed in the nineteenth century to help visualize microbial cells. Without staining, many microbial cells are transparent and difficult to see. The Gram-positive and Gram-negative designation also provides some insight into the structure of the microbial cell wall. The cell wall of a Gram-positive bacterium is approximately 90% peptidoglycan, whereas the cell wall of a Gram-negative bacterium is approximately 10% peptidoglycan (Madigan *et al.*, 2012). Interestingly, the Gram designation also provides information about the taxonomy of bacteria. Gram-positive bacteria are generally found in phyla Actinobacteria and Firmicutes. Some relatively well-known Gram-positive bacteria include genera *Bacillus* (rod), *Clostridium* (rod), and *Streptococcus* (coccus), all of which are in phylum Firmicutes. The GC content of a cell can be used to generally distinguish between Actinobacteria and Firmicutes. The GC content represents

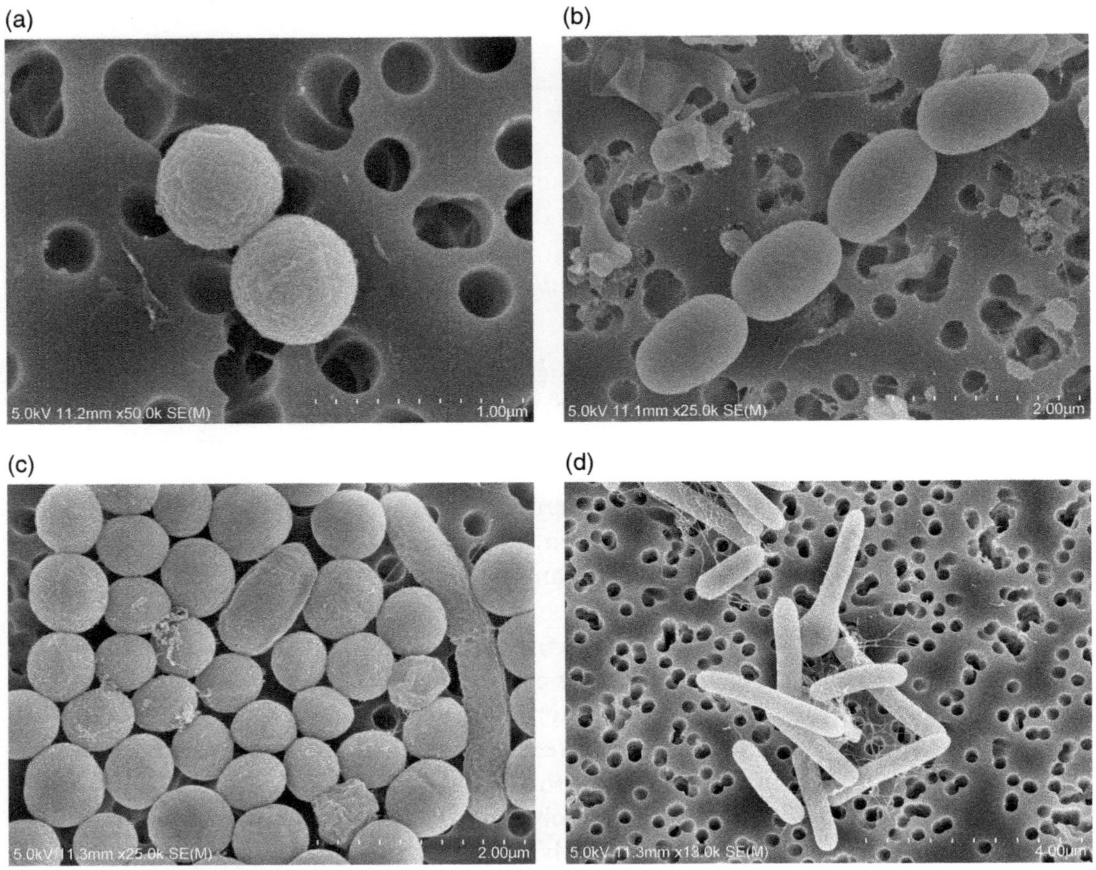

Figure 1.1 Scanning electron micrograph showing some of the contrasting morphologies observed in microbial cells including the widely observed coccus (a: *Staphylococcus cohnii*) and rod (b: *Bacillus subtilis*) shapes, which can vary and occur in association with other morphologies (c). Microbial cells can be observed as several other shapes including spiral (spirilla), curved rods (vibrio), and club shaped (coryneform), the latter of which is observed in *Sporosarcina contaminans* (d)

the proportion of the bacterial genome that comprises GC base pairs, rather than adenine–thymine base pairs, and is presented as high GC (>50% GC content) or low GC (<50% GC content). Phylum Actinobacteria includes high GC bacteria, while phylum Firmicutes includes low GC bacteria.

Microbes can also be identified based on their function. In fact, one of the reasons that microbes are so important to life on Earth is the vast diversity of functions that they carry out. Microbes decompose organic material, utilize carbon dioxide, fix nitrogen, and help plants and animals to acquire nutrients, and they contribute many other vital functions that keep habitats stable. These functions are the result of microorganisms competing for that which is essential to microbial life, water, energy, and nutrients, where the energy source is often organic (contains carbon) and the nutrients include several essential elements such as nitrogen, phosphorus, potassium, sulfur, and calcium. As seen with other organisms, the ability of microbes to function is greatly influenced by their environment: oxygen availability, temperature, chemistry (particularly pH and Eh), and light (Ball, 1997). Thus, humans live in habitats that are the result of several processes, many of which are carried out by microorganisms. To effectively use microorganisms as physical evidence, we must therefore understand where microbes live, what their function is within that habitat, and how their environment affects their function and survival.

Microbes use several strategies to acquire energy and nutrients, referred to here as metabolic strategy. Some microbes even have the ability to change their metabolic strategy depending on environmental conditions. A microbe that is restricted to a single metabolic strategy is known as obligate. An obligate aerobe is a microbe that requires oxygen, for example. A microbe that can change metabolic strategies is termed facultative. A facultative anaerobe is a microbe that can maintain metabolic function regardless of oxygen availability. Understanding these metabolic strategies is important for forensic microbiology because it forms the foundation for interpreting microbial evidence or identifying a particular microorganism. For example, one would expect to see decomposing remains associated with chemoorganotrophic microbes, like the Enterobacteriaceae (e.g., Benbow *et al.*, 2015), because a dead body is a high quality source of water, energy, and nutrients (Carter *et al.*, 2007b).

The metabolic terms defined in Table 1.2 are presented to help understand the diversity of metabolic strategies used by microbes and show that they are regularly described by their metabolic strategy. Phototrophs, such as those in the phylum Cyanobacteria, use light to conduct photosynthesis. Chemotrophs, such as those in class Gammaproteobacteria, consume organic compounds (e.g., human remains, plant detritus) to generate energy. Because of this they are often referred to as decomposers. These decomposers represent the bulk of microorganisms associated with decomposing remains and trace evidence. It is important to accept that some taxa have physiologies that are not easily described using the terminology in Table 1.2 because of the vast array of metabolic strategies employed by microorganisms (e.g., Slonczewski and Foster, 2011; Madigan *et al.*, 2012)

1.2.2 Enzyme activity

Enzymes are important in microbial ecology. Microbial activity and metabolism can be measured in many different ways (e.g., carbon mineralization, calorimetry), but these metrics are often the result of enzyme-substrate reactions, if not a direct measure of potential enzyme activity (e.g., Carter *et al.*, 2008). Microbial enzymes can be classified in many ways, for example, as hydrolase, lyase, oxidoreductase, and transferase enzymes (Table 1.3). They can also be classified as intracellular and extracellular, which are important distinctions to consider; intracellular enzymes react with substrates within

Table 1.2 Glossary of terms commonly used to describe the habitat preferences and metabolic strategies of microorganisms

Aerobe	An organism that lives in the presence of oxygen
Aerotolerant	An organism that can live in aerobic and anaerobic conditions
Anaerobic	An organism that lives in the absence of oxygen
Autotrophy	Using carbon dioxide as the only source of carbon
Chemoautotrophy	Oxidizing chemical compounds to obtain energy while acquiring carbon only from carbon dioxide
Chemolithotrophy	Oxidizing inorganic compounds to obtain energy
Chemoorganotrophy	Oxidizing organic compounds to obtain energy
Chemotrophy	Oxidizing chemical compounds to obtain energy
Extremotroph, -phile	An organism that lives in extreme environmental conditions such as temperature and pH
Facultative	Not required, optional
Fermentation	Obtaining energy by using an organic compound as both an electron donor and an electron acceptor
Halotroph, -phile	An organism that lives in high salt concentrations
Heterotrophy	Using organic compounds as sources of carbon
Hyperthermotroph, -phile	An organism with an optimal growth temperature of ≥80°C
Mesotroph, -phile	An organism that lives at moderate temperature with an optimal growth temperature from 15 to 40°C
Metabolism	All reactions, anabolic and catabolic, in a cell
Microaerotroph, -phile	An organism that requires low levels of oxygen
Obligate	Required
Photoautotrophy	Using light as an energy source while acquiring carbon only from carbon dioxide
Photoheterotrophy	Using light as an energy source while using organic compounds as sources of carbon
Phototroph	An organism that obtains energy from light
Psychrotroph, -phile	An organism that lives at low temperature with an optimal growth temperature of <15°C
Respiration	Obtaining energy by oxidizing chemical compounds through a series of reactions to a terminal electron acceptor
Thermotroph, -phile	An organism that lives at high temperature with an optimal growth temperature from 45 to 80°C

See also Madigan *et al.* (2012) and Slonczewski and Foster (2011).

the microbial cell, while extracellular enzymes are released from microbial cells so that the enzyme–substrate product can then be transported across the cell membrane into the cell. Microorganisms commonly release extracellular enzymes into their habitat so that they can use the products to acquire the resources necessary for survival. This degradative and overall enzyme profile can be used to determine the range of substrates

Table 1.3 General classification of enzymes commonly associated with microorganisms

Enzyme	Examples	Function
Hydrolase	Amylase, chitinase, lipase, peptidase, phosphatase, phosphodiesterase, protease, sulfatase, urease	Catalyzes the hydrolysis of chemical bonds
Lyase	Aldehyde lyase, amino acid decarboxylase, cyclase, dehydratase	Catalyzes an elimination reaction and oxidation to break chemical bonds
Oxidoreductase	Alcohol oxidoreductase, amino acid oxidoreductase, ammonia monooxygenase, glucose oxidase, nitrite oxidoreductase, methane monooxygenase, monoamine oxidase, peroxidase	Catalyzes the transfer of electrons from the electron donor (reductant) to the electron acceptor (oxidant)
Transferase	Amino transferase, CoA transferase, methyltransferase, polymerase, kinase	Catalyzes the transfer of a functional group from one molecule (donor) to another (acceptor)

that can be used by a microbe (e.g., Pechal *et al.*, 2013), known as community level physiological profiling (Degens and Harris, 1997). This potential substrate use is then modified by resource availability/quality, decomposer community, and physicochemical environment (Swift *et al.*, 1979; Killham and Prosser, 2007).

The release of extracellular enzymes into a habitat does introduce some complexity into interpreting the ecology of a microbe. Once extracellular enzymes are released from a microbial cell, they can function independently of the microbial cell. This ability means that microbial habitats can contain free enzymes capable of reacting with substrates. As a result, a microbial cell might take up the product of a reaction that used an enzyme it did not release. To further complicate matters, these free enzymes do not necessarily react with substrates soon after release from the microbial cell. They can be bound to organic material and inorganic surfaces, such as soil minerals, and even accumulate in habitats so that the measurement of enzyme activity might provide a misleading metric of microbial activity (Carter *et al.*, 2007a). This phenomenon must be researched in greater detail before we truly understand postmortem microbial ecology.

1.3 Microorganisms and their habitats

Microorganisms and their activity are greatly influenced by their environment. Temperature, relative humidity, pH, oxidation–reduction potential (Eh), and oxygen availability are all environmental parameters that play a role in defining the microbial community. For example, the *Psychrobacter* sp. observed by Carter *et al.* (2015) in gravesoil during the winter in Nebraska, United States, is a taxon that prefers cold temperature. Many of these environmental parameters, such as ambient temperature and precipitation, can be measured with relative ease (see Chun *et al.*, 2015), and these quantitative measures can be used to understand the ecology of microbes associated with a given sample or habitat. In addition, some of these chemical measures can be used as physical evidence in their own right. For example, these measures have allowed us to observe that fly (Diptera) larval masses in decomposing remains are slightly acidic-reducing environments of high temperature (Chun *et al.*, 2015). As a result we would expect to see larval masses select for bacteria and archaea that can grow in a habitat of neutral pH with little/no oxygen at approximately 35°C.

1.3.1 Oxygen and moisture

Microorganisms are regularly classified as aerobes or anaerobes. Thus, the presence or absence of oxygen is a useful index for classifying microorganisms. Obligate aerobes like *Pseudomonas aeruginosa* and *Mycobacterium tuberculosis* require oxygen as it serves as the terminal electron acceptor in the electron transport chain. In contrast, obligate anaerobes like *Clostridium* spp. cannot tolerate oxygen, and they must employ another metabolic strategy for survival; *Clostridium* spp. are well-known fermenters of carbohydrates, amino acids, and ethanol. Fermentation is a metabolic strategy where an organic compound is used as the terminal electron acceptor. Fermentation is an important process to understand as it drives putrefaction, which is essentially the fermentation of human remains.

Facultative anaerobes, such as *Staphylococcus aureus* can grow with or without oxygen. Many microbes can switch between oxidative phosphorylation (aerobic respiration) and anaerobic fermentation. Others may respire in the absence of oxygen using alternative inorganic electron acceptors such as nitrate, sulfate, or oxidized metals. For example, a denitrifier from the order Rhizobiales would reduce nitrate, instead of oxygen, to either nitrous oxide (N_2O) or free nitrogen (N_2) (Metcalf *et al.*, 2016). The switch between aerobic and anaerobic metabolism by a microorganism can occur within a few hours (Killham and Prosser, 2007), easily within the early postmortem period. However, the energy yield of anaerobic metabolism is low because it results in the accumulation of incompletely oxidized products. As a result, anaerobic and microaerobic habitats tend to result in a slower decomposition process, such as the formation of adipocere (Forbes, 2008).

The influence of oxygen is also related to the microbial need for moisture as moisture and oxygen availability are coupled in many habitats. Moisture concentration is negatively correlated to the concentration of oxygen. Moisture is necessary for microbial life, but many microbes have contrasting moisture preferences. Some microorganisms, like the Actinobacteria, Firmicutes, and Proteobacteria detected on dry human skin (Grice and Segre, 2011), are able to thrive in habitats of low moisture content, whereas some microorganisms, like those observed on vertebrate remains in postmortem submersion interval studies (Benbow *et al.*, 2015), require an aquatic habitat to maintain growth. Because of this, many microorganisms have potential as microbial trace evidence because they can survive on human skin as well as the surfaces on which they come into contact (Lax *et al.*, 2014).

An excellent example of the relationship between oxygen and moisture was presented by Carter *et al.* (2010): the addition of moisture to a sufficiently aerated habitat can lead to anaerobic conditions as air diffuses more slowly through water. This effect is enhanced by the introduction of organic material available for decomposition, such as human remains. In this scenario, the presence of human remains would stimulate microbial activity because it is a highly attractive food source. This microbial activity would consume air more rapidly than it can be replenished, which is significantly slower in a wet environment (Carter *et al.*, 2010). This effect is particularly pronounced when decomposing remains are buried in soil, but the relationship also holds true for remains decomposing on the surface. The remains will remain an anaerobic hotspot (because air cannot be replenished rapidly enough to meet demand) until moisture levels drop sufficiently to retard enzyme activity. At this point, often referred to as advanced decay (Carter *et al.*, 2007b), it becomes difficult for bacteria to compete and the fungi can proliferate (Carter and Tibbett, 2003; Sagara *et al.*, 2008).

1.3.2 Temperature

Microorganisms and their activity are significantly influenced by temperature. Microbial activity increases as temperature increases and slows with decreasing temperatures. Doubling of microbial activity with a 10°C increase in temperature is not uncommon (e.g., Carter and Tibbett, 2006) up to approximately 40°C. As temperature increases from 40°C, it becomes too hot for many microorganisms. Similarly, a lot of microbial activity ceases at 0°C. Several studies (Putman, 1978; Carter *et al.*, 2008; Pechal *et al.*, 2013) have reported an increase in postmortem microbial activity with an increase in temperature, which is one of the reasons why decomposition is more rapid during warmer months. Although insects and scavengers often consume the majority of remains, microorganisms make a significant contribution to taphonomy (Lauber *et al.*, 2014).

Microbes have preferred temperature ranges. Microbes are typically characterized as one of the three temperature regimes: psychrotrophic, mesotrophic, or thermotrophic. Psychrotrophs live in low temperature habitats, such as glaciers, and have an optimal growth temperature <15°C. Mesotrophs live at moderate temperature and grow optimally between 15 and 40°C. Thermotrophs live at high temperature and grow optimally between 50 and 80°C in environments such as thermal vents and hot springs. Some microbes, particularly archaea, are recognized as hyperthermotrophs. These microbes have an optimal growth temperature ≥80°C.

One way to analyze the influence of temperature on biological activity is by using the accumulated degree day (ADD), which is essentially the sum of median daily temperature and represents physiological time (e.g., Pechal *et al.*, 2014). This metric is commonly used, and it has been incorporated into recent postmortem microbiome studies

(Metcalf *et al.*, 2013; Pechal *et al.*, 2014; Hauther *et al.*, 2015; Metcalf *et al.*, 2016), yet one important issue requires resolution: the use of a minimum developmental threshold. The minimum developmental threshold is a variable used to compensate for the cessation of biological activity when calculating ADD. For example, if an organism is known to cease function at 6°C, then these six degrees are subtracted from each degree day. One problem that forensic microbiology currently faces is the establishment of a minimum developmental threshold for microorganisms. Vass *et al.* (1992) used 0°C based on the rationale that most microbial activity ceases at 0°C. However, this is not always the case, as microsites might remain slightly above 0°C or free extracellular enzymes might remain active below 0°C. Establishing a minimum developmental threshold for forensic microbiology might be straightforward for studies using a single taxon or simple microbial communities comprising only a few taxa. However, the complex microbial communities that are currently the focus of postmortem microbiology and trace microbiology could make the issue quite difficult to resolve.

1.4 Competition for resources

At this point we hope that you can appreciate the usefulness of identifying a microbe, for example, as a psychrotrophic obligate aerobe. This brief description refers to a microorganism that would be found in a cold, well-aerated habitat. However, this categorization does not provide insight into the ability of a microbe to compete for nutrients. This competition is an issue of growth kinetics where the growth rate of a microbe is a function of substrate concentration and availability (Figure 1.2). Figure 1.2 shows that

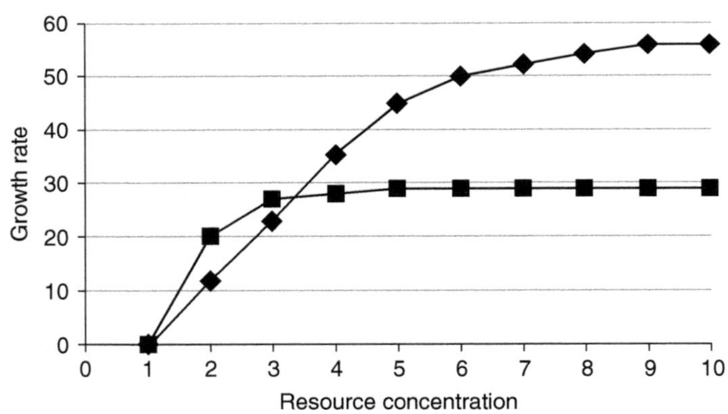

Figure 1.2 The growth rate of microorganisms is regulated, in part, by the availability of resources. Zymogenous (◆) microbes are able to use resources and multiply rapidly so they can dominate a habitat in which resources are abundant. In contrast, autochthonous (■) microbes tend to represent the basal, resting community that uses resources and multiplies slowly and forms the foundation of a microbial community

the microbial population represented by a diamond symbol will outcompete the microbial population represented by the square symbol at high resource concentration. This organism is termed zymogenous. Zymogenous organisms are associated with a high growth rate and a high substrate affinity. Zymogenous microbes will dominate habitats following a disturbance, such as digging in the soil, or when habitats contain a high quality source of energy. In contrast, the square population will outcompete the other populations at low substrate concentrations. This organism is termed autochthonous. Autochthonous organisms are associated with a low growth rate and substrate affinity. Autochthonous microbes will dominate environments with low resource availability or low resource quality and likely represent the foundational, basal microbial community for a given habitat. These populations might occur in the less accessible microenvironments where C availability is low so that, in reality, a scale of autochthony and zymogeny exists. Furthermore the scale changes over time so that microbial succession will occur as a resource is being decomposed.

The autochthony–zymogeny concept can be viewed as analogous to the r-K continuum used in plant and animal ecology. K-strategists and oligotrophic microbes are adapted to growth under conditions of low energy and nutrient availability. Oligotrophy also refers to low nutrient content and the use of unusual C resources and volatile organic acids. In contrast, copiotrophs are adapted to an excess of nutrients. Recall that microorganisms are physiologically versatile and flexible; facultative anaerobes metabolize most efficiently as aerobes, but they can survive when oxygen is insufficient for aerobic metabolism. Another aspect of flexibility is the ability to metabolize more than one substrate, often simultaneously. This is reflected in the diversity of enzymes associated with microbes. Also, microorganisms can adapt to the repeated introduction of compounds such as pesticides (Jayachandran *et al.*, 1998) and muscle tissue (Carter and Tibbett, 2008), which results in more rapid decomposition.

1.5 The ecology of some forensically relevant bacteria

In recent years unprecedented insight into the ecology of microorganisms relevant to medicolegal death investigation and trace evidence analysis has been gained (Fierer *et al.*, 2010; Nagasawa *et al.*, 2013; Song *et al.*, 2013; Lax *et al.*, 2014; Tridico *et al.*, 2014; Bouslimani *et al.*, 2015; Lax *et al.*, 2015). This research has identified many bacteria from several phyla, but there are three phyla that have been observed regularly: Actinobacteria, Firmicutes, and Proteobacteria. The remainder of this chapter will provide an introduction to the characteristics of some microorganisms from these phyla.

1.5.1 Actinobacteria

The phylum Actinobacteria is both morphologically and metabolically diverse. The taxonomy of this phylum is somewhat confusing as it has only one class (Actinobacteria) but several subclasses (www.bacterio.net). Actinobacteria comprises 10 orders and 58 families. These Gram-positive, high GC bacteria can be observed as cocci, rods, and filaments while conducting respiration or fermentation. The Actinobacteria are primarily aerobic and are common inhabitants of soil, skin, and aquatic environments. Many of the more familiar genera in this phylum are members of subclass Actinobacteriales, including *Corynebacterium*, *Micrococcus*, and *Mycobacterium* (Brown, 2015). As with most phyla, there are several ways to categorize the Actinobacteria. For this chapter, phylum Actinobacteria is organized into three groups: coryneform bacteria, propionic acid bacteria, and filamentous bacteria.

1.5.1.1 Coryneform bacteria

Although the term coryneform indicates club-shaped cells (Figure 1.1d), the coryneform bacteria are also rod-shaped. The coryneform bacteria are commonly associated not only with human skin and mucosae but are also observed in soils and foods (Table 1.4). Common genera in this group include *Arthrobacter*, *Corynebacterium*, and *Kurthia* (Madigan *et al.*, 2012). *Corynebacterium* is a diverse group of greater than 100 species that includes decomposers as well as pathogenic bacteria (e.g., *Corynebacterium diphtheriae*). *Corynebacterium glutamicum* is of great economic interest because it is used to generate monosodium glutamate. Coryneform bacteria have been observed in association with decomposing remains and present an interesting trend in this context; their abundance can remain relatively consistent over time (Carter *et al.*, 2014). This persistence during decomposition will likely lead to great forensic interest because it might indicate their ability to survive in dynamic habitats, which is an important characteristic for trace evidence.

Table 1.4 General characteristics of two genera recognized as coryneform bacteria

	Corynebacterium	***Arthrobacter***
Morphology	Gram-positive rods, club shaped	Gram-variable, rod–coccus–rod developmental cycle (e.g., *Arthrobacter globiformis*)
Oxygen	Aerobic or facultative anaerobe	Mostly aerobic, but capable of anaerobic growth with nitrate (e.g., *Arthrobacter globiformis*)
Metabolism	Lactic acid fermentation	Mostly respiratory; can decompose herbicides, pesticides, nicotine, caffeine, and phenol
Habitat	Soil, food, human skin	Primarily upper layers of soil
Other	Of significant interest to industry (*Corynebacterium glutamicum*) and public health (*Corynebacterium diphtheriae*)	Highly resistant to desiccation and starvation (no spore formation)

From Brown (2015), Madigan *et al.* (2012), and Slonczewski and Foster (2011).

1.5.1.2 Propionic acid bacteria

The propionic acid bacteria are anaerobic or aerotolerant, and their morphology can be rod or filamentous. Common genera in this group include *Propionibacterium* and *Eubacterium*. The ecology of these bacteria is interesting because of their ability to ferment lactic acid. Typically the lactic acid bacteria, such as *Lactobacillus* (see Section 1.5.2.2), are very effective competitors that are able to inhibit the proliferation of other bacteria due to their ability to decrease the pH of a habitat. This is not the case with the propionic acid bacteria. In fact, the propionic bacteria decompose lactic acid to generate propionic acid, acetic acid, and carbon dioxide. A succession then occurs whereby propionic acid bacteria succeed lactic acid bacteria. This metabolic strategy, where propionic bacteria use a fermentation product released by other bacteria, is known as secondary fermentation (see Madigan *et al.*, 2012). The propionic acid bacteria are also among the many Gram-positive taxa present in the human oral cavity. The propionic bacteria thrive on human skin and in the ducts of sebaceous glands and are associated with acne, although they do not necessarily cause it. They also contribute to some components of body odor. One species in particular, *Propionibacterium acnes*, is ubiquitous on the human skin and is generally harmless.

1.5.1.3 Filamentous bacteria

The filamentous bacteria are a large group of phylogenetically related aerobic Gram-positive bacteria that form filamentous hyphae that are similar to fungal hyphae. This ability makes these bacteria particularly interesting because they essentially function as multicellular bacteria. Filamentous bacteria have cells to generate and release vast amounts of spores that are sometimes called arthrospores (Brown, 2015). These spores are not as durable as endospores released by *Bacillus* spp. and *Clostridium* spp. (see following text); however, they are resistant to desiccation (Killham and Prosser, 2007). Commonly observed filamentous bacteria include genera *Streptomyces* and *Actinomyces*.

Filamentous bacteria are primarily soil dwellers and are important decomposers because they can generate extracellular enzymes to decompose polysaccharides, proteins, and lipids. They prefer alkaline and neutral soils that are well drained and aerobic. They are able to proliferate in well-drained soils because, like fungi, their filaments can bridge gaps in water and nutrients, an ability that single-celled bacteria do not possess. The hallmark of these bacteria, however, is their prolific production of antibiotics, such as streptomycin, spectinomycin, tetracycline, erythromycin, clindamycin, and chloramphenicol. It remains to be seen if these antibiotics play a relevant role in postmortem or trace microbiology.

1.5.2 Firmicutes

The phylum Firmicutes are a large and diverse group of bacteria comprising 5 classes, 9 orders, and 39 families (www.bacterio.net). Firmicutes are usually decomposers (heterotrophic) found in soils, sediments, and in association with skin, mucosae, and gastrointestinal tract. Members of the class Bacilli are primarily aerobic, while many of the other taxa in this phylum are usually anaerobic but often aerotolerant. Firmicutes are able to use a wide range of carbon sources, and therefore their fermentation products are similarly diverse, which might explain the vast diversity of volatile organic compounds that are released during decomposition (e.g., Statheropoulos *et al.*, 2005; Vass, 2012; Forbes *et al.*, 2016). Here we have divided Firmicutes into three main groups: sporulating firmicutes, lactic acid bacteria, and non-sporulating, non-lactic acid Firmicutes.

1.5.2.1 Sporulating Firmicutes

These bacteria are characterized by their ability to form endospores. These structures are highly heat-, chemical-, and radiation-resistant, thick-walled, and metabolically inactive. Endospores allow these organisms to endure unfavorable conditions such as extreme temperature, desiccation, or nutrient depletion. For example, endospore-forming Firmicutes can be selectively isolated from an environment by heating the sample to 80°C for 10 minutes. This exposure will kill active cells but will not harm the endospore. The endospore-forming Firmicutes are a large group of bacteria that are commonly found in association with soil, food, dust, and animals. Key genera in this group include *Bacillus*, *Clostridium*, *Heliobacterium*, and *Sporosarcina* (Figure 1.3). The common endospore formers can be grouped into two categories: aerobic (*Bacillus* and relatives) and anaerobic (*Clostridium* and relatives).

The aerobic endospore formers are found in the class Bacilli. Many Bacilli generate extracellular hydrolytic enzymes that can break down complex polymers, including polysaccharides, nucleic acids, and lipids, which explains their association with decomposing remains (Chun *et al.*, 2015). One of the most well-documented bacteria in this

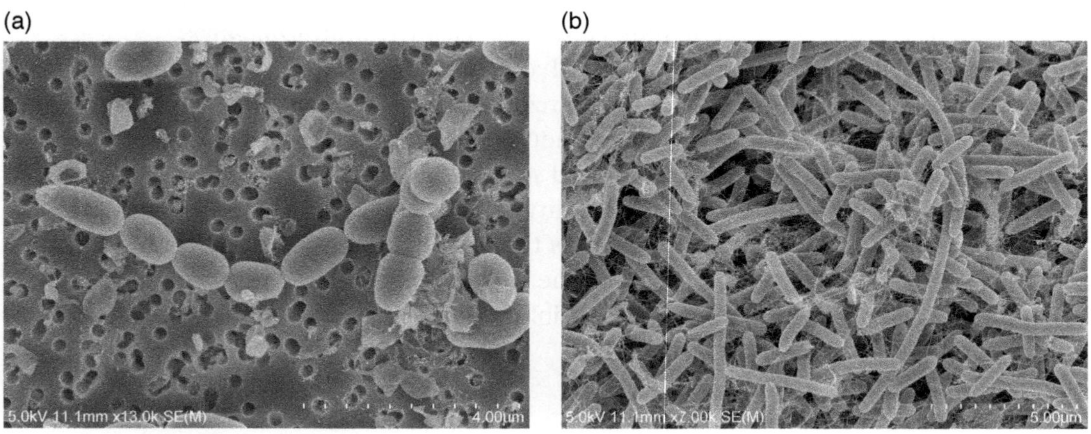

Figure 1.3 Scanning electron micrograph of endospore-forming Firmicutes *Bacillus subtilis* (a) and *Sporosarcina contaminans* (b) collected from the skin of a decomposing swine carcass on Oahu, Hawaii

group is *Bacillus anthracis*, the causative agent of anthrax. This species has the ability to produce endospores that can withstand harsh environments and have the ability to be industrialized. These characteristics have allowed terrorists to mass-produce and distribute the endospores as a biological weapon via simple methods, such as the postal service, as seen in 2001 (Government Accountability Office (GAO), 2005). Like many pathogens, this bacterium is an opportunist and will remain as an endospore until an ideal host or environment presents itself, such as a *B. anthracis* spore that is introduced to the human body through inhalation, ingestion, or through a break in the skin (cutaneous). The endospore then germinates and becomes an active microbial cell when it is introduced to these favorable environments. A number of aerobic endospore formers were formerly members of genus *Bacillus* but have since been reassigned to new genera such as *Paenibacillus*, *Lysinibacillus*, and *Geobacillus*. These genera have also been observed with decomposing remains (Chun *et al.*, 2015).

The anaerobic spore formers, class Clostridia, commonly reside in the soil and the animal gut where they primarily live in anaerobic microsites. Clostridia are fermenters since they are unable to use an electrochemical gradient and inorganic final electron acceptor. This trait has allowed further classification since there is variation in the fermentation substrates of *Clostridium* spp. For example, members of the class Clostridia are well-known fermenters of amino acids and fatty acids, which are in high abundance in decomposing vertebrate remains. Consequently, these microbes are regularly observed in postmortem microbial communities (Pechal *et al.*, 2014; Damann *et al.*, 2015; Metcalf *et al.*, 2016). Other Clostridia with medicolegal relevance include *Clostridium botulinum*, *Clostridium tetani*, and *Clostridium difficile* (Caplan and Koontz, 2001).

1.5.2.2 Lactic acid Firmicutes

The lactic acid bacteria are fermenters that produce lactic acid. This group of bacteria (e.g., *Enterococcus*, *Lactobacillus*, *Lactococcus*, *Streptococcus*) are usually Gram-positive rods or cocci. One way to further classify the lactic acid bacteria is based on the pattern of products formed from their fermentation. The lactic acid bacteria are either homofermentative (only produce lactic acid) or heterofermentative (produce lactic acid and other compounds like ethanol and carbon dioxide). All lactic acid bacteria grow anaerobically but many are aerotolerant. The genera *Streptococcus* and *Lactococcus* are homofermentative and play an important role in dairy production. *Enterococcus* is typically grouped among the fecal bacteria. *Lactobacillus* can be homo- or heterofermentative and is more tolerant of acidic habitats than most other lactic acid bacteria; lactobacilli can grow at pH 4. They are often responsible for the final stages of most lactic acid fermentations, when the pH is too acidic for other bacteria. Because of this, they are able to outcompete many other bacteria because their production of lactic acid inhibits the growth of their microbial competitors.

1.5.2.3 Non-sporulating, non-lactic acid Firmicutes

The non-sporulating Firmicutes, such as *Sarcina*, *Staphylococcus*, and *Erysipelothrix*, are catalase positive, which separates them from *Streptococcus* and other Gram-positive cocci and rods. Oftentimes, *Micrococcus* is included among the non-sporulating Firmicutes even though it belongs to phylum Actinobacteria. *Staphylococcus* and *Micrococcus* are both aerobic bacteria that use respiration to generate energy. Staphylococci are common in humans and other animals. *Micrococcus* can be isolated from skin, but they are also common in soil and dust. In general, these Gram-positive cocci are tolerant of low moisture content and high salt concentrations (e.g., 7.5% NaCl in growth medium will select for these taxa), which makes them ideal as trace evidence since the human skin is a relatively salty environment.

1.5.3 Proteobacteria

The phylum Proteobacteria is by far the largest and most diverse of the bacterial phyla. Many bacteria commonly cultured from soil, water, and animals are species of Proteobacteria. Proteobacteria show a high metabolic diversity with autotrophic, heterotrophic, chemolithotrophic, chemoorganotrophic, and phototrophic taxa. They are equally diverse in their relationship to oxygen; obligate anaerobes, obligate aerobes, microaerophilic organisms, and facultative anaerobes are known. 16S rRNA gene sequences divide the phylum Proteobacteria into five classes: Alpha-, Beta-, Gamma-, Delta-, and Epsilon-. However, all of the aforementioned characteristics appear to be widely dispersed within these classes, making Proteobacteria difficult to organize. Brown (2015) groups Proteobacteria according to class, in which there is value (Table 1.5), but here we have grouped these bacteria by considering function as done by Madigan *et al.* (2012).

1.5.3.1 Enteric Proteobacteria

The enteric Proteobacteria are a major group of highly related bacteria of major medical importance that are found in class Gammaproteobacteria, family Enterobacteriaceae. These bacteria are facultatively anaerobic, Gram-negative, non-sporulating rods that have relatively simple nutrient requirements, can ferment sugars, and are often grouped by their metabolic strategy (Table 1.6). These bacteria are regularly found in soil and water but are most known for their presence in the digestive tracts of humans and several other animals. For example, *Escherichia* spp. are nearly universal inhabitants of human and other animal intestinal tracts. *Proteus* spp. are typically a frequent cause of urinary tract infections and have been shown to play a significant role in attracting flies to decomposing remains (Zheng *et al.*, 2013). Many enteric bacteria can be pathogenic, such as *Escherichia coli*, and the enterics are sometimes called coliforms because of their relatively close relationship with *E. coli*. Enteric bacteria have been observed in several recent studies (e.g., Lax *et al.*, 2014; Carter *et al.*, 2015), and it will be interesting to observe their use as physical evidence because they are relevant to cause of death, postmortem interval, and trace evidence.

Three genera from this group have recently garnered great attention because of their consistent association with human remains. These bacteria are *Proteus*, *Ignatzschineria* (Figure 1.4), and *Wohlfahrtiimonas*. *Proteus mirabilis* plays an important role in attracting

Table 1.5 Some general characteristics of the classes that comprise phylum Proteobacteria classes

Class	Morphology	Metabolism	Habitat	Forensically significant divisions
Alpha-	Rod, coccus, spirilla, appendaged	Phototrophy, heterotrophy, methylotrophy, autotrophy	Aerobic and anaerobic aquatic environments	Nitrifiers, methanotrophs, methylotrophs, sulfur, and iron oxidizers
Beta-	Rod, coccus, spirilla, filamentous	Aerobic, facultative anaerobic, heterotrophy, chemolithotrophy	Organic-rich environments (soils, wastewater, sediments)	Sulfur-, iron-, manganese-, methane-oxidizers, nitrifiers
Gamma-	Rod, coccus, spirilla, vibrio, filamentous	Aerobic, facultative anaerobic, microaerotrophic, obligate anaerobic; heterotrophy, chemoautotrophy, autotrophy	Ubiquitous; psychro-, meso-, and thermophiles	Methanotrophs, methylotrophs, sulfur and iron oxidizers, pseudomonads, enterics
Delta-	Straight or curved rod	Anaerobic sulfate and sulfur reduction	Varies with phenotype (i.e., sulfate-reducers in sulfur-rich anaerobic environments)	Sulfate and sulfur reducers, sulfur oxidizers, nitrifiers
Epsilon-	Helically curved rods with flagella	Microaerotrophy, heterotrophy, chemoautotrophy	Intestinal symbionts, deep sea	Sulfate and sulfur reducers

From Brown (2015), Madigan *et al.* (2012), and Slonczewski and Foster (2011).

Table 1.6 Grouping of enteric bacteria by metabolic product

Group	Metabolism	Taxa
Mixed acid fermenters	Forms acetic, lactic, and succinic acids; ethanol, CO_2, and H_2; equal amounts of CO_2 and H_2 are produced	*Escherichia, Proteus, Salmonella, Yersinia*
2,3-butanediol producers	Less acid formed, main products are butanediol, ethanol, CO_2, and H_2; more CO_2 is produced than H_2	*Enterobacter, Klebsiella, Serratia*

From Brown (2015), Madigan *et al.* (2012), and Slonczewski and Foster (2011).

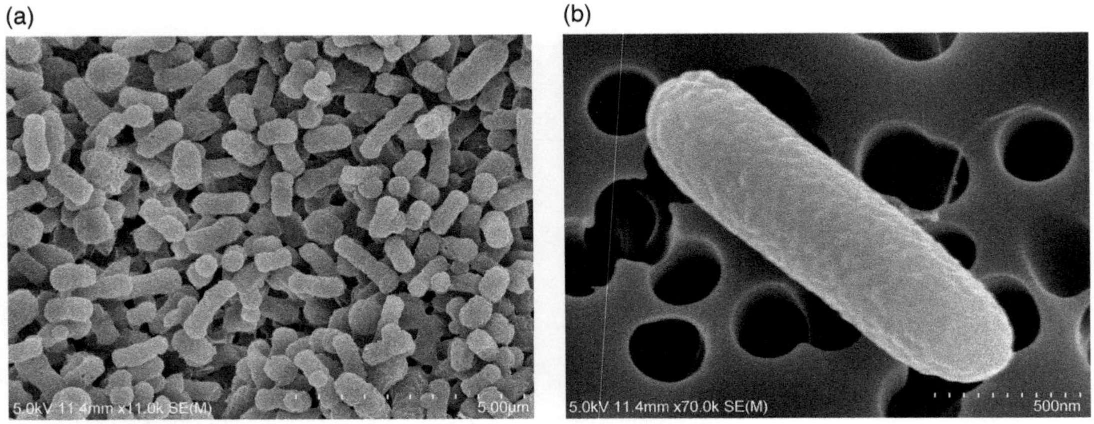

Figure 1.4 Scanning electron micrographs of enteric Proteobacteria *Ignatzschineria indica* collected from the skin of a decomposing swine carcass on Oahu, Hawaii presented as a colony (a) and as a single cell on a filter (b)

forensically important insects to resources (Ma *et al.*, 2012; Zheng *et al.*, 2013). This bacterium can colonize resources rapidly through a process called swarming where the cells grow and spread rapidly, ultimately forming biofilm. *Ignatzschineria* and *Wohlfahrtiimonas* also have a strong association with insects. Most likely as a consequence of this interaction, they are also regularly found in association with decomposing remains (e.g., Chun *et al.*, 2015). The forensic value of these microbes is not yet fully understood, but these bacteria are already an excellent example in the importance of understanding insect–microbe interactions (Chapter 11).

1.5.3.2 *Pseudomonas* and the pseudomonads

Pseudomonas is a genus within Gammaproteobacteria and the term pseudomonad refers to other Proteobacteria that share similar characteristics: Gram-negative chemoorganotrophic rods with polar flagella, such as *Burkholderia* and *Xanthomonas* (Madigan *et al.*, 2012). Common in most aerobic environments, the pseudomonads typically use aerobic respiration for energy. However, many pseudomonads can grow under anoxic conditions, often using nitrate as an alternative electron acceptor. The pseudomonads can thrive at relatively low nutrient levels because of their simple nutrient requirements and ability to use a vast array of different organic compounds. Since they tend to lack the hydrolytic enzymes necessary for catalysis of high molecular compounds, pseudomonads tend to use lower molecular weight organic compounds. Pseudomonads are also effective decomposers of xenobiotics, such as herbicides and pesticides. Because of their ability to capitalize on low nutrient environments, grow anoxically, and vary their electron acceptor to what is available, these organisms are well equipped to thrive in a decomposition environment.

1.5.3.3 Nitrifying Proteobacteria

Nitrification is a process where ammonia (NH_3) is oxidized to nitrite (NO_2^-), which is then oxidized to nitrate (NO_3^-). This process is critical for microbial function in all habitats and probably explains some of the pH changes observed in gravesoil (Vass *et al.*, 1992; Carter and Tibbett, 2008). These bacteria are widespread in soil and water and proliferate where the concentration of ammonia is high, such as in gravesoil (e.g., Anderson *et al.*, 2013) and bodies of water that receive an input of sewage or wastewater.

All aerobic ammonia oxidizers are members of the class Betaproteobacteria except for *Nitrosococcus*, which is in Gammaproteobacteria. Most nitrifying bacteria are obligate chemolithotrophs and obligate aerobes that oxidize inorganic nitrogen. However, *Nitrobacter* spp. are able to grow via chemoorganotrophy on acetate or pyruvate as the carbon source. Phylogenetically, nitrifying bacteria are found in Alpha-, Beta-, Gamma-, and Deltaproteobacteria. However, the genus *Nitrospira* forms its own phylum and represents nitrifying bacteria in metabolism only, even though it has been suggested that it is the most abundant nitrifying bacterial taxon in nature (Madigan *et al.*, 2012).

1.5.3.4 Methanotrophic and methylotrophic Proteobacteria

Methylotrophs are bacteria that can grow using C_1 compounds such as methanol, formate, and formamide. Thus, all organic compounds in the cell must be synthesized from C_1 precursors. Methanotrophs are methylotrophs that can also oxidize methane (CH_4). Methane is found in anoxic environments throughout the planet, including the mammalian digestive tracts, decomposing remains, water-treatment areas, and landfills. A methane cycle almost certainly exists in any coupled anaerobic/aerobic system making it a likely inhabitant of a postmortem microbial community. The main consumers of methane are Gammaproteobacteria and Alphaproteobacteria. These consumers are aerobic and present in several soil, water, and digestive systems, often located at the interface of methane and oxygen-rich systems. They conduct chemoautotrophy using methane released from the anaerobic environment to generate energy in the aerobic environment. Because of this, methanotrophs play an important role in the carbon cycle as they convert methane into cellular material and carbon dioxide.

1.5.3.5 Sulfur- and iron-oxidizing Proteobacteria

Sulfide oxidizers are found among the Alpha-, Beta-, Gamma-, Delta-, and Epsilonproteobacteria (Plante, 2007). However, the Gammaproteobacteria are the most commonly observed. These bacteria are typically found in black hydrogen sulfide-rich soils and sediments, such as those associated with decomposing remains (Vass *et al.*, 1992). Ecosystems in which hydrogen sulfide is produced tend to be microaerobic or anaerobic. Because of this, sulfur oxidizers must position themselves at the interface of sulfur-rich and oxygenated habitats, in a similar way as the methanotrophs. These bacteria oxidize hydrogen sulfide and other reduced sulfur compounds for energy with oxygen or nitrate as electron acceptors. They use either carbon dioxide or organic compounds as carbon sources. Sulfur oxidizers form the foundation of a food web, just like photosynthesizers in other ecosystems. Sulfur and hydrogen sulfide provide energy for various autotrophic sulfur or sulfide oxidizers, and the oxidation of these substrates generates sulfuric acid (H_2SO_4).

1.5.3.6 Sulfate- and sulfur-reducing Proteobacteria

Sulfate and sulfur reducers proliferate in sulfur-rich environments. The most obvious indicator of these organisms is the characteristic odor associated with decomposing vertebrate remains often encountered during forensic investigations. These bacteria use organic compounds or H_2 in anaerobic respiration. Sulfate and sulfur are electron acceptors, while hydrogen, lactate, and pyruvate are often the electron donors. Hydrogen sulfide is the end product of both reductions. The genera of dissimilative sulfate and sulfur reducers can be placed in one of the two groups: (i) those that can oxidize acetate and other fatty acids to carbon dioxide, such as *Desulfobacter*, *Desulfonema*, and *Desulfosarcina* and (ii) those that cannot, such as *Desulfovibrio*, *Desulfuromonas*, *Desulfotomaculum*, and *Desulfobulbus*. Dissimilative simply means that the sulfate or sulfur is used for energy, not for growth. Sulfate-reducing bacteria are widespread and often form communities with sulfur oxidizers. *Desulfosarcina* and *Desulfonema* are actually chemolithotrophs with carbon dioxide serving as the sole C source. Several other taxa are also able to reduce sulfur, including *Proteus*, *Campylobacter*, *Pseudomonas*, and *Salmonella*.

1.6 Archaea and microbial eukaryotes

The amount of space dedicated to the domains Archaea and Eukaryota does not accurately reflect their importance to forensic microbiology. Yet, there are major gaps in our understanding of these organisms. Here we will briefly discuss the fungi, other microbial eukaryotes, and domain Archaea.

Fungi have received the most research attention of the members of this section. Fungi are regularly associated with decomposing remains (Hitosugi *et al.*, 2006; Sagara *et al.*, 2008; Chimutsa *et al.*, 2015; Olakanye *et al.*, 2015; Metcalf *et al.*, 2016) and have value as trace evidence (Hawksworth and Wiltshire, 2011; Wiltshire *et al.*, 2014; Young *et al.*, 2015, 2016). The fungi associated with decomposing remains tend to proliferate after the flush of bacterial and insect activity or when they can colonize the skin (Hitosugi *et al.*, 2006; Ishii *et al.*, 2006). These fungi apparently proliferate in response to the availability of ammonia and nitrate, so they probably have a relationship with the nitrifying Proteobacteria.

The Archaea are virtually unstudied from a forensic perspective. They are currently organized into five phyla with the majority of Archaea representing phylum Crenarchaeota or phylum Euryarchaeota. These organisms, like bacteria, are small (0.5–5.0 μm) prokaryotes that are widely distributed around the planet. However, the Archaea are more ecologically diverse than the bacteria and eukaryotes because they inhabit several extreme environments. Some archaea are hyperthermophiles that can thrive at hot temperatures like those found in hot springs and hydrothermal vents. Other archaea are psychrophiles that can live in association with sea ice and glacial lakes. Others are tolerant to extreme pressure, acidity, alkalinity, and salinity, and it will be valuable to determine if these microbes have forensic value due to their extremely diverse habitat ranges. For example, some archaea are methanogens, generating methane from carbon dioxide and hydrogen. These archaea might be of value to the investigation of decomposing remains.

1.7 Conclusions

Modern advances in microbiology and microbial ecology allow for the study of prokaryotes and eukaryotes at an unprecedented resolution. This knowledge is potentially helpful for forensic science because it allows us to better understand microorganisms and their environment. Some key environmental parameters that significantly affect microorganisms are moisture, oxygen, and temperature. These parameters must be considered when interpreting microbial evidence. Also critical is the understanding that all microbes are in competition for energy and nutrients. Understanding how microbes compete to survive is vital for a robust interpretation of microbial evidence. In terms of taxonomy, many recent studies have highlighted the forensic value of bacterial phyla Actinobacteria, Firmicutes, and Proteobacteria. We expect that these taxa will continue to be explored as physical evidence, and we are hopeful that significant attention will be paid to Archaea and microbial eukaryotes, as many of these organisms have already shown some potential to serve as bioindicators of past events.

Acknowledgments

We kindly thank R. W. Carter and C. Carter for the helpful comments on the manuscript. We also thank T. Carvalho and the University of Hawaii at Manoa Biological Electron Microscope Facility for assistance with scanning electron micrographs.

References

Almirall, J. and Trejos, T. (2016) Analysis of glass evidence, in *Forensic Chemistry: Fundamentals and Applications* (ed J.A. Siegel), John Wiley & Sons, Ltd., Chichester, UK, pp. 228–272.

Anderson, B., Meyer, J. and Carter, D.O. (2013) Dynamics of ninhydrin-reactive nitrogen and pH in gravesoil during the extended postmortem interval. *Journal of Forensic Sciences*, **58**, 1348–1352.

Ball, A.S. (1997) *Bacterial Cell Culture: Essential Data*, John Wiley & Sons, Ltd., Chichester, UK.

Benbow, M.E., Pechal, J.L., Lang, J.M. *et al.* (2015) The potential of high-throughput metagenomic sequencing of aquatic bacterial communities to estimate the postmortem submersion interval. *Journal of Forensic Sciences*, **60**, 1500–1510.

Bisbing, R.E. (2016) The forensic comparison of soil and geological microtraces, in *Forensic Chemistry: Fundamentals and Applications* (ed J.A. Siegel), John Wiley & Sons, Ltd., Chichester, UK, pp. 273–317.

Bouslimani, A., Porto, C., Rath, C.M. *et al.* (2015) Molecular cartography of the human skin surface in 3D. *Proceedings of the National Academy of Sciences of the United States of America*, **112**, E2120–E2129.

Brown, J.W. (2015) *Principles of Microbial Diversity*, American Society for Microbiology Press, Washington, DC.

Brzezinski, J.L. and Craft, D.L. (2012) Characterization of microorganisms isolated from counterfeit toothpaste. *Journal of Forensic Sciences*, **57**, 1365–1367.

Budowle, B., Schutzer, S.E., Breeze, R.G. *et al.* (eds) (2011) *Microbial Forensics*, 2nd edn, Academic Press, Burlington, MA.

Caplan, M.J. and Koontz, F.P. (2001) *Cumitech 35 Postmortem Microbiology*, American Society for Microbiology Press, Washington, DC.

Carter, D.O. and Tibbett, M. (2003) Taphonomic mycota: fungi with forensic potential. *Journal of Forensic Sciences*, **48**, 168–171.

Carter, D.O. and Tibbett, M. (2006) Microbial decomposition of skeletal muscle tissue (*Ovis aries*) in a sandy loam soil at different temperatures. *Soil Biology and Biochemistry*, **38**, 1139–1145.

Carter, D.O. and Tibbett, M. (2008) Does repeated burial of skeletal muscle tissue (*Ovis aries*) in soil affect subsequent decomposition? *Applied Soil Ecology*, **40**, 529–535.

Carter, D.O., Yellowlees, D. and Tibbett, M. (2007a) Autoclaving can kill soil microbes yet enzymes can remain active. *Pedobiologia*, **51**, 295–299.

Carter, D.O., Yellowlees, D. and Tibbett, M. (2007b) Cadaver decomposition in terrestrial ecosystems. *Naturwissenschaften*, **94**, 12–24.

Carter, D.O., Yellowlees, D. and Tibbett, M. (2008) Temperature affects microbial decomposition of cadavers (*Rattus rattus*) in contrasting soils. *Applied Soil Ecology*, **40**, 129–137.

Carter, D.O., Yellowlees, D. and Tibbett, M. (2010) Moisture can be the dominant environmental parameter governing cadaver decomposition in soil. *Forensic Science International*, **200**, 60–66.

Carter, D.O., Knight, R. and Metcalf, J.L. (2014) *A preliminary survey of a postmortem skin microbiome on Oahu*. Proceedings of 66th Annual Meeting of the American Academy of Forensic Sciences, vol. **20**, Seattle, WA, February 17–22, 2014, American Academy of Forensic Sciences, Colorado Springs, CO, pp. 361–362.

Carter, D.O., Metcalf, J.L., Bibat, A. and Knight, R. (2015) Seasonal variation of postmortem microbial communities. *Forensic Science, Medicine, and Pathology*, **11**, 202–207.

Chimutsa, M., Olakanye, A.O., Thompson, T.J.U. and Komang Ralebitso-Senior, T. (2015) Soil fungal community shift evaluation as potential cadaver decomposition indicator. *Forensic Science International*, **257**, 155–159.

Chun, L., Miguel, M., Junkins, E.N. *et al.* (2015) An initial investigation in the ecology of culturable postmortem bacteria. *Science and Justice*, **55**, 394–401.

Damann, F.E., Williams, D.E. and Layton, A.C. (2015) Potential use of bacterial community succession in decaying human bone for estimating postmortem interval. *Journal of Forensic Sciences*, **60**, 844–850.

Degens, B.P. and Harris, J.A. (1997) Development of a physiological approach to measuring the catabolic diversity of soil microbial communities. *Soil Biology and Biochemistry*, **9–10**, 1309–1320.

Fierer, N., Lauber, C.L., Zhou, N. *et al.* (2010) Forensic identification using skin bacterial communities. *Proceedings of the National Academy of Sciences*, **107**, 6477–6481.

Finley, S.J., Pechal, J.L., Benbow, M.E. *et al.* (2016) Microbial signatures of cadaver gravesoil during decomposition. *Microbial Ecology*, **71**, 524–529.

Forbes, S.L. (2008) Decomposition chemistry in a burial environment, in *Soil Analysis in Forensic Taphonomy: Chemical and Biological Effects of Buried Human Remains* (eds M. Tibbett and D.O. Carter), CRC Press, Boca Raton, FL, pp. 203–223.

Forbes, S.L., Troobnikoff, A.N., Ueland, M. *et al.* (2016) Profiling the decomposition odour at the grave surface before and after probing. *Forensic Science International*, **259**, 193–199.

Government Accountability Office (GAO) (2005) Anthrax Detection: Agencies Need to Validate Sampling Activities in Order to Increase Confidence in Negative Results. GAO, Washington, DC.

Grice, E.A. and Segre, J.A. (2011) The skin microbiome. *Nature Reviews Microbiology*, **9**, 244–253.

Guo, J., Fu, X., Liao, H. *et al.* (2016) Potential use of bacterial community succession for estimating postmortem interval as revealed by high throughput sequencing. *Scientific Reports*, **6**, 24197.

Hauther, K.A., Cobaugh, K.L., Meadows-Jantz, L. *et al.* (2015) Estimating time since death from postmortem human gut microbial communities. *Journal of Forensic Sciences*, **60**, 1234–1240.

Hawksworth, D.L. and Wiltshire, P.E.J. (2011) Forensic mycology: the use of fungi in criminal investigations. *Forensic Science International*, **206**, 1–11.

Hitosugi, M., Ishii, K., Yaguchi, T. *et al.* (2006) Fungi can be a useful forensic tool. *Legal Medicine*, **8**, 240–242.

Houck, M. (2016) Textiles, in *Forensic Chemistry: Fundamentals and Applications* (ed J.A. Siegel), John Wiley & Sons, Ltd., Chichester, UK, pp. 40–74.

Hug, L.A., Baker, B.J., Anantharaman, K. *et al.* (2016) A new view of the tree of life. *Nature Microbiology*, **1**, 1–6.

Iancu, L., Carter, D.O., Junkins, E.N. and Purcarea, C. (2015) Using bacterial and necrophagous insect dynamics for post-mortem interval estimation during cold season: novel case study in Romania. *Forensic Science International*, **254**, 106–117.

Ishii, K., Hitosugi, M., Kido, M. *et al.* (2006) Analysis of fungi detected in human cadavers. *Legal Medicine*, **8**, 188–190.

Jayachandran, K., Stolpe, N.B., Moorman, T.B. and Shea, P.J. (1998) Application of ^{14}C most probable number technique to enumerate atrazine-degrading microorganisms in soil. *Soil Biology and Biochemistry*, **30**, 523–529.

Killham, K. and Prosser, J.I. (2007) The prokaryotes, in *Soil Microbiology, Ecology, and Biochemistry*, 3rd edn (ed E.A. Paul), Academic Press, Burlington, MA, pp. 119–144.

Kirkbride, P. (2016) Paint and coatings examination, in *Forensic Chemistry: Fundamentals and Applications* (ed J.A. Siegel), John Wiley & Sons, Ltd., Chichester, UK, pp. 75–134.

Lauber, C.L., Metcalf, J.L., Keepers, K. *et al.* (2014) Vertebrate decomposition is accelerated by soil microbes. *Applied and Environmental Microbiology*, **80**, 4920–4929.

Lax, S., Smith, D.P., Hampton-Marcell, J. *et al.* (2014) Longitudinal analysis of microbial interaction between humans and the indoor environment. *Science*, **345**, 1048–1052.

Lax, S., Hampton-Marcell, J.T., Gibbons, S.M. *et al.* (2015) Forensic analysis of the microbiome of phones and shoes. *Microbiome*, **3**, 21.

Locard, E. (1930a) The analysis of dust traces. Part III. *American Journal of Political Science*, **1**, 496–514.

Locard, E. (1930b) The analysis of dust traces. Part II. *American Journal of Political Science*, **1**, 401–418.

Locard, E. (1930c) The analysis of dust traces. Part I. *American Journal of Political Science*, **1**, 276–298.

Ma, Q., Fonseca, A., Liu, W. *et al.* (2012) *Proteus mirabilis* interkingdom swarming signals attract blow flies. *The ISME Journal*, **6**, 1356–1366.

MacCallum, W.G. and Hastings, T.W. (1899) A case of acute endocarditis caused by *Micrococcus zymogenes* (nov. spec.), with a description of the microorganism. *Journal of Experimental Medicine*, **4**, 521–534.

Madigan, M., Martinko, J., Stahl, D. and Clark, D. (2012) *Brock Biology of Microorganisms*, 13th edn, Pearson, San Francisco, CA.

Metcalf, J.L., Wegener Parfrey, L., Gonzalez, A. *et al.* (2013) A microbial clock provides an accurate estimate of the postmortem interval in a mouse model system. *eLife*, **2**, e01104.

Metcalf, J.L., Xu, Z.Z., Weiss, S. *et al.* (2016) Microbial community assembly and metabolic function during mammalian corpse decomposition. *Science*, **351**, 158–162.

Nagasawa, S., Motani-Saitoh, H., Inoue, H. and Iwase, H. (2013) Geographic diversity of *Helicobacter pylori* in cadavers: forensic estimation of geographical origin. *Forensic Science International*, **229**, 7–12.

Olakanye, A.O., Thompson, T. and Komang Ralebitso-Senior, T. (2015) Soil fungal community shift evaluation as a potential cadaver decomposition indicator. *Forensic Science International*, **257**, 155–159.

Pechal, J.L., Crippen, T.L., Tarone, A.M. *et al.* (2013) Microbial community functional change during vertebrate carrion decomposition. *PLoS One*, **8**, e79035.

Pechal, J.L., Crippen, T.L., Benbow, M.E. *et al.* (2014) The potential use of bacterial community succession in forensics as described by high throughput metagenomic sequencing. *International Journal of Legal Medicine*, **128**, 193–205.

Plante, A.F. (2007) Soil biogeochemical cycling of inorganic nutrients and metals, in *Soil Microbiology, Ecology, and Biochemistry* (ed E.A. Paul), Academic Press, Burlington, MA, pp. 389–432.

Putman, R.J. (1978) Patterns of carbon dioxide evolution from decaying carrion. 1. Decomposition of small mammal carrion in temperate systems. *Oikos*, **31**, 47–57.

Sagara, N., Yamanaka, T. and Tibbett, M. (2008) Soil fungi associated with graves and latrines: toward a forensic mycology, in *Soil Analysis in Forensic Taphonomy: Chemical and Biological Effects of Buried Human Remains* (eds M. Tibbett and D.O. Carter), CRC Press, Boca Raton, FL, pp. 67–108.

Slonczewski, J.L. and Foster, J.W. (2011) *Microbiology: An Evolving Science*, 2nd edn, W. W. Norton and Company, New York, NY.

Song, S.J., Lauber, C., Costello, E.K. *et al.* (2013) Cohabiting family members share microbiota with one another and with their dogs. *eLife*, **2**, e00458.

Statheropoulos, M., Spiliopoulou, C. and Agapiou, A. (2005) A study of volatile organic compounds evolved from the decaying human body. *Forensic Science International*, **153**, 147–155.

Swift, M.J., Heal, O.W. and Anderson, J.M. (1979) *Decomposition in Terrestrial Ecosystems*, Blackwell Scientific, Oxford, UK.

Tridico, S.R., Murray, D.C., Addison, J. *et al.* (2014) Metagenomic analyses of bacteria on human hairs: a qualitative assessment for applications in forensic science. *Investigative Genetics*, **5**, 16.

Vass, A.A. (2012) Odor mortis. *Forensic Science International*, **222**, 234–241.

Vass, A.A., Bass, W.M., Wolt, J.D. *et al.* (1992) Time since death determinations of human cadavers using soil solution. *Journal of Forensic Sciences*, **37**, 1236–1253.

Wiltshire, P.E.J., Hawksworth, D.L., Webb, J.A. and Edwards, K.J. (2014) Palynology and mycology provide separate classes of probative evidence from the same forensic samples: a rape case from southern England. *Forensic Science International*, **244**, 186–195.

Woese, C.R., Kandler, O. and Wheelis, M.L. (1990) Towards a natural system of organisms: proposal for the domains Archaea, Bacteria, and Eucarya. *Proceedings of the National Academy of Sciences of the United States of America*, **87**, 4576–4579.

Young, J.M., Weyrich, L.S., Breen, J. *et al.* (2015) Predicting the origin of soil evidence: high throughput eukaryote sequencing and MIR spectroscopy applied to a crime scene scenario. *Forensic Science International*, **251**, 22–31.

Young, J.M., Weyrich, L.S. and Cooper, A. (2016) High throughput sequencing of trace quantities of soil provides reproducible and discriminative fungal DNA profiles. *Journal of Forensic Sciences*, **61**, 478–484.

Zheng, L., Yu, Z., Crippen, T.L. *et al.* (2013) Bacteria mediate oviposition by the black soldier fly, *Hermetia illucens* (L.), (Diptera: Stratiomyidae). *Scientific Reports*, **3**, 2563.

CHAPTER 2

History, current, and future use of microorganisms as physical evidence

Zachary M. Burcham and Heather R. Jordan

Department of Biological Sciences, Mississippi State University, Starkville, MS, USA

2.1 Introduction

2.1.1 Why and how are microorganisms used in forensic science?

What is forensic microbiology? What differentiates forensic microbiology from molecular epidemiology? The answer lies in focus and application. Molecular epidemiology has been used for years to trace outbreaks of microbial diseases. However, what distinguishes forensic microbiology from molecular epidemiology is that forensic microbiology data must hold up, not only to the scrutiny of scientists in the health care community, but also to the scrutiny of judges and juries. Nonetheless, work done to date on microbial epidemiology will provide an invaluable starting point for the additional work that needs to be done to make microbial forensics ready for its day in court. Forensic microbiology is thus a marriage between the application of scientific method and the criminal justice system.

Forensic microbiology is a relatively new field. In fact, the push for the increased utility of microorganisms in forensic science intensified as the fear of biological terrorism and biocrimes increased after the anthrax attacks in the United States in 2001. Since then, the use of microbes has been seen as the next frontier in forensic science. It was not until recently that the ability to study microorganisms at the genomic level led to the power to accurately and consistently use them in forensics. Microorganisms are expected to provide insights into the origin of biocrimes, cause of death, and time since death, and also to provide evidence in other crimes such as sexual assault and medical malpractice.

In this chapter, we will (i) introduce methods for microbial identification including traditional, culture-based methods followed by newer molecular technologies targeting microbial nucleic acids, (ii) show how all of these methods have been studied to answer

Forensic Microbiology, First Edition. Edited by David O. Carter, Jeffery K. Tomberlin, M. Eric Benbow and Jessica L. Metcalf.

questions about postmortem interval and cause of death, (iii) describe how microbial methods facilitate investigations of biocrimes, and (iv) look at how microbes are utilized as trace evidence in violent crimes, such as homicide and sexual assault, medical malpractice, and agricultural contamination. Overall, the chapter encompasses the importance of microbial evidence in past, current, and future criminal cases.

2.2 Methods for identification

2.2.1 Classical microbiology

The oldest form of microbial identification is by observing and biochemically testing a microorganism in pure culture (Maccallum and Hastings, 1899). Various growth media have been developed and refined. These methods range from being able to provide nutrients to a broad spectrum of microorganisms, to selective and differential media for the isolation of targeted genera or species (Murray *et al.*, 2009). Growth media are usually found in either an agar or a broth form (Figure 2.1), and the medium chosen as well as additives for enrichment is dependent upon the desired microorganisms to be isolated (see Chapter 6 for more details on culturing).

Once a microorganism has been obtained in pure culture, it can be analyzed phenotypically and/or genotypically. The phenotype is the physical appearance based upon the genetic makeup of the organism, and the genotype is the specific genetic foundation leading to the phenotype (Slonczewski and Foster, 2013). Historically, microbiologists did not possess the tools to study genotypes intensively; therefore, we will only focus on

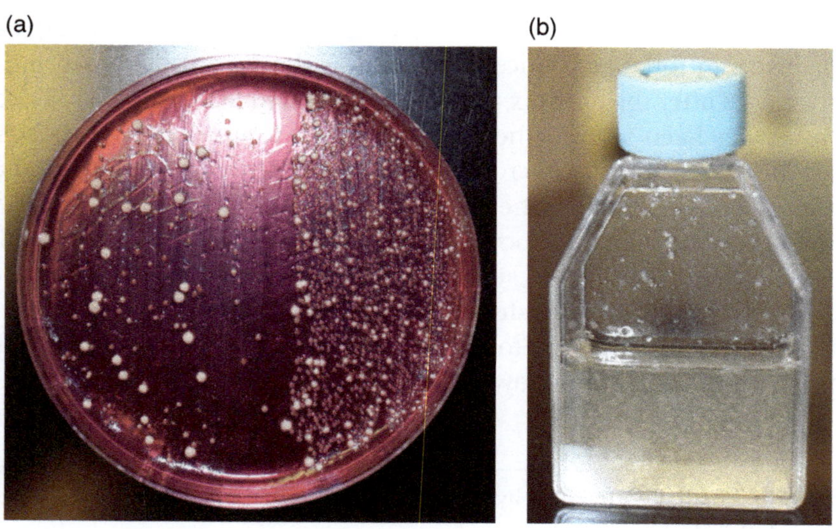

Figure 2.1 Examples of bacteria isolated on agar (a) and broth (b) growth media

phenotypes in this section, as genotyping and genome studies will be discussed in subsequent Section 2.2.2.

The phenotype of a microorganism can be described in a number of ways. Observing the color, colony morphology, pattern of growth on agar and in broth, growth requirements, and antibiotic susceptibility are just a few examples. The biochemical processes that the organism is capable of performing such as creating the catalase enzyme that is responsible for breaking down hydrogen peroxide to oxygen and water can also be used to identify some microbes (Murray *et al.*, 2009; Slonczewski and Foster, 2013).

Although classical, culture-based methods have and continue to advance our knowledge of microorganisms, there are restrictions that limit our ability to analyze all organisms this way. Not every microorganism is able to be cultured in a standard laboratory setting; in fact, it has been estimated that only <1 to 2% of microbial species are culturable (Amann *et al.*, 1995; Wilson *et al.*, 1997). Other limiting factors may be that some organisms are fastidious with specific environmental and nutritional requirements or prolonged growth times, and the organisms present must be live in order to be cultivated. When combining these limitations, the ability to obtain a usable culture that provides strong evidence in a forensic case can be difficult. The use of culturing and long-term organism storage is discussed in greater detail in Chapter 6.

2.2.2 Genomics and strain typing

Traditionally, molecular methods such as sequencing of a microbe have been performed on culturable isolates grown in a laboratory setting. Sequencing allows detection of targeted genes as well as entire genomes of specific microorganisms. However, most microbes grow as a mixed microbial community, and potential biases become possible from the study of microbial genetic diversity, gene expression, and identification from a single-species environment. Microbial community analysis using next-generation sequencing targeting microbial nucleic acids has recently been utilized as a method to study genetic material recovered from mixed microbial populations in naturally occurring environments. This ability to provide information specific for species identification to the strain level from both live and dead microorganisms that are nonculturable and in mixed populations has greatly advanced the field of microbiology (Woese *et al.*, 2000). A few common sequencing techniques include 16S or 18S ribosomal RNA (rRNA) gene sequencing, next-generation sequencing such as pyrosequencing, and pulsed field gel electrophoresis (Gu *et al.*, 2015; Harmsen and Karch, 2004; Kripalani-Joshi and Law, 1994). From these, each DNA sequence can be compared against known sequences archived in online databases. A list of sequencing methods and their application can be found in Table 2.1. Non-DNA-based analytical technologies can also be utilized, such as studying microbial chemistry and matrix using mass spectrometry (Seng *et al.*, 2009). We will only provide a cursory overview of these methods; Chapters 3–5 describe sampling methods, data generation, and analyses in greater detail.

Table 2.1 Sequencing methods and applications

Method	Target	Application	Method summary	References
16/18S ribosomal RNA (rRNA) gene	rRNA gene	Phylogenetic determination	Highly conserved rRNA genes containing slight mutations across species are sequenced for taxonomy.	Wimberly *et al.* (2000), Schroeder *et al.* (2001)
Single-nucleotide polymorphisms	Point mutations	Strain Identification	Single-nucleotide mutations in conserved regions of DNA is sequenced for identification of strains within species.	Nachman (2001)
Internal transcribed spacers	Noncoding DNA	Species Identification	Highly variable regions of noncoding DNA between rRNA genes are sequenced for species identification.	Lafontaine and Tollervey (2001), Baldwin (1992)
Whole genome shotgun	Entire genome	Genome Assembly	Entire genomic DNA is broken apart, sequenced, and aligned based on overlapping regions to recreate the full original genome.	Ventner *et al.* (2004)

These genomic methods have been used in forensic science to create genetic profiles for a single species and also for the identification of entire microbial communities. Genetic profiling can provide ancestral information on the microbial communities involved that can suggest if the strains are found in a certain geographic location or a laboratory, or provide profiles used to compare DNA found at a crime scene to a suspect or a victim. Narrowing down the locations is useful when investigating a criminal case so that time and energy can be concentrated on the correct tasks. We encourage the readers to review Chapters 3 and 4 for a focused discussion on experimental design and research considerations associated with sampling and sequencing microbial communities within the forensic context.

Sequencing of the 16S rRNA gene was one of the first targets for identifying taxa/organisms in the domains of bacteria and archaea species and determining their phylogenetic relationships. The 16S rRNA gene codes for a single-stranded RNA that is a component of the 30S small subunit of prokaryotic ribosomes. Similarly, the eukaryotic homolog that can be used for identification is the 18S rRNA that codes for the 40S small subunit of eukaryotic ribosomes. Both the 16S and the 18S rRNA genes play an important role in the formation of the 70S ribosome and the 80S ribosome, respectively, which are subsequently used in translation for protein synthesis (Schroeder *et al.*, 2001; Wimberly *et al.*, 2000). Because of their importance for survival, large parts of the gene sequence in both ribosomes are highly conserved across taxonomic phyla. While the gene is highly conserved, there are variable regions with interspecific polymorphisms or mutations that can be used for identification. Universal primers are used to locate and begin translation at the conserved regions and then continue translating the variable regions. Other regions of DNA used for taxonomic analyses are the noncoding regions of DNA between the genes for ribosomal RNA called the internal transcribed spacer (ITS) such as between the 16S and the 23S in bacteria and archaea (Lafontaine and Tollervey, 2001). These regions have a high mutation rate since they are nonessential for survival, and this allows them to be compared across similar species (Baldwin, 1992). It is important to note that since prokaryotic mitochondria are present inside eukaryotes due to an endosymbiotic relationship, mitochondrial 16S rRNA from the host will be present in the microbial community and needs to be excluded. The resulting sequences are then compared to online sequence databases for taxa-referenced matches from either the 16S, 18S, or ITS. In forensic science, this method can be useful for quickly determining microorganisms associated with death from an infection, using the presence of microorganisms to determine if the individuals have been to a location specific for those strains, or to investigate if a person was intentionally infected. The intentional use of a pathogen to cause harm to an individual is called a biocrime and is discussed later in the chapter.

While all microorganisms of a given species may have closely related DNA sequences up to 99.9% similarity, they are not exactly the same. This can arise from a phenomenon called single-nucleotide polymorphisms (SNPs) where a nucleotide has mutated into a different nucleotide base. The mutation may not affect the overall protein created, but it can be used as a marker for specific strains in a species or help determine the ancestry (Nachman, 2001). The ability to determine the exact strain of a species is known as strain typing and is extremely useful in biocrime cases. The method of strain typing was widely utilized after the 2001 anthrax attacks in the United States to determine that the *Bacillus anthracis* spores used were likely from the Ames strain found at the government's biodefense lab in Frederick, Maryland (Van Ert *et al.*, 2007).

The identification of a microorganism is often a crucial component of an investigation, and the ability to identify microbes has improved significantly. However, there are multiple reasons why an organism can be misidentified after sequencing: (i) microbial communities exchange DNA with each other constantly and have the potential to exchange a genetic forensic marker; (ii) pathogenic strains and nonpathogenic strains can have different sequences; and (iii) databases can provide incorrect or incomplete organism identification, as database accuracy has been shown to vary (Budowle, 2004; Budowle *et al.*, 2009; LeClerc *et al.*, 1996). Another drawback of using the newer

Table 2.2 Advantages and limitations of culture-based versus molecular-based microbiological methods

Classical microbiology		Molecular microbiology	
Advantages	**Disadvantages**	**Advantages**	**Disadvantages**
• Inexpensive • Often specific • Distinguishes between viable and nonviable organisms • Can determine phenotype	• Low sensitivity • Training required • Requires dedicated laboratory • Often slow results • Often requires pure culture	• Sensitive • Specific • Distinguishes between strains depending upon target • Rapid	• Expensive • Requires strict quality control • Requires specialized laboratory • Requires highly trained personnel • Cannot distinguish between viable and nonviable • Requires isolation of DNA or RNA • Requires personnel skilled in bioinformatics for analysis

identification methods comes from the fact that some of the methods are not yet validated. Extensive amounts of quality assurances must be performed to validate the information obtained from them. The methods must go through rigorous validation by having strict guidelines of procedure and equipment calibration that is consistent across the forensic field and that complies with the current chain of custody laws. The quality of information is one of the most important aspects in being able to use data as evidence in a criminal case that cannot be thrown out or discredited. A comparison of culture- and genomic-based methods can be seen in Table 2.2. While these identification methods may not be perfect, they are important steps in the ability to use microorganisms in forensics. One of the research areas where these analyses have been implemented is in determining postmortem interval (PMI).

2.3 Estimating PMI

Current forensic analyses of body decomposition for estimating the PMI—the time elapsed since death occurred—are mainly derived from taphonomy (Sorg and Haglund, 1996), or in some cases using organisms (e.g., insects) that have colonized a set of human remains (Amendt *et al.*, 2011; Tomberlin *et al.*, 2011). Time of death is an extremely critical part of the evidence that can include or exclude witnesses and suspects. Investigators will use invertebrates (see Chapter 11) to help narrow the PMI estimation window by noting their abundance and current life stage (e.g., larval and egg) (Rodriguez and Bass, 1983) or the succession of the arthropod community (Goff and Flynn, 1991). Another commonly used method relies on using the stage of decomposition to estimate the PMI (Megyesi *et al.*, 2005). The stages of decomposition follow an order of events that occur over decomposition time. Stages of decomposition can be described as fresh, discolorization, bloat, active decay, and skeletonization (Wilson-Taylor, 2013), although some researchers have more elaborate descriptors associated with insect colonization and succession (Payne, 1965; Wilson-Taylor, 2013). Each stage is marked by certain characteristics such as mottled skin, livor mortis, rigor mortis in the fresh and discolorization stage, distention (bloat), purging of body fluids (active), and exposure and drying of the skeleton (skeletonization) that can be useful tools for narrowing the PMI (see Figure 2.2). However, multiple factors, such as temperature, humidity/moisture, oxygen availability, sunlight exposure, burial, cause and manner of death, and body composition, affect how quickly decomposition takes place (Carter *et al.*, 2007; Mann *et al.*, 1990). Environmental factors, such as temperature and moisture, apparently have the most significant impacts on the speed of the decomposition process. Studies have shown that hot, humid environments will cause decay to take place at a drastically quicker pace (within 4 minutes), as compared to cold dry environments, which may take multiple days to reach the same decomposition stage (Carter *et al.*, 2007; Mann *et al.*, 1990; Vass, 2001). This is because moisture is available for microbes and the enzymatic temperature preferred is around the natural body temperature that they are accustomed to inhabiting (Janaway *et al.*, 2009).

While these factors have been studied extensively, it was only recently that another major player in the decomposition process, microorganisms, has found new appreciation. The utilization of microbes in medicolegal death investigation has been overlooked

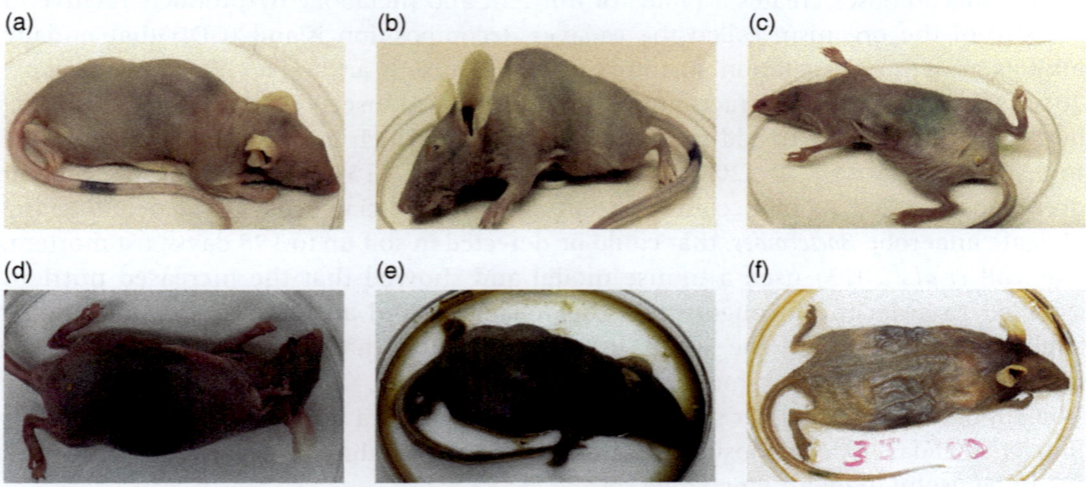

Figure 2.2 Mouse model displaying stages of decomposition. Fresh stage immediately after sacrifice (a), 3–4 hours after death showing signs of discolorization and livor mortis (b), and 5–7 hours after death showing rigor mortis (c). The remaining stages include bloat (d), active decay, shown here with purging of body fluids (e), and drying of tissues later in the stage (f), and natural mummification (g) or skeletonization (not shown). Mouse mass and length varied between 16–22 g and 7.5–8.5 cm, respectively

largely because of their difficulty to visualize and culture. Now that next-generation methods of sequencing are available, it has become more apparent that these microscopic organisms are a driving force in decomposition processes (Gunn and Pitt, 2012; Lauber *et al.*, 2014). Prokaryotic and eukaryotic organisms present in the environment around the corpse (e.g., soil), on the corpse, and inside the corpse begin breaking down tissues and fats for nutrition once the immune system ceases to function (Janaway *et al.*, 2009). They are also able to transmigrate and colonize previously uninhabitable areas of the body in search for nutrients and space to proliferate (Janaway *et al.*, 2009). Once it was discovered that microbes play such a significant role in decomposition, the question raised was whether microbial presence, abundance, and transmigration could be used to estimate PMI. We will introduce studies aimed toward answering this question by analyzing the soil and cadaver microbiota, but this will be discussed in greater detail in Chapters 8 and 10.

2.3.1 Microbial succession

2.3.1.1 Soil microbial communities

Cadaver decomposition results in the release of organic and ammonia-rich nutrients, altering both the pH and nutrient content of the soil stimulating microbial activity (Cobaugh *et al.*, 2015). Research has shown that cadaver-released nutrients yield higher amounts of microbial biomass carbon and nitrogen, respiration, ammonium, and sulfide (Carter *et al.*, 2007; Cobaugh *et al.*, 2015; Hopkins *et al.*, 2000). The area in which a cadaver decomposes creates a radius of nutrient and metabolic by-products relative to the size of the organism called the cadaver decomposition island (CDI) that initially inhibits growth of vegetation surrounding the carcass (Carter *et al.*, 2007; Weiss *et al.*, 2016). The cadaver also attracts animal scavengers and insects that defecate and/or die in the area continuing to add to the nutrient influx, which increases microbial activity. In a study by Carter *et al.* (2007) and Cobaugh *et al.* (2015), it was found that human decomposition led to high concentrations of host-associated bacteria, including the obligate anaerobe *Bacteroides*, that could be detected in soil up to 198 days post mortem. (Metcalf *et al.*, 2013) used a mouse model and showed that the increased nutrient load led to a decline in oligotrophic Acidobacteria and an increase in copiotrophic Alphaproteobacteria. In the study, it was also found that the microbial eukaryotic community at all sites became dominated by a nematode, *Oscheius tipulae*, in the family Rhabditidae during the later stages of decomposition. Data from these demonstrate the impact of cadaver decomposition in nutrient pulses, leading to organismal shifts that might be useful in identifying grave soil.

Recent data suggest that exogenous soil microbial communities also significantly influence the rate of decomposition processes (Lauber *et al.*, 2014), and that cadavers placed on or in a soil substrate with intact microbial communities reach advanced stages of decomposition much more rapidly than those in sterilized soil or in the absence of soil (Lauber *et al.*, 2014). Taxonomic and functional succession of microbial populations in soils below decomposing human cadavers showed no change in total bacterial abundances; however, distinct changes in both function and community composition associated with decomposition from active to advanced decay were found (Cobaugh *et al.*, 2015). Additionally, Lauber *et al.* (2014) showed significant differences in microbial communities associated with cadaver skin and sterile versus untreated soils; however, the authors were unable to differentiate whether the decomposer community was influenced more from the soil or the original cadaver microbial community. Nevertheless, these data provide strong evidence that the substrate on which carrion decomposes influence microbial decomposer community, and these data may be informative for identifying predictor populations for estimating PMI.

Environmental parameters such as temperature, moisture, soil type, and soil texture can affect the rate at which decomposition occurs and the ability of decomposer organisms to function (Carter *et al.*, 2007, 2008, 2010; Hopkins *et al.*, 2000). Burial also affects the quantity of organic material left from a cadaver in the soil since breakdown is mainly dependent on microbes, and scavengers are not able to ingest the tissues and displace the organic material (Carter *et al.*, 2007). It is likely that these parameters will affect postmortem microbial communities.

For instance, grave soil microbial communities, and corresponding cadaver decomposition, are influenced by season (Carter *et al.*, 2015). Decomposition has been observed to be approximately three times slower in winter than in summer, with increased insect activity also observed in summer. The authors found that eukaryotic microbiomes were not significantly different overall between seasons, but that bacterial communities showed a significant difference (Carter *et al.*, 2015). Whether these parameters would be universal across geographical locations, climates, microbial communities, or cadavers is not known. However, this underscores the importance of understanding these relationships in order to develop a method to estimate PMI for use in the criminal justice system.

2.3.1.2 Mammalian microbial communities

Another important consideration is the cadaver-associated microbial communities and their impact on decomposition. Several studies have begun to reveal the complex dynamics associated with decomposing mammals, including mice and swine carrion, as well as human cadavers to determine the utility of these cadaver-associated microbial communities for determining PMI. Mammals contain a large community of microorganisms that live inside and outside of their body. These organisms outnumber the host cells by 10-fold and can provide benefits, opportunistic harm, or have no effect at all on the host (Bengmark, 1998). The microorganisms usually live in a specific niche on the host, for instance, bacteria that prefer to live in the nasal passages or others that prefer the colon. In a healthy individual, organisms remain in their preferred niche; but if a dramatic event happens, they can change the host's environment and immune system that allows the organisms to transmigration to other areas of the body (Janaway *et al.*, 2009). This drastic change can be caused by multiple factors, such as illness, but permanent host environmental changes will also take place when the host dies and begins to decompose (Clark *et al.*, 1997; Janaway *et al.*, 2009). If we want to understand microbe transmigration after death in humans, then we need to start by studying systems in animal surrogate models (see limitations discussed in Chapter 3). This information will provide context into our own microbiota and aid in determining whether microbiota transmigration can be a useful tool in forensics for estimating PMI.

Melvin *et al.* (1984) performed a study investigating postmortem transmigration of the microbiota through the small intestine of mice. The authors sacrificed mice and removed a segment of their small intestine and left them to incubate at 4, 25, or 37°C for 3 days. At each temperature, transmigration was measured from the gut tissue to the saline by culture and scanning electron microscopy of aerobic, anaerobic, and mycotic species. They found that the dominant organisms were *Staphylococcus* spp., followed by coliforms, fungi, and lastly anaerobes. Within the 4°C model, *Staphylococcus* spp. did not appear until 66–68 hours, coliforms/fungi did not appear until 68–72 hours, and coliforms/anaerobes were rare at 72 hours. The low temperature slowed the rate of transmigration as compared to the 25 and 37°C models. In the 25°C model, *Staphylococcus* spp. appeared in 5–6 hours, coliforms/fungi in 8–10 hours, and coliforms/anaerobes in 12–16 hours (Melvin *et al.*, 1984). This transmigration occurred drastically quicker than the 4°C model, but still slower than the 37°C model that provided the fastest transmigration. The 37°C model showed *Staphylococcus* spp. transmigration in as little as 2–3 hours, coliforms/fungi in 4–5 hours, and coliforms/anaerobes in 6–8 hours. This result is not surprising since these models are measuring organisms that live as commensals inside a mouse body whose natural internal body temperature is around 37°C during life. This study showed that transmigration and decomposition by microbes is temperature dependent, and species preferring oxygen dominate host decomposition first, with anaerobic species dominating at later stages of decomposition. Melvin *et al.* (1984) suggested that this information may be useful in estimating the PMI of a known cadaver if the temperature in which decomposition is taking place is known, but the surface of a mouse's small intestine differs from humans. To further test if microbiota of the human host can be used to determine PMI, a model closer to the human anatomy, such as swine, which has a high similarity to a human digestive tract, skin, and immune system, should be assessed (Meurens *et al.*, 2012). Postmortem bacterial transmigration is discussed in detail in Chapter 8.

More recent studies using molecule-based methods of analysis further demonstrated a significant change in microbial communities during decomposition. For example, sequencing of 16S rRNA isolated from skin of swine cadavers was used to assess community composition and succession with respect to decomposition, and it was found that succession was dependent upon season and insect activity (Pechal *et al.*, 2013, 2014). A follow-up study showed that at the time of the swine cadaver placement in the forest (2–3 hours after death), 70% of the microbes were of the phylum Proteobacteria and 20% were of the phylum Firmicutes. Samples of day 1 and day 3 contained about 40%

Proteobacteria and 40% Firmicutes. Samples of day 5 contained about 5% Proteobacteria and 95% Firmicutes (Pechal *et al.*, 2014). These results correlated with the previous study published by Melvin *et al.* (1984) as the aerobic microbes (Proteobacteria) dominated decomposition early and were almost completely replaced by anaerobic species (Firmicutes) by day 5.

Metcalf *et al.* (2013) documented changes of microbial communities associated with mouse decomposition, demonstrating a similar trend of microbial succession in microbial communities associated with decomposing mouse cadavers. Microbial communities associated with the decomposing cadavers became increasingly differentiated from starting communities over time, in the abdominal cavity, gravesoil, and skin sites where similar taxa became abundant at each sample site in the later stages of decomposition. Further, the eukaryotic nematode *O. tipulae* dominated microbial eukaryotic communities during late-stage decomposition.

Since not all forensic cases consist of a cadaver found on or in soil, some research has been performed looking at microbial succession in aquatic environments. While the literature for this field is sparse, Dickson *et al.* (2011) determined that swine head decomposition takes place at a slower rate when submerged in salt water during winter as compared to fall. Further, the marine bacterial phyla and genera that colonize the host were able to be identified but did not provide reliable evidence to determine the postmortem submersion interval (PMSI) with enough accuracy. Another study used algal community growth rates as an indicator of PMSI of pig decomposition. Pigs and ceramic tiles were completely submerged and regularly sampled for periphyton growth. Algal growth rates were greater on pigs compared with tiles, and there was a strong correlation between algal growth rate and time on pigs and tile substrates (Haefner *et al.*, 2004). Both studies strengthen the correlation between temperature and decomposition by microorganisms along with providing a foundation for future research using microorganisms in forensic investigations for determining the length of time a victim has been submerged in water. With animal surrogate models showing consistent data with the potential to estimate PMI, the next step is to perform studies in human models in order to gather information for use in forensic investigations.

2.3.1.3 Studies involving human microbial communities

The use of human samples in microbial forensic investigations is still in the beginning stages of research development. Soil and animal surrogate studies have shown the potential for use of microorganisms as indicators of time since death or placement into an environment occurred. Next, researchers will need to set up experiments at decomposition research areas or "body farms" to study the effects of microbial succession in human models. In one of only a few studies, Hyde *et al.* (2013) used pyrosequencing to measure microbial community richness and diversity from oral, rectal, and internal organs of two human cadavers during stages of pre- and postbloat. Results of this study showed that richness increased from the upper gastrointestinal tract (mouth, stomach, and small intestine) to the lower (colon and rectal/fecal). Firmicutes were the most prevalent among all body sites, though organisms from the phylum Proteobacteria and also *Bacteroides* spp. were present. Samples from the lower gastrointestinal tract and body cavity shared *Clostridium, Lactobacillus, Eggerthella,* and *Bacteroides,* while *Streptococcus, Prevotella,* and *Veillonella* were found in the upper gastrointestinal tract. Firmicutes were found in the oral cavity during prebloat, while Proteobacteria were found during postbloat (Hyde *et al.*, 2013). Hyde *et al.* (2015) recently conducted a study determining microbial succession on several body sites from human cadavers in an outdoor setting. Microbial succession was evident as in previous studies. Additionally, bacteria associated with flies, such as *Ignatzschineria* and *Wohlfahrtiimonas,* were common at bloat and purge and until tissues began to dehydrate or were removed (Hyde *et al.*, 2015).

The future direction for investigating the utility of microbes for measuring PMI will involve sampling from victims in current and future forensic cases with various manners of death and locations of death or discovery. Since current methods of estimating PMI will be used in these forensic investigations, it is imperative to estimate the PMI from the same cadavers using microbial succession so that the data can be compared for accuracy and reliability. The resulting data will also be validated by experiments conducted at decomposition facilities across the world to gain a database of decomposition in multiple environments and the changes of microbial activity associated with them. Microbial estimations of PMI along with current methods will allow investigators to construct an investigative time line to make a stronger case against assailants or release potential false witnesses.

2.4 Cause of death

When a person dies, one of the first objectives is to determine the cause and manner of death. For example, if a drug overdose is suspected, then traces of the drug, or its metabolites, can usually be found in the urine, blood, or tissues; lacerations and bruises may be present if an assault took place; even microorganisms can give clues to the cause of death. Microbes can be used in some cases to determine whether an individual died as a result of natural causes, such as disease, or intent to harm by poisoning or biological attack (Gunn and Pitt, 2012).

2.4.1 Natural causes

2.4.1.1 Disease

Often, an individual dies due to an infection while in a hospital setting. If the medical staff was unable to identify the infectious agent while the patient was alive, diagnostic tests such as serological or genomic analysis may be ordered at autopsy that could give insight into the causative organism. Many times, the etiologic agent is identified antemortem; however, final confirmatory testing of target organs for presence and pathogen concentrations may be conducted postmortem (Rodriguez *et al.*, 2005). This can be beneficial for treating other patients that may be suspected of having the same infection. This information can also be used to determine whether the patient traveled to an area that may be a "hotspot" for this organism or ingested contaminated food and allow officials to either warn the public or quarantine a suspected contaminated region or food source.

One example is sudden infant death syndrome (SIDS), which is one of the leading causes of postneonatal infant mortality in the world. While the exact mechanism of SIDS in unknown, it has been suggested that microorganisms may be playing a crucial role in many of the cases (Gleeson and Cripps, 2004). Studies have shown that viral infections caused by human herpesvirus-6, Epstein–Barr virus, and cytomegalovirus were present among tissues from infants that died from SIDS (Alvarez-Lafuente *et al.*, 2008). SIDS has also been associated with the bacterial organisms *Staphylococcus aureus, Escherichia coli, Streptococcus* spp., and *Haemophilus influenzae* (Blackwell, 2004). While infections are not attributed to all cases of SIDS, proving infection may help alleviate some of the stress on the parents from law enforcement officials since parents are typically investigated for the death of the child, and may also provide the grieving families with some closure during this tragic time.

However, investigators must be hesitant when attributing death from a SIDS case to microbial infection. Human bodies naturally carry numerous bacterial cells that will start to take advantage of the lack of an immune system when the host dies and migrate throughout the organs and tissues that may have previously been sterile during life in search for nutrients (Janaway *et al.*, 2009). This migration can lead to the organism being misdiagnosed as the cause of death since it will be present in the tissue. It is also important to avoid contamination of any of the samples with organisms from the environment. Organisms like *E. coli* can cause deadly infections, but are also ubiquitous, including living nondetrimentally inside and outside the body; therefore, *E. coli* could be suspected as being the agent leading to death while actually being a contaminant from improper sample care. Investigators must be able to differentiate ubiquitous, possibly contaminating organisms from the organism responsible for death.

2.4.1.2 Drowning

When a cadaver is discovered in a body of water or its shore, one of the first suspicions is that the victim drowned. Since the external and internal signs of drowning, such as large amounts of fluid in the lungs and ruptured alveoli, can only suggest drowning took place without complete certainty, there has been a push to find other ways to strengthen the methods of analyzing drowning as a cause of death (Lucci *et al.*, 2008). Whether physical evidence of an altercation is present or not, microbial communities in the water may be able to provide evidence to drowning being the cause of death as well as the location that the drowning took place. One suggested method is to investigate the microorganisms present in the lungs or organs as a result of the inhalation and diffusion of water through the body. The "gold standard" for using microorganisms to determine drowning has been by analyzing diatoms, but other methods utilizing fecal microorganisms have also been proposed (Lin *et al.*, 2014; Lucci *et al.*, 2008).

Diatoms are one of the most common microorganisms used to establish drowning as the cause of death (Timperman, 1972). Diatoms are phytoplanktons composed of a silica cell wall that is able to resist degradation from acid, enzymes, and temperature for a long period of time, which allows them to be present in discovered cadavers (Timperman, 1972; Lin *et al.*, 2014). Diatoms have been found in lung tissue, blood, bone marrow, and internal organs (Ludes *et al.*, 1996). Diatoms are typically only present in cases where the individual drowned in a natural environment, such as a lake, river, or sea. Individuals who drown in treated water (e.g., swimming pool) generally present an absence of diatoms, which is most likely due to the water treatment process (Lin *et al.*, 2014). Diatoms found in the individual can also be compared to those found in the environment to confirm the location that the drowning took place (Lin *et al.*, 2014).

Fecal coliforms and fecal streptococci have also been studied as possible indicator organisms for death by drowning. Since these organisms are smaller than diatoms, it is speculated that they would be able to migrate into the bloodstream faster (Lucci *et al.*, 2008). These organisms are considered ubiquitous since they have been found in seawater, fresh water, and are present inside the gastrointestinal tract (Lucci *et al.*, 2008). However, they are not naturally found in the blood, so migration into the bloodstream along with the drowning medium (e.g., water) may be an indicator of drowning (Lucci *et al.*, 2008). Fecal microorganisms have been shown to provide a possible mechanism of confirming drowning cases. While one might suspect that the bacterial flora may migrate into the bloodstream after death even if the submerged decedent died from factors unrelated to drowning, it appears that the fecal coliforms and streptococci tests are specific enough to confirm drowning in freshwater and seawater without false positives (Lucci *et al.*, 2008).

2.4.2 Biocrimes

Creating biological weapons can be accomplished with little to no expertise in the microbiological field and without the need for large amounts of space. Although the amount of a pathogen needed to create massive causalities would be difficult, creating panic could be achieved with infection of only a small amount of cases (Henderson, 1999). The National Center for Biotechnology Information (NCBI) defines a biocrime as "an assault crime, except, instead of a gun or knife, the weapon is a pathogen or a toxin" (Schutzer *et al.*, 2005). Biocrimes can target individuals specifically or an area's agriculture or water supply.

2.4.2.1 Biocrimes involving humans

One, now infamous, biocrime took place in 1994 when a gastroenterologist knowingly injected his girlfriend with a mixed blood sample that contained human immunodeficiency virus (HIV) and hepatitis C while telling her that the shot contained vitamin B.

In 1995, the victim tested positive for HIV, sparking an investigation even though she had been HIV negative through 1994. The woman's past sexual partners were tested for HIV, and all were negative. There was also no record of any kind of incident or accident that may have caused her to come in contact with infected fluids. During the investigation, one of the boyfriend's logbooks indicated that he had drawn blood from an HIV-positive patient in 1994. Investigators then obtained blood from the victim, infected patient, and other HIV-positive individuals in the area so that the victim's viral phylogeny could be compared against the patient and the community. At the end of the analyses, the infected patient and woman's HIV strain were similar enough to be concluded that they were related, and the doctor ended up being charged with second-degree attempted murder (Metzker *et al.*, 2002).

In 2001, an unknown person sent endospores formed by the bacterium *B. anthracis* through mail to individuals in the United States (Higgins *et al.*, 2003). These spores cause the disease anthrax. This attack led to heightened awareness of biocrimes, especially throughout the news media. The anthrax investigation took a microbiological approach when some of the letters were preserved that allowed the *B. anthracis* used to be identified as the Ames strain (Higgins *et al.*, 2003). The strain was identified by locating the SNPs present in the samples, and comparing them with known strains across laboratories in the United States (Van Ert *et al.*, 2007). Although the Federal Bureau of Investigation (FBI) declares the strain could not be precisely narrowed down to a single, specific laboratory source, the perpetrator was concluded to be a laboratory worker based on a combination of factors. Unfortunately, the suspect committed suicide before a court case could be held.

2.4.2.2 Cases involving agriculture

Animal and plant agricultural products have the potential to be targeted for a bioterrorism attack. While creating enough of a pathogen to cause massive causalities in humans can be difficult to achieve, a massive attack on animals and crops can prove much easier. Agricultural pathogens are usually not infectious to humans, preventing detection in the early stages of dispersal, making it significantly harder to contain. The mass dispersal also allows the attackers to require only a small amount of the pathogen as an initial inoculation with potential for travel to surrounding areas. While a massive attack on agriculture would not cause famine across a nation in many cases, trade sanctions could be affected leading to the loss of billions of dollars. The most common plant pathogens are fungi, which can be controlled by fungicides, but animals and some plants can also be affected by bacterial and viral organisms (Wheelis *et al.*, 2002). These infections are harder to treat and usually result in the product being quarantined or eradicated. As in cases involving humans, if an attack is suspected, the microorganism can be isolated and identified using various genetic techniques. The genetic code may be able to provide information that may lead to determining if the strain is related to an environmental isolate or if the pathogen is completely diverse from the communities. This might suggest that the pathogen was obtained in a lab setting or came from exotic environment (Wheelis *et al.*, 2002). For more detailed discussion on biocrimes, please see Budowle *et al.* (2003).

2.5 Trace evidence

2.5.1 Human

Trace evidence is physical evidence that can be transferred from contact between objects. Locard first recognized the concept of trace evidence in the context of forensic science, which he summarized as follows: "Any action of an individual and the violent actions of a crime cannot occur without leaving a trace" (Inbau, 1934). Trace evidence helps put pieces of the investigative puzzle together, is used to identify location of both the decedent and witness(es), and is essential for excluding or including witnesses and unlocking circumstances leading to death. Evidence is carefully collected in the form of hair or fibers from the victim's or suspect's clothing, or from the crime scene (Figure 2.3). Understanding the microbiological associations between people and their surroundings has become a powerful tool for providing evidentiary data, especially with respect to location. For instance, bacterial communities, fungi, and fungal spores may be picked up by any object contacting them and are subject to similar evidentiary considerations as hair or fibers as they are proxy indicators for the environment from whence they came (Wiltshire, 2010). While placing a suspect at a crime scene is not conclusive evidence, it is nonetheless an important factor of consideration in an investigation.

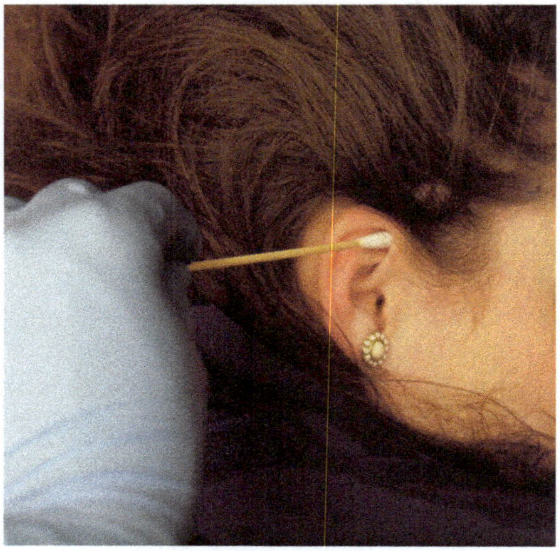

Figure 2.3 Enactment of collecting microbial trace evidence from a cadaver with a sterile cotton swab

One of the first cases reporting the involvement of microbiological case evidence was when a young woman's body was found dumped on a patch of stinging nettles (*Urtica dioica*) (Hawksworth and Wiltshire, 2011). Spores from two fungal species associated with dead stinging nettles were found both at the crime scene and in the suspect's car, demonstrating a link between the suspect and the place where the body had been dumped. In another case, a woman claimed to have been raped in a wooded strip of land near her home (Hawksworth and Wiltshire, 2011). The suspect denied the claim and stated that the act was consensual and occurred in an area of turf in a park, and that he had never been in the wooded area at any time. Botanical and mycological profiles from analyses of shoe and clothing samples matched that of the wooded area. Confronted with this evidence, the suspect subsequently confessed.

Investigations of the human microbiome have increased our understanding of the ecology and structural composition of the skin microbiota and have shown a high degree of inter- and intradiversity between individuals and body locations. These communities are also relatively stable and recover just hours after body washing (Costello *et al.*, 2009; Fierer *et al.*, 2008; Grice and Segre, 2011; Leake *et al.*, 2016). Additionally, the presence and composition of microbial communities associated with multiple inanimate objects such as door knobs, light switches, and smart phones have been assayed, and data have shown specific microbial signatures matching individuals to these objects (Lax *et al.*, 2014; Luongo *et al.*, 2016; Meadow *et al.*, 2014).

In a study by Fierer *et al.* (2010), next-generation sequencing was used to compare bacterial communities from samples taken from personal computer keyboards and compared to public or differing individual keyboards. Results showed that the bacterial communities were unique enough to identify who had used each keyboard, and could not be linked to the public or other specific individuals, as those had differential bacterial signatures. In a very interesting companion study, bacterial communities taken from computer mice that had not been touched for more than 12 hours were compared against a database of 270 hand samples acquired over a period of 2 years from those who had never touched the mouse (Fierer *et al.*, 2010). Data showed that bacteria from the mouse matched bacteria on every owner's hand more significantly than the samples from the other 270 hands. Another study, by Lax *et al.* (2015), found that microbial communities taken from shoes clustered to distinct groups according to geographical location and surface type and that bacterial taxa often increased in abundance on the shoe soles while walking through a particular space. Although still being refined, results from all of these studies demonstrate the utility of such an approach for complementary or, in the absence of human DNA or fingerprint evidence, independent confirmation of forensic evidence to determine whether skin's unique bacterial community structure could be recovered from touched surfaces at crime scenes and potentially matched back to the individual from whence they came. Please see Chapters 13–15 for more detail on microorganisms as spatial evidence.

2.5.2 Nonhuman animals and food

2.5.2.1 Agroterrorism

The close relationships between public health, agricultural, and veterinary science with respect to epidemiological and forensic investigations are well recognized, as each seeks to determine the origin and source of an outbreak even though subsequent response pathways may differ. The ability to track plant and animal pathogens, or instances of agroterrorism to their source, requires analyses at levels from whole plant/animal to the landscape (2014). Proactive epidemiological approaches are used to determine intensity or geographical extent of a disease against natural background, and reconstructive methodologies enable understanding of temporal sequences (Monke, 2007). Both of these approaches are used to determine whether a disease outbreak was a natural event or the result of direct human action. Ideally, an investigator would be able to provide evidence that the pathogen was absent before the outbreak, or present, but not under conducive conditions for infection (Elbers and Knutsson, 2013; Martensson *et al.*, 2013). However, as one might imagine, proving the absence before an outbreak is difficult, if not impossible. Currently, surveillance systems are being developed containing DNA fingerprints for microbes involved in foodborne or zoonotic infections (those that can be transmitted between nonhuman animals and humans) in order to trace strains and to clarify the chains of transmission (Gronvall *et al.*, 2014; The White House, 2014). Although these surveillance systems are relatively new, the associated databases are growing rapidly and being continuously updated. For instance, the Global Early Warning and Response System for Major Animal Diseases, including Zoonoses (GLEWS) and PulseNet of the Centers for Disease Control are international efforts for early warning of animal and plant outbreaks (2013, 2015).

2.5.2.2 Animal abuse cases

Microbiological trace evidence is often collected during cases of animal cruelty. In these cases, samples for microbial analyses are collected to determine evidence of homicides from poisoning, gunshot, or drowning. Even if a death visually appears to be a clear case of intentional drowning (an animal recovered from water with a rope and weight attached to its neck), other manners of death must still be ruled out (Munro and Munro, 2008). To do so, an investigator may search for the presence of diatoms, plant material or mud in the respiratory pathway, or specific lung pathology such as alveolar overdistention or narrowed alveolar capillaries (Munro and Munro, 2008).

As with human cases of homicide, the body is considered evidence (Gerdin and McDonough, 2013). Photographs, whole-body radiographs, histology, and necropsy are critical to differentiate homicide from death from diseases, natural causes, or unintentional body obstructions (Gerdin and McDonough, 2013). External examination is undertaken to look for matted hair, projectile wounds, or gross evidence of neglect or malnutrition. Other evidence may also be collected for testing of hair fibers, DNA of the animal or suspected pathogen (biocrimes), ballistics, or for forensic entomology (Gerdin and McDonough, 2013).

2.6 Other medicolegal aspects

2.6.1 Sexual assault

Generally in cases of sexual assault, an investigator interviews a living victim who can provide a statement toward the events leading to the crime and evidence pertinent to the investigation. However, there may be instances, especially in cases of child sexual abuse, where individuals cannot speak for themselves, or cannot reconstruct events leading to the crime, and investigators must rely on collected evidence for assault determination. Collected body swab or exogenous fluid specimens, as well as bed sheets, blankets, clothing, hair, and fibers may be analyzed using nucleic acid amplification tests (NAATs), culture, or other point-of-care tests. These tests are used for the detection of sexually transmitted pathogens, such as *Neisseria gonorrhoeae*, *Chlamydia trachomatis*, or *Trichomonas vaginalis*, or serum sample evaluation for HIV, hepatitis B, and syphilis, all causative agents of sexually transmitted diseases (STDs) (Frieden *et al.*, 2015; Jaureguy *et al.*, 2016). A number of STDs from infectious agents can be transmitted during a sexual assault. Generally, testing for an STD does not have forensic value, but is often a concern of sexual assault victims, and in some instances may be important in establishing the transmission of an infectious agent in a biocrime.

A recent study testing the utility of bacterial communities for the identification of sexual offenders found unique bacterial communities associated with human hairs (Tridico *et al.*, 2014). Although no correlation was found between scalp hair microbiota and sex, niche-specific taxa associated with female pubic hairs enabled individual discrimination between males and females. The authors concluded that these data suggest that metagenomic analysis might be of evidentiary value in sexual assault cases in the absence of associative evidence (Tridico *et al.*, 2014). See Chapter 15 for a more detailed discussion of bacterial communities associated with human hairs.

2.6.2 Medical malpractice

Generally, establishing negligence by the medical care provider whose conduct falls below an established standard of care can show legal liability for medical malpractice. Often, medical malpractice can include acts of misdiagnosis, or mistakes during medical procedures such as inadvertent infection due to inappropriate carelessness in cleaning and maintaining buildings, equipments, and supplies. DNA sequencing methods have provided a cost-effective means to obtain comprehensive information about the contaminating source to demonstrate causation of an injury or ailment, or are also used to refute such claims.

In a study of malpractice in a criminal case, molecular evidence was used to convict a gastroenterologist of attempted second-degree murder by injecting his former girlfriend's blood of an HIV-infected patient under his care (Metzker *et al.*, 2002). This was the first known use of phylogenetic analysis as evidence in a US criminal proceeding. Another study used maximum likelihood phylogenetic analysis and nucleotide sequencing to reject a hypothesis of transmission between a Baltimore surgeon and his patients (Holmes *et al.*, 1993).

2.6.3 Nosocomial infections and antibiotic resistance

Hospital-acquired infections are of major concern, causing an estimated $30 billion in direct medical cost annually (Scott Douglas, 2009). A study by the Center for Disease Control (CDC) estimated a total of 721,800 infections in US hospitals in 2011 (Magill *et al.*, 2014). However, proving whether the infection originated at the hospital at all, was due to failure to observe aseptic technique, or was inevitable despite sufficient standard of care is often a challenge. A breakthrough in nosocomial source tracking of hospital outbreaks came when whole genome sequencing was used to investigate an outbreak of carbapenem-resistant *Klebsiella pneumonia* that occurred in 2011 (Snitkin *et al.*, 2012). Genomic comparisons provided evidence that three transmission events occurred from a single patient who was discharged 3 weeks prior to clinical presentation of the next infected patient. The investigation showed the bacterial source in a respirator the index patient had used as well as in sink drains in the patient room. Although the respirator and sinks had been cleaned, the disinfecting agent had failed. Results of the investigation demonstrated that genetic evidence could provide litigable insights into chains of transmission to facilitate the control of nosocomial infections and could help prove or refute a plaintiff's claim of a hospital-acquired infection due to malpractice.

2.6.4 Food safety and environmental contamination

2.6.4.1 Environmental contamination

Animal production has shifted dramatically over the past decade from small family farms toward large commercial confinement operations (Rogers and Haines, 2005). Animal agriculture results in the production of copious amounts of manure, much of which is ultimately used as a fertilizer for crops or spread onto land. Human or zoonotic pathogens can be shed in animal wastes or found associated with animal carcasses and may include *Mycobacterium* spp., rotaviruses, hepatitis E virus, *Salmonella* spp., *E. coli* O157:H7, and diseases caused by zoonotic pathogens include salmonellosis, tuberculosis, leptospirosis, infantile diarrheal disease, Q Fever, trichinosis, cryptosporidiosis, and giardiasis to name a few (Rogers and Haines, 2005). A list of some human and zoonotic pathogens and their source are listed in Table 2.3.

The survival of pathogenic organisms in the environment varies widely depending on the pathogen, environmental conditions, and the chemical, physical, and biological composition of the matrix in question. Considering the many potential exposure routes following release of zoonotic pathogens from animal rearing facilities, identification of the risks associated with pathogens coming from these facilities require technologies that enable the measurement of pathogens in air, water, meat and produce, soil and sediments, and feces of various animals among others. Methods include classical cultivation

Table 2.3 Examples of human and zoonotic bacterial, parasitic, and viral pathogens transmitted through environmental contamination

Pathogen	Mode of transmission	Disease	Host range	Reservoir
Bacteria				
Clostridium botulinum	Ingestion of contaminated food, inhalation, contamination of wounds	Botulism including symptoms of paralysis, respiratory failure, death	Humans, fowl, fish, cattle, dogs	Soil, aquatic sediments, birds, mammals, fish, honey, vegetables
Clostridium tetani	Contamination of wounds	Tetanus including symptoms of renal failure, muscle spasms, respiratory failure, cardiac arrest, death	Humans, domestic and wild animals	Soil, feces, sewage, aquatic sediments, humans
Mycobacterium ssp.	Inhalation, passive infection, ingestion	Tuberculosis, swimming pool granuloma, contact dermatitis	Humans, water birds, fish, shellfish, herbivores	Humans, cattle, swine, birds, fish
Enterohemorrhagic *Escherichia coli*	Consumption of contaminated water, fruits, vegetables, raw milk or meat; person-to-person	Travelers diarrhea, acute renal failure, infantile diarrheal disease	Humans, cattle, sheep, some poultry	Healthy cattle
Listeria monocytogenes	Ingestion of contaminated food; direct contact with diseased animals; mother to fetus	Listeriosis including symptoms of fever, chills, malaise, spontaneous abortion, death	Humans and other mammals, fish, crustaceans, insects	Soil, manure, decaying vegetable matter, fresh and frozen poultry, raw milk, cheese, silage, water
Salmonella spp.	Ingestion of contaminated food or water	Salmonellosis including symptoms of diarrhea, fever, abdominal cramps, nausea, vomiting	Humans, cattle, swine, poultry, horses, rodents, reptiles, amphibians	Humans, cattle, swine, poultry, horses, rodents, reptiles, amphibians
Bacillus anthracis	Inhalation, direct contact with diseased animals	Anthrax, respiratory and/or abdominal distress, skin lesions, death	Humans, cattle, sheep, goats, horses, pigs	Soil, dried or processed hides

(*Continued*)

Table 2.3 (Continued)

Pathogen	Mode of transmission	Disease	Host range	Reservoir
Campylobacter jejuni	Ingestion of contaminated food; direct contact with diseased animals	Campylobacteriosis including symptoms of inflammatory, often bloody diarrhea, dysentery, cramps, skin ulceration, fever, pain, pneumonic disease, death	Humans, pigs, cattle, poultry, wild birds	Pigs, poultry, cattle, sheep, wild birds
Parasites				
Trichinella spp.	Consumption of raw or undercooked meat	Fever, edema, diarrhea, neurological and cardiac complications	Domestic and sylvatic mammals, birds, reptiles	Pigs, horses, bears, game animals
Giardia lamblia	Contact with contaminated soil and water; fecal–oral route	Nausea, fever, chills, diarrhea, weight loss	Humans and other mammals	Humans and other animals
Cryptosporidium parvum	Fecal–oral route through ingestion of contaminated food, water, ruminants	Gastroenteritis, abdominal pain, cramps, fever, pancreatitis	Humans and other mammals, fish, reptiles, birds	Environment and many mammalian species
Viruses				
Rotavirus	Fecal–oral route from person-to-person or from contact with contaminated food or water	Diarrhea, fever, chills, vomiting, enterocolitis	Humans and experimentally infected animals	Humans, calves, pigs, dogs, foals, cats, some birds
Hepatitis E	Fecal–oral; foodborne; bloodborne, vertical transmission	Jaundice, malaise, anorexia, fever, diarrhea, abdominal pain	Humans, swine, nonhuman primates	Humans, monkeys, swine, boar, rats, bear, cows, sheep, camels, horses

The table includes information of mode of transmission, disease signs and symptoms, host range, and natural reservoirs. Data obtained from searching through pathogen safety data sheets found at: http://www.phac-aspc.gc.ca/lab-bio/res/psds-ftss/index-eng.php.

approaches, and more recently, identification and quantification of agents via the detection of surface antigens or nucleic acids. The hindrances of low infectious doses, detection of viability, detection of multiple agents in a limited sample, and contamination of outcompeting species on growth media have been largely overcome through the advent of next-generation sequencing technologies where considerable advances in emerging nucleic acids and sensor technologies are allowing cost-effective and rapid analysis reducing analyses (Ho *et al.*, 2005).

Difficulty in quantifying pathogens at relevant concentrations in environmental matrices, a lack of epidemiological data, and the large number of assays required to measure all of the zoonotic pathogens shed in livestock feces make it difficult to monitor pathogenic organisms in waste streams to determine safe levels of pathogens in the environment (Rogers and Haines, 2005). Instead, indicator organisms for fecal pollution, including coliform bacteria, fecal coliforms, *E. coli*, and/or Enterococci have been monitored in lieu of overt pathogens (Cabral, 2010). This method detects fecal bacteria as an indicator for fecal contamination, and although their presence does not necessarily mean pathogens are present, it indicates a potential health hazard. While this surrogate method of monitoring has been in place for over 100 years, recent experiments have shown that these indicators are not reliable surrogates for many pathogens (Cabral, 2010).

2.6.4.2 Plant pathogens

Plants are susceptible to over 50,000 different pathogens, primarily fungi, viruses, bacteria, and nematodes (Madden and Wheelis, 2003; Pimentel *et al.*, 2000). There have been no documented cases, as yet, of the deliberate use of pathogens to attack crops or other plants; however, accidental introduction of exotic pathogens account for roughly 65% of crop losses that amount to an estimated cost of $137 billion annually (Pimentel *et al.*, 2000). To that end, the examination of natural or accidental plant pathogen or pest introductions can provide insight into the possible impacts of a successful deliberate attack. Plant pathogens of high risk for the United States are designated as select agents under the Code of Federal Regulations, title 7, part 331, by the Agricultural Bioterrorism Protection Act of 2002 and the United States Department of Agriculture (USDA) Animal and Plant Health Inspection Service (APHIS). Standard operating procedures are enforced during initial field assessments of disease and subsequent investigation of the epidemiology surrounding transmission of the etiological agent.

2.6.4.3 Foodborne illness

It is estimated that 48 million foodborne illnesses occur each year in the United States alone. More than 250 pathogens and toxins can lead to foodborne illness and foodborne toxicity, respectively, of which nearly all can lead to disease outbreaks (Prevention Centers for Disease Control 2014). A better sense of the relationship between contaminated foods and illnesses supports food safety along the entire food production chain from farm to table and is vital for identifying opportunities to improve food safety. Trends identified by laboratory-confirmed foodborne diseases are monitored through surveillance programs such as Foodborne Disease Active Surveillance Network (FoodNet) including infections caused by the bacteria *Campylobacter*, *E. coli* O157, *Listeria*, *Salmonella*, *Shigella*, *Vibrio*, and *Yersinia*, and the parasites *Cryptosporidium* and *Cyclospora*. These data allow local authorities and policy makers to make decisions regarding the best agricultural practices and to monitor progress in reaching national goals (USDHS, 2014) for reducing foodborne illness.

2.7 Needs that must be met for use in chain of custody

Forensic microbiology faces certain challenges in order to ensure that the procedures for chain of custody and sample collection are followed (Figure 2.4). This entails proving that the evidence was collected and stored legally, without tampering and under the

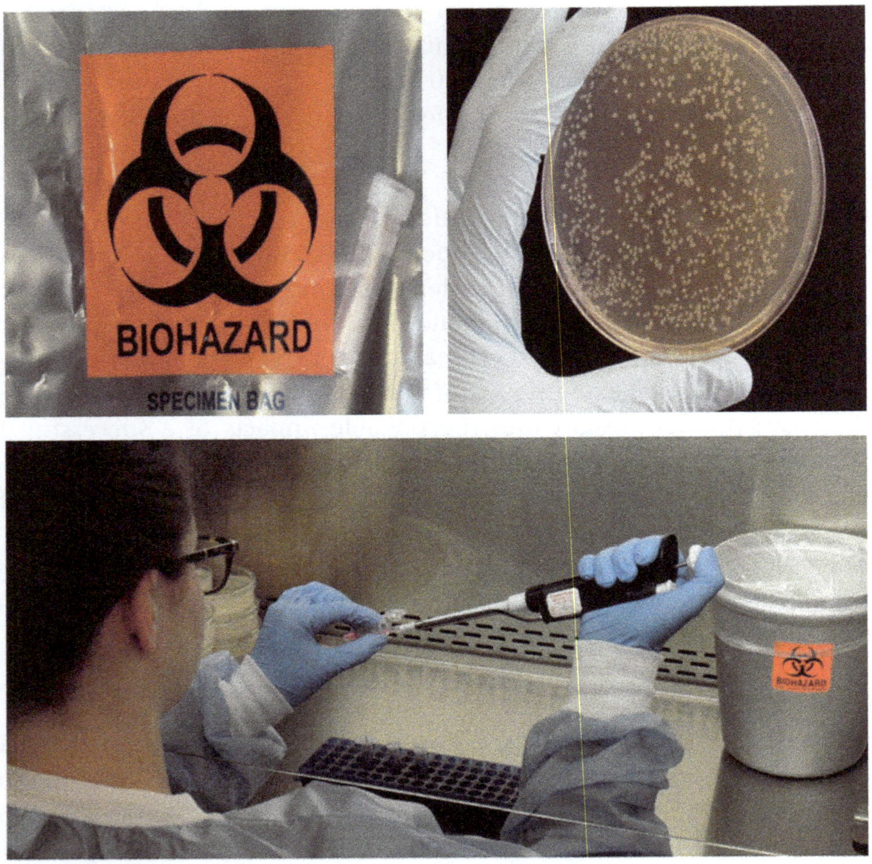

Figure 2.4 Enactment of proper aseptic conditions to ensure samples avoid contamination and meet the proper chain of custody needs

correct conditions to avoid sample contamination and/or degradation (Evans, 2003). Chain of custody follows legal protocol and documents every person who comes in contact with the sample (Evans, 2003). This is particularly important because microorganism sample collection can differ from other, more routine samples collected from a crime scene. For instance, one must take into account the history of the specimen, the environmental context, physical and chemical composition, and pathogenicity (i.e., toxicity and/or biosafety level) in order to assess the level of necessary containment, especially when a biocrime is suspected. Whether analysis of live microbes is required or if nucleic acid isolation is necessary should also be considered; in all instances, aseptic technique should be employed to prevent exogenous microbial or genetic contamination. This is especially true if microbial species are used as trace evidence for the inclusion or exclusion of a suspect.

Live organisms must be placed in or onto a nutrient medium sufficient for growth of the species present. Once grown, nucleic acids from a particular strain of interest can be isolated for whole genome sequencing for comparisons to other known strains. Such information could prove useful when tracking sources of biocrimes, medical malpractice, or with contamination leading to large outbreaks. However, most microorganisms cannot be easily cultured in a laboratory setting so obtaining an accurate representation of the microbial community from live cultured samples may prove difficult (Wade, 2002). Alternatively, microbial community profiling using genetic-based methods and a taxonomically informative marker provides a more robust analysis since genes of organisms present can be used for identification without the biases often encountered through live culture (Wilson *et al.*, 1997).

Additionally, new molecular tests and methods for data analysis are being published every month. As such, standardized methods for sampling and analytical methods for microbial analyses will need to be established (e.g., Pechal *et al.*, 2014) across fields and disciplines to maintain consistency and reproducibility of results, especially for use in court.

2.8 Summary

Microbial forensics for legal proceedings involves systematic, evidence-based data collection, objective observation, and rigorous, reproducible analyses suitable for use in a court of law. Unlike public health investigations, forensic microbiology requires detailed characterization of microbial isolates and special processing procedures for producing legal evidence to link the causative agent with a specific individual or group. Detailed genomic analyses and other specific physical and chemical characterizations are critical parts of forensic analyses, as are chain of custody forms that are needed to sustain the validity of these analyses in legal proceedings.

Culture-based and molecular microbiological techniques have been used for years to trace outbreaks of microbial diseases. Microbial forensics will continue to evolve and improve as the availability of basic scientific information concerning microbial genetics, evolution, physiology, and ecology grows. With this basic knowledge and with ever-improving cutting-edge technologies currently available, it has become feasible to move toward an understanding of microbe community–host interactions as a system rather than as a collection of component parts.

This chapter was composed to provide the reader not only with a historical understanding of the use of microbiology as evidence but also to give the reader an appreciation of the exciting, dynamic future of forensic microbiology. With this goal in mind, we have introduced concepts and background to further make explicit the interconnectedness of topics both within a chapter and between chapters. The remaining chapters will emphasize specific applications of relevant concepts, methods, and data evaluation from several subspecialties in the field in order to give the reader a broad understanding of the scientific substance behind the discipline, both in theory and in practice.

Acknowledgments

The authors thank the anonymous reviews provided on earlier drafts of this chapter. Heather Jordan recognizes the National Institute of Justice (2014-DN-BX-K243) for providing financial support leading toward the preparation of the chapter. Points of view in the chapter are those of the authors and do not necessarily represent the official position or policies of the U.S. Department of Justice.

References

Alvarez-Lafuente, R., Aguilera, B., Suarez-Mier, M.A. *et al.* (2008) Detection of human herpesvirus-6, Epstein-Barr virus and cytomegalovirus in formalin-fixed tissues from sudden infant death: a study with quantitative real-time PCR. *Forensic Science International*, **178** (2–3), 106–111.

Amann, R., Ludwig, W. and Schleifer, K.H. (1995) Phylogenetic identification and in situ detection of individual microbial cells without cultivation. *Microbiological Reviews*, **59**, 143–169.

Amendt, J., Richards, C.S., Campobasso, C.P. *et al.* (2011) Forensic entomology: applications and limitations. *Forensic Science, Medicine, and Pathology*, **7**, 379–392.

Baldwin, B.G. (1992) Phylogenetic utility of the internal transcribed spacers of nuclear ribosomal DNA in plants: an example from the compositae. *Molecular Phylogenetics and Evolution*, **1** (1), 3–16.

Bengmark, S. (1998) Ecological control of the gastrointestinal tract. The role of probiotic flora. *Gut*, **42**, 2–7.

Blackwell, C. (2004) Infection, inflammation, and SIDS. *FEMS Pathogens and Disease*, **42** (1), 1–2.

Budlowe, B. (2004) SNP typing strategies. *Forensic Science International*, **146** (Suppl), S139–S142.

Budowle, B., Schutzer, S.E., Einseln, A. *et al.* (2003) Public health. Building microbial forensics as a response to bioterrorism. *Science*, **301** (5641), 1852–1853.

Budowle, B., Eisenberg, A.J. and vandal, A. (2009) Validity of low copy number typing and applications to forensic science. *Croatian Medical Journal*, **50** (3), 207–217.

Cabral, J.P. (2010) Water microbiology. Bacterial pathogens and water. *International Journal of Environmental Research and Public Health*, **7**, 3657–3703.

Carter, D.O., Yellowlees, D. and Tibbett, M. (2007) Cadaver decomposition in terrestrial ecosystems. *Naturwissenschaften*, **94**, 12–24.

Carter, D.O., Yellowlees, D. and Tibbett, M. (2008) Temperature affects microbial decomposition of cadavers (*Rattus rattus*) in contrasting soils. *Applied Soil Ecology*, **40**, 129–137.

Carter, D.O., Yellowlees, D. and Tibbett, M. (2010) Moisture can be the dominant environmental parameter governing cadaver decomposition in soil. *Forensic Science International*, **200**, 60–66.

Carter, D.O., Metcalf, J.L., Bibat, A. and Knight, R. (2015) Seasonal variation of postmortem microbial communities. *Forensic Science, Medicine, and Pathology*, **11**, 202–207.

Centers for Disease Control and Prevention (2013) *Pulsenet* [Online], https://www.cdc.gov/features/dspulsenetfoodborneillness/ (accessed December 7, 2016).

Clark, M.A., Worrell, M.B. and Pless, J.E. (1997) Postmortem changes in soft tissue, in *Forensic Taphonomy: The Postmortem Fate of Human Remains* (eds W.D. Haglund and M.H. Sorg), CRC Press, Boca Raton, pp. 151–164.

Cobaugh, K.L., Schaeffer, S.M. and DeBruyn, J.M. (2015) Functional and structural succession of soil microbial communities below decomposing human cadavers. *PLoS One*, **10**, e0130201.

Costello, E.K., Lauber, C.L., Hamady, M. *et al.* (2009) Bacterial community variation in human body habitats across space and time. *Science*, **326**, 1694–1697.

Dickson, G.C., Poulter, R.T., Maas, E.W. *et al.* (2011) Marine bacterial succession as a potential indicator of postmortem submersion interval. *Forensic Science International*, **209**, 1–10.

Douglas Scott II, R. (2009) The Direct Medical Costs of Healthcare-Associated Infections in U.S. Hospitals and the Benefits of Prevention. Centers for Disease Control and Prevention.

Elbers, A. and Knutsson, R. (2013) Agroterrorism targeting livestock: a review with a focus on early detection systems. *Biosecurity and Bioterrorism*, **11** (Suppl 1), S25–S35.

Evans, M.M. (2003) Maintaining the chain of custody: evidence handling in forensic cases. *AORN Journal*, **78** (4), 563–569.

Fierer, N., Hamady, M., Lauber, C.L. and Knight, R. (2008) The influence of sex, handedness, and washing on the diversity of hand surface bacteria. *Proceedings of the National Academy of Sciences of the United States of America*, **105**, 17994–17999.

Fierer, N., Lauber, C.L., Zhou, N. *et al.* (2010) Forensic identification using skin bacterial communities. *Proceedings of the National Academy of Sciences of the United States of America*, **107**, 6477–6481.

Frieden, T.R., Jaffe, H.W. and Cono, J. (2015) Sexually Transmitted Diseases Treatment Guidelines, in *Morbidity and Mortality Weekly Report* (ed S.A. Rasmussen), Center for Surveillance, Epidemiology, and Laboratory Services, Centers for Disease Control and Prevention, US Department of Health and Human Services, Atlanta.

Gerdin, J.A. and McDonough, S.P. (2013) Forensic pathology of companion animal abuse and neglect. *Veterinary Pathology*, **50**, 994–1006.

Gleeson, M. and Cripps, A.W. (2004) Development of mucosal immunity in the first year of life and relationship to sudden infant death syndrome. *FEMS Immunology and Medical Microbiology*, **42** (1), 21–33.

Global Early Warning System (2015) GLEWS+ The joint FAO-OIE-WHO Global Early Warning System for health threats and emerging risks at the human-animal-ecosystems interface [Online], http://www.glews.net/ (accessed October 20, 2016).

Goff, M.L. and Flynn, M.M. (1991) Determination of postmortem interval by arthropod succession: a case study from the Hawaiian Islands. *Journal of Forensic Science*, **36**, 607–614.

Grice, E.A. and Segre, J.A. (2011) The skin microbiome. *Nature Reviews. Microbiology*, **9**, 244–253.

Gronvall, G., Boddie, C., Knutsson, R. and Colby, M. (2014) One health security: an important component of the global health security agenda. *Biosecurity and Bioterrorism*, **12**, 221–224.

Gu, Y., Mao, X., Zha, L. *et al.* (2015) Development of a candidate method for forensic microbial genotyping using multiplex pyrosequencing combined with a universal biotinylated primer. *Forensic Science International*, **246**, e1–e6.

Gunn, A. and Pitt, S.J. (2012) Microbes as forensic indicators. *Tropical Biomedicine*, **29**, 311–330.

Haefner, J.N., Wallace, J.R. and Merritt, R.W. (2004) Pig decomposition in lotic aquatic systems: the potential use of algal growth in establishing a postmortem submersion interval (PMSI). *Journal of Forensic Science*, **49**, 330–336.

Harmsen, D. and Karch, H. (2004) 16s rDNA for diagnosing pathogens: a living tree. *ASM News*, **70**, 19–24.

Hawksworth, D.L. and Wiltshire, P.E. (2011) Forensic mycology: the use of fungi in criminal investigations. *Forensic Science International*, **206** (1–3), 1–11.

Henderson, D.A. (1999) The looming threat of bioterrorism. *Science*, **283**, 1279–1282.

Higgins, J.A., Cooper, M., Schroeder-Tucker, L. *et al.* (2003) A field investigation of *Bacillus anthracis* contamination of U.S. Department of Agriculture and other Washington, D.C. buildings during the anthrax attack of October 2001. *Applied Environmental Microbiology*, **69**, 593.

Ho, C.K., Robinson, A., Miller, D.R. and Davis, M.J. (2005) Overview of sensors and needs for environmental monitoring. *Sensors*, **5**, 4–37.

Holmes, E.C., Zhang, L.Q., Simmonds, P. *et al.* (1993) Molecular investigation of human immunodeficiency virus (HIV) infection in a patient of an HIV-infected surgeon. *Journal of Infectious Diseases*, **167**, 1411–1414.

Hopkins, D.W., Wiltshire, P.E.J. and Turner, B.D. (2000) Microbial characteristics of soils from graves: an investigation at the interface of soil microbiology and forensic science. *Applied Soil Ecology*, **14**, 283–288.

Hyde, E.R., Haarmann, D.P., Lynne, A.M. *et al.* (2013) The living dead: bacterial community structure of a cadaver at the onset and end of the bloat stage of decomposition. *PLoS One*, **8**, e77733.

Hyde, E.R., Haarmann, D.P., Petrosino, J.F. *et al.* (2015) Initial insights into bacterial succession during human decomposition. *International Journal of Legal Medicine*, **129**, 661–671.

Inbau, F.E. (1934) Scientific evidence in criminal cases. *Journal of Criminal Law and Criminology*, **25** (3), 500–516.

Janaway, R.C., Percival, S.L. and Wilson, A. (2009) Decomposition of Human Remains, in *Microbiology and Aging: Clinical Manifestations* (ed S.L. Percival), Humana Press, Totowa.

Jaureguy, F., Chariot, P., Vessieres, A. and Picard, B. (2016) Prevalence of *Chlamydia trachomatis* and *Neisseria gonorrhoeae* infections detected by real-time PCR among individuals reporting sexual assaults in the Paris, France area. *Forensic Science International*, **266**, 130–133.

Kripalani-Joshi, S. and Law, H.Y. (1994) Identification of integrated Epstein-Barr virus in nasopharyngeal carcinoma using pulse field gel electrophoresis. *International Journal of Cancer*, **56**, 187–192.

Lafontaine, D.L. and Tollervey, D. (2001) The function and synthesis of ribosomes. *Nature Reviews. Molecular Cell Biology*, **2**, 514–520.

Lauber, C.L., Metcalf, J.L., Keepers, K. *et al.* (2014) Vertebrate decomposition is accelerated by soil microbes. *Applied and Environmental Microbiology*, **80**, 4920–4929.

Lax, S., Smith, D.P., Hampton-Marcell, J. *et al.* (2014) Longitudinal analysis of microbial interaction between humans and the indoor environment. *Science*, **345**, 1048–1052.

Lax, S., Hampton-Marcell, J.T., Gibbons, S.M. *et al.* (2015) Forensic analysis of the microbiome of phones and shoes. *Microbiome*, **3**, 21.

Leake, S.L., Pagni, M., Falquet, L. *et al.* (2016) The salivary microbiome for differentiating individuals: proof of principle. *Microbes and Infection*, **18**, 399–405.

LeClerc, J.E., Li, B., Payne, W.L. and Cebula, T.A. (1996) High mutation frequencies among *Escherichia coli* and Salmonella pathogens. *Science*, **274** (5290), 1208–1211.

Lin, C.Y., Yen, W.C., Hsieh, H.M. *et al.* (2014) Diatomological investigation in sphenoid sinus fluid and lung tissue from cases of suspected drowning. *Forensic Science International*, **244**, 111–115.

Lucci, A., Campobasso, C.P., Cirnelli, A. and Lorenzini, G. (2008) A promising microbiological test for the diagnosis of drowning. *Forensic Science International*, **182** (1–3), 20–26.

Ludes, B., Coste, M., Tracqui, A. and Mangin, P. (1996) Continuous river monitoring of the diatoms in the diagnosis of drowning. *Journal of Forensic Sciences*, **41** (3), 425–428.

Luongo, J.C., Barberan, A., Hacker-Cary, R. *et al.* (2016) Microbial analyses of airborne dust collected from dormitory rooms predict the sex of occupants. *Indoor Air* doi: 10.1111/ina.12302.

Maccallum, W.G. and Hastings, T.W. (1899) A case of acute endocarditis caused by Micrococcus Zymogenes (Nov. Spec.), with a description of the microorganism. *Journal of Experimental Medicine*, **4**, 521–534.

Madden, L.V. and Wheelis, M. (2003) The threat of plant pathogens as weapons against U.S. crops. *Annual Review of Phytopathology*, **41**, 155–176.

Magill, S.S., Edwards, J.R., Bamberg, W. *et al.* and Emerging Infections Program Healthcare-Associated Infections & Antimicrobial Use Prevalence Survey Team (2014) Multistate point-prevalence survey of health care-associated infections. *New England Journal of Medicine*, **370**, 1198–1208.

Mann, R.W., Bass, W.M. and Meadows, L. (1990) Time since death and decomposition of the human body: variables and observations in case and experimental field studies. *Journal of Forensic Science*, **35**, 103–111.

Martensson, P.A., Hedstrom, L., Sundelius, B. *et al.* (2013) Actionable knowledge and strategic decision making for bio- and agroterrorism threats: building a collaborative early warning culture. *Biosecurity and Bioterrorism*, **11** (Suppl 1), S46–S54.

Meadow, J.F., Altrichter, A.E. and Green, J.L. (2014) Mobile phones carry the personal microbiome of their owners. *PeerJ*, **2**, e447.

Megyesi, M.S., Nawrocki, S.P. and Haskell, N.H. (2005) Using accumulated degree-days to estimate the postmortem interval from decomposed human remains. *Journal of Forensic Science*, **50**, 618–626.

Melvin, J.R., Jr, Cronholm, L.S., Simson, L.R., Jr and Isaacs, A.M. (1984) Bacterial transmigration as an indicator of time of death. *Journal of Forensic Science*, **29**, 412–417.

Metcalf, J.L., Wegener Parfrey, L., Gonzalez, A. *et al.* (2013) A microbial clock provides an accurate estimate of the postmortem interval in a mouse model system. *eLife*, **2**, e01104.

Metzker, M.L., Mindell, D.P., Liu, X.M. *et al.* (2002) Molecular evidence of HIV-1 transmission in a criminal case. *Proceedings of the National Academy of Sciences of the United States of America*, **99**, 14292–14297.

Meurens, F., Summerfield, A., Nauwynck, H. *et al.* (2012) The pig: a model for human infectious diseases. *Trends in Microbiology*, **20**, 50–57.

Monke, J. (2007) Agroterrorism: Threats and Preparedness. CRS Report for Congress.

Munro, M.C. and Munro, R. (2008) *Animal Abuse and Unlawful Killing: Forensic Veterinary Pathology*, Elsevier, Edinburgh.

Murray, P.R., Rosenthal, K.S. and Pfaller, M.A. (2009) *Medical Microbiology*, 6 edn, Mosby/Elsevier, Philadelphia.

Nachman, M.W. (2001) Single nucleotide polymorphisms and recombination rate in humans. *Trends in Genetics*, **17**, 481–485.

Payne, J.A. (1965) A summer carrion study of the baby pig Sus scrofa Linnaeus. *Ecology*, **46**, 592–602.

Pechal, J.L., Crippen, T.L., Tarone, A.M. *et al.* (2013) Microbial community functional change during vertebrate carrion decomposition. *PLoS One*, **8**, e79035.

Pechal, J.L., Crippen, T.L., Benbow, M.E. *et al.* (2014) The potential use of bacterial community succession in forensics as described by high throughput metagenomic sequencing. *International Journal of Legal Medicine*, **128**, 193–205.

Pimentel, D., Lach, L., Zuniga, R. and Morrison, D. (2000) Environmental and economic costs of nonindigenous species in the United States. *Bioscience*, **50**, 53–65.

Prevention Centers for Disease Control (2014) *Estimating Foodborne Illness: An Overview* [Online]. Centers for Disease Control and Prevention, http://www.cdc.gov/foodborneburden/estimates-overview.html (accessed October 20, 2016).

Rodriguez, W.C. and Bass, W.M. (1983) Insect activity and its relationship to decay rates of human cadavers in East Tennessee. *Journal of Forensic Science*, **28**, 423–432.

Rodriguez, A.F., Ballesteros, S., Ory, F. *et al.* (2005) Virological analysis in the diagnosis of sudden children death: a medico-legal approach. *Forensic Science International*, **4633**, 1–7.

Rogers, S. and Haines, J. (2005) *Detecting and Mitigating the Environmental Impact of Fecal Pathogens Originating from Confined Animal Feeding Operations: Review*, National Risk Management Research Library, Cincinnati, OH.

Schroeder, J.M., Booton, G.C., Hay, J. *et al.* (2001) Use of subgenic 18S ribosomal DNA PCR and sequencing for genus and genotype identification of acanthamoebae from humans with keratitis and from sewage sludge. *Journal of Clinical Microbiology*, **39**, 1903–1911.

Schutzer, S.E., Budowle, B. and Atlas, R.M. (2005) Biocrimes, microbial forensics and the physician. *PLoS Medicine*, **2** (12), e337.

Seng, P., Drancourt, M., Gouriet, F. *et al.* (2009) Ongoing revolution in bacteriology: routine identification of bacteria by matrix-assisted laser desorption ionization time-of-flight mass spectrometry. *Clinical Infectious Diseases*, **49**, 543–551.

Slonczewski, J.L. and Foster, J.W. (2013) *Microbiology: An Evolving Science*, W.W. Norton and Company, New York.

Snitkin, E.S., Zelazny, A.M., Thomas, P.J. *et al.* (2012) Tracking a hospital outbreak of carbapenem-resistant Klebsiella pneumoniae with whole-genome sequencing. *Science Translational Medicine*, **4**, 148ra116.

Sorg, M.H. and Haglund, W.D. (1996) *Forensic Taphonomy: The Postmortem Fate of Human Remains*, CRC Press, Boca Raton, FL.

The White House (2014) Fact Sheet: *Global Health Security Agenda* [Online], http://www.globalhealth.gov/global-health-topics/global-health-security/GHS Agenda.pdf (accessed October 20, 2016).

Timperman, J. (1972) The diagnosis of drowning. A review. *Forensic Science*, **1**, 397–409.

Tomberlin, J.K., Mohr, R., Benbow, M.E. *et al.* (2011) A roadmap for bridging basic and applied research in forensic entomology. *Annual Review of Entomology*, **56**, 401–421.

Tridico, S.R., Murray, D.C., Addison, J. *et al.* (2014) Metagenomic analyses of bacteria on human hairs: a qualitative assessment for applications in forensic science. *Investigative Genetics*, **5**, 16.

USDHS (2014) *Food Safety* [Online]. Office of Disease Prevention and Health Promotion: Department of Health and Human Services, http://www.healthypeople.gov/2020/topics-objectives/topic/food-safety?topicid=14 (accessed October 20, 2016).

Van Ert, M.N., Easterday, W.R., Simonson, T.S. *et al.* (2007) Strain-specific single-nucleotide polymorphism assays for the *Bacillus anthracis* Ames strain. *Journal of Clinical Microbiology*, **45**, 47–53.

Vass, A.A. (2001) Beyond the grave-understanding human decomposition. *Microbiology Today*, **28**, 190–192.

Ventner, J.C., Remington, K., Heidelberg, J.F. *et al.* (2004) Environmental genome shotgun sequencing of the Sargasso Sea. *Science*, **304** (5667), 66–74.

Wade, W. (2002) Unculturable bacteria—the uncharacterized organisms that cause oral infections. *Journal of the Royal Society of Medicine*, **95** (2), 81–83.

Weiss, S., Carter, D.O., Metcalf, J.L. and Knight, R. (2016) Carcass mass has little influence on the structure of gravesoil microbial communities. *International Journal of Legal Medicine*, **130**, 253–263.

Wheelis, M., Casagrande, R. and Madden, L.V. (2002) Biological attack on agriculture: low-tech, high-impact bioterrorism. *BioScience*, **52** (7), 569–576.

Wilson, M.J., Weightman, A.J. and Wade, W.G. (1997) Applications of molecular ecology in the characterization of uncultured microorganisms associated with human disease. *Reviews in Medical Microbiology*, **8**, 51–103.

Wilson-Taylor, R.J. (2013) Time Since Death Estimation and Bone Weathering: The Postmortem Interval, in *Forensic Anthropology: An Introduction* (eds M.T.A. Tersigni-Tarrant and N.R. Shirley), CRC Press, Boca Raton, FL.

Wiltshire, P.E.J. (2010) Forensic Ecology, in *Crime Scene to Court: The Essentials of Forensic Sciences*, 3rd edn (ed P. White), RSC Publishing, Cambridge.

Wimberly, B.T., Brodersen, D.E., Clemons, W.M., Jr *et al.* (2000) Structure of the 30S ribosomal subunit. *Nature*, **407**, 327–339.

Woese, C.R., Olsen, G.J., Ibba, M. and Soll, D. (2000) Comparisons of complete genome sequences allow the most objective and comprehensive descriptions possible of a lineage's evolution. *Microbiology and Molecular Biology Reviews*, **64**, 202–236.

CHAPTER 3

Approaches and considerations for forensic microbiology decomposition research

M. Eric Benbow[1,2] and Jennifer L. Pechal[1]

[1] *Department of Entomology, Michigan State University, East Lansing, MI, USA*

[2] *Department of Osteopathic Medical Specialties, Michigan State University, East Lansing, MI, USA*

3.1 Introduction

The study of vertebrate decomposition for applications in the forensic sciences is often challenging, but can provide invaluable scientific data. There are some unique challenges faced by scientists studying human death or other aspects of human violence and justice. Often studies of the decomposition of vertebrate remains, especially if the remains are human, are linked to ethical, cultural, and social complexities. Many vertebrate decomposition studies are conducted in the field and often with human and nonhuman (hereafter referred to as animal) remains that are allowed to decompose in a variety of habitats and circumstances. If humans are not available (e.g., there is not a donation program), animal surrogates are often employed to understand how a carcass decomposes, or in the case of wildlife forensic research, the specific species (e.g., bear, elephant, and bison) of animal is used for direct application (Huffman and Wallace, 2012). Inferences are then made from the surrogate decomposition process to that of humans. Whether human or other animal remains are used in such research, it is often difficult to acquire the necessary number of identical or similar subjects for executing a robust experimental design to achieve strong statistical inference and generalizable conclusions. Further, field research in carcass decomposition can be unpleasant. Research into decomposing remains has the obvious negative attributes of repulsive odors and the natural human aversion to evaluating the processes and conditions of death. However, the data generated by studies of death and decomposition provide important information to the forensic sciences and to the broader body of scientific knowledge that ranges from molecular genetics of cell regulation to ecosystem level plant community change and functional processes (Benbow *et al.*, 2015).

Forensic Microbiology, First Edition. Edited by David O. Carter, Jeffery K. Tomberlin, M. Eric Benbow and Jessica L. Metcalf.

These challenges are not exclusive to vertebrate remains decomposition research; indeed, a well-developed experimental design is the foundation of scientific research. Therefore, in this chapter the basic concepts related to developing a strong experimental design are reviewed in general for research while specifically considering the unique challenges of using cadavers for research in field settings. Using examples from the literature, the chapter provides an overview of studies that provide new information related to pattern or processes during decomposition, and how hypothesis-driven research provides a strong, mechanistic approach for investigations with forensic applications. The focus of the chapter will be on research using human remains during the decomposition process; however, many of the same principles and considerations are relevant to animal systems in cases of abuse, negligence, or poaching. While the focus of the chapter is on human remains research, the guiding considerations related to experimental design and strong statistical inference hold for other aspects of forensic microbiology, such as soil microbial communities and research within the built environment (e.g., Chapters 13 and 14) in addition to the other chapters of this book. Further, the goal of the chapter is not to cover all aspects of experimental design or the statistical theory behind developing robust and defendable statistical approaches for analyzing data from different experimental scenarios. For readers interested in additional detail and background of the design and analyses of carrion research, please refer to two excellent resources by Schoenly *et al.* (2015) and Moreau *et al.* (2015).

3.2 Challenges of human remains research

For studying forensic microbiology during decomposition, outside of wildlife forensic research, the most direct and realistic approach is to use human cadavers. However, although human remains are the best way to represent past, present, or future human death investigations and circumstances, there are several drawbacks. One of the most common limitations for using human cadavers for research is access and secured research facilities. There are few active human remains donation programs in the United States that are linked to facilities where the cadavers are placed outside for field research focused on decomposition: University of Tennessee at Knoxville, Western Carolina University, Texas State University-San Marcos, Colorado Mesa University, and Sam Houston State University. The only other facility outside of the United States is located at the University of Technology, Sydney, Australia. The few facilities that have the infrastructure, permits, and capacity to handle donated human cadavers limit the ability of acquiring human remains for field research in many regions of the world. Further, the donation program cannot predict the availability or characteristics of the cadavers that it receives. Having subjects that are treated in consistent ways and are as similar as possible is important for proper experimental design and replication. However, this is a rare, if not impossible, task when using donated human cadavers.

Often donated cadavers represent a wide range of individuals that differ in many ways, including sex, ancestry, manner of death, and the degree and types of long-term antemortem medication or medical procedures used at the end of life. For proper experimental design, it is necessary to acquire subjects that are as similar as possible to control for outside factors like medication. As an example relevant to microbiology, researchers would assume that the degree and types of antibiotics that a person has taken would impact the antemortem microbiome, as well as the initial communities of the postmortem microbiome (Crippen *et al.*, 2015). However, there is no research that we are aware of that has investigated the effects of antemortem antibiotic usage and the postmortem microbiome. The same is true for many other aspects of the human postmortem microbiome, including how such factors of ancestry, sex, age, geographic area of residence, overall diet and weight, health status, and the treatment of the body after death (e.g., autopsies, cold/frozen storage time, and transportation methods) affect the microbial communities of the recently dead. Donated human cadavers can represent a highly variable initial microbiome that will need to be much better understood and modeled with error rates for broad use in the forensic sciences. However, there are alternatives to using donated human cadavers for this important research. Here, a brief discussion of two alternatives is provided: human remains that can be accessed as part of routine death investigations and the use of animal surrogates.

3.3 Human remains research during death investigations

A potential alternative to human cadaver donation programs is to acquire research access to human remains that have expired for different reasons and that are part of routine death investigations. This scenario is dependent on collaborations with medical examiner offices or other agencies or departments that provide the medical or forensic evaluations of human or animal deaths as part of their charge and professional responsibilities. There are several limitations to this approach of acquiring access to human remains, including that the manners of death can range from natural, homicide, and suicide, to accidental. Another factor to consider includes the inherent variation in the human attributes of the remains; these may include attributes of ancestry, sex, age, medical history, manner of death, geographic region of domicile or death event, and toxicity screens, as examples. Thus, the investigator still does not have control over the types of human remains that will be available for study; however, in this scenario the collaborative investigative office can provide samples or data (e.g., microbial swabs from the body) that have been acquired from many sets of human remains. Based on preliminary analyses and an evaluation of the manner of death and the conditions of the remains (as described earlier), a researcher can identify and choose which subjects to use for a specific forensic microbiology question or hypothesis.

As an example, this approach is being used in the authors' current research with a collaboration with a medical examiner. Ultimately, this collaboration will result in microbial swab samples collected from different body locations (e.g., eyes, mouth, and rectum) of over 1000 bodies resulting from various manners of death (Pechal *et al.*, 2015). Ultimately, the entire pool of subjects will be evaluated from a total of 200–300 decedents to (i) test if postmortem microbial communities, or the postmortem microbiomes, can be used to estimate the time since death; (ii) determine if the communities change (or go through species succession) in synchrony among locations of the body; and (iii) if the attributes of the subjects themselves influence the overall human postmortem microbiome, or just specific areas of the body. Thus, while it is not possible to *a priori* select which subjects will be used, a database of subjects, their attributes, and circumstances of death will be analyzed and used to identify a set of subjects to develop a strong statistical experimental design where certain attributes can be selected and included in model development.

There are several advantages to this approach: there is often a continuous availability of human remains that represent a different population compared to that of donated remains programs. This sample of human remains represents common manners of death, including suspicious deaths. Thus, a wide range of manners and circumstances of death can be surveyed, including a range of postmortem intervals (PMIs) and body discovery circumstances. These cases are part of an investigation, and so additional information about the handling of the body, and some medical history or toxicology results and autopsy reports may be available. Further, since these cases are part of an investigation, any microbial data that collected could potentially be used as evidence in the distant future, as forensic microbiology develops into a scientifically justified discipline for use in legal investigations.

There are several disadvantages beyond the considerations presented earlier (and human attribute variation) when working with investigative teams. A major disadvantage is that there are very few investigative offices that are currently working with forensic researchers in this capacity; this makes it difficult to employ a new collection strategy or protocol that goes beyond current investigator responsibilities. A second potential disadvantage is that when a collaborative agreement is reached, it will often require a shift in perspective of the investigative office to better understand and appreciate the nuances and rigor of the research, while the researchers will need to better understand and appreciate the political and economically complex atmosphere of many investigative offices. Last, in many death investigations, there can be initially very little known about the identity of the human remains, the PMI, manner of death, and other circumstances of the recovery conditions. Further, appropriate measures and protocols for quality control and quality assurance will necessarily be written into the collaborative agreement, and training sessions and critical oversight of the research as part of the investigation are required. Additional attention to policy development for ethical consideration will likely vary by investigative office. When human remains from either a donation program or through death investigations are not possible, nonhuman surrogates are often the next best systems for forensic decomposition research.

3.4 Human surrogates in research

By far, the most common nonhuman animal surrogate or proxy for humans in forensic decomposition research is the domesticated swine (*Sus scrofa* L.). The reasons for this are several and have been described for forensic entomology research (Catts and Goff, 1992). While one study (Schoenly *et al.*, 2007) showed that there were no significant differences in the arthropod communities that arrived and colonized swine versus human remains when the carcasses were between 23 and 27 kg at death, there is still controversy over how well swine represent humans: the authors of that study noted that there was no true replicate human cadavers used in their investigation. Nevertheless, swine have been the most widely used surrogates for human decomposition for many studies from around the world (Tomberlin *et al.*, 2012). However, in a recent literature search Tomberlin *et al.* (2012) quantified the number of several other animals that have been used in research as proxies for humans in various parts of the world, including rats (*Rattus* spp.), chickens (*Gallus gallus domesticus* L.), and different species of primates (*Presbytis* sp.), among others (Table 3.1) (Tomberlin *et al.*, 2012). Additionally, Catts and Goff (1992) reviewed several earlier decomposition studies that used animals ranging in size from lizards and toads to elephants.

There are many arguments for the close similarity of swine and humans. Swine can be similar in terms of size and mass. They can also have similar fat distribution to the average torso of an adult human and the dermis is structurally similar between the two species. Swine have similar internal microbial communities (Schoenly *et al.*, 2006), and the body hair density is within the range of most humans (Dent *et al.*, 2004).

Table 3.1 Animal surrogates from publications that have been used in global research for making inferences of human remains decomposition

Common name	Scientific name	Number of publications
Swine	*Sus* spp.	37
Rodent	*Rattus* spp.	8
Chicken	*Gallus gallus domesticus*	2
Rabbit	*Oryctolagus* spp.	3
Raccoon	*Procyon lotor*	1
Alligator	*Alligator mississippiensis*	3
Monkey	*Presbytis cristata*	1
Domestic cat	*Felis catus*	2
Black bear	*Ursus americanus*	2
White-tailed deer	*Odocoileus virginianus*	1
Domestic goat	*Capra aegagrus hircus*	1

Source: Data from figure 4 in Tomberlin *et al.* (2012).

The arguments against swine as ideal surrogates for human decomposition research, especially for postmortem microbiology, come from understood differences between the two species: for example, swine have a different life style and overall diet compared to humans (e.g., swine feed on what is given, often in proximity to soil); they are indeed very different species that occupy disparate habitats. However, given the logistical constraints and ethical issues surrounding the use of human remains in decomposition research, most researchers accept that animal surrogates are useful and necessary for advancing forensic science, including dimensions of microbiology. Depending on the type of remains and the location and conditions of the research, several aspects should be considered while developing an experimental design.

3.5 Considerations for field studies

Once the subject (i.e., human or animal surrogate remains) is identified and has been acquired for research, there are several aspects of field studies that should be considered. The authors highly recommend reading Chapter 4 for in-depth suggestions, approaches, and considerations for field studies, but a brief discussion of the main factors that should be considered when developing a research project are given here. There are logistical, financial, and practical limitations that one must consider when developing a field study based on the study objective(s): How many field sites do I need? What is the appropriate area or size of the habitat patch or location? How far apart should I place or collect my samples? What is my sample size and frequency? How will the topography (slope or gradient) influence adjacent sample collection locations? What is the directionality of flow in an aquatic habitat? Once a good understanding of these factors, among other abiotic and biotic factors as discussed in Chapter 4, is decided upon the researchers can appropriately prepare for their collections.

Microbial communities are unique from other field research samples (e.g., entomological), and so efforts must be made to maintain sterility during aseptic sample collections and reduce the potential for cross-contamination among samples. Sterility of individual samples may be maintained in the field by having premade sampling kits available during sampling periods with individual sterile containers (e.g., 2.0 mL tubes) and collection devices (e.g., swabs or biopsy punches). The resource type (e.g., soil, insect, carcass, dung, and water column) will ultimately dictate the collection devices and containers. Researchers may also want to consider tools and materials that one may find outside of a scientific supplier; for example, a sterilized galvanized nail is an excellent homogenization tool for soft-bodied insects, while a piece of sterilized steel pipe can provide an alternative for a standardized soil corer. However, one should anticipate opportunities for vandalism and other unexpected setbacks during field investigations that are not within a fenced and locked facility. For instance, vertebrate scavengers, such as vultures, bears, and foxes, may remove the resource if adequate protection measures (large wire cages) are not taken. Additionally, if field experiments are not frequently visited due to a particular objective of the study, there can be other potential dangerous animals that may use the cage as a barrier against larger predators (e.g., venomous snakes residing in the same cage as the carcass). Contingency plans, such as alternative sites for the experiment or extra sample collection kits, are essential to consider when establishing the field experiment in order to minimize the aforementioned potential pitfalls or limitations. However, field research can be unpredictable and unexpected; thus, researchers may need to approach their project with an open mind, adaptability, and innovation to overcome any expected situations in the field.

3.6 Descriptive and hypothesis-driven research

A strong experimental design rests on how a research question is developed, and one of the most important elements of this is to determine whether the study will include an experimental component that tests competing hypotheses or not and be more descriptive in nature. In many ways, more descriptive studies describe the characterization of *patterns* in decomposition. In the case of forensic microbiology, most of the recent studies using high-throughput genomic sequencing provide a survey of the microbial taxa that are occurring on or in the cadaver or in the soil associated or impacted by the decomposition (Parkinson *et al.*, 2009; Hyde *et al.*, 2013, 2014; Metcalf *et al.*, 2013; Pechal *et al.*, 2013, 2014; Can *et al.*, 2014; Lauber *et al.*, 2014; Weiss *et al.*, 2015). In these research projects, the pattern often does not include an experimental manipulation that can be tested and compared to a control condition; however, these pattern-oriented studies provide important information for additional studies.

Many pattern-oriented studies are predominately descriptive (i.e., lacking a manipulated independent variable) but still provide valuable information about the processes and mechanisms of decomposition, especially when characterized over time. For instance, Weiss *et al.* (2015) evaluated the effect of carcass size on the soil microbial community changes beneath decomposing swine cadavers, noting that time since placement in the field was significantly more important on the microbial (i.e., bacterial, archaeal, and eukaryotic) community structure than the effect of carcass mass at death. Similarly, Pechal *et al.* (2014) evaluated the Bacteria and Archaea epinecrotic community (i.e., microbial consortia on surfaces of carrion) succession on swine carcasses in the field and developed statistical models identifying indicator taxa that were associated with the amount of time of decomposition. These results were similar to those produced by Metcalf *et al.* (2013), where the bacterial, archaeal, and eukaryotic epinecrotic and soil microbial communities were surveyed over time in a laboratory setting. These example studies, and the others cited earlier, are immensely valuable as they provide important and pioneering information on the pattern of microbial community succession on both human surrogate cadavers and the soil impacted by their decomposition. However, much of that information was descriptive in nature and did not test or evaluate specific hypotheses in an experimental context that would test the mechanistic processes occurring within the changing microbial communities during decomposition.

When manipulation experiments are developed, they are conceived to test an explicit hypothesis. New discoveries and novel processes and patterns that occur in nature are often (but not always) revealed through experimental manipulations of replicate experimental units—in the case of cadaver decomposition and forensic microbiology, the experimental units are usually considered the decomposing sets of remains. Often, the study design is created in a way that tests a null hypothesis against one or more competing or alternative hypotheses (Montgomery, 2009). To identify processes, a subcomponent of the mechanism of interest is controlled or in some way manipulated to test the response of pattern or process that reveals how that change occurred. Several carcass decomposition studies have manipulated swine carcasses as a means to evaluate changes in the epinecrotic microbial community succession (Carter *et al.*, 2015; Weiss *et al.*, 2015). For instance, Pechal *et al.* (2013) prevented approximately 95–99% of primary arthropod colonizers (see Chapter 11 on entomological aspects of decomposition) access to three replicate swine carcasses for 5 days of decomposition. This approach was taken in order to compare to the community succession of microbes on control carcasses that were immediately and consistently colonized by these arthropods during that same 5-day period (Figure 3.1). The investigators identified significant changes in the microbial communities due to the exclusion of arthropod colonizers to these replicate sets of remains, and speculated that the process of epinecrotic microbial succession was at least partially dependent on arthropod activity on the carcasses.

The authors suggested that these changes in successional patterns were through direct inoculation of new (exogenous) microbial taxa through regurgitation and defecation (normal activities of some flies when they contact a cadaver) that added to the community and shifted interspecies interactions. Alternatively, blow fly (Diptera: Calliphoridae) larvae, for example, that grow and consume the carcass tissue were mechanically changing the communities through feeding activities or physical movement. While this study was a relatively simple manipulation experiment, it tested the hypothesis that arthropod colonization of decomposing remains will have a significant impact on the epinecrotic community successional patterns.

Manipulation studies such as this provide new information that can be used for developing new hypotheses that can more explicitly be used to understand deeper and more mechanistic processes of decomposition. In the aforementioned example, follow-up manipulation studies could be developed to test the general hypothesis that if only certain arthropod taxa (e.g., blow flies) are excluded from carcasses, the pattern of epinecrotic microbial succession will shift compared to controls and to other treatments where, for example, other arthropod taxa (e.g., beetle (Coleoptera) larvae) are directly introduced to carcasses while all other taxa are excluded. This is not to suggest that nonmanipulation experiments are not hypothesis driven, but from an experimental design approach,

Figure 3.1 Pechal *et al.* (2013) conducted a manipulation experiment where necrophagous arthropods were either allowed to (a) naturally interact with and colonize the carcasses or (b) were excluded from accessing and interacting with the carcasses for 5 days of decomposition using mesh netting that was sealed at the bottom edges with plastic and soil. The exclusion treatments were 99% effective at preventing any arthropods from accessing the treatment (excluded) carcasses. Inadvertent arthropod access into the exclusion treatments was monitored by small glue traps (i.e., white rectangle attached to the wire case in panel (b) placed inside the mesh cages. The small cages made of chicken wire in both (a) and (b) were antivertebrate scavenging cages to prevent access of vertebrates to the carcasses. Both worked at 100% efficiency in preventing vertebrate scavenging

manipulations allow for more explicit testing of mechanisms that produce patterns and processes. Many other process-oriented experiments could be developed to address increasingly higher resolution understanding of microbial mechanisms during decomposition. However, any such studies require strong and rigorous experimental design.

3.7 Experiment design

A strong experimental design rests on how the research question is developed. Attention should be focused on independent experimental units, adequate replication, and statistical power to achieve increased inference strength from the resulting data set. Better understanding of the naturally occurring microbial community variability within an ecosystem (e.g., human cadaver) will lead to an increased ability to extrapolate information from the existing data set and apply it to remains found in similar circumstances. For instance, if a researcher characterized microbial community succession from three areas (e.g., eyes, nose, and buccal cavity) of a decomposing carcass placed in a mixed-grass prairie once every other day for 10 days, would this specific data set be useful to forensic practitioners investigating a death scene, and attempting to estimate a PMI in a similar ecoregion? The answer is not entirely clear, as the resulting data would statistically be limited to that prairie during that time period because of the lack of replication or independent experimental units, and the limited sampling frequency and intensity would potentially miss important microbial community changes denoting shifts in decomposition.

An alternative experimental design to characterize microbial community succession after death that would increase the inference strength and potential applicability to forensic practitioners would be as follows: place three sets of remains (subsamples to understand within experimental unit variation) in a mixed-grass prairie (experimental unit); collect and identify microbes from three locations on the body (samples) at 8-hour intervals; and with the same design and sampling scheme, simultaneously conducted this experiment in another mixed-grass prairie with similar biotic and abiotic conditions located at an independent distance away (e.g., 10 km). This experimental design has independent experimental units ($N=2$ mixed-grass prairies) that represent the minimal number of replicates (Moreau *et al.*, 2015; Schoenly *et al.*, 2015), and the sampling frequency necessary for characterizing microbial community variation throughout decomposition among individuals. One must take into consideration the level of inference, study goals or objectives, and variability one wishes to accomplish with the study to inform their experimental design.

An experimental unit is the smallest division to which a treatment can be applied. There are a variety of treatments that may be of interest for research, but consider the following scenario as an example of an experimental unit: a researcher hypothesizes microbial community structure on decomposing rodent carcasses in a temperate forest has greater microbial species diversity than rodent carcasses decomposing in a prairie. The experimental unit is the habitat type (forest or prairie); thus, the replicates for the study would be the number of forests and prairies the researcher places carcasses within; the number of carcasses represent subsamples and not true replicates (see simple pseudoreplication definition later) (Moreau *et al.*, 2015; Schoenly *et al.*, 2015). The experimental units to test for treatment effects must be independent in time and space; otherwise, there can be confounding interactive effects among the samples that may bias the results. This is especially relevant in terms of distances between replicate carcasses (Moreau *et al.*, 2015; Schoenly *et al.*, 2015).

Furthermore, randomization of the treatments is important. There is an inherent, and still largely underexplored, diversity among naturally occurring microbial communities. For example, an estimated 10^9 microorganism cells and tens of thousands of species can be detected in a single gram of soil (Torsvik and Ovreas, 2002; Kassen and Rainey, 2004; Achtman and Wagner, 2008). Despite recent advances and increased cost effectiveness of sequencing capabilities (e.g., gene amplicon-based sequencing or metagenomics) and computation power, the temporal and spatial complexity of microorganisms during the decomposition process remains largely unknown. Therefore, researchers should consider the natural microbial variability (biodiversity) when designing their experiments.

There are many experimental design approaches that can be implemented to randomize samples within the treatments that will account for variability (Moreau *et al.*, 2015; Schoenly *et al.*, 2015), such as complete randomization, randomized block, nested designs, and split-plot designs. Each of these examples is summarized as follows:

- Complete randomization is a simple one-factor experiment that tests a single response variable (e.g., CO_2 production) after undergoing testing of the desired treatment(s) and control; each individual (e.g., blow fly larva) or sample (e.g., soil type) is randomly assigned to the treatment(s) or control.
- A randomized block design takes into consideration that a known factor, such as sex, will affect the response to a treatment. For example, to test the size of flies that have fed on different concentrations of antibiotics, the individuals are first separated by sex (block) and then randomly assigned to the different antibiotic treatments. Multifactor experiments will use more advanced experimental design, such as nested and split-plot, to appropriately test hypotheses.
- Nested designs use a hierarchical approach to test the effects of a fixed factor (i.e., treatment) on individuals randomly assigned to each treatment. There are typically two- and three-factor nested designs with each factor being similar but not identical. For example, multiple soil samples collected from four sites from two different fields (a treated field with decomposing carcasses and a control field) represent a nested design because there is no interaction among any of the sites collected in the treated field to any sites collected in the control field.

Split-plot designs are necessary when the level(s) of certain factors are more difficult to change during the experiment and have at least two sizes of experimental units. For example, to test microbial community richness in soil samples below four carcass types (e.g., rat, chicken, rabbit, and swine) in two habitats (agricultural field and forested area), one would consider using a split-plot design. Equal-sized plots would be established in each habitat type (whole plot). Each whole plot is divided into four equal plots (split-plot) and the four carcass types are randomly assigned to each split-plot. Each of these designs, among other experimental design approaches, has its own set of pros and cons for statistical analysis, power, inference, and detection of statistically significant differences that the researcher(s) will need to take into consideration for their project (Moreau *et al.*, 2015; Schoenly *et al.*, 2015).

A common experimental design flaw in ecological and forensic field studies is pseudoreplication (for an excellent review of this topic in forensic entomology, please see Michaud *et al.* (2012)). Pseudoreplication, as defined by Hurlbert (1984), is "the use of inferential statistics to test for treatment effects with data from experiments where either treatments are not replicated (though samples may be) or replicates are not statistically independent." There are three common types of pseudoreplication that can occur in laboratory or field experiments. *Simple pseudoreplication* occurs when there is only a single experimental unit per treatment (Figure 3.2); many researchers incorrectly classify the

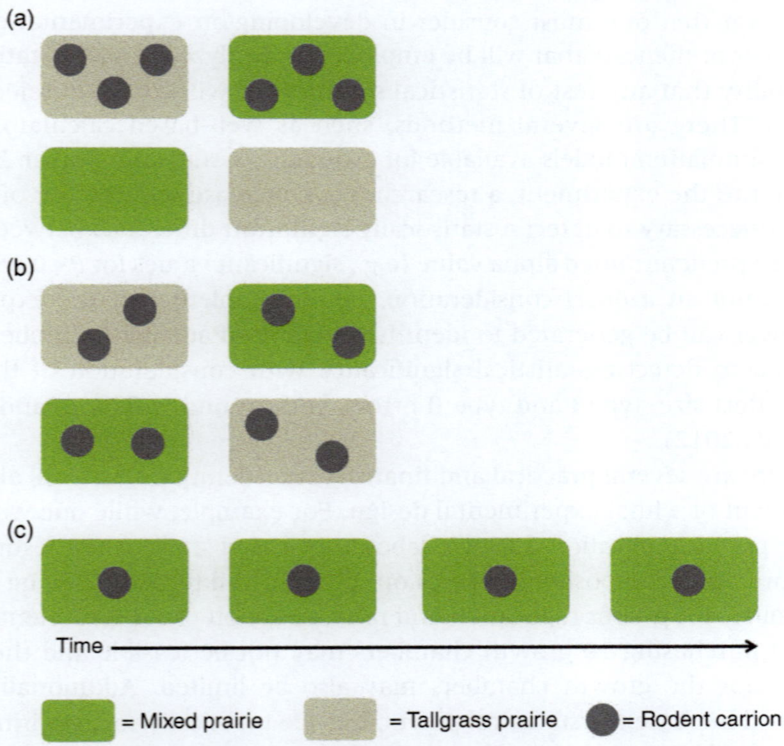

Figure 3.2 Examples of common types of pseudoreplication: (a) simple, (b) sacrificial, and (c) temporal. The larger rectangles represent the experimental unit with the circles denoted as one set of rodent remains that is sampled. In the case of temporal pseudoreplication, the same set of remains is sampled repeatedly over time and each sampling event is incorrectly considered a replicate sample from that carcass

samples within the experimental unit as "replicates." This type of pseudoreplication is neither surprising nor necessarily a fatal experimental flaw, as true replication with experimental units is often impossible or undesirable when very large-scale systems (e.g., cities, forests, watersheds, and rivers) are the experimental unit. Another type is *sacrificial pseudoreplication*, which is when the experimental design involves true replication of treatments, but the replicate data are pooled prior to statistical analysis or if two or more samples collected from each experimental unit are treated as independent replicates (Figure 3.2). The third type of pseudoreplication is *temporal pseudoreplication*; this differs from simple pseudoreplication only in that the multiple samples from each experimental unit are not taken simultaneously, but rather each independent experimental unit is sampled sequentially over each of several dates (Figure 3.2). In decomposition research, this can be the sequential sampling of a carcass that is incorrectly considered replicate samples of an individual set of remains. Each project will be designed based on the goals and practical and financial considerations, but researchers aware of the pitfalls of pseudoreplication for statistical analysis and interpretation will make continued efforts to avoid experimental designs with these limitations.

Another factor that one must consider in developing an experimental design is the statistical power of the tests that will be employed to analyze the data. Statistical power is the probability that any test of statistical significance will correctly reject a null hypothesis (H_0). There are several methods, such as web-based calculators, software programs, or simulation models available for estimating a statistical power. Based on the estimates prior to the experiment, a researcher can increase the number of samples (or observations) necessary to detect a statistically significant difference between or among treatments at a predetermined alpha value (e.g., significant values for $P<0.05$). However, if power was not an *a priori* consideration, upon completion of the experiment the statistical power can be generated to identify whether an adequate number of samples were collected to detect a statistical significance with consideration of the following parameters: effect size, type I and type II errors, background variation, and sample size (Michaud *et al.*, 2012).

Finally, there are several practical and financial considerations that will also influence the development of a final experimental design. For example, while one would want to avoid simple pseudoreplication during a laboratory-based study when testing the effect of temperature on decomposing carcasses on different soil types by having access to 10 growth chambers for proper replication and randomization of the treatments, the financial reality of purchasing 10 growth chambers may not be feasible and the logistics of finding space for the growth chambers may also be limited. Additionally, increased sample size will increase the statistical power, but the practicality of acquiring 50 (45 kg) nondiseased, similarly processed swine carcasses (or sets of human remains) for simultaneously deposition into a field is highly unlikely. Therefore, researchers can take alternative sampling approaches, such as using a complete randomized block design for several trials in the previously mentioned growth chamber example to account the variation within the growth chamber. Also, researchers may choose to composite the samples by collecting a single sample from different areas of an individual carcass and pooling them for an overall sample. This would be compared to collecting single samples from different areas of a carcass and evaluating differences among the areas. For researchers interested in the nucleic acids of a sample (DNA or RNA), the composite step may occur during the sampling period, or samples may be combined in equal quantities post extraction for further downstream applications (e.g., amplicon-based and metagenomic sequencing or quantitative polymerase chain reaction).

There are many considerations for developing a study with certain limitations, some of which have been addressed in the chapter, but others may occur on a project-by-project basis. In any case, if the investigator is not confident of a strong experimental design and wishes to avoid some of the issues discussed here, it is highly recommended that the investigator seek professional statistical advice and consultation.

3.8 Validation studies

Another consideration for forensic microbiology research is the development of validation studies. Validation studies test and confirm primary studies that have shown significant patterns or processes of microbial communities during decomposition. When a primary research project has been undertaken with a well-developed experimental design and has robust statistical power and inference, the results should be broadly relevant. However, to determine how well a specific study can be used to make inference or understanding in other, more realistic and variable contexts, validation studies are a powerful tool for confirming results for broader applications. Within the focus of the chapter, an example would be the blind validation of using microbial community succession as a biological indicator of the PMI.

In such an example, there have been several primary studies that have shown promise for this technique in estimating the PMI, but each was conducted in a certain habitat or in the laboratory (Metcalf *et al.*, 2013; Pechal *et al.*, 2014). These studies have modeled the community succession in a way to predict what microbial taxa are most important for estimating a PMI for the circumstance and conditions of each study. To evaluate how well those models can be applied in other situations, a validation study would sample the microbial communities from new sets of remains in several different environmental conditions, representing the variation in death scene circumstances that are commonly encountered by law enforcement investigators. Ideally, the samples would be collected from a new investigator and provided to a "blind" investigator who would only be provided the published statistical models. The blind investigator would process the samples and use the models to estimate the PMI. If the results of the models correctly estimate the PMI within the defined prediction interval, then the study would have validated the earlier modeling of the primary studies. If the validation models fail to agree with the primary model estimates, then the study would not validate the primary studies. In this way, validation studies inform the general discipline of researchers on how well the findings of primary studies can be extrapolated or used in future applications that represent a broad range of circumstances. This is an important consideration for any research in the forensic sciences, and should become a mandated activity for the rigorous evaluation of what kind of biological evidence can be used in death investigations and evaluated within the larger subject of jurisprudence.

Acknowledgments

We would like to thank the editors and two reviewers for a critical and helpful review of the chapter. The authors would also like to recognize the National Institute of Justice (2010-DN-BX-K243 and 2014-DN-BX-K008) for providing financial support leading toward the preparation of this chapter. Points of view in the chapter are those of the authors and do not necessarily represent the official position or policies of the US Department of Justice.

References

Achtman, M. and Wagner, M. (2008) Microbial diversity and the genetic nature of microbial species. *Nature Reviews Microbiology*, **6**, 431–440.

Benbow, M.E., Tomberlin, J.K. and Tarone, A.M. (2015) *Carrion Ecology, Evolution, and Their Applications*, CRC Press, Taylor & Francis, Boca Raton, FL.

Can, I., Javan, G.T., Pozhitkov, A.E. and Noble, P.A. (2014) Distinctive thanatomicrobiome signatures found in the blood and internal organs of humans. *Journal of Microbiological Methods*, **106**, 1–7.

Carter, D., Metcalf, J., Bibat, A. and Knight, R. (2015) Seasonal variation of postmortem microbial communities. *Forensic Science, Medicine, and Pathology*, **11**, 202–207.

Catts, E. and Goff, M.L. (1992) Forensic entomology in criminal investigations. *Annual Review of Entomology*, **37**, 253–272.

Crippen, T.L., Benbow, M.E. and Pechal, J.L. (2015) Microbial interactions during carrion decomposition, in *Carrion Ecology, Evolution and Their Applications* (eds M.E. Benbow, J.K. Tomberlin and A.M. Tarone), CRC Press, Boca Raton, FL, pp. 31–63.

Dent, B.B., Forbes, S.L. and Stuart, B.H. (2004) Review of human decomposition processes in soil. *Environmental Geology*, **45**, 576–585.

Huffman, J.E. and Wallace, J.R. (2012) *Wildlife Forensics: Methods and Applications*, vol. **6**, Wiley Blackwell, Oxford.

Hurlbert, S.H. (1984) Pseudoreplication and the design of ecological field experiments. *Ecological Monographs*, **54**, 187–211.

Hyde, E.R., Haarmann, D.P., Lynne, A.M. *et al.* (2013) The living dead: bacterial community structure of a cadaver at the onset and end of the bloat stage of decomposition. *PLoS One*, **8**, e77733.

Hyde, E., Haarmann, D., Petrosino, J. *et al.* (2014) Initial insights into bacterial succession during human decomposition. *International Journal of Legal Medicine*, **129** (8), 661–671.

Kassen, R. and Rainey, P.B. (2004) The ecology and genetics of microbial diversity. *Annual Review of Microbiology*, **58**, 207–231.

Lauber, C.L., Metcalf, J.L., Keepers, K. *et al.* (2014) Vertebrate decomposition is accelerated by soil microbes. *Applied and Environmental Microbiology*, **80**, 4920–4929.

Metcalf, J.L., Wegener Parfrey, L., Gonzalez, A. *et al.* (2013) A microbial clock provides an accurate estimate of the postmortem interval in a mouse model system. *eLife*, **2**, e01104.

Michaud, J.-P., Schoenly, K.G. and Moreau, G. (2012) Sampling flies or sampling flaws? Experimental design and inference strength in forensic entomology. *Journal of Medical Entomology*, **49**, 1–10.

Montgomery, D. (2009) *Design and Analysis of Experiments*, Wiley, Hoboken, NJ.

Moreau, G., Michaud, J.-P. and Schoenly, K.G. (2015) Experimental design, inferential statistics, and computer modeling, in *Carrion Ecology, Evolution and Their Applications* (eds J.K. Tomberlin and M.E. Benbow), CRC Press, Boca Raton, FL, pp. 205–229.

Parkinson, R., Dias, K.-R., Horswell, J. *et al.* (2009) Microbial community analysis of human decomposition on soil, in *Criminal and Environmental Soil Forensics* (eds K. Ritz, L. Dawson and D. Miller), Springer, Dordrecht, pp. 379–394.

Pechal, J.L., Crippen, T.L., Tarone, A.M. *et al.* (2013) Microbial community functional change during vertebrate carrion decomposition. *PLoS One*, **8**, e79035.

Pechal, J.L., Crippen, T.L., Benbow, M.E. *et al.* (2014) The potential use of bacterial community succession in forensics as described by high throughput metagenomic sequencing. *International Journal of Legal Medicine*, **128**, 193–205.

Pechal JL, Schmidt CJ and Benbow ME 2015. *The first use of postmortem microbiomes in human death investigations*, p. 919. In, Proceedings of the 67th Annual Meeting of the American Academy of Forensic Science, February 15–21, 2015, Orlando, FL. American Academy of Forensic Science.

Schoenly, K.G., Haskell, N.H., Mills, D.K. *et al.* (2006) Using pig carcasses as model corpses to teach concepts of forensic entomology and ecological succession. *The American Biology Teacher*, **68**, 402–410.

Schoenly, K.G., Haskell, N.H., Hall, R.D. and Gbur, J.R. (2007) Comparative performance and complementarity of four sampling methods and arthropod preference tests from human and porcine remains at the Forensic Anthropology Center in Knoxville, Tennessee. *Journal of Medical Entomology*, **44**, 881–894.

Schoenly, K.G., Michaud, J.-P. and Moreau, G. (2015) Design and analysis of field studies in carrion ecology, in *Carrion Ecology, Evolution and Their Applications* (eds M.E. Benbow, J.K. Tomberlin and A.M. Tarone), CRC Press, Boca Raton, FL, pp. 129–148.

Tomberlin, J., Byrd, J., Wallace, J. and Benbow, M. (2012) Assessment of decomposition studies indicates need for standardized and repeatable research methods in forensic entomology. *J Forensic Research*, **3**, 147.

Torsvik, V. and Ovreas, L. (2002) Microbial diversity and function in soil: from genes to ecosystems. *Current Opinion in Microbiology*, **5**, 240–245.

Weiss, S., Carter, D., Metcalf, J. and Knight, R. (2015) Carcass mass has little influence on the structure of gravesoil microbial communities. *International Journal of Legal Medicine*, **130** (1), 253–263.

CHAPTER 4

Sampling methods and data generation

Jennifer L. Pechal[1], M. Eric Benbow[1,2] and Tawni L. Crippen[3]

[1] *Department of Entomology, Michigan State University, East Lansing, MI, USA*

[2] *Department of Osteopathic Medical Specialties, Michigan State University, East Lansing, MI, USA*

[3] *Southern Plains Agricultural Research Center, Agricultural Research Service, United States Department of Agriculture, College Station, TX, USA*

4.1 Introduction

The study of forensic microbiology is an inherent blend of forensic science and microbiology, and both disciplines have recently been undergoing rapid advancements in technology that are allowing for exciting new research avenues. However, the integration of two different disciplines poses challenges because experts in two different fields are required to become competent, if not proficient, in the conceptual and technical background and tools for sampling and data generation of the other discipline. For instance, most microbiologists have not been trained in how to approach death scenes or interpret data within a forensic context. Likewise, forensic scientists often have not received in-depth training in classical microbiology or the recent molecular advances that are somewhat revolutionizing these fields. Therefore, this chapter provides an introduction to sampling methods and data generation that are an integration of techniques from laboratory-based microbiology, field and environmental microbial ecology, and considerations within a forensic context. The focus of the chapter is on the methods and approaches for collecting, processing, analyzing, and the handling of samples and data related to microbial communities associated with the decomposition of carrion/human remains. While many important topics are covered concerning sampling methods and data generation, the chapter is not comprehensive, but rather serves as an introduction of the basic principles necessary for research in forensic microbiology. For additional information on specific methodologies related to forensic microbiology, please see other chapters (e.g., microbiology at autopsy (Chapter 7) or within the built environment (Chapter 13)) that cover more detailed surveys of sampling and data generation methods.

Forensic Microbiology, First Edition. Edited by David O. Carter, Jeffery K. Tomberlin, M. Eric Benbow and Jessica L. Metcalf.

4.2 Materials

4.2.1 Financial considerations

Often, one of the most important constraints to any research or application of a given discipline is financing, and forensic microbiology is no exception. Financial limitations will often dictate the experimental design of most research projects. As discussed in Chapter 3, the number of replicates, the number of samples, and how often they are collected in a study all require consideration for the cost of each sample, in addition to the other expenses necessary to successfully complete a project (e.g., high-throughput sequencing or server space). While the cost of the expendable supplies (e.g., swabs or DNA extraction reagents) and service activities, such as various forms of genomic sequencing or purchasing specific bacterial cultures, are often considered early on in the development of research, the costs associated with personnel and time are often miscalculated. Therefore, a realistic estimate of the costs associated with acquiring permanent equipments, expendable supplies, contractual services, personnel, and other costs necessary to undertake the research are critical for successful research projects that reflect a rigorous and scientifically defendable experimental design.

Microbiological research has historically been expensive, and new advances in genomic technology have continued to be costly especially when the research is new to a scientist without funding. This is particularly true in situations when a researcher or collaborative group is required to generate preliminary data necessary to secure competitive funding, which is often the circumstance for many new researchers in forensic microbiology. Additionally, to best represent real-world scenarios and natural circumstances that are relevant to forensics, researches in many aspects of forensic microbiology are conducted in the field rather than in the laboratory. Moving such research from controlled laboratory conditions into more realistic settings poses new and sometimes unexpected costs to the research budget.

For instance, a laboratory microbiologist may be aware but may not anticipate the costs associated with doing field research on multiple human cadavers, such as the equipment (e.g., vehicle) to transport the cadavers or the environmental instrumentation necessary for measuring and monitoring weather conditions, or the material and supply costs for building cages to go around the cadavers to prevent undesired scavenging events by vertebrates. The same potential under appreciation of the cost, time, and personnel expertise necessary for evaluating microbial samples using high-throughput sequencing (e.g., amplicon-based or metagenomic sequencing) may fall on an environmental scientist that focuses on real-world scenarios of forensic importance. In both scenarios, the best path to a successful multidisciplinary project is early, frequent, and clear communication of budget expectations for each aspect of a research project. For the remaining portions of the chapter, each set of sampling activities comes with a cost that should be considered when planning and developing a research plan.

4.2.2 Terrestrial settings

4.2.2.1 Environmental conditions

There are six primary terrestrial biomes (tundra, taiga, temperate forest, desert, grassland, mangroves, and tropical rain forest) that can be divided into 14 major habitat types with 867 smaller ecoregions according to the World Wildlife Fund. Each ecosystem has unique and important biological functions, such as nutrient cycling, but also has various abiotic characteristics (e.g., temperature, pH, terrain, and plant diversity). The variation in biotic and abiotic diversity encountered across ecoregions may present challenges that forensic practitioners must consider when assessing naturally occurring microbial communities associated with decomposition. For example, in North America, there are 15 Level I terrestrial ecoregions ranging from the Arctic Cordillera in northern Canada to tropical humid forests in the lower parts of Florida and southern gulf coasts of Mexico, and including various plains, mountains, deserts, and forests ecoregions distributed throughout the continent (Commission for Environmental Cooperation Working Group, 1997). Here, we will not provide an exhaustive material list for each specific ecoregion, but rather provide insights and considerations for those materials that may be able to be universally used across the terrestrial biomes, as there are no currently accepted forensic microbiology standard practices and research guidelines for the natural environment (Pechal *et al.*, 2014b).

Cadavers can be discovered in almost any terrestrial location (biomes) under numerous circumstances that can either completely conceal the remains by burial or leave the remains fully or partially exposed. Thus, researchers conducting decomposition studies to characterize microbial communities after death should document abiotic conditions throughout the duration of an experiment since temperature, pH, and other physiochemical properties influence microbial community dynamics from an ecosystem to microhabitat scale. There are numerous instruments available to measure abiotic conditions, but as previously discussed the financial circumstances will be important for frequency and accuracy of data collected. Local meteorological conditions, such as temperature, humidity, and rainfall, are important abiotic factors that can influence the rate of decomposition and the development of consumers feeding on the remains (e.g., Calliphoridae (Diptera) larvae). However, these conditions can be altered at a microhabitat scale based on local vegetation or topography; thus, it is important to document the changes in abiotic factors throughout decomposition for each replicate vertebrate carcass, when financially and logistically feasible.

Data loggers are available that document the following environmental conditions: temperature, humidity, barometric pressure, moisture content, light, precipitation, and wind. These instruments range in data collection capabilities (single vs. multiparameter), size (1.3 cm to 0.6 m), and price ($30 to >$5000). Additional tools may be necessary for measuring biotic factors that influence the decomposition site in specific habitats. For example, in a forested area, the amount of light availability can influence the microbial communities; thus, it is important to measure forest canopy density and light intensity parameters. Additionally, researchers may consider using remote-sensing tools and available databases, such as the normalized difference vegetation index (NDVI), enhanced vegetation index (EVI), or Web Soil Survey (WSS), to quantify the land cover at and surrounding the research location (Lillesand *et al.*, 2014). Finally, it is a good practice to have photographic documentation and written notation of the characteristics of the local habitat where the remains are decomposing.

4.2.2.2 Consumer collections

Many consumers utilize and interact with decomposing remains for food, shelter, progeny rearing, and/or a mating location; there are many excellent texts available that discuss in depth the role and community dynamics of invertebrates and vertebrates during decomposition (see Byrd and Castner (2001), Tomberlin and Benbow (2015), and Benbow *et al.* (2015a) for further details). Primarily, insects and other invertebrates have the most contact with decomposing remains, and therefore may be of the most interest for assessing their microbiology. Flies can be mechanical vectors of numerous microorganisms (e.g., bacteria, fungi, archaea, and protists); they can inoculate the remains with microbes from other sources of decomposing organic material, such as manure found at a livestock facility, and can acquire microbes from the remains for potential distribution back into the ecosystem. Additionally, specific species of blow flies (Diptera: Calliphoridae) and beetles (Coleoptera) can produce antimicrobial peptides, which can ultimately alter the microbial community composition on the resource (Rozen *et al.*, 2008; Čeřovský *et al.*, 2010). Better understanding of the microbial constituents present during decomposition through surveying will increase the empirical data of microbes associated with decomposition and lead to a better mechanistic understanding of the decomposition process in natural environments.

There are several entomological trapping methods widely used depending on the species of interest to be collected (Figure 4.1). Examples of these methods include but are not limited to flight intercept trap, Malaise trap, light trap, beating sheet, sifting (Berlese funnel), and carbon dioxide trap. Forensic entomologists typically use a subset of methods for collecting the insect communities (Schoenly *et al.*, 2007). Flying insects commonly associated with decomposing remains, such as blow flies and flesh flies (Diptera: Sarcophagidae), can be collected either using active trapping methods (sweep nets) or through passive trapping methods (baited, inverted cone traps or glue traps). Glue traps need to be attached to structures close to the remains, such as a wooden stake or rebar that can be placed near the remains during a field study; alternatively, glue traps can be placed in the field using a "tent method" where clothespins are attached at each of the corners, the trap is bent in the middle, the sticky side is placed up, and the trap is placed on the ground surface near the body to collect incoming flying insects. Hand collections using forceps are useful for collecting less mobile insects, such as fly eggs, larvae, pupae, and beetles, while pitfall traps are a passive sampling method that can be

Figure 4.1 Examples of sampling methodologies for collecting microbial communities from various substrates. (a) Inverted cone trap for collecting insects. A bait resource, in this instance the white arrow pointing to a "stink jar" consisting of decomposed liver and water, is placed at the bottom opening as an attractant for necrophagous insects. (b) An antiscavenging cage placed over a swine carcass the field is an excellent place to attached passive trapping methods, such as glue traps (see (c) for an up close view), for collecting flying invertebrates attracted to carrion. (d) Swabs can be effectively used to collect the epinecrotic microbial communities from cadavers. (e) In aquatic habitats, stable structures placed in an aqueous habitat, such as these unglazed ceramic tiles attached to a brick, create a standardized area for biofilm formation and can be left for weeks to months or up to a year depending on the scope of the study

used to collect those ground-dwelling invertebrates attracted to the carcasses over time. Furthermore, there are soil and leaf-litter dwelling invertebrates that may be of interest to researchers. Samples can be collected using trowels, soil corers, or hand collections with further isolation of the invertebrates using sieves and/or Berlese funnels occurring in the laboratory. The soil and leaf litter material may be of interest; thus following the previously described methods, an investigator can collect soil samples for further microbial analysis. However, the samples should be standardized by soil depth, core size, or weight collected.

Gloves should be worn when collecting samples in order to minimize cross-contamination and exposure to potential pathogens. Insect samples can be placed in sterile glass vials, conical tubes, or microcentrifuge tubes depending on the number of specimens collected. Glue traps can be carefully collected, closed cylindrically with the sticky side facing the interior to protect the insects and prevent collapse, and then placed in polyethylene bags. Litter/soil samples also can be stored in sterile glass vials, conical tubes, microcentrifuge tubes, or polyethylene bags. For additional details about proper aseptic collection methods, please see Section 4.3. All of the collected samples (e.g., insects, leaf litter, or soil) for microbial analyses should be preserved in 96–100% molecular grade ethanol, placed in a solution specifically for stabilizing and protecting RNA/DNA (e.g., RNAProtect®), or immediately frozen (≤ –20°C). Finally, and most importantly, sample containers should be appropriately labeled using the following information: date, time, sample identification number, case number (if applicable), geographic location (GPS coordinates), and the collector's name or initials.

4.2.3 Aquatic settings

In addition to the materials outlined earlier for terrestrial cadaver decomposition settings, in aquatic settings, several other sets of supplies and instrumentation are important for a successful research project that accounts for the unique aspects of aquatic habitats. For an excellent review of the different habitat types and importance of carrion decomposition in aquatic settings that are important for research in these habitats, we recommend Wallace (2015). Here, we provide a review of the most common aquatic habitat types and the materials and instrumentation that are important for studying vertebrate decomposition in water systems.

Aquatic settings, or habitats, can generally be grouped into freshwater and marine, but there is some interface of these habitats (brackish habitats) in estuaries and some saltwater inland lakes. In addition, these habitats can be flowing, relatively stagnant, or not flowing. Flowing aquatic habitats are streams, rivers, large lakes, or oceanic currents, while no-, or little, flowing habitats are ephemeral ponds, wetlands, marshes, and lakes. Some parts of oceans have little flow, but most of the oceans and large lakes have dynamic water movement in the form of vertical and horizontal currents. These habitats, whether flowing or not, are functionally important to microbial communities, because nutrient availability, oxygen content, ion concentrations, sedimentation, and many other variables are substantially different in aqueous environments. To account for these abiotic variables, there are numerous materials and instruments that are necessary and different from those used in terrestrial settings.

When cadavers are discovered or studied in aquatic settings, there are several water quality variables that should be measured because they are important factors to decomposition, especially microbial communities. These water quality condition variables are discussed in more detail elsewhere (Hauer and Lamberti, 2007; Wallace, 2015), but the primary factors that can be measured either as spot, one-time measurements or unattended with continuously measured intervals (e.g., 30 minutes, 1 hour, and 24 hours) are the following: dissolved oxygen, temperature, pH, redox potential, various ions (e.g., chloride and iron), conductivity, turbidity, total dissolved solids, clarity, and suspended solids. Additionally, many of these variables can change over small areas, and so it is important to note where the water quality parameters are measured. For instance, in a nonflowing, shallow wetland dissolved oxygen can vary by an order of magnitude with near 100% saturation at the surface of the water to complete anoxia where the water and bottom sediments meet. Temperature is also highly variable depending on depth and shading. If the habitat is a flowing water setting, the velocity, depth, and volume of the water are often important, in addition to the variables mentioned earlier. Streamflow velocity is known to affect how microbial community biofilms form on decomposing carrion in streams (Dickson *et al.*, 2010; Benbow *et al.*, 2015b) along with other variables, such as oxygen, temperature, and sedimentation (turbidity) that affect aquatic biofilms (Battin and Sengschmitt, 1999; Battin *et al.*, 2003; Lang *et al.*, 2015).

In order to measure these variables in aquatic habitats, there are many options that are too numerous and different to review in the chapter but can be found in other resources (Hauer and Lamberti, 2007). The key considerations for choosing the correct instrument will be related to the research question and any financial constraints. There are multiparameter probes that are capable of measuring several water condition variables simultaneously either for a single spot assessment or with the capability of unattended sampling. These probes are often quite expensive (e.g., over $15,000 for the housing unit, all probes, and connecting cables) and can often be cost prohibitive. There are also smaller, more cost-effective probes that do not have the attended sampling capability and often can only measure three to four water quality condition variables (usually dissolved oxygen, temperature, conductivity, and pH). These probes are often a suitable, yet limited, alternative to multiparameter, unattended probes for aquatic research.

Further, if the microbial communities of the water column are an important research question, there are several potential avenues to filter the microbial and associated sediments onto filters, measuring the volume of water that is filtered, and matching this to water quality condition. These filtering strategies can range from small handheld syringes to larger, battery-powered vacuum pumps, to automated storm water samplers that pump water from the source and then store the water in climate-controlled conditions for later filtering. The filters necessary for capturing the microbes of interest will vary in composition, pore size, and diameter, as discussed later in the chapter. Examples of the materials and supplies used for measuring aquatic water quality conditions, particulate organic carbon, and microbial communities can be found in the book titled *Methods in Stream Ecology* (Hauer and Lamberti, 2007). It is an excellent book on aquatic sampling techniques, which provides specific methods and literature for sampling abiotic and biotic components of aquatic habitats that can be modified for forensic microbiological applications.

An additional consideration for forensic microbiology research in aquatic settings is the materials and supplies necessary for securing a carcass in place in the aquatic habitat. Carcasses often go through a stage of bloating from the accumulation of microbially generated gases (Chapter 10). The bloating causes the carcass to float, and in flowing systems, to move downstream. There are numerous ways to consider securing the carcasses, ranging from small cages, stakes, rope, or other creative ways of tying a carcass down like zip ties. The composition of these carcass security materials can have an effect on the microbial communities and should be taken into consideration. For instance, a metal cage will quickly oxidize in a stream, generating new chemical conditions that may affect the carcass and any associated microbial communities. It is recommended that nonmetal materials be used in aquatic forensic microbiology research.

4.3 Sample collection techniques

The use of proper aseptic (sterile) techniques to eliminate contamination of collected microbial samples that will undergo sequencing analysis is the most vital protocol consideration in any experiment. As microbial forensic techniques are developed, it is imperative that the collection of samples for future forensic DNA/RNA analysis be efficient and nondestructive and that contamination by surroundings or personnel is mitigated. Inadvertent contamination of a sample with extraneous microbial DNA/RNA is unfortunately quite possible given the environment in which samples are taken, and if it goes undetected contamination sequencing reads will be incorrectly identified as legitimate results. Therefore, aseptic sample collection techniques should always be taken seriously and extra caution is always warranted. Strict protocols should be established, practiced by all personnel, and consistently adhered to during sample collection and manipulation. Preferentially, the same person should always be assigned to conduct the same job to maintain consistency in sample collection. It is vital that all new staff be trained on and practice sterile techniques prior to their utilization in the field. They should apprentice with a skilled practitioner and be supervised at length until comfortable with sterile techniques before they are allowed to work independently. Records of who made the collection, what was collected and how, where, when, and any other relevant data should be recorded at the time of sample collection: an approach analogous to maintaining the chain of custody.

All materials that will come into contact with the sample must be sterile, and manipulations of the sample must be designed to avoid contact with any nonsterile surroundings. Personnel collecting samples should be required to wear personal protective equipment (PPE) (Zheng *et al.*, 2013), such as booties, gloves, aprons/suits, sleeves, safety glasses, surgical masks, and hair covers. PPE will prevent both contamination of the sample with extraneous microbes and possible passage of infectious pathogens to personnel. However, just donning PPE does not automatically prevent contamination of a sample, it only acts as a sterility barrier; it is up to the person wearing the PPE to be purposefully aware of their actions in order to avoid contact with nonsterile environments or prevent contamination of themselves. All supplies for field collection should be labeled and prepackaged for convenience, and this should be done in a sterile manner in the laboratory prior to use in the field.

An absolute sterility barrier cannot be accomplished in the field without prohibitively impeding routine manipulations and sample collections; therefore, procedures should be designed to significantly reduce the probability of contamination. One primary consideration is to arrange your supplies within your work area to be easily accessible, with wide spacing so that you avoid reaching over one sample to get to another. If items in use are tightly packed, you will inevitably brush a sterile surface with a nonsterile one or inadvertently drop nonsterile items (e.g., dust, skin cells, and dirt) into sterile containers. Whenever possible, hold items, such as collection tubes and swabs, in your sterilely gloved hands or sterile tube holders/racks and avoid setting them onto nonsterile surfaces. Particular sample collection procedures will vary with the environment or host from which collections are made. Careful preplanning and consideration of field conditions likely to be encountered, along with in-the-field simulated practice and consulting experienced personnel, are essential components to designing and maintaining aseptic techniques for your experiment. Once a successful design is established, all precautions should be scrupulously maintained over replicate experiments to avoid sample inconsistency and possible corruption.

4.4 Sample preservation, storage, and handling techniques

A sample must be taken in such a fashion that it is representative of the microbial community. For example, skin microbes from healthy humans can usually be collected by taking swabs, scrapes, or punch biopsies. Each of these collection methods has pros and cons as far as convenience, cost, and applicability in the field are concerned; however, according to Grice *et al.* (2008), each method yields nearly identical microbial profiles. Therefore, the choice is that of the researcher with respect to experimental design.

From a forensic perspective, minimal disturbance of evidence is desired; therefore, noninvasive methods such as swabbing with a sterile cotton swab or forensic collection swabs made from rayon, polyester, or foam are preferred. As the number of microbes on the epidermal surface may be low, special care must be taken to adequately sample in order to capture a usable quantity of microbes that assures sufficient high-purity DNA for polymerase chain reaction (PCR) and subsequent sequencing. This is determined by systematic experimentation where all sampling is standardized in terms of the type of swab used, the time spent swabbing (e.g., 30 seconds and 1 minutes), and the size and location of the area swabbed. Microbes are generally present in larger quantities in the gastrointestinal (GI) tract and body orifices (e.g., buccal cavity) making collection from those areas easier. In general, 0.25–0.50 g of microbial GI material per sample is sufficient for DNA isolation. Replicate samples should always be collected if possible and not just the minimal triplicate to assure statistical relevance, but also additional samples for possible replacement of corrupted or low-yield samples.

Sterility factors and temperature need to be controlled and standardized for field collection. Dependent on the experimental design, the amount of time spent collecting samples prior to processing in the lab could be significant enough to affect microbial species viability; therefore, immediate, short-term sample storage in the field, such as placing samples on ice packs in a cooler, should be a consideration in the experimental design.

When working in the field, one must also be cognizant of other animals that exploit carrion, both from an aspect of being a competitor for the rich food supply offered and also being harmful to personnel making field collections. Always determine what other fauna exist in your ecoregion of interest, from insects to larger predators that would utilize carrion. Take appropriate protective action for your personnel while in the field and protect your study resource (the carrion) from exploitation by other necrophagous fauna. This usually requires utilization of a physical barrier such as a sturdy, protective cage made from heavy gauge fencing (wire utility panels) to exclude larger predators, covered by a layer of small-diameter fencing (chicken wire) or plastic heavy duty bird netting to defend against the long necks of vultures or the resourceful hands of raccoons. This larger sized opening barrier or mesh can be layered by a finer netting (i.e., reduced mesh size) to exclude insects, if warranted.

In the study of microbes associated with carrion, insects that feed on carrion and soil under and surrounding the carrion may be of interest as these communities contribute to the decomposition of carcasses (e.g., Damann *et al.*, 2012; Metcalf *et al.*, 2013; Zheng *et al.*, 2013; Pechal *et al.*, 2014a, b; Carter *et al.*, 2015; Weiss *et al.*, 2015). The collection of microbial samples from different life stages of flies of forensic importance has been previously described in the work of Zheng *et al.* (2013). If collecting soil microbes associated with carrion, the evaluation of the placement of replicate carrion specimens is essential. Appropriate distances and considerations of inclines to avoid contamination resulting from leaching from another carcass must be assessed before placement. The carrion is generally a rich contributor of nutrients to soil inhabitants, and the soil beneath the carrion contains a mix of microbes from both the original soil community and from the carrion, which was placed on top (Parkinson *et al.*, 2009; Carter *et al.*, 2010). Because of this, the diversity and density of soil microbes varies greatly, particularly with soil depth and the quantity of soil sampled (Ranjard *et al.*, 2001; Kakirde *et al.*, 2010). The physical composition of soil, such as its chemical characteristics, content of organic matter, and pH, will influence the overall microbial structure and biomass (Hassink *et al.*, 1993). In general, moist soils around neutral pH with more labile carbon (available for microbial metabolism) and high nitrogen mineralization tend to have a higher microbial biomass; whereas dry arid, low nutrient value, and extreme pH soils have lower biomass (Hassink *et al.*, 1993; Hoyle *et al.*, 2006; Cookson *et al.*, 2008). Therefore to enhance reproducibility, it is important to consistently sample the appropriate quantity of soil (which will vary with the density of microbes present) and document from which soil layer it was collected.

Sampling from different soil depths will have a significant effect on analysis of archaeal and bacterial communities (Ranjard *et al.*, 2001; Nicol *et al.*, 2003). Bacterial density is generally higher near the surface of the soil rather than in the subsurface (Ranjard *et al.*, 2001; Parkinson *et al.*, 2009; Kakirde *et al.*, 2010). Archaeal communities can also shift as a result of sampling methodologies. In a study by Pesaro and Widmer (2002) along four soil depth profiles from the surface soil layer (0–9 cm) to a bottom soil layer (50–100 cm depth) in a terrestrial system of forest Haplumbrept soil, elemental decreases in organic carbon and total nitrogen paralleled the disappearance of Euryarchaeota and Group I.1b Crenarchaeota at increased soil depths. In a study by Kemnitz *et al.* (2007) of temperate acidic forest soil, the ratio of Archaea to bacteria increased with soil depth while the absolute abundance decreased. Therefore, the quantity of sample collected must be adjusted to accommodate the desired biomass within a soil type and depth. Soil collected within 5 cm from the surface usually requires 5–10 g of soil, whereas subsurface soil may require a larger volume in order to collect sufficient biomass for replicate DNA/RNA extraction. Such differences in sample volume need to be recorded for analyzing microbe density. Each sample should be collected using a separate sterile device, such as a tube/pipe (4–6 cm × 2–4 cm for surface volumes) or, if not possible, the device should be thoroughly cleaned between samples with a sterilizing agent, such as ethanol, bleach, or Lysol™ (Kakirde *et al.*, 2010).

There are constraints imposed on experimental design by natural ecosystem complexity, seasonal abiotic influences, and land usage (Lauber *et al.*, 2013). A researcher must decide how many samples and experimental replications are feasible, cost effective, and necessary to adequately test his/her scientific hypothesis. To add to the complexity, microbes are not static, but are rapidly reproducing, evolving, and often mobile organisms making accurate characterization of the community structure challenging. Additionally, locally microenvironments can exist, such as pockets of anaerobiosis in seemingly aerobic environments (Swift *et al.*, 1979). Due to the natural biological, temporal, and spatial heterogeneity in samples, larger sample sizes ($n>3$) provide more powerful statistical validation of a study. However, the cost of sequencing so many individual samples may preclude a truly comprehensive experimental design; therefore, composite samples (pooling of multiple individual samples) can be used to reduce the costs, although a minimal triplicate replication should be conducted even for composites. Composite samples can sometimes be preferred over individual samples in many experimental designs to obtain a better mean microbial diversity from the site; however, composite analysis will not garner information about microbial microenvironments (Wallenius, 2011). Therefore, the appropriate experimental design to answer the hypothesis needs to be well thought out prior to sampling.

Collection of soil from underneath the cadaver offers its own challenges. The cadaver must be rolled on its side or completely lifted from the ground using sterile techniques to avoid contamination of the soil or the personnel doing the collecting under the remains. This may take a team effort with large cadavers, and this becomes even more challenging in longitudinal studies (collecting the same type of data on the same subject at multiple points) as the decomposition process proceeds and the cadaver no longer rolls in a single piece due to disarticulation. Again all personnel should be in proper PPE with sampling implements (e.g., sterile tube, shovel, spatula, and soil corer) ready to recover the desired quantity of soil once the cadaver has been moved. One should randomize the collection locations under or adjacent to the body, in such a way as to avoid resampling from previously sampled sites. Sampling from the same location can disturb the natural microbial community structure for successive sampling or lead to the collection of microbes from different soil layers in different samples. Additionally, one must be careful to place the remains back into the exact same spot covering all the soil beneath; otherwise, exposure to the sun, ultraviolet light, or other environmental elements will affect the composition of soil microbes.

In aquatic environments, the aforementioned considerations for aseptic techniques and personnel protection need to be considered but altered due to the aqueous nature of the habitat. The methods to remove microbial communities from an aquatic habitat will depend on the location, size, and substrate. Epilithic biofilms can be directly collected from hard substrates (rock surfaces) using sterile tubular samplers, which would provide a standardized sampling area, but this method can be logistically challenging since it would be difficult to sterilize the brush component of the apparatus in the field between sampling locations or would require replicate, sterile brushes be brought into the field. As an alternative, researchers have collected the epilithic biofilms from a standardized area (10 cm × 10 cm) using sterile, dehydrated sponges or sterile razor blades. These methods will allow for characterization of the naturally occurring microbial communities, but there are other methods available for assessing biofilm succession and community dynamics in aquatic habitats (Hauer and Lamberti, 2007). Specifically, using artificial substrates, such as unglazed clay tiles, for biofilm development and accumulation in this ecosystem also provides a standardized sampling area for microbe collections. Finally, invertebrate communities, macrophytes, and water column samples can also be collected using various standardized methods. Hess samplers and drop kick nets provide a set perimeter for quantitative sampling of benthic invertebrates, while macrophytes can be collected using a predetermined area or sample size. Water column samples can be standardized based on filtrate volume, filter diameter size, material, and porosity (no greater than 20 μm for the final filter pore size); all of these sampling factors will be in response to the classification of the water body (e.g., oligotrophic, mesotrophic, or eutrophic). Oligotrophic water bodies are considered pristine, low production systems with low nutrient content; thus, an increase in filtrate volume may be required to adequately collect all representatives of the microbial communities. On the other hand, eutrophic water bodies are highly productive systems with a high nutrient and biomass; thus, an initial filtering of the sample with a larger porosity (45 μm) may be necessary to prevent saturation of the filter with debris and not an adequate representation of the microbial communities. Individuals should not conduct fieldwork in aquatic environments alone; there are potential dangers that can unexpectedly occur in a flowing habitat, such as flooding events, rapid currents, or submersion. Additionally, depending on location, a water source is likely to have potentially dangerous wildlife (e.g., bears, wolves, crocodiles, hippopotamuses, snakes) within a close proximity (<1 km).

The analysis of the microbiota for experimental investigation requires that the sampling procedure does not alter the microbial composition in the sample, but this may be easier said than done. First, there is a deficiency of studies that answer the many questions on exactly how different collection, storage, and processing procedures might affect the actual abundance and composition of microbes within a sample or the assessment of that sample. For example, the use of different primer pairs to analyze the 16S rRNA in different studies introduces biases into the interpretation of the results due to specificity of primer–template hybridizations for different species of bacteria. This makes it more difficult to compare between studies, although large-scale efforts such as the Earth Microbiome Project (http://www.earthmicrobiome.org/) aim to standardize methods across research labs to allow for large meta-analyses.

As a result of this lack of specific knowledge, the current state of decomposition microbial exploration is that no universally accepted or validated method to process or store samples exists. Even more vexing to the applied side of decomposition microbiology has been the conflicting results reported on the efficacy of various storage methods. For instance, Roesch *et al.* (2009) tested the bacterial community structure in replicate human fecal samples that were frozen immediately at −80°C or frozen after 12, 24, 48, or 72 hours at room temperature (about 25–27°C). They determined that some taxa changed across time with the most change occurring between 12 and 24 hours. The average bacterial community abundance across all subjects changed from about 3 to 10%, generally increasing with increased time at room temperature prior to freezing. However, the variability between subjects was high, making analysis of the specific genera change caused by storage difficult; but within the same subject, some of the most abundant genera found, *Bacteroides* and *Clostridium*, decreased over time, whereas other bacteria from the family Enterobacteriaceae increased. The variability in the community structure from person to person was hypothesized by the authors to be due to the nutrient value of the faeces from different people differentially affecting bacterial growth rates when left at room temperature. Cardona *et al.* (2012) found that DNA and RNA started to fragment after 24 hours at room temperature, but degradation also occurred in frozen samples within 1 hour after defrosting. This degradation altered the relative abundance of bacterial taxa in the final community analysis. However, Lauber *et al.* (2010) surveyed bacterial communities harvested from soil, human faeces, and human skin stored at 20, 4, −20, and −80°C for 3 or 14 days. They determined that the phylogenetic structure, community diversity, and relative abundances of most taxa were maintained over 14 days regardless of storage temperature or time. Consideration of how difficult microbial samples can be to obtain depend on the field conditions from which they are taken, which is not always compatible with immediate freezing or processing of the sample. Such constraints add to the complexity of developing a universal methodology across all forensic microbiology applications (see Song *et al.*, 2016).

In general, it is preferable to process samples immediately (within 2 hours) without any freezing or changes in oxygenation, but that is rarely logistically possible when samples are collected in the field (Rochelle *et al.*, 1994). So field samples should be protected from extreme temperature changes, changes in pH, and changes in ionic strength of a solution, if applicable (Wilfinger, 1997). One should also consider if samples contain aerobes where transport or storage in limited volume, airtight containers could affect community composition by restricting air exchange, as would storage of anaerobic samples in containers with excessive oxygen. Rochelle *et al.* (1994) demonstrated the sensitivity of anaerobic and aerobic bacteria from marine sediments to sample handling which affected oxygenation in the sample storage container. Similar considerations would apply to samples taken from top layers of soil with mixed anaerobia and aerobic bacterial communities. When immediate processing is impractical, the most common storage method includes some level of freezing. Storage at −20°C is recommended. Samples destined for RNA or DNA extractions can be stored in RNAlater® solution for approximately 1 week at room temperature, or at −80°C for long term per the manufacturer's instructions. The majority of published work has used freezing during their sample collection and processing, despite some studies that suggest freezing changes in the microbial community structure. Samples destined for DNA/RNA extractions have been stored at −20°C for up to 2 weeks, or at −80°C for up to 6 months reportedly without significant effect on bacterial community structure (Kakirde *et al.*, 2010; Lauber *et al.*, 2010; Carroll *et al.*, 2012). Once a sample is frozen, it is important to keep it frozen until DNA/RNA extraction can be performed and do not thaw it until immediately before the extraction. Repeated freeze–thaw cycles should be avoided as this can cause fragmentation, which can significantly affect characterization of microbial diversity (Männistö *et al.*, 2009; Cardona *et al.*, 2012).

Finally, during sample collection one must be adaptable and more prepared than when sampling in a laboratory due to unforeseen circumstances that become apparent during sampling events. It is a good practice to have all sampling materials in a unified location that can be easily transported and have maneuverability in the field (e.g., plastic boxes or polyethylene bags containing all of the supplies that are placed in a backpack or a tool box). Furthermore, a checklist of all materials and supplies is highly recommended so nothing is forgotten in the laboratory prior to sampling in the field.

4.5 Data considerations

Collection of metadata associated with each sample is a highly individualized practice with unique preferences for data organization and collection of information. For example, the following is a protocol used for collecting microbial communities from an arctic lake using filters (courtesy of Pechal and Benbow):

Sampling Equipment
Each sampling kit (1 per sampling location, 2 per lake) will consist of sterile

1. 60 mL Luer lock syringe (1)
2. Swin-Lok plastic filter holder with preloaded 25 mm MCE filter (1)
3. prelabeled polyethylene bag (1)

Sampling Protocol
At each sample location,

1. Record the date, time, name of the lake, the lake location (latitude/longitude), general geographic characteristic of the sampling location, and water chemistry.
2. Attach the plastic filter holder (cartridge) to the syringe wearing gloves; if the cartridge is loose, gently turn clockwise to tighten.
3. Fill the syringe with water, making note of the volume (50–60 mL). Push the collected water through the filter (filtrate).
4. When it becomes difficult to push the sample through the filter, the filter is close to being saturated and you want to stop filtration, so as to not tear or rupture the filter.
5. Repeat four to five times until the filter is saturated and record the final filtrate volume (mL).
6. Remove cartridge from the syringe and place in the sterile, prelabeled polyethylene bag.
7. Repeat points 2–6 for the remaining locations.
8. Keep all samples as cold as possible during transport from field to laboratory, with final storage at −20°C.

The following example data collection sheet is for terrestrial decomposition research using swine carrion as a model for human decomposition (courtesy of Pechal and Benbow):

Decomposition Research
Report Sheet

Date: ____________________ **Researchers Present:** ______________
Time In: ____________________ ______________
Time Out: ____________________ **Geographic Location:** ______________

Remains ID: ________

Decomposition Progression: Fresh ___ Bloat ___ Early ___ Advanced ___ Dry ___

Details: ____
____

Temperature Measurements: Ambient (1 m): _______ Ambient (5 cm): _______

Ground (10 cm): _______ Body Surface: _______ Under body: _______

Larval Mass 1: _______ Location: ______________________________
Larval Mass 2: _______ Location: ______________________________
Larval Mass 3: _______ Location: ______________________________
Notes: ____
____

"Top" of Carcass (Air side) "Bottom" of Carcass (Soil side)

However, researchers should follow the minimum information about any (x) sequence (MIxS) for submission of metagenomic/metatranscriptomic sequences for proper collection of metadata. These guidelines established by the Genomic Standards Consortium (GSC) consist of Minimum Information about a Genome Sequence (MIGS), Minimum Information about a Metagenome Sequence (MIMS), and Minimum Information about a MARKer gene Sequence (MIMARKS) standards and describe 14 environments. There are mandatory attributes and descriptors for each sample that is submitted to a data archiving center (e.g., NCBI or EBI); for an example of the list of the requirements, please see Tables 4.1 and 4.2, and visit the GSC (http://gensc.org/) for an extended list of available detailed descriptors.

Table 4.1 Metadata information for submission of resulting sequencing data for each sample as determined by the MIMARKS-compliant metadata template

Structured comment name	Item name	Definition
investigation_type	Investigation type	This field is either eukaryote, bacteria, virus, plasmid, organelle, metagenome, miens-survey or miens-culture.
project_name	Project name	Name of the project within which the sequencing was organized.
lat_lon	Geographic location (latitude and longitude)	The geographical origin of the sample as defined by latitude and longitude. The values should be reported in decimal degrees and in WGS84 system
geo_loc_name	Geographic location (country and/or sea, region)	The geographical origin of the sample as defined by the country or sea name followed by specific region name. Country or sea names should be chosen from the INSDC country list (http://insdc.org/country.html), or the GAZ ontology (v1.446) (http://bioportal.bioontology.org/visualize/40651)
collection_date	Collection date	The time of sampling, either as an instance (single point in time) or interval. In case no exact time is available, the date/time can be right truncated, that is, all of these are valid times: 2008-01-23T19:23:10+00:00; 2008-01-23T19:23:10; 2008-01-23; 2008-01; 2008; Except: 2008-01; 2008 all are ISO8601 compliant.
biome	Environment (biome)	In environmental biome level are the major classes of ecologically similar communities of plants, animals, and other organisms. Biomes are defined based on factors such as plant structures, leaf types, plant spacing, and other factors like climate. Examples include: desert, taiga, deciduous woodland, or coral reef.
feature	Environment (feature)	Environmental feature level includes geographic environmental features. Examples include: harbor, cliff, or lake.
material	Environment (material)	The environmental material level refers to the matter that was displaced by the sample, prior to the sampling event. Environmental matter terms are generally mass nouns. Examples include: air, soil, or water.
env_package	Environmental package	MIGS/MIMS/MIMARK extension for reporting of measurements and observations obtained from one or more of the environments where the sample was obtained.
seq_meth	Sequencing method	Sequencing method used; For example, Sanger, Illumina, ABI-solid.

For a list of full descriptors, please see The Genomic Standards Consortium (GSC) website at http://gensc.org.

Table 4.2 Metadata information for submission of resulting sequencing data for each sample as determined by the MIMARKS-compliant metadata template

Environmental package	Structured comment name	Item name	Definition
Air	Alt	Altitude	The altitude of the sample is the vertical distance between Earth's surface above sea level and the sampled position in the air
Host-associated	None	—	—
Human-associated	None	—	—
Human-oral	None	—	—
Human-gut	None	—	—
Human-skin	None	—	—
Human-vaginal	None	—	—
Microbial mat/ biofilm	Depth	Depth	Depth is defined as the vertical distance below surface, e.g. for microbial mat samples depth is measured from mat surface. Depth can be reported as an interval for subsurface samples.
Microbial mat/ biofilm	Elev	Elevation	The elevation of the sampling site as measured by the vertical distance from mean sea level.
Miscellaneous	None	—	—
Plant-associated	None	—	—
Sediment	Depth	Depth	Depth is defined as the vertical distance below surface, e.g. for sediment samples depth is measured from sediment surface. Depth can be reported as an interval for subsurface samples.
Sediment	Elevation	Elevation	The elevation of the sampling site as measured by the vertical distance from mean sea level.
Soil	Depth	Depth	Depth is defined as the vertical distance below surface, e.g. for soil samples depth is measured from soil surface. Depth can be reported as an interval for subsurface samples.
Soil	Elevation	Elevation	The elevation of the sampling site as measured by the vertical distance from mean sea level.
Wastewater sludge	None	—	—
Water	Depth	Depth	Depth is defined as the vertical distance below surface, e.g. for water samples depth is measured from water surface. Depth can be reported as an interval for subsurface samples.

For a list of full descriptors, please see The Genomic Standards Consortium (GSC) website at http://gensc.org.

4.6 Conclusions

Learning how to properly and efficiently sample microbial communities in various biomes/habitats, from numerous substrates, and under suboptimal conditions is vital for forensic purposes. Currently, there is a lack of standardization and established protocols for sampling naturally occurring microbiological communities of forensic importance, but not directly related to bioterrorism/security, food safety, or microbiological protocols commonly used by forensic practitioners during autopsies. Decomposition of remains can occur in a number of environments and under a number of circumstances; thus, protocols need to be developed that are consistent, yet flexible to adapt to the sampling across all possible death scenarios. The use of microorganisms for forensic applications is a rapidly advancing field, with most articles documenting microbial successional changes over time on decomposing remains (Metcalf *et al.*, 2013; Pechal *et al.*, 2014b; Hyde *et al.*, 2015). However, there are also studies documenting microbial community structure in insects associated with decomposing remains (e.g., blow flies (Singh *et al.*, 2015) and black soldier flies (Diptera: Stratiomyidae) (Zheng *et al.*, 2013)), soil communities (Lauber *et al.*, 2014; Cobaugh *et al.*, 2015), and aquatic habitats (Pechal and Benbow, 2015). Basic surveys of microbial communities from the aforementioned sample "types" (e.g., soil vs. insects) are imperative for informing future hypothesis-driven research with strong experimental design (see Chapter 3 for further details). From this chapter, the intention was to demonstrate the importance for researches and practitioners alike to have continued vigilance in their aseptic sampling techniques, sample storage methods, proper metadata collection, and standardization of samples for all microbial analyses.

Acknowledgments

The authors would like to acknowledge the generous donations of individuals who support research occurring at anthropological decomposition research facilities; without these generous gifts, this type of research would not have been possible. The authors would also like to recognize the National Institute of Justice (2010-DN-BX-K243 and 2014-DN-BX-K008) for providing financial support leading toward the preparation of this chapter. Points of view in this document are those of the authors and do not necessarily represent the official position or policies of the U.S. Department of Justice. Mention of trade names, companies, or commercial products in this publication is solely for the purpose of providing specific scientific information and does not imply recommendation or endorsement of the products by the U.S. Department of Agriculture.

References

Battin, T.J. and Sengschmitt, D. (1999) Linking sediment biofilms, hydrodynamics, and river bed clogging: evidence from a large river. *Microbial Ecology*, **37** (3), 185–196.

Battin, T.J., Kaplan, L.A., Newbold, J.D. *et al.* (2003) Effects of current velocity on the nascent architecture of stream microbial biofilms. *Applied and Environmental Microbiology*, **69** (9), 5443–5452.

Benbow, M., Tomberlin, J. and Tarone, A. (eds) (2015a) *Carrion Ecology, Evolution and Their Applications*, CRC Press, Boca Raton, FL.

Benbow, M.E., Pechal, J.L., Lang, J.M. *et al.* (2015b) The potential of high-throughput metagenomic sequencing of aquatic bacterial communities to estimate the postmortem submersion interval. *Journal of Forensic Sciences*, **60** (6), 1500–1510.

Byrd, J.H. and Castner, J.L. (2001) *Forensic Entomology: The Utility of Arthropods in Legal Investigations*, CRC Press, Boca Raton, FL.

Cardona, S., Eck, A., Cassellas, M. *et al.* (2012) Storage conditions of intestinal microbiota matter in metagenomic analysis. *BMC Microbiology*, **12** (1), 158.

Carroll, I.M., Ringel-Kulka, T., Siddle, J.P. *et al.* (2012) Characterization of the fecal microbiota using high-throughput sequencing reveals a stable microbial community during storage. *PLoS One*, **7** (10), e46953.

Carter, D.O., Yellowlees, D. and Tibbett, M. (2010) Moisture can be the dominant environmental parameter governing cadaver decomposition in soil. *Forensic Science International*, **200** (1–3), 60–66.

Carter, D.O., Metcalf, J.L., Bibat, A. and Knight, R. (2015) Seasonal variation of postmortem microbial communities. *Forensic Science and Medical Pathology*, **11** (2), 202–207.

Čeřovský, V., Ždárek, J., Fučík, V. *et al.* (2010) Lucifensin, the long-sought antimicrobial factor of medicinal maggots of the blowfly *Lucilia sericata*. *Cellular and Molecular Life Sciences*, **67** (3), 455–466.

Cobaugh, K.L., Schaeffer, S.M. and DeBruyn, J.M. (2015) Functional and structural succession of soil microbial communities below decomposing human cadavers. *PLoS One*, **10** (6), e0130201.

Commission for Environmental Cooperation Working Group. (1997) Ecological Regions of North America—Toward a Common Perspective, p. 71. In Montreal CfEC (ed.), Commission for Environmental Cooperation: Montréal, Canada.

Cookson, W., Murphy, D. and Roper, M. (2008) Characterising the relationships between soil organic matter components and microbial function and composition along a tillage disturbance gradient. *Soil Biology and Biochemistry*, **40** (3), 763–777.

Damann, F.E., Tanittaisong, A. and Carter, D.O. (2012) Potential carcass enrichment of the University of Tennessee Anthropology Research Facility: a baseline survey of edaphic features. *Forensic Science International*, **222** (1), 4–10.

Dickson, G.C., Poulter, R.T.M., Maas, E.W. *et al.* (2010) Marine bacterial succession as a potential indicator of postmortem submersion interval. *Forensic Science International*, **209** (1), 1–10.

Grice, E.A., Kong, H.H., Renaud, G. *et al.* (2008) A diversity profile of the human skin microbiota. *Genome Research*, **18** (7), 1043–1050.

Hassink, J., Bouman, L., Zwart, K. *et al.* (1993) Relationships between soil texture, physical protection of organic matter, soil biota, and C and N mineralization in grassland soils. *Geoderma*, **57** (1), 105–128.

Hauer, F.R. and Lamberti, G.A. (eds) (2007) *Methods in Stream Ecology*, Academic Press, New York.
Hoyle, F., Murphy, D. and Fillery, I. (2006) Temperature and stubble management influence microbial CO_2-C evolution and gross transformation rates. *Soil Biology and Biochemistry*, **38** (1), 71–80.
Hyde, E.R., Haarmann, D.P., Petrosino, J.F. *et al.* (2015) Initial insights into bacterial succession during human decomposition. *International Journal of Legal Medicine*, **129** (3), 661–671.
Kakirde, K., Parsley, L. and Liles, M. (2010) Size does matter: application-driven approaches for soil metagenomics. *Soil Biology and Biochemistry*, **42** (11), 1911–1923.
Kemnitz, D., Kolb, S. and Conrad, R. (2007) High abundance of Crenarchaeota in a temperate acidic forest soil. *FEMS Microbiology Ecology*, **60** (3), 442–448.
Lang, J.M., McEwan, R.W. and Benbow, M.E. (2015) Abiotic autumnal organic matter deposition and grazing disturbance effects on epilithic biofilm succession. *FEMS Microbiology Ecology*, **91** (6), fiv060.
Lauber, C.L., Zhou, N., Gordon, J.I. *et al.* (2010) Effect of storage conditions on the assessment of bacterial community structure in soil and human-associated samples. *FEMS Microbiology Letters*, **307** (1), 80–86.
Lauber, C., Ramirez, K., Aanderud, Z. *et al.* (2013) Temporal variability in soil microbial communities across land-use types. *The ISME Journal*, **7** (8), 1641–1650.
Lauber, C.L., Metcalf, J.L., Keepers, K. *et al.* (2014) Vertebrate decomposition is accelerated by soil microbes. *Applied and Environmental Microbiology*, **80** (16), 4920–4929.
Lillesand, T., Kiefer, R. and Chipman, J. (eds) (2014) *Remote Sensing and Image Interpretation*, John Wiley & Sons, New York.
Männistö, M.K., Tiirola, M. and Haggblom, M.M. (2009) Effect of freeze-thaw cycles on bacterial communities of arctic tundra soil. *Microbial Ecology*, **58** (3), 621–631.
Metcalf, J.L., Wegener Parfrey, L., Gonzalez, A. *et al.* (2013) A microbial clock provides an accurate estimate of the postmortem interval in a mouse model system. *eLife*, **2**, e01104.
Nicol, G.W., Glover, L.A. and Prosser, J.I. (2003) Spatial analysis of archaeal community structure in grassland soil. *Applied and Environmental Microbiology*, **69** (12), 7420–7429.
Parkinson, R.A., Dias, K.-R., Horswell, J. *et al.* (2009) Microbial community analysis of human decomposition on soil, in *Criminal and Environmental Soil Forensics* (eds K. Ritz, L. Dawson and D. Miller), Springer, Dordrecht, pp. 379–394.
Pechal, J.L. and Benbow, M.E. (2015) Microbial ecology of the salmon necrobiome: evidence salmon carrion decomposition influences aquatic and terrestrial insect microbiomes. *Environmental Microbiology*, **18**, 1511–1522.
Pechal, J., Benbow, M., Crippen, T. *et al.* (2014a) Delayed insect access alters carrion decomposition and necrophagous insect community assembly. *Ecosphere*, **5** (4), art45.
Pechal, J.L., Crippen, T.L., Benbow, M.E. *et al.* (2014b) The potential use of bacterial community succession in forensics as described by high throughput metagenomic sequencing. *International Journal of Legal Medicine*, **128** (1), 193–205.
Pesaro, M. and Widmer, F. (2002) Identification of novel Crenarchaeota and Euryarchaeota clusters associated with different depth layers of a forest soil. *FEMS Microbiology Ecology*, **42** (1), 88–98.
Ranjard, L., Poly, F., Lata, J.C. *et al.* (2001) Characterization of bacterial and fungal soil communities by automated ribosomal intergenic spacer analysis fingerprints: biological and methodological variability. *Applied and Environmental Microbiology*, **67** (10), 4479–4487.
Rochelle, P., Cragg, B., Fry, J. *et al.* (1994) Effect of sample handling on estimation of bacterial diversity in marine-sediments by 16s ribosomal-RNA gene sequence-analysis. *FEMS Microbiology Ecology*, **15** (1–2), 215–225.

Roesch, L., Casella, G., Simell, O. *et al.* (2009) Influence of fecal sample storage on bacterial community diversity. *The Open Microbiology Journal*, **3** (1), 40–46.

Rozen, D.E., Engelmoer, D.J.P. and Smiseth, P.T. (2008) Antimicrobial strategies in burying beetles breeding on carrion. *Proceedings of the National Academy of Sciences of the United States of America*, **105** (46), 17890–17895.

Schoenly, K.G., Haskell, N.H., Hall, R.D. and Gbur, J.R. (2007) Comparative performance and complementarity of four sampling methods and arthropod preference tests from human and porcine remains at the forensic anthropology center in Knoxville, Tennessee. *Journal of Medical Entomology*, **44** (5), 881–894.

Singh, B., Crippen, T.L., Zheng, L. *et al.* (2015) A metagenomic assessment of the bacteria associated with *Lucilia sericata* and *Lucilia cuprina* (Diptera: Calliphoridae). *Applied Microbiology and Biotechnology*, **99** (2), 869–883.

Song, S.J., Amir, A., Metcalf, J.L. *et al.* (2016) Preservation methods differ in fecal microbiome stability, affecting suitability for field studies. *mSystems*, **1** (3), e00021-16.

Swift, M.J., Heal, O.W. and Anderson, J.M. (1979) The influence of the physio-chemical environment on decomposition processes, in *Decomposition in Terrestrial Ecosystems*, vol. **5** (eds D.J. Anderson, P. Greig-Smith and F.A. Pitelka), University of California Press, Berkeley, CA, pp. 220–225.

Tomberlin, J.K. and Benbow, M.E. (eds) (2015) *Forensic Entomology: International Dimensions and Frontiers*, CRC Press, Boca Raton, FL.

Wallace, J.R. (2015) Aquatic vertebrate carrion decomposition, in *Carrion Ecology, Evolution and Their Applications* (eds M.E. Benbow, J.K. Tomberlin and A.M. Tarone), CRC Press, Boca Raton, FL, pp. 247–271.

Wallenius K. (2011). Microbiological characterisation of soils: evaluation of some critical steps in data collection and experimental design. Ph.D., University of Helsinki: Helsinki.

Weiss, S., Carter, D., Metcalf, J. and Knight, R. (2015) Carcass mass has little influence on the structure of gravesoil microbial communities. *International Journal of Legal Medicine*, **130** (1), 253–263.

Wilfinger, W. (1997) Effect of pH and ionic stregth on the spectrophotometric assessment of nucleic acid purity. *BioTechniques*, **22** (3), 474–481.

Zheng, L., Crippen, T.L., Singh, B. *et al.* (2013) A survey of bacterial diversity from successive life stages of black soldier fly (Diptera: Stratiomyidae) by using 16S rDNA pyrosequencing. *Journal of Medical Entomology*, **50** (3), 647–658.

CHAPTER 5

An introduction to metagenomic data generation, analysis, visualization, and interpretation

Baneshwar Singh[1], Tawni L. Crippen[2] and Jeffery K. Tomberlin[3]
[1] *Department of Forensic Science, Virginia Commonwealth University, Richmond, VA, USA*
[2] *Southern Plains Agricultural Research Center, Agricultural Research Service, United States Department of Agriculture, College Station, TX, USA*
[3] *Department of Entomology, Texas A&M University, College Station, TX, USA*

5.1 Introduction

As discussed in virtually every chapter in this book, the emergence of next-generation sequencing technology in the last decade has revolutionized the field of forensic microbiology and decomposition ecology. Because of such technological developments, we now recognize that on average approximately 3.9×10^{13} bacteria live on and within us, which is almost equal to the total number of human cells in our body (Sender *et al.*, 2016). The taxonomic diversity of microbes (especially those that are rare in abundance) in skin, gut, and saliva vary significantly between individuals and thus can serve as a marker to establish individualization (Franzosa *et al.*, 2015; Oh *et al.*, 2014; Schloss, 2014).

Sequencing the enormity of individual microbes that exist in a given environmental community is no longer a limiting factor. In reality, managing the sheer volume of sequences generated through such analyses turns out to be the greater challenge. In concert with the utilization of these new technologies in scientific investigations, the number of microbial sequences entered into public databases has increased exponentially, and with the current rate of sequencing, it is predicted that most microbial taxa will have been described by the end of this decade (Yarza *et al.*, 2014). Thus the magnitude of sequence data that will be generated by these platforms creates huge challenges for scientists. This chapter will focus on the current status of different next-generation sequencing methods, 16S ribosomal DNA (16S rDNA), and whole-community shotgun sequence data analysis, visualization, and interpretation of results (see Figure 5.1). Please note that these methods, while used at the time this chapter was developed,

---

Forensic Microbiology, First Edition. Edited by David O. Carter, Jeffery K. Tomberlin, M. Eric Benbow and Jessica L. Metcalf.

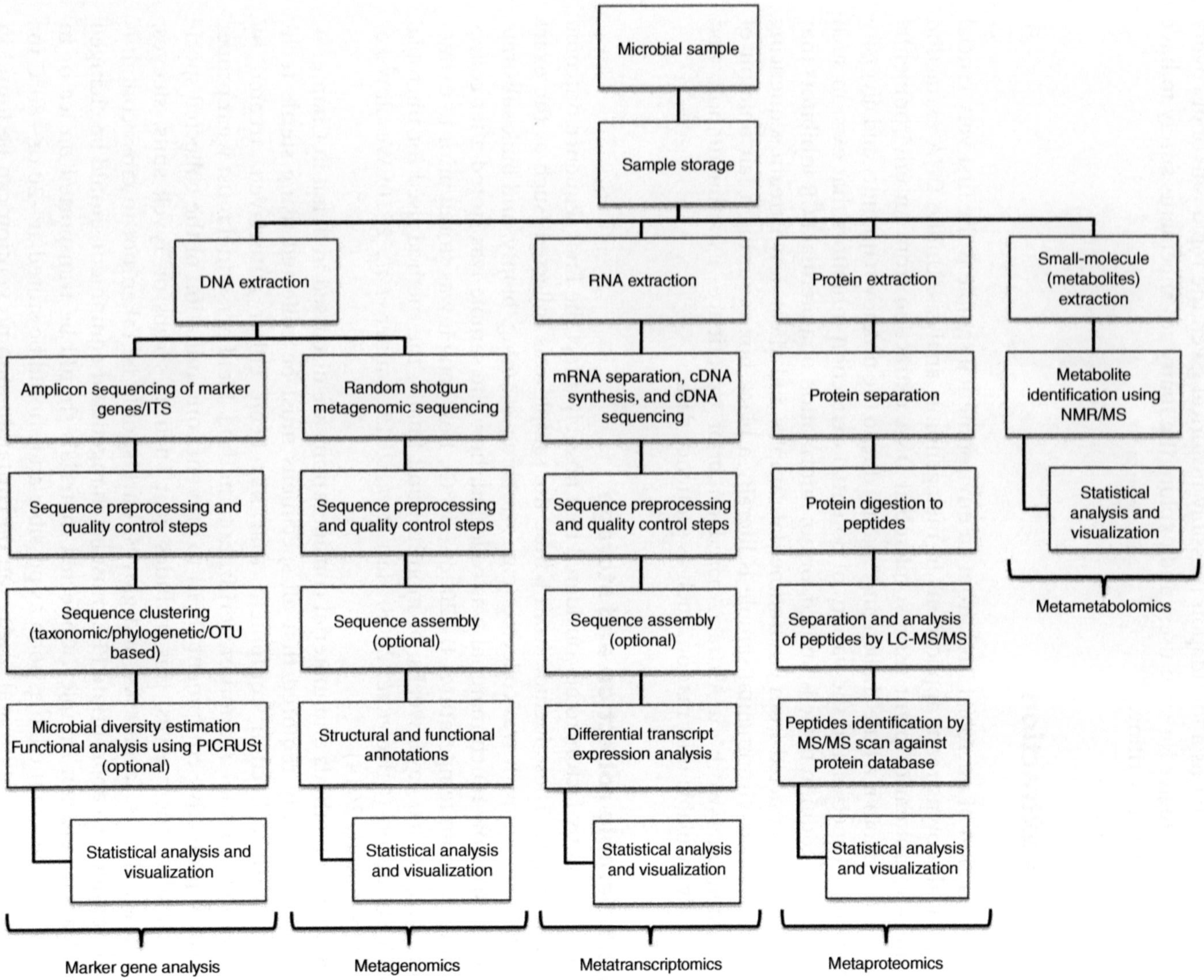

Figure 5.1 Flow diagram of a typical multi-omics workflow

could have evolved and diversified or possibly be no longer in use. The reader is encouraged to use this chapter as a foundation (possibly historical given the rapid rate of discovery of new technologies for such uses) and explore the many advancements surely to have occurred since its writing.

5.2 DNA extraction

The extraction of microbial DNA from an environmental sample is the first very crucial step for all downstream applications in metagenomic analysis. Unlike DNA extraction from pure animal or plant tissues, microbial DNA extraction from an environmental sample (e.g., soil) is a little challenging, mainly due to the unknown quantity and diversity of microbial species and the variety of potential extraction inhibitors that exist in such samples. In addition there is limited, or no, information on potential PCR inhibitors that could be co-extracted from environmental samples and affect downstream sequencing analyses. An environmental sample is literally a black box requiring a fair amount of "educated guesswork" to extract, based on prior experience, experimentation, and publically available information, such as sequence data.

5.2.1 Sample collection and storage

There are many factors to be considered that possibly affect the final sequence outcome from a sample. These include from where the sample was collected (such as the exact depth in soil), how it was collected (e.g., swabbing, scraping, biopsy, and bulk substrate collection), how much material was collected, how the sample was stored after collection (e.g., room temperature, 4, −20, or −80°C), how long it was stored prior to extraction (e.g., hours, days, weeks, or months), and finally, the method used for microbial DNA extraction (Grice *et al.*, 2008; Hart *et al.*, 2015; Lauber *et al.*, 2010; Wesolowska-Andersen *et al.*, 2014).

Sampling methods and material considerations are discussed in detail in Chapter 4, but it is generally required that all specimens must be collected using sterile technique, with collection equipment that has been either autoclaved, treated by flame-95% ethanol or another sterilization method. Field personnel must wear proper personal protective equipment (PPE) to prevent contamination of the collected specimens and of themselves; this includes gear such as scrubs or Tyvek suits, sleeves, booties, hairnets, facemasks, and gloves. Care should be taken not to cross contaminate samples, so appropriate PPE must be changed and equipment should be changed or cleaned between samples. Collected materials should be transported on ice or in liquid nitrogen to the lab as soon as possible and should be stored at −20 or −80°C for later processing. Additional discussion on contamination prevention can be found in Chapter 4.

5.2.2 Extraction methods

The choice of a DNA extraction method can significantly impact the interpretation of generated microbial structural and functional results (Henderson *et al.*, 2013; Wagner Mackenzie *et al.*, 2015; Wesolowska-Andersen *et al.*, 2014). More than 100-fold differences in absolute microbial number and significant differences in microbial community composition were observed when different DNA extraction methods were applied to cow and sheep rumen samples (Henderson *et al.*, 2013). Thus, for an unbiased estimate of microbial composition in a genomic analysis, choosing a DNA extraction method that is capable of recovering both prokaryotic and eukaryotic DNA efficiently is vital, especially if a researcher plans to perform shotgun sequencing on the extracted DNA. It is also crucial to consider that DNA extraction from Gram-positive, unlike Gram-negative, bacteria requires a harsher lysis step as the cell wall must be disrupted in order to gain access to wanted DNA (Harrison *et al.*, 1991); however, this additional step complicates method development as it is very important to choose an approach that is unbiased toward recovery of either Gram-positive or Gram-negative bacterial DNA.

Microbial DNA can be extracted from an environmental sample either directly (entire sample lysed during cell-lysis step) or indirectly (microbial cells first get separated, and then separated cells get lysed during cell-lysis step) (Delmont *et al.*, 2011; Ogram *et al.*, 1987; Parachin *et al.*, 2010). Although both methods have some advantages and disadvantages (Singh and Crippen, 2015), they yield very similar microbial diversity (Delmont *et al.*, 2011); hence, the choice of a particular method depends on the experimental design and downstream applications. For example, if a researcher wants to construct large-insert clone libraries (insert size >12 kb), then an indirect method may be a better choice as direct methods can cause shearing of genomic DNA (DNA fragments are usually <12 kb in size) during intensive cell-lysis stages. But if the purpose is to use DNA for 16S rDNA, or even shotgun, sequencing where small sample sizes (<1 g) are adequate, then a direct method is generally preferred mainly because it is less laborious and time consuming and produces high DNA yield.

The main disadvantage of a direct method is its inability to efficiently remove PCR inhibitors from some environmental samples. Although several traditional methods (e.g., hydroxyapatite columns, cesium chloride density centrifugation, chromatographic separation, chemical flocculation, gel extraction) are available for the removal of inhibitors after DNA extraction, the developments of commercial DNA purification kits (e.g., MOBIO PowerClean® DNA Clean-Up kit) that are capable of eliminating common PCR inhibitors (e.g., humic substances, polyphenolics, polysaccharides, heparin) from difficult samples have helped in popularization of the direct DNA extraction method in recent years (see Table 5.1). The majority of direct DNA extraction methods use either silica spin-column-based approaches or organic phenol–chloroform-based approaches for the separation of contaminants from microbial DNA.

The silica spin-column-based approach usually first resuspends the sample in an enzyme-based lysis buffer in order to break open the cells and release the cellular

Table 5.1 Commonly used DNA extraction methods for microbiome studies

Sr. #	DNA extraction method	References
1	MoBio PowerSoil® DNA Isolation Kit	Fierer *et al.* (2010), Gilbert *et al.* (2014), Human Microbiome Project Consortium (2012), and Metcalf *et al.* (2015)
2	Zymo ZR Fecal DNA extraction kit	Gajer *et al.* (2012)
3	Qiagen QIAamp® DNA stool Mini Kit	Mirsepasi *et al.* (2014)
4	MP Biomedicals FastSpin Soil DNA kit	Eren *et al.* (2015)
5	Phenol: chloroform-based DNA isolation	Pechal *et al.* (2013), Singh *et al.* (2014a), and Zheng *et al.* (2013)

components from inside the cell (e.g., proteins, DNA, phospholipids). The buffer may also contain stabilizers to help preserve the nucleic acids once released from the protective environment inside the cell. Some types of cells, such as bacterial cells, are not easily lysed and require harsher treatment. Heating of the sample may facilitate rupture of particularly tough cell walls. Additionally, zirconium or glass beads can be added to the sample initially, followed by rapid homogenization with a bead beater or extended vortexing (Yuan *et al.*, 2012). A binding buffer is then used at a specific pH and salt concentration to bind the DNA to the silicon dioxide gel filter of the spin column while allowing other cellular components to be centrifuged through and separated from the sample. DNA absorption to the filter is most efficient in high ionic strength buffers with a pH at or below the pK_a of the filter silanol groups. The flow-through of the column is discarded, and the filter is washed and centrifuged to remove remaining impurities in an effort to leave only nucleic acid bound to the silica gel filter. Finally an elution buffer is added, which releases the nucleic acid from the filter, and through centrifugation it is harvested into a collection tube. While the purity of such preparations is generally very high, a significant amount of DNA from the sample is often not recovered during the procedure; regardless of whether it is unbound to the filter and is discarded in the flow-through or whether it remains attached to the filter during the elution step, it is lost to the researcher.

By contrast phenol–chloroform extraction is a liquid–liquid extraction technique for isolating DNA and RNA by hydrophobicity at neutral pH (Chomczynski and Sacchi, 1987). This method takes longer than a silica-based system and requires a fume hood as volatile, hazardous chemicals are used, which will also require proper disposal. The technique can be challenging, as it involves extraction in which phenol–chloroform denatures proteins within the sample and separates them into the organic phase. When mixed with an aqueous solution and the pH kept neutral (or slightly alkaline pH 7–8), nucleic acids (both DNA and RNA) will partition into the aqueous phase along with other contaminants such as salts; whereas if the pH is acidic, DNA is retained in the organic phase and RNA partitions into the aqueous phase. The aqueous phase is collected and mixed with pure chloroform, which functions to help remove phenol residues. These residues reduce the DNA purity and interfere with later PCR procedures. DNA is insoluble in alcohols such as ethanol and isopropanol; therefore mixing of collected aqueous phase with alcohol (often at ice cold temperatures) causes the DNA to aggregate together, which can then be collected by centrifugation. The precipitation can be improved by increasing the solutions' ionic strength, such as by adding sodium acetate. With practice this method will result in samples of high DNA purity. Its advantage generally lies in the more efficient recovery of nucleic acids, and so it is often used for difficult samples with low DNA content in order to improve yield.

For a shotgun metagenomic approach, obtaining very high quality and quantity of target DNA from an environmental sample during DNA extraction is extremely important. Some samples (e.g., water, skin) yield very low quantities of target DNA during extraction irrespective of the method used, which may not be sufficient for shotgun metagenomic sequencing. In this situation, DNA yield can be increased by using a non-PCR-based multiple displacement amplification (MDA) method (Peura *et al.*, 2015; Shoaib *et al.*, 2008). If a sample is collected from an environment that includes abundant nontarget host DNA, then fractionation or selective lysis steps help to minimize collection of nontarget DNA. In some instances, blocking primers can be used to reduce amplification of nontarget host DNA during PCR amplification of marker gene (Gilbert *et al.*, 2014).

5.3 DNA sequencing

5.3.1 Amplicon sequencing of marker (16S rDNA/18S rDNA/ITS) loci

Amplification and sequencing of targeted marker loci (e.g., 16S rDNA for bacteria/archaea, 18S rDNA for eukaryotes, and internal transcribed spacer (ITS) for fungi) is currently the most common culture-independent molecular method for the detection and characterization of the microbial community structure from a particular environment. With this method, a marker locus of a target community is amplified directly from the extracted DNA using specific universal PCR primers, and the amplified product is then sequenced in parallel (no traditional cloning step is needed before sequencing) on a next-generation sequencing platform of choice (see Section 5.3.3). The choice of target loci depends on the sampled community and the experimental hypothesis.

The 16S rRNA gene (1541 bp in *Escherichia coli*) encodes a ribosomal small subunit (SSU) and is present in all bacteria and archaea. This is the marker of choice for bacterial and archaeal marker gene studies because (i) it consists of both conserved (good for universal primer designing) and variable (good for taxonomic identification) regions, (ii) several reference databases (e.g., Ribosomal Database Project (RDP) (Cole *et al.*, 2009), Greengenes (DeSantis *et al.*, 2006), Silva (Quast *et al.*, 2013), GenBank (Benson *et al.*, 2005)) are available for alignment and identification of bacteria and archaea directly from DNA sequences, and (iii) it can be amplified from a degraded sample, as usually it is present in multiple copies (1–15 per cell). The 18S rRNA gene (1798 bp in *Saccharomyces cerevisiae*) is a homolog of 16S rRNA gene in eukaryotes, and like the 16S rRNA gene, the 18S rRNA gene is also present in multiple copies and includes both variable and conserved regions (Mankin *et al.*, 1986). However, variable regions of 18S rRNA gene are not as variable as those of the 16S rRNA gene and hence of less use in classification of sequences at lower taxonomic levels (e.g., genus) (Schoch *et al.*, 2012). Because the variable region of 18S rRNA gene is comparatively less complex, its alignment is easier, and hence 18S rRNA gene is preferred in phylogenetic studies of eukaryotes (Schoch *et al.*, 2012). The ITS region is a transcribed intergenic noncoding spacer in the nuclear rRNA cistron. The ITS locus is present in multiple copies, which makes it an easy target for amplification, even in the presence of small quantities of DNA (Schoch *et al.*, 2012). In bacteria and archaea genomes, the ITS is located between the 16S and 23S rRNA genes. In the eukaryotic genome, the ITS is split into two parts: ITS1 and ITS2; ITS1 is located between the 18S rRNA gene and the 5.8S rRNA gene, and ITS2 is located between the 5.8S rRNA gene and the 25S (in plants) or the 28S (in animals) rRNA gene. The ITS in bacteria and archaea is the analog of ITS1 in eukaryotes. Because hypervariable ITS1 and ITS2 are flanked by conserved regions, it is easy to design universal primers for amplification of these targets. The size of ITS region varies greatly in different eukaryotes (e.g., ITS1 length varies from 791 bp (*Coccidula rufa*) to 2572 bp (*Exochomus quadripustulatus*) in different species of ladybird beetles (Coleoptera: Coccinellidae)). In the widely recognized eukaryotic model yeast organism *S. cerevisiae*, the ITS1 and ITS2 lengths are 361 and 232 bp, respectively (Korabecna, 2007; von der Schulenburg *et al.*, 2001).

The majority of next-generation sequencing platforms generate short reads (usually <400 bp in length) that do not cover all variable and conserved regions of a target locus (16S rDNA/18S rDNA/ITS). As a result, researchers usually sequence a region of the target locus for estimation of microbial diversity. Unfortunately, no single region of a target locus is ideal for all samples, and hence the choice of a region of a target locus depends on sample type and experimental design (Chakravorty *et al.*, 2007; Kumar *et al.*, 2011). Klindworth *et al.* (2013) evaluated 512 primers based on sequence obtained from SILVA database and concluded that primer pair S-D-Bact-0341-b-S-17 (5′- CCT ACG GGN GGC WGC AG-3′) and S-D-Bact-0785-a-A-21 (5′-GAC TAC HVG GGT ATC TAA TCC-3′) was the best for overall coverage of bacteria and archaea. This primer pair covers variable regions 3 and 4 of the 16S rDNA and generates a 464 bp amplicon (appropriate for MiSeq® paired-end 300 bp reads using v3 reagents).

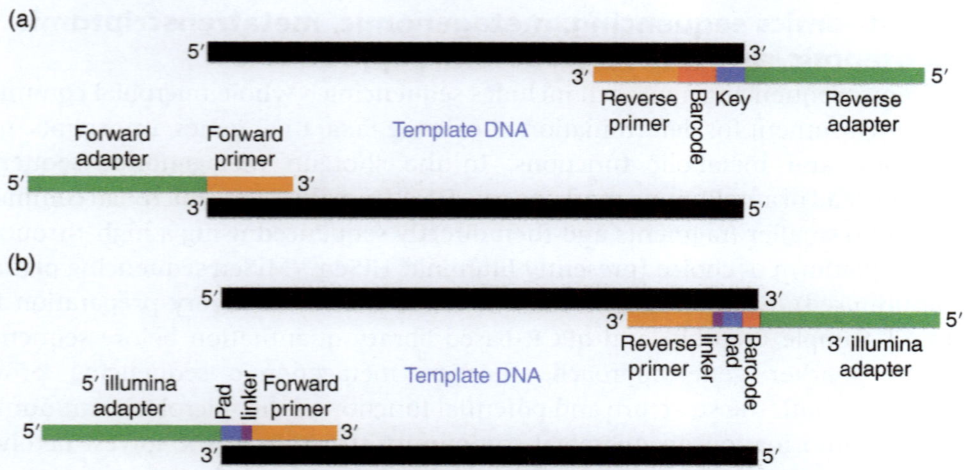

Figure 5.2 Schematic showing barcoded fusion primer design for (a) Ion Torrent PGM/454 and (b) MiSeq® sequencing platforms

Target region primers should be chosen based on what works best for a particular targeted community (host/environment) and to what database or other studies one plans to compare the results. For example, the Earth Microbiome Project (http://press.igsb.anl.gov/earthmicrobiome/emp-standard-protocols/) recommends primer pair 515fB (5′-GTG YCA GCM GCC GCG GTA A-3′) and 806rB (5′-GGA CTA CNV GGG TWT CTA AT-3′) for sequencing 16S rDNA variable region 4 and primer pair Euk_1391F (5′-GTA CAC ACC GCC CGT C-3′) and EukBR (5′-TGA TCC TTC TGC AGG TTC ACC TAC-3′) for sequencing of 18S rDNA variable region 9 (Caporaso *et al.*, 2012). For more detailed information on marker gene primers and their utility, readers are advised to refer to Smith and Peay (2014), Singh and Crippen (2015), and Walters *et al.* (2016).

An amplicon library for the target loci can be constructed either by using a barcoded fusion primer approach (Dowd *et al.*, 2008), where the primer pair of choice are fused with adapter sequences and barcode sequences for amplification of a target region of interest (see Figure 5.2), or by using commercial amplicon library preparation kits, where barcode or index sequence is added either by an enzymatic ligation (Ion Torrent PGM sequencing) or by a second sequential PCR (MiSeq sequencing: https://www.Illumina®.com/content/dam/Illumina®-support/documents/documentation/chemistry_documentation/16s/16s-metagenomic-library-prep-guide-15044223-b.pdf) of the first PCR-amplified product. Detail protocol for amplicon library preparation using fusion primer approach is available at Earth Microbiome Project website (http://press.igsb.anl.gov/earthmicrobiome/emp-standard-protocols/) and at https://github.com/SchlossLab/MiSeq®_WetLab_SOP/blob/master/MiSeq®_WetLab_SOP_v4.md, and readers are advised to consult these links for more information.

5.3.2 Multi-omics sequencing: metagenomic, metatranscriptomic, metaproteomic, and metametabolomic approaches

A multi-omics sequencing approach includes sequencing a whole microbial community from the environment for determination of existing taxa, their genes, transcripts, metabolic profiles, and metabolic functions. In the shotgun metagenomic sequencing approach, instead of amplifying a marker gene, DNA from the entire microbial community is sheared into smaller fragments and then directly sequenced using a high-throughput sequencing platform of choice (presently Illumina® HiSeq™/MiSeq sequencing platforms are typically used). Several commercial kits are available for library preparation (e.g., Nextera XT Sample Prep kit) and qPCR-based library quantitation before sequencing. Unlike the marker gene approach, shotgun metagenomic sequencing provides information on both the structure and potential function of the microbial community. It also provides information on microbial community structure at the lowest taxonomic levels (species or strain) because the microbial classification is based on hundreds of microbial loci rather than the one or two loci used in the marker gene-based approach (Segata *et al.*, 2012). The relative abundance of microbes obtained using a whole-genome shotgun sequencing approach is also more accurate because the relative abundance calculations in this approach are based on a single-copy gene rather than the multi-copy ribosomal RNA gene, as in the marker gene approach. Most importantly, whole-genome shotgun sequencing helps to avoid biases associated with degenerate PCR primers, PCR amplification, and databases used for the hierarchical classification of microbes (Lee *et al.*, 2012). This can aid the discovery of new organisms not detected with a marker gene approach.

Despite these advantages, the shotgun metagenomic approach possesses many challenges that must be considered before and after sequencing. Before sequencing, an understanding of microbial diversity and relative abundance of the community being sampled is important. For example, a microbial community in a soil sample is generally more complex than those found in human skin samples; hence for proper coverage and meaningful results, more sequence data will be needed from the complex communities (e.g., soil). Shotgun metagenomic approaches typically generate more sequences of dominant species than those species present at <1% relative abundance. In this scenario, if sequencing depth (i.e., number of reads achieved from a single sample during sequencing) is not sufficient, then there is a strong possibility that rare taxa will not be sequenced at all. The higher the sequencing depth is, the more reliable the sequencing outcome is. The high sequencing depth that is required to recover rare taxa substantially increases cost of sequencing and complicates sequence data analysis (Sharpton, 2014). Sometimes, even deep sequencing fails to yield meaningful results from complex community samples or samples where host DNA sequences outnumber actual microbial DNA sequences. In these situations alternative methods, such as cell sorting by flow cytometry to enrich the target species, may serve as a cost-efficient approach.

The metatranscriptomic approach also uses shotgun sequencing, but instead of genomic DNA, cDNA is used for direct shotgun sequencing. This approach helps with determining active spatial and temporal expression of RNA pools. Because RNA is easily degraded, immediate processing and storage of samples in reagents and vials that are RNase-free is critical. To maintain the integrity of RNA, the use of gloves and nuclease-free items such as plasticware and RNaseZap® (Thermo Fisher Scientific Inc., Waltham, MA, USA) for cleaning of work areas and pipettors is required. Generally, flash freezing of samples in liquid nitrogen or dry ice immediately after collection followed by storage at −80°C helps to protect RNA from degradation and results in high total RNA yield during RNA extractions (Simister *et al.*, 2011). Nevertheless, the use of liquid nitrogen or dry ice during field sampling is often difficult; hence in these situations, use of a stabilizer, such as RNA*later*® (Life Technologies, Carlsbad, CA, USA) is recommended for preservation of the sample prior to RNA extraction (Simister *et al.*, 2011). These stabilizers can protect samples from degradation for almost a week at room temperature and indefinitely at −20°C.

Whenever possible, small molecules (metabolites) and biomacromolecules (DNA/RNA/proteins) should be extracted simultaneously from the same microbial sample, because splitting the sample for individual extraction introduces artifacts (Roume *et al.*, 2013). If the plan is to perform only metagenomic and metatranscriptomic studies, then DNA should be extracted along with RNA for a better understanding of transcript abundance with respect to gene or taxon abundance. There are some commercial kits that can extract both DNA and RNA together (e.g., AllPrep DNA/RNA Mini Kit™ (Qiagen Inc., Valencia, CA, USA), PowerMag® Microbiome RNA/DNA Isolation Kit (MO BIO Laboratories, Inc., Carlsbad, CA, USA)), but these kits have not been widely used in multi-omics studies, and their usefulness in these applications is still in its infancy. Recently, Roume *et al.* (2013) developed a method for co-isolation of small molecules (metabolites) and biomacromolecules (RNA/DNA/proteins) from a single microbial sample. Although this method looks promising for samples with high microbial load (e.g., fecal samples), its efficacy for samples with low microbial load (e.g., skin samples) is still not determined. For metatranscriptomic studies, total RNA can be extracted by using either an organic method (e.g., TRIzol® Isolation Reagent; Thermo Fisher Scientific Inc.) or commercial RNA extraction kits (e.g., RNeasy RNA extraction kit; Qiagen). Although organic methods generally result in slightly higher RNA yield than spin-column-based commercial kits, it is still not preferred in many laboratories because of hazardous chemical use (e.g., phenol, chloroform) during the extraction process (Simister *et al.*, 2011).

After RNA extraction, contaminant DNA is removed by DNase treatment. Because more than 95% of total extracted RNA is ribosomal RNA, it is very important to efficiently remove rRNA and enrich for mRNA that is the main focus of metatranscriptome analysis (Nilsen, 2012). This is the most significant methodological challenge in metatranscriptomics because prokaryotic mRNA does not have poly-A tails as observed in eukaryotic mRNA, and commercial kits that use an oligo-dT-based approach for conversion of mRNA to cDNA do not work for prokaryotic mRNA. There are several alternative methods, however, such as mRNA separation and elution by gel electrophoresis (McGrath *et al.*, 2008), subtractive hybridization (Stewart *et al.*, 2010), and enzymatic digestion of rRNA (Nilsen, 2012). Commercial subtractive hybridization kits, such as MICROB*Express*™ Bacterial mRNA Enrichment Kit, RiboMinus™ Transcriptome Isolation Kit, Bacteria (Thermo Fisher Scientific Inc.), and Ribo-Zero™ rRNA Removal Kit (Illumina Inc., San Diego, CA, USA), are also available for the efficient removal of rRNA from the total RNA. Although all commercial subtractive hybridization kits are designed to efficiently remove rRNA from total RNA, published research indicates that the Ribo-Zero rRNA removal Kit is more productive than the MICROB*Express* Bacterial mRNA Enrichment Kit or mRNA ONLY (Giannoukos *et al.*, 2012).

Unfortunately, removal of rRNA also results in the substantial loss of usable mRNA, and hence it is important to amplify usable mRNA, such as by using commercially available MessageAmp™ II-Bacteria RNA Amplification Kit (Thermo Fisher Scientific Inc.). This kit works in four steps: (i) polyadenylation, (ii) synthesis of single-stranded cDNA by reverse transcription, (iii) synthesis of second strand cDNA, and (iv) synthesis of antisense aRNA by *in vitro* transcription. The single-stranded aRNA is then converted into cDNA by a cDNA synthesis kit (e.g., Universal RiboClone® cDNA Synthesis System; Promega Corporation, Madison, WI, USA), and new synthesized cDNA is then used for a random shotgun sequencing with a sequencing platform of choice. Fortunately, over time, sequencing costs have been decreasing rapidly, and in the future getting enough coverage of mRNA transcript without any rRNA removal steps might be possible.

The second major methodological challenge in metatranscriptomic analyses is the measurement of transcript abundance. Traditionally relative transcript abundance is measured within a sample, but this is not very useful in determining the absolute transcript expression level. This challenge can be overcome by adding known quantity of artificial mRNA (~1 kb in size) into the microbial sample before RNA extraction and then standardization of mRNA transcript based of abundance of the artificial mRNA level (Gifford *et al.*, 2011). This method assumes that RNA is extracted at 100% efficiency, and both artificial and natural mRNA degrade at the same rate. It is also important that the sequence of the artificial mRNA be unique and very different from the target environmental mRNA sequence.

Metaproteomics, or whole-community proteomics (WCP), is the identification and quantitation of all proteins of microbial communities that are present in an environment at a particular time. Determination of temporal or spatial differences in relative abundance of proteins from different treatment and control groups helps researchers develop a better understanding of "who is doing what" (i.e., how a microbial community physiologically responds to environmental disturbances at a protein expression level). Metaproteomic studies in general include the following four steps: (i) proteins are extracted from an environmental sample by an appropriate protein extraction method (Keiblinger *et al.*, 2012), (ii) extracted proteins are digested by proteolytic enzymes (e.g., trypsin), (iii) digested peptides are separated and analyzed based on their charge and mass on liquid chromatography–tandem mass spectrometry (LC-MS/MS), and (iv) peptide fragments are then identified by matching MS/MS spectra against a protein database such as UniProt (Bateman *et al.*, 2015) by using software such as Sipros/ProRata (Wang *et al.*, 2013) and the MetaProteomeAnalyzer (Muth *et al.*, 2015). The MetaProteomeAnalyzer (Muth *et al.*, 2015) also includes options for graphic visualization of end results.

Compared with metagenomic, metatranscriptomic, and single-cell proteomic studies, metaproteomic and metametabolomic studies are more challenging (Morgan and Huttenhower, 2014; Muth *et al.*, 2013). The challenge is mainly because metaproteomic samples contain highly heterogeneous proteins at varying levels of relative abundance. There is not a standardized protein extraction method for recovery of different proteins (e.g., hydrophilic, hydrophobic), and protein databases are incomplete for a majority of environments, which leads to low success in identification of peptides and proteins. Similarly, metametabolomic studies that focus on small metabolites from an environment at a given point of time face several challenges, such as the lack of a single analytical tool to analyze different kinds of metabolites. To our knowledge, Global Natural Products Social Molecular Networking (GNPS) (https://gnps.ucsd.edu/ProteoSAFe/static/gnps-splash.jsp) is the only pipeline that can analyze major part of metametabolomic data.

5.3.3 Next-generation sequencing platforms

Over the last 10 years, several next-generation sequencing platforms have been developed. Among them, the Roche 454-pyrosequencer (Roche Diagnostic Corp., Indianapolis, IN, USA), the Illumina HiSeq/MiSeq sequencer (Illumina Inc.), the Ion PGM™/Ion Proton™ System (Thermo Fisher Scientific Inc.), and the PacBio® RS II sequencing system (Pacific Biosciences of California, Inc., Menlo Park, CA, USA) have gained popularity. Each sequencing platform has its own advantages and disadvantages (see Table 5.2), but the choice of sequencing platform for a particular project depends on the goal of the project, required read length, sequence accuracy, sequence depth, and sequencing cost.

Until recently, the 454-pyrosequencer and Ion PGM had an advantage over Illumina's MiSeq/HiSeq sequencing platforms for marker gene-based metagenomic sequencing, mainly because of longer read length. However, in recent years Illumina's MiSeq/HiSeq sequencing platforms have greatly improved their technologies, and as a result sequencing reads of up to 300 bp (total 600 bp in paired-end sequencing) and 250 bp (total 500 bp in paired-end sequencing) can be obtained with the MiSeq and HiSeq platforms, respectively, which is comparable with 454-pyrosequencing and Ion PGM platforms. Improved read length along with easy operational steps, lower cost per sample, lower error rate, and higher sequence depth (see Table 5.2) than other comparable sequencing platforms (e.g., 454, Ion PGM™/Ion Proton™) have made Illumina's/HiSeq sequencer the platform of choice for metagenomic analysis in recent years (Salipante *et al.*, 2014). The MiSeq sequencing platform is generally preferred over the HiSeq platform for marker gene-based metagenomic analysis because MiSeq generates longer read lengths than HiSeq, and the amount of data generated per run in MiSeq platform is sufficient for smaller projects (e.g., 25 million 300 bp reads per run).

Table 5.2 Comparison of next-generation sequencing platforms

Sequencing instruments	Method	Read length (bp)	Error (%); primary error	Output (million reads/run)	Remarks
454	Emulsion PCR and Pyrosequencing	700	1; indels	1	Longer read but comparatively high cost/Mb data and high error rate compared to MiSeq®/HiSeq™. Ok for marker gene-based metagenomic sequencing
Ion PGM	Emulsion PCR and semiconductor Sequencing	400	≈1; indels	0.4–0.5 (314 chip)	Faster and reasonably longer read than MiSeq/HiSeq but low throughput than to MiSeq/HiSeq. Also too much hands-on time and high error rate associated with indels, OK for marker gene-based metagenomic sequencing but strict size selection is a problem
				2–3 (316 chip)	Same as 314 chip
				4–5.5 (318 chip)	Same as 314 chip
Ion proton	Emulsion PCR and semiconductor sequencing	200	≈1, indels	60–80 (PI chip)	Higher throughput than Ion PGM but lower throughput than HiSeq, shorter reads than Ion PGM/MiSeq/HiSeq/PacBio
MiSeq	Bridge amplification and reversible dye terminator sequencing	300 + 300	≈0.1; substitution	25	High throughput, low error rate, and comparable read length in paired-end sequencing make this platform best among all for marker gene-based metagenomic sequencing
HiSeq 2500	Bridge amplification and reversible dye terminator sequencing	250 + 250	≈0.1; substitution	600 (rapid run v2 kit)	Same as MiSeq but with much higher throughput and slightly shorter read length. Good for shotgun Meta-omic sequencing
		125 + 125		4000	
PacBio® RS II	Single molecule real-time (SMRT) sequencing	8500 bp	≈11; indels	0.05 (per smart cell per run)	No amplification step, longest read among all sequencing platforms (up to 20 kb) but with low throughput and very high error rate in a single pass. Good for shotgun sequence assembly in meta-omic analysis using hybrid approach. Also works better for hard to sequence DNA template (e.g., AT/GC rich DNA, highly repetitive sequences, sequence with long homonucleotide stretch, etc.)

For shotgun sequencing, NextSeq®/HiSeq sequencing platforms are preferred over other high-throughput sequencing platforms mainly because NextSeq/HiSeq generates a much higher sequence depth per run at a much lower cost. The PacBio RS II sequencing system, unlike other sequencing platforms, is a next-generation sequencing system that avoids the amplification step and looks very promising for metagenomic sequencing projects. It yields the longest read length (up to 20 kb) for a low cost and is suitable for labs with few samples. The major problem with this platform is that it has a very high error rate (>11%) (Fichot and Norman, 2013; Kuczynski *et al.*, 2012). However, Pacific Biosystems recently incorporated a circular consensus sequencing (CCS) approach in their sequencing platform, which significantly reduces the error by repeating the sequencing of the same fragment (consensus accuracy >99.999%; http://www.pacb.com/blog/a-closer-look-at-accuracy-in-pacbio/). However, even with the current high error rate, it is very useful as a hybrid approach where the longer read length obtained from PacBio RS II sequencing system can be combined with shorter less erroneous reads obtained from NextSeq/HiSeq sequencing system for assembly and annotation of metagenomics sequencing data. Nevertheless, studies utilizing this hybrid approach are relatively uncommon.

5.4 Marker gene data analysis, visualization, and interpretation

5.4.1 Data analysis pipelines

Marker gene data obtained from different sequencing platforms can be analyzed either by using software from the manufacturer of the sequencing platforms (e.g., Ion Reporter™ software for data from Ion PGM/Ion Proton system or MiSeq Reporter Software for data from MiSeq sequences) or by using external data analysis pipelines such as mothur (Schloss *et al.*, 2009), QIIME (Caporaso *et al.*, 2010b), RDPipeline (Cole *et al.*, 2009), and CloVR (Angiuoli *et al.*, 2011). A manufacturer's built-in software is generally adequate for obtaining taxonomic profiles of the samples, but they are not very common in scientific genomic literature because they often omit important quality control steps (e.g., chimera removal) during sequence preprocessing steps (see Section 5.4.2). They also have limited options for user-based customizations. All external sequence analysis pipelines are capable of analyzing marker gene metagenomic data completely, but selection of a particular pipeline depends on the user's bioinformatics preference and skill, as well as the size of the dataset. RDPipeline (Cole *et al.*, 2014) is a web-based data analysis pipeline that is very user friendly and works well with smaller datasets. Many tools under RDPipeline are now available for download, which makes it suitable for faster analysis in some situations. Currently QIIME (Caporaso *et al.*, 2010b) and mothur (Schloss *et al.*, 2009) are the two most common pipelines available for marker gene-based metagenomic analysis of large datasets. Both pipelines yield almost similar results, can be customized by the user based on their requirements, and are capable of handling large marker gene datasets from different sequencing platforms in an efficient manner.

The QIIME (Caporaso *et al.*, 2010b) pipeline has incorporated "original implementation" of many freely available software packages for analysis and visualization of the final outputs. The tool mothur (Schloss *et al.*, 2009) utilizes a slightly edited version of these softwares for data analysis and yields output that can be further analyzed and visualized in (R Development Core Team, 2011) and other freely available visualization softwares. CloVR (Angiuoli *et al.*, 2011) is a software wrapper that facilitates push-button QIIME and mothur data analysis and can analyze a wide range of data quickly. The main limitation of this pipeline is its lack of flexibility for user customization of analysis steps.

5.4.2 Preprocessing of sequence data

Raw sequence data are preprocessed before hierarchical classification of sequences to remove low quality sequences (e.g., sequence with ambiguous base calls during the process of assigning nucleotide identification, primer and barcode mismatches, unusually long and short sequence reads) and to minimize errors (e.g., indels, homopolymers) from sequencing steps (Huse *et al.*, 2007; Schloss *et al.*, 2011). Although built-in software associated with sequencing platforms is capable of converting specialized file format (e.g., SFF) to basic file format (e.g., FASTA/FASTAQ), investigators should use the original file format (e.g., SFF file in 454, FASTAQ file in MiSeq/HiSeq, etc.) as a starting file for the sequence preprocessing as this provides better error correction. For example, processing of SFF files (compared with FASTA files) from the 454 sequencing platform (either by Mothur or QIIME pipelines) not only helps with the removal of low quality sequence reads but also reduces noise from the flowgrams (plots illustrating signal intensity corresponding to nucleotide incorporation during 454 or Ion PGM sequencing) using several noise correction software packages (see Table 5.3). Noise correction software that was developed for 454 flowgrams may also work with Ion PGM flowgrams; however, it is not recommended that investigators use this approach. This is because the implementation is very time consuming (authors' personal experience) and the algorithms have not been validated using Ion PGM data (Quince *et al.*, 2011). Recently, Golan and Medvedev (2013) developed a new algorithm for noise correction of Ion PGM data, which looks promising, but no study has vetted its use. Because the MiSeq/HiSeq platforms do not produce flowgrams, no noise correction steps are needed for MiSeq/HiSeq data.

The next quality control step (optional) in sequence preprocessing is the identification and removal of chimeric sequences. Chimeras (a sequence made up of two or more parents) are an artifact of the PCR amplification step and constitute anywhere from 5 to 45% of all raw sequence reads (Haas *et al.*, 2011; Schloss *et al.*, 2011). Removal of chimeric sequences before sequence classification and diversity estimation is important as the presence of such sequences increases diversity and gives a researcher a false impression that they have discovered new taxa. But how to remove only chimeric sequences and not true biological sequences is a very challenging question at the moment, because it is very difficult to differentiate chimeric sequences from true biological sequences. To avoid any loss of true biological sequences, many researchers avoid this step during sequence preprocessing. Several chimera detection software packages are freely

Table 5.3 List of commonly used software for multi-omics data analyses

Sr. #	Name of software	Utility	Reference
1	Mothur	Marker gene data analysis pipeline	Schloss *et al.* (2009)
2	QIIME	Marker gene data analysis pipeline	Caporaso *et al.* (2010b)
3	RDPipeline	Marker gene data analysis pipeline	Cole *et al.* (2009)
4	CloVR	Marker gene data analysis pipeline	Angiuoli *et al.* (2011)
5	PyroNoise	Flowgram noise correction for 454 data	Quince *et al.* (2009)
6	PyroTagger	Flowgram noise correction for 454 data	Kunin and Hugenholtz (2010)
7	Denoiser	Flowgram noise correction for 454 data	Reeder and Knight (2010)
8	AmpliconNoise	Flowgram noise correction for 454 data	Quince *et al.* (2011)
9	FlowgramFixer	Flowgram noise correction for Ion PGM data	Golan and Medvedev (2013)
10	Decipher	Chimeric sequence detection	Wright *et al.* (2012)
11	ChimeraSlayer	Chimeric sequence detection	Haas *et al.* (2011)
12	Perseus	Chimeric sequence detection	Quince *et al.* (2011)
13	Uchime	Chimeric sequence detection	Edgar *et al.* (2011)
14	RDP classifier	Taxonomic identification	Wang *et al.* (2007)
15	RTAX	Taxonomic identification	Soergel *et al.* (2012)
16	Muscle	Multiple sequence alignment	Edgar (2004)
17	MAFFT	Multiple sequence alignment	Katoh and Standley (2013)
18	Infernal	Multiple sequence alignment	Nawrocki *et al.* (2009)
19	PyNAST	Multiple sequence alignment	Caporaso *et al.* (2010a)
20	FastTree	Phylogenetic analysis using maximum likelihood (ML) method	Price *et al.* (2010)
21	Clearcut	Phylogenetic analysis using neighbor-joining (NJ) method	Sheneman *et al.* (2006)
22	EMPeror	Multivariate data visualization in 3D	Vazquez-Baeza *et al.* (2013)
23	rgl package in R	Multivariate data visualization in 3D	https://cran.r-project.org/web/packages/rgl/index.html
24	PICRUSt	Functional profile prediction based on 16S rDNA sequence data	Langille *et al.* (2013)
25	MEGAN	Shotgun metagenome sequence analysis	Huson *et al.* (2007)
26	CAMERA	Shotgun metagenome sequence analysis	Seshadri *et al.* (2007)
27	IMG/M	Shotgun metagenome sequence analysis	Markowitz *et al.* (2012)
28	MG-RAST	Shotgun metagenome sequence analysis	Glass *et al.* (2010)
29	MetaPathways	Shotgun metagenome sequence analysis	Konwar *et al.* (2015)
30	HUMAnN	Gene prediction and characterization of metabolic pathways	Abubucker *et al.* (2012)

(Continued)

Table 5.3 (Continued)

Sr. #	Name of software	Utility	Reference
31	ShortBRED	Targeted community functional profiling from metagenomic sequences	Kaminski *et al.* (2015)
32	MetaTrans	Metatranscriptomics analysis	Martinez *et al.* (2016)
33	GNPS	Metabolomics analysis	https://gnps.ucsd.edu/ProteoSAFe/static/gnps-splash.jsp
34	ngs_backbone	Quality control of shotgun metagenomic sequences	Blanca *et al.* (2011)
35	FASTAX-Toolkit	Quality control of shotgun metagenomic sequences	http://hannonlab.cshl.edu/fastx_toolkit/
36	ABBA	Guided metagenomic assembly	Salzberg *et al.* (2008)
37	MetaVelvet	De novo metagenomic assembly	Namiki *et al.* (2012)
38	Meta-IDBA	De novo metagenomic assembly	Peng *et al.* (2011)
39	MetaORFA	De novo metagenomic assembly	Ye and Tang (2009)
40	PhyloPythiaS	Sequence composition-based taxonomic profiling	Patil *et al.* (2012)
41	PhymmBL	Sequence composition-based taxonomic profiling	Brady and Salzberg (2009)
42	TACOA	Sequence composition-based taxonomic profiling	Diaz *et al.* (2009)
43	NBC	Sequence composition-based taxonomic profiling	Rosen *et al.* (2011)
44	AMPHORA2	Sequence similarity-based taxonomic profiling	Wu and Scott (2012)
45	AmphoraNet	Sequence similarity-based taxonomic profiling (web version of AMPHORA2)	Kerepesi *et al.* (2014)
46	MetaPhlAn2	Shotgun sequence similarity-based taxonomic profiling	Truong *et al.* (2015)
47	Qiita	Multi-omics data analysis	https://qiita.ucsd.edu/
48	STAMP	Statistical analysis	Parks *et al.* (2014)
49	MetaStats	Statistical analysis	White *et al.* (2009)
50	LEfSe	Statistical analysis	Segata *et al.* (2011)
51	SourceTracker	Statistical analysis	Knights *et al.* (2011)

available (see Table 5.3) and are capable at removing most of chimeric sequences. ChimeraSlayer (Haas *et al.*, 2011) and Decipher (Wright *et al.*, 2012) require a chimera-free reference database for the detection of chimeric sequences, whereas Perseus (Quince *et al.*, 2011) can detect chimera *de novo* (no chimera-free reference database is required) and assumes that the parent sequences will be more common than chimeric sequence in products from a single PCR. Uchime (Edgar *et al.*, 2011) detects chimeras either *de novo* or using a reference database, and in both situations it is the fastest among all chimera detection software packages (Edgar *et al.*, 2011). Uchime can detect even those chimeras that have originated from more than two parents (Edgar *et al.*, 2011). Decipher (Wright *et al.*, 2012) is computationally very demanding and is not recommended for large datasets. Because no reference database is 100% chimera-free and complete, it is sometimes better to use a *de novo* approach in chimera detection than a reference database approach.

5.4.3 Sequence clustering approaches

After the preprocessing steps, the remaining sequences are clustered based on either taxonomy/phylogeny/operational taxonomic units (OTU) or a combination of more than one approach before diversity estimates can be made.

5.4.3.1 Taxonomic identification

Taxonomic identification of sequences are based on similarity of new sequences to the known sequences from existing reference databases, such as the RDP database (Cole *et al.*, 2014), Greengene database (DeSantis *et al.*, 2006), SILVA database (Quast *et al.*, 2013), National Center for Biotechnology Information (NCBI) taxonomy database (http://www.ncbi.nlm.nih.gov/taxonomy), and UNITE database (Tedersoo *et al.*, 2011). To perform this task, software packages such as RDP classifier (Wang *et al.*, 2007) and RTAX (Soergel *et al.*, 2012) are available and can be used either as a standalone application or as a part of the pipeline (RTAX is available only in QIIME pipeline). Taxonomic classification of sequences helps researchers in qualitative and quantitative comparison of their study with other similar studies. Usually at higher taxonomic levels (e.g., phylum, class), taxonomic classification of sequences works very well, but at lower taxonomic level (e.g., species) some sequences remain unclassified, especially for samples that are from an environmental source that is relatively less studied.

5.4.3.2 Phylogenetic clustering

In phylogenetic clustering, sequences are clustered based on phylogenetic relationships between DNA sequences, and the resulting phylogenetic tree is used for diversity estimations. Phylogenetic clustering also helps with identifying close relatives of unclassified sequences resulting from a taxonomic clustering method. To make a phylogenetic tree, sequences are first aligned either *de novo* (e.g., Muscle (Edgar, 2004), MAFFT (Katoh and Standley, 2013)) or by using existing reference alignments (e.g., Infernal (Nawrocki *et al.*, 2009), SILVA reference file (Quast *et al.*, 2013), PyNAST (Caporaso *et al.*, 2010a)), and then aligned sequences are used as an input file for construction of a phylogenetic tree using phylogenetic methods such as neighbor-joining (NJ) method, maximum parsimony (MP), maximum likelihood (ML), or Bayesian methods. Phylogenetic tree construction software, such as FastTree (Price *et al.*, 2010) for a ML tree and Clearcut (Sheneman *et al.*, 2006) for a NJ tree are preferred over traditional packages, such as PAUP* (Swofford, 2003) and PHYLIP (Felsenstein, 2005). This is mainly because they are much faster than traditional packages for the construction of a phylogenetic tree from a huge metagenomic dataset. Also reference-based alignment of marker gene sequences is preferred over *de novo* alignment, which is computationally more challenging.

5.4.3.3 OTU-based clustering

In OTU-based clustering methods, sequences are clustered based on genetic similarity between each other, and each genetic similarity level generally represents a taxonomic level. For example, genetic similarity at ≥97, ≥95, ≥90, and ≥85% hypothetically represent species, genus, family, and order, respectively (Cressman *et al.*, 2010). There are three options for OTU-based clustering: *de novo*, closed reference, and open reference. *De novo* method cluster sequences based on genetic similarities between each other, whereas closed-reference method cluster sequences based on reference sequence database. Because a *de novo* method is independent of reference sequences, it clusters all sequences and hence all sequences are utilized in downstream applications, whereas closed-reference method clusters only those sequences that hit the reference sequences of the reference sequence database (remaining sequences are discarded from the downstream analysis). The closed-reference method works well when microbial diversity of sampled environment is well represented in the reference datasets (e.g., closed-reference method discards only ~5% of all sequences from human gut microbiome samples, whereas same OTU picking approach discards ~50% of sequences from soil samples mainly because of lack of representative soil sample sequence in the reference database) (Goodrich *et al.*, 2014). The open-reference clustering method is a hybrid approach where sequences are first clustered by closed-reference method, and then those sequences that did not hit reference sequences are clustered *de novo*. This approach helps in detection of both known and unknown (novel) diversity and hence preferred over other OTU picking methods in the majority of situations. Kopylova *et al.* (2016) have evaluated the performance of different commonly used OTU picking methods, and readers are advised to read this paper for more information on this topic.

5.4.4 Microbial diversity estimations

Once sequences are clustered, the next step is to utilize them for community diversity estimations. Before diversity estimation, it is important to normalize the sequence datasets so that each sample has the same number of reads. This is very important because no next-generation sequencing platform generates equal yields in all samples, and unequal sequencing depths across samples can lead to biased diversity estimation (Gihring *et al.*, 2012). One approach to solve this problem is to calculate the simple proportion of taxa/OTU within each sample (i.e., make total sequence count for each sample to 100 and then calculate relative abundance for each taxa or OTU). Many researchers do not recommend this approach mainly because it does not address heteroscedasticity and is prone to clustering of samples based on sequencing depth (Goodrich *et al.*, 2014; McMurdie and Holmes, 2014). The second approach is to normalize samples by random subsampling of all samples to the one with lowest number of reads. If some samples have extremely low numbers of sequence reads compared with most samples, then it is preferable to discard these samples from the analysis rather than decreasing the number of reads of all other samples to an extremely low read level. There is no clear-cut guideline for sequence count per sample that will be considered extremely low, but for sequence generated from Illumina MiSeq sequencing platform, samples with sequence read <1000 are generally discarded from the analysis. The main problem with this approach is the loss of precious sequence data from high-sequence count samples. To avoid problems associated with previous approaches, McMurdie and Holmes (2014) have proposed an alternative mixed model-based approach for this purpose.

To estimate community richness, nonparametric richness estimators, such as the Chao (Chao, 1984; Chao and Bunge, 2002) and the ACE, (Chao, 2005) are commonly used. Fortunately, these are available for use in different data analysis pipelines. Estimation of species richness by nonparametric estimators is based on the number of rare and abundant taxa; hence to avoid biased richness estimation using these methods, it is important to make sure that the rare taxa present in the sample are real and not an artifact of the sequencing reaction. The mothur pipeline includes a parametric richness estimator, CatchAll (Bunge *et al.*, 2012), which is based on the best-fit model, and is less sensitive to outliers that are commonly observed in next-generation metagenomic datasets. All popular metagenomic pipelines (e.g., mothur, QIIME) have incorporated different methods that calculate α- (within-sample microbial diversity, e.g., Shannon index and inverse-Simpson index) and β-diversity (between-sample microbial diversity, e.g., Jaccard similarity coefficient, Yue and Clayton θ similarity coefficient, Bray–Curtis similarity coefficient, etc.) indices based on relative abundance of OTU or taxonomy. Distances calculated based on β-diversity estimators are then utilized for multivariate analysis using principle coordinate analysis (PCoA) or nonmetric multidimensional scaling (NMDS), either within the data analysis pipelines or by using the Vegan package (Oksanen *et al.*, 2012) in (R Development Core Team, 2011). Alternatively, phylogenetic information from the sequence data (i.e., phylogenetic tree) can also be used in the calculation of α-diversity (e.g., phylogenetic diversity) and β-diversity (e.g., UniFrac (Lozupone and Knight, 2005)). UniFrac distance can be calculated by using either a weighted (quantitative method, i.e., based on relative abundances of OTUs) or unweighted (qualitative, i.e., based on only presence and absence of OTUs) approach, and calculated UniFrac distance by either methods can be used for unweighted pair group method with arithmetic mean (UPGMA) clustering of communities or for multivariate analysis, such as PCoA and NMDS. The choice of the ordination method depends on the data type and if assumptions of a method are satisfied or not. In general, NMDS is preferred over other ordination methods in microbiome data analysis, because it is a nonparametric ordination method with no major assumptions. The quality of an NMDS plot depends on the R^2 (higher is better) and stress (lower is better) values. Generally, the stress value of NMDS plots decreases with an increase in the number of dimensions (i.e., two-dimensional (2D) NMDS plots generally have higher stress value than three-dimensional (3D) NMDS plots) and increases with an increase in number of samples. But in general, NMDS plots with stress values ≤ 0.2 are considered as a reasonable representation of data (http://www.mothur.org/wiki/454_SOP). Stress value can be reduced by different data transformation methods (e.g., logarithmic transformation, square root transformation, arcsine transformation), but before selection of a particular data transformation method, it is important to ensure that the selected method is appropriate for that data type. Multivariate results obtained from different ordination methods can be visualized either within a data analysis pipeline (e.g., QIIME) or by using external visualization softwares such as gplots and rgl packages in (R Development Core Team, 2011) and EMPeror (Vazquez-Baeza *et al.*, 2013).

5.5 Multi-omics data analysis, visualization, and interpretation

Multi-omics data analysis is more complicated and challenging than marker gene-based analyses mostly because of incomplete coverage, high volume of data, and high computational requirements. Although there are some easy-to-use pipelines for shotgun metagenome sequence analysis and metatranscriptomics (see Table 5.3); the same is not true for other multi-omics (e.g., metaproteomic, metametabolomic) data analyses.

In general, the shotgun metagenomic data analysis pipeline includes sequence quality control steps (preprocessing steps), sequence assembly steps (optional) (guided/*de novo*), taxonomic profiling steps, and gene prediction and metabolic profiling steps. Steps for metatranscriptomic analysis are almost similar to shotgun metagenomic analysis (e.g., sequence preprocessing including removal of leftover rRNA, tRNA, and host RNA, sequence assembly, taxonomic profiling, gene prediction, and metabolic profiling), but in metatranscriptomic analysis, differential transcript expression between different samples is also calculated (e.g., CuffDiff (Trapnell *et al.*, 2013)).

5.5.1 Sequence preprocessing

Quality control steps for the sequence analyses are generally sequencing platform specific because raw data from different platforms come as different files. These sequence files are preprocessed for quality check, and during this process adapter, linkers, barcodes, and low quality region of the sequences are trimmed out (see Section 5.4.2). Apart from platform-specific quality assurance software (e.g., FastQC, Pyrobayes), there are other software (e.g., FASTX toolkit, ngs_backbone, etc.) that can be used for preprocessing of sequences obtained from different sequencing platforms.

5.5.2 Sequence assembly

Whether to assemble metagenomic data into contiguous sequences (i.e., contigs—a series of overlapping short DNA sequences, which make a map that reconstructs the sequence of a region of a chromosome) or use unassembled reads directly for taxonomic and functional annotations depends on community complexity, sequencing depth, read length, and purpose of the experiment. Genome assembly, whenever possible, helps in the development of longer contigs, which ultimately increases accuracy in taxonomic and functional annotations and helps in identification of entire gene sequence and associated regulatory control regions. However genome assembly in many situations is impractical and very time consuming. Hence, many researchers prefer metagenomic analysis using softwares that skip genome assembly stages (e.g., MetaPhlAn, HUMAnN).

Sequences can be assembled either by a guided approach (based on a reference genome) or by *de novo* approach. In the guided approach, a closely related reference genome (i.e., a reference genome that differs from the true genome by small polymorphisms) is used for the assembly of the sequence. However if the closely related taxa are not available as a reference genome, then differences in true genome and reference genome (e.g., large indels, large polymorphism) can lead to fragmented genome assembly. Guided genome assembly can be done by using software such as ABBA (Salzberg *et al.*, 2008) and AMOScmp (Chaisson and Pevzner, 2008). These softwares process data at a much faster rate (in hours) and require much less memory; hence it is possible to assemble a genome on regular desktop or laptop computers. *De novo* assembly of metagenomic data, on the other hand, is more challenging, mainly because the majority of currently available genome assemblers were designed for assembly of DNA fragments from a single organism with the assumption that all DNA fragments came from the same individual, but that is not true for metagenomic sequence data. In recent years several metagenomic assembly software packages were released (see Table 5.3) that assume non-clonality of DNA fragment before assembly and can assemble almost complete genomes of at least some dominant members. However, metagenomic assembly is still in its infancy, and the accuracy of assembled genomes is not guaranteed in the majority of cases, especially when a selected sample is from a complex community and sequencing coverage is low.

Because metagenomic assembly is performed from sequences obtained from a mixture of communities, there is always a chance of forming a chimeric assembly. Unlike a marker gene-based method, there is no specifically designed software for identification of chimeric assembly during metagenomic analysis. However formation of chimeric assemblies can be minimized by changing parameters during sequence assembly steps such as the use of very high-sequence identity threshold (>95% sequence identity), G + C skew changes, and codon usage changes and by blasting of open reading frames against the chimera-free databases.

5.5.3 Taxonomic profiling

One of the major goals of whole-genome shotgun sequencing is to identify all microbes (at the lowest taxonomic level) and their potential function from an environmental sample. Sequence reads/contigs/scaffolds can be assigned to different clusters or bins, based on their taxonomic origins. The success and accuracy of taxonomic identification of sequences depends on the read or contig length. Usually longer sequence read/contig length (>800 bp) results in a more accurate taxonomic clustering at lower taxonomic levels (e.g., species or genus level) than shorter reads. Sequence clustering can be done based either on sequence composition (i.e., *de novo* clustering based on shared sequence characteristics such as GC content, oligonucleotide frequency, codon usage, and read coverage) or based on sequence similarity to a reference dataset (e.g., genus, family, order, etc.). Because sequence clustering based on composition of the sequence does not require a reference dataset, it is much faster and requires less computational resources than similarity-based clustering (Sharpton, 2014). Similarity-based approaches generally provide higher taxonomic resolution and more accurate sequence classification compared with composition-based approaches. However, if prior genomic information about an environment is limited, then a composition-based approach may perform better (Sharpton, 2014). There are several freely available software packages that can cluster metagenomic sequences based on composition and similarity (see Table 5.3). Among sequence similarity-based clustering software packages, AMPHORA2 (Wu and Scott, 2012), AmphoraNet (Kerepesi *et al.*, 2014), and MetaPhlAn (Segata *et al.*, 2012) are much faster than others, mainly because instead of a whole-genome search, the software performs a search only on those parts of the genome that are more informative for taxonomic clustering. For example, MetaPhlAn (Segata *et al.*, 2012) utilizes limited numbers of clade-specific (clade—a common ancestor and all its descendants classified together on the basis of phylogenetic relationships) marker genes for the direct identification of metagenomic sequences.

5.5.4 Gene prediction and metabolic profiling

In the multi-omics data analysis, it is important to understand not only which bacterial taxa are there but also the function of the bacterial community. This information can be obtained by identifying genes and their expression levels, as well as by identifying metabolic pathways. Sequence annotations can be done by using either assembled or unassembled data. Both approaches have advantages and disadvantages, and the choice of a particular approach depends on the experimental design and sample type. Use of an assembled genome during sequence annotation helps in more accurate identification of genes and its neighboring regulatory regions, but it can also cause bias toward those sequences that get assembled relatively easily and are more abundant in number. On the other hand, use of an unassembled genome is more sensitive to rare sequences, but it is comparatively less accurate for functional annotations. Metagenomic sequences are generally scanned against reference genomes for gene prediction, and then predicted genes are functionally annotated using reference functional databases such as KEGG (Kanehisa and Goto, 2000), COG (Tatusov *et al.*, 2003), SEED (Overbeek *et al.*, 2005), EggNOG (Powell *et al.*, 2014), PFAM (Finn *et al.*, 2014), TIGRFAMs (Haft *et al.*, 2013), MetaCyc (Caspi *et al.*, 2014), and PhyloFacts (Afrasiabi *et al.*, 2013). Several widely used multi-omics analysis pipelines such as HUMAnN (Abubucker *et al.*, 2012), MG-RAST (Glass *et al.*, 2010), and MetaPathways (Konwar *et al.*, 2015) use unassembled sequences as an input for gene prediction and characterization of metabolic pathways. Once gene or metabolic profiles are obtained from each sample, then relative abundances of these profiles can be analyzed similar to taxonomic profiling.

5.6 Statistical analysis

The decreasing costs associated with next-generation sequencing have allowed researchers to increase the number of replicates in metagenomic studies, which ultimately produces more robust statistical analyses. Selection and use of a particular statistical test will ultimately depend on the experimental design and hypothesis being investigated. For example, ordination methods, such as PCoA and NMDS, as discussed in Section 5.4.4, serve as a good exploratory technique in visualizing how samples from different treatments are grouped together in 2D or 3D scale. But whether spatial separation between different treatment groups is statistically significant or not can only be determined by a statistical test of significance such as analysis of molecular variance (AMOVA) or multiresponse permutation procedure (MRPP). If different treatment groups have significantly different microbial composition from each other, then either indicator species analysis (ISA) (Dufrene and Legendre, 1997) or random forest (Breiman, 2001) modeling approach can help identify indicator taxa or predictor taxa that are responsible for the significant difference between treatment groups. Indicator taxa or predictor taxa from the previous analysis then can be used in development of predictive models (Metcalf *et al.*, 2013; Pechal *et al.*, 2013) such as generalized additive models (GAMs) and generalized linear models (GLMs). Although most of the multi-omics data can be analyzed by general statistical analysis software (e.g., R Development Core Team (2011),

SPSS (SPSS Inc., Chicago, IL, USA), and SAS (SAS Institute Inc., Cary, NC, USA)), there are many (see Table 5.3) that are specifically developed for statistical analysis of multi-omics datasets.

5.7 Major challenges and future directions

The development of next-generation sequencing platforms in the last decade has revolutionized the field of microbial ecology. Forensic scientists have embraced this technology and performed several excellent studies, mostly based on a marker gene approach (Fierer *et al.*, 2010; Metcalf *et al.*, 2013; Pechal *et al.*, 2013; Singh *et al.*, 2014b, c; Tridico *et al.*, 2014) investigating the utilization of differential microbial community structure in forensic analyses. As discussed previously, although marker gene-based approaches provide information on "who is there," it does not provide information on "what they are doing" or on "who is doing what." Such integrated multi-omics approaches could help forensic scientists develop new, more precise, and robust forensic tools (e.g., trace evidences identification, human identification), but the number of studies using multi-omics approaches on forensic samples are currently very limited (Khodakova *et al.*, 2014). This is mainly because the acquisition and analysis of several multi-omics data types simultaneously is very challenging. Although the cost of next-generation sequencing is decreasing, it is still beyond the reach of many small labs to perform integrated multi-omics analysis of a complex forensic samples (e.g., soil (Aitkenhead-Peterson *et al.*, 2012, 2015), insects (Singh *et al.*, 2015; Tomberlin *et al.*, 2012; Zheng *et al.*, 2013)). However these technologies are constantly evolving, and hopefully in the near future, forensic scientists will have appropriate multi-omics tools to conduct more comprehensive analysis of microbial communities and their relevance to forensic science.

Even with no further technological development, there are still many unanswered questions in forensic microbiology that today's tools and technologies are capable of answering. Such investigations still require thoughtful hypotheses with strong experimental design strategies and global collaborations. The impacts of many other factors (e.g., drugs, scavengers) on bacterial succession during decomposition are still unknown and require further investigation. For example, although it is well established that microbial communities change with time on human cadavers and that this information can help in the prediction of post-mortem interval (PMI) (Metcalf *et al.*, 2013, 2015; Pechal *et al.*, 2013), the method is still too cumbersome for common usage in most forensic laboratories and lacks a global perspective that incorporates geographical variations affecting microbial community structure. International studies designed to incorporate worldwide microbial geosignatures could significantly assist forensic scientists to implement this information for the benefit of forensic investigations.

References

Abubucker, S., Segata, N., Goll, J. *et al.* (2012) Metabolic reconstruction for metagenomic data and its application to the human microbiome. *PLoS Computational Biology*, **8**, e1002358.

Afrasiabi, C., Samad, B., Dineen, D. *et al.* (2013) The PhyloFacts FAT-CAT web server: ortholog identification and function prediction using fast approximate tree classification. *Nucleic Acids Research*, **41**, W242–W248.

Aitkenhead-Peterson, J.A., Owings, C.G., Alexander, M.B. *et al.* (2012) Mapping the lateral extent of human cadaver decomposition with soil chemistry. *Forensic Science International*, **216**, 127–134.

Aitkenhead-Peterson, J.A., Alexander, M.B., Bytheway, J.A. *et al.* (2015) Applications of soil chemistry in forensic entomology, in *Forensic Entomology: International Dimensions and Frontiers* (eds J.K. Tomberlin and M. Eric Benbow), CRC Publishers, Boca Raton.

Angiuoli, S.V., Matalka, M., Gussman, A. *et al.* (2011) CloVR: a virtual machine for automated and portable sequence analysis from the desktop using cloud computing. *BMC Bioinformatics*, **12**, 356.

Bateman, A., Martin, M.J., O'Donovan, C. *et al.* (2015) UniProt: a hub for protein information. *Nucleic Acids Research*, **43**, D204–D212.

Benson, D.A., Karsch-Mizrachi, I., Lipman, D.J. *et al.* (2005) GenBank. *Nucleic Acids Research*, **33**, D34–D38.

Blanca, J.M., Pascual, L., Ziarsolo, P. *et al.* (2011) ngs_backbone: a pipeline for read cleaning, mapping and SNP calling using next generation sequence. *BMC Genomics*, **12**, 285.

Brady, A. and Salzberg, S.L. (2009) Phymm and PhymmBL: metagenomic phylogenetic classification with interpolated Markov models. *Nature Methods*, **6**, 673–676.

Breiman, L. (2001) Random forests. *Machine Learning*, **45**, 5–32.

Bunge, J., Woodard, L., Bohning, D. *et al.* (2012) Estimating population diversity with CatchAll. *Bioinformatics*, **28**, 1045–1047.

Caporaso, J.G., Bittinger, K., Bushman, F.D. *et al.* (2010a) PyNAST: a flexible tool for aligning sequences to a template alignment. *Bioinformatics*, **26**, 266–267.

Caporaso, J.G., Kuczynski, J., Stombaugh, J. *et al.* (2010b) QIIME allows analysis of high-throughput community sequencing data. *Nature Methods*, **7**, 335–336.

Caporaso, J.G., Lauber, C.L., Walters, W.A. *et al.* (2012) Ultra-high-throughput microbial community analysis on the Illumina HiSeq and MiSeq platforms. *ISME Journal*, **6**, 1621–1624.

Caspi, R., Altman, T., Billington, R. *et al.* (2014) The MetaCyc database of metabolic pathways and enzymes and the BioCyc collection of Pathway/Genome Databases. *Nucleic Acids Research*, **42**, D459–D471.

Chaisson, M.J. and Pevzner, P.A. (2008) Short read fragment assembly of bacterial genomes. *Genome Research*, **18**, 324–330.

Chakravorty, S., Helb, D., Burday, M. *et al.* (2007) A detailed analysis of 16S ribosomal RNA gene segments for the diagnosis of pathogenic bacteria. *Journal of Microbiological Methods*, **69**, 330–339.

Chao, A. (1984) Nonparametric-estimation of the number of classes in a population. *Scandinavian Journal of Statistics*, **11**, 265–270.

Chao, A. (2005) Species estimation and applications, in *Encyclopedia of Statistical Sciences* (eds S. Kotz, N. Balakrishnan, C.B. Read and B. Vidakovic), John Wiley & Sons, Inc, Hobone, pp. 7907–7916.

Chao, A. and Bunge, J. (2002) Estimating the number of species in a stochastic abundance model. *Biometrics*, **58**, 531–539.

Chomczynski, P. and Sacchi, N. (1987) Single-step method of RNA isolation by acid guanidinium thiocyanate-phenol-chloroform extraction. *Analytical Biochemistry*, **162**, 156–159.

Cole, J.R., Wang, Q., Cardenas, E. *et al.* (2009) The Ribosomal Database Project: improved alignments and new tools for rRNA analysis. *Nucleic Acids Research*, **37**, D141–D145.

Cole, J.R., Wang, Q., Fish, J.A. *et al.* (2014) Ribosomal Database Project: data and tools for high throughput rRNA analysis. *Nucleic Acids Research*, **42**, D633–D642.

Cressman, M.D., Yu, Z., Nelson, M.C. *et al.* (2010) Interrelations between the microbiotas in the litter and in the intestines of commercial broiler chickens. *Applied and Environmental Microbiology*, **76**, 6572–6582.

Delmont, T.O., Robe, P., Clark, I. *et al.* (2011) Metagenomic comparison of direct and indirect soil DNA extraction approaches. *Journal of Microbiological Methods*, **86**, 397–400.

DeSantis, T.Z., Hugenholtz, P., Larsen, N. *et al.* (2006) Greengenes, a chimera-checked 16S rRNA gene database and workbench compatible with ARB. *Applied and Environmental Microbiology*, **72**, 5069–5072.

Diaz, N.N., Krause, L., Goesmann, A. *et al.* (2009) TACOA—taxonomic classification of environmental genomic fragments using a kernelized nearest neighbor approach. *BMC Bioinformatics*, **10**, 56.

Dowd, S., Callaway, T., Wolcott, R. *et al.* (2008) Evaluation of the bacterial diversity in the feces of cattle using 16S rDNA bacterial tag-encoded FLX amplicon pyrosequencing (bTEFAP). *BMC Microbiology*, **8**, 125.

Dufrene, M. and Legendre, P. (1997) Species assemblages and indicator species: the need for a flexible asymmetrical approach. *Ecological Monographs*, **67**, 345–366.

Edgar, R.C. (2004) MUSCLE: multiple sequence alignment with high accuracy and high throughput. *Nucleic Acids Research*, **32**, 1792–1797.

Edgar, R.C., Haas, B.J., Clemente, J.C. *et al.* (2011) UCHIME improves sensitivity and speed of chimera detection. *Bioinformatics*, **27**, 2194–2200.

Eren, A.M., Sogin, M.L., Morrison, H.G. *et al.* (2015) A single genus in the gut microbiome reflects host preference and specificity. *ISME Journal*, **9**, 90–100.

Felsenstein J. 2005. PHYLIP (Phylogeny Inference Package) Distributed by the author. Seattle: Department of Genome Sciences, University of Washington.

Fichot, E.B. and Norman, R.S. (2013) Microbial phylogenetic profiling with the Pacific Biosciences sequencing platform. *Microbiome*, **1**, 10.

Fierer, N., Lauber, C.L., Zhou, N. *et al.* (2010) Forensic identification using skin bacterial communities. *Proceedings of the National Academy of Sciences of the United States of America*, **107**, 6477–6481.

Finn, R.D., Bateman, A., Clements, J. *et al.* (2014) Pfam: the protein families database. *Nucleic Acids Research*, **42**, D222–D230.

Franzosa, E.A., Huang, K., Meadow, J.F. *et al.* (2015) Identifying personal microbiomes using metagenomic codes. *Proceedings of the National Academy of Sciences of the United States of America*, **112**, E2930–E2938.

Gajer, P., Brotman, R.M., Bai, G. *et al.* (2012) Temporal dynamics of the human vaginal microbiota. *Science Translational Medicine*, **4**, 132ra52.

Giannoukos, G., Ciulla, D.M., Huang, K. *et al.* (2012) Efficient and robust RNA-seq process for cultured bacteria and complex community transcriptomes. *Genome Biology*, **13**, R23.

Gifford, S.M., Sharma, S., Rinta-Kanto, J.M. and Moran, M.A. (2011) Quantitative analysis of a deeply sequenced marine microbial metatranscriptome. *ISME Journal*, **5**, 461–472.

Gihring, T.M., Green, S.J. and Schadt, C.W. (2012) Massively parallel rRNA gene sequencing exacerbates the potential for biased community diversity comparisons due to variable library sizes. *Environmental Microbiology*, **14**, 285–290.

Gilbert, J.A., Jansson, J.K. and Knight, R. (2014) The Earth Microbiome project: successes and aspirations. *BMC Biology*, **12**, 69.

Glass, E.M., Wilkening, J., Wilke, A. *et al.* (2010) Using the metagenomics RAST server (MG-RAST) for analyzing shotgun metagenomes. *Cold Spring Harbor Protocols*, **2010**, pdb. prot5368.

Golan, D. and Medvedev, P. (2013) Using state machines to model the Ion Torrent sequencing process and to improve read error rates. *Bioinformatics*, **29**, i344–i351.

Goodrich, J.K., Di Rienzi Sara, C., Poole Angela, C. *et al.* (2014) Conducting a microbiome study. *Cell*, **158**, 250–262.

Grice, E.A., Kong, H.H., Renaud, G. *et al.* (2008) A diversity profile of the human skin microbiota. *Genome Research*, **18**, 1043–1050.

Haas, B.J., Gevers, D., Earl, A.M. *et al.* (2011) Chimeric 16S rRNA sequence formation and detection in Sanger and 454-pyrosequenced PCR amplicons. *Genome Research*, **21**, 494–504.

Haft, D.H., Selengut, J.D., Richter, R.A. *et al.* (2013) TIGRFAMs and genome properties in 2013. *Nucleic Acids Research*, **41**, D387–D395.

Harrison, S.T., Dennis, J.S. and Chase, H.A. (1991) Combined chemical and mechanical processes for the disruption of bacteria. *Bioseparation*, **2**, 95–105.

Hart, M.L., Meyer, A., Johnson, P.J. and Ericsson, A.C. (2015) Comparative evaluation of DNA extraction methods from feces of multiple host species for downstream next-generation sequencing. *PLoS One*, **10**, e0143334.

Henderson, G., Cox, F., Kittelmann, S. *et al.* (2013) Effect of DNA extraction methods and sampling techniques on the apparent structure of cow and sheep rumen microbial communities. *PLoS One*, **8**, e74787.

Human Microbiome Project Consortium (2012) A framework for human microbiome research. *Nature*, **486**, 215–221.

Huse, S.M., Huber, J.A., Morrison, H.G. *et al.* (2007) Accuracy and quality of massively parallel DNA pyrosequencing. *Genome Biology*, **8**, R143.

Huson, D.H., Auch, A.F., Qi, J. and Schuster, S.C. (2007) MEGAN analysis of metagenomic data. *Genome Research*, **17**, 377–386.

Kaminski, J., Gibson, M.K., Franzosa, E.A. *et al.* (2015) High-specificity targeted functional profiling in microbial communities with ShortBRED. *PLoS Computational Biology*, **11**, e1004557.

Kanehisa, M. and Goto, S. (2000) KEGG: kyoto encyclopedia of genes and genomes. *Nucleic Acids Research*, **28**, 27–30.

Katoh, K. and Standley, D.M. (2013) MAFFT multiple sequence alignment software version 7: improvements in performance and usability. *Molecular Biology and Evolution*, **30**, 772–780.

Keiblinger, K.M., Wilhartitz, I.C., Schneider, T. *et al.* (2012) Soil metaproteomics—comparative evaluation of protein extraction protocols. *Soil Biology and Biochemistry*, **54**, 14–24.

Kerepesi, C., Banky, D. and Grolmusz, V. (2014) AmphoraNet: the webserver implementation of the AMPHORA2 metagenomic workflow suite. *Gene*, **533**, 538–540.

Khodakova, A.S., Smith, R.J., Burgoyne, L. *et al.* (2014) Random whole metagenomic sequencing for forensic discrimination of soils. *PLoS One*, **9**, e104996.

Klindworth, A., Pruesse, E., Schweer, T. *et al.* (2013) Evaluation of general 16S ribosomal RNA gene PCR primers for classical and next-generation sequencing-based diversity studies. *Nucleic Acids Research*, **41**, e1.

Knights, D., Kuczynski, J., Charlson, E.S. *et al.* (2011) Bayesian community-wide culture-independent microbial source tracking. *Nature Methods*, **8**, 761–763.

Konwar, K.M., Hanson, N.W., Bhatia, M.P. *et al.* (2015) MetaPathways v2.5: quantitative functional, taxonomic and usability improvements. *Bioinformatics*, **31**, 3345–3347.

Kopylova, E., Navas-Molina, J.A., Mercier, C., *et al.* (2016). Open-source sequence clustering methods improve the state of the art. *mSystems*, 1–16. doi: 10.1128/mSystems.00003-15

Korabecna, M., ed. (2007). The variability in the fungal ribosomal DNA (ITS1, ITS2, and 5.8S rRNA gene): its biological meaning and application in medical mycology. Badajoz: Formatex Research Center. pp. 783–787.

Kuczynski, J., Lauber, C.L., Walters, W.A. *et al.* (2012) Experimental and analytical tools for studying the human microbiome. *Nature Reviews. Genetics*, **13**, 47–58.

Kumar, P.S., Brooker, M.R., Dowd, S.E. and Camerlengo, T. (2011) Target region selection is a critical determinant of community fingerprints generated by 16S pyrosequencing. *PLoS One*, **6**, e20956.

Kunin, V. and Hugenholtz, P. (2010) PyroTagger: a fast, accurate pipeline for analysis of rRNA amplicon pyrosequence data. *Open J*, 1–8.

Langille, M.G., Zaneveld, J., Caporaso, J.G. *et al.* (2013) Predictive functional profiling of microbial communities using 16S rRNA marker gene sequences. *Nature Biotechnology*, **31**, 814–821.

Lauber, C.L., Zhou, N., Gordon, J.I. *et al.* (2010) Effect of storage conditions on the assessment of bacterial community structure in soil and human-associated samples. *FEMS Microbiology Letters*, **307**, 80–86.

Lee, C.K., Herbold, C.W., Polson, S.W. *et al.* (2012) Groundtruthing next-gen sequencing for microbial ecology-biases and errors in community structure estimates from PCR amplicon pyrosequencing. *PLoS One*, **7**, e44224.

Lozupone, C. and Knight, R. (2005) UniFrac: a new phylogenetic method for comparing microbial communities. *Applied and Environmental Microbiology*, **71**, 8228–8235.

Mankin, A.S., Skryabin, K.G. and Rubtsov, P.M. (1986) Identification of ten additional nucleotides in the primary structure of yeast 18S rRNA. *Gene*, **44**, 143–145.

Markowitz, V.M., Chen, I.M., Chu, K. *et al.* (2012) IMG/M: the integrated metagenome data management and comparative analysis system. *Nucleic Acids Research*, **40**, D123–D129.

Martinez, X., Pozuelo, M., Pascal, V. *et al.* (2016) MetaTrans: an open-source pipeline for metatranscriptomics. *Science Reports*, **6**, 26447.

McGrath, K.C., Thomas-Hall, S.R., Cheng, C.T. *et al.* (2008) Isolation and analysis of mRNA from environmental microbial communities. *Journal of Microbiological Methods*, **75**, 172–176.

McMurdie, P.J. and Holmes, S. (2014) Waste not, want not: why rarefying microbiome data is inadmissible. *PLoS Computational Biology*, **10**, e1003531.

Metcalf, J.L., Wegener Parfrey, L., Gonzalez, A. *et al.* (2013) A microbial clock provides an accurate estimate of the postmortem interval in a mouse model system. *eLife*, **2**, e01104.

Metcalf, J.L., Xu, Z.Z., Weiss, S. *et al.* (2015) Microbial community assembly and metabolic function during mammalian corpse decomposition. *Science*, **351**, 158–162.

Mirsepasi, H., Persson, S., Struve, C. *et al.* (2014) Microbial diversity in fecal samples depends on DNA extraction method: easyMag DNA extraction compared to QIAamp DNA stool mini kit extraction. *BMC Research Notes*, **7**, 50.

Morgan, X.C. and Huttenhower, C. (2014) Meta'omic analytic techniques for studying the intestinal microbiome. *Gastroenterology*, **146**, 1437–1448.

Muth, T., Benndorf, D., Reichl, U. *et al.* (2013) Searching for a needle in a stack of needles: challenges in metaproteomics data analysis. *Molecular BioSystems*, **9**, 578–585.

Muth, T., Behne, A., Heyer, R. *et al.* (2015) The MetaProteomeAnalyzer: a powerful open-source software suite for metaproteomics data analysis and interpretation. *Journal of Proteome Research*, **14**, 1557–1565.

Namiki, T., Hachiya, T., Tanaka, H. and Sakakibara, Y. (2012) MetaVelvet: an extension of Velvet assembler to de novo metagenome assembly from short sequence reads. *Nucleic Acids Research*, **40**, e155.

Nawrocki, E.P., Kolbe, D.L. and Eddy, S.R. (2009) Infernal 1.0: inference of RNA alignments. *Bioinformatics*, **25**, 1335–1337.

Nilsen, T.W. (2012) Removing rRNA from deproteinized, phenol-extracted total RNA by enzymatic digestion. *Cold Spring Harbor Protocols*, **2012**, 1187–1188.

Ogram, A., Sayler, G.S. and Barkay, T. (1987) The extraction and purification of microbial DNA from sediments. *Journal of Microbiological Methods*, **7**, 57–66.

Oh, J., Byrd, A.L., Deming, C. *et al.* (2014) Biogeography and individuality shape function in the human skin metagenome. *Nature*, **514**, 59–64.

Oksanen, J., Blanchet, G.F., Kindt, R., *et al.* (2012). Vegan: community ecology package. R package version 2.0-4, https://cran.r-project.org and https://github.com/vegandevs/vegan (accessed November 30, 2016).

Overbeek, R., Begley, T., Butler, R.M. *et al.* (2005) The subsystems approach to genome annotation and its use in the project to annotate 1000 genomes. *Nucleic Acids Research*, **33**, 5691–5702.

Parachin, N.S., Schelin, J., Norling, B. *et al.* (2010) Flotation as a tool for indirect DNA extraction from soil. *Applied Microbiology and Biotechnology*, **87**, 1927–1933.

Parks, D.H., Tyson, G.W., Hugenholtz, P. and Beiko, R.G. (2014) STAMP: statistical analysis of taxonomic and functional profiles. *Bioinformatics*, **30**, 3123–3124.

Patil, K.R., Roune, L. and McHardy, A.C. (2012) The PhyloPythiaS web server for taxonomic assignment of metagenome sequences. *PLoS One*, **7**, e38581.

Pechal, J.L., Crippen, T.L., Benbow, M.E. *et al.* (2013) The potential use of bacterial community succession in forensics as described by high throughput metagenomic sequencing. *International Journal of Legal Medicine*, **128**, 193–205.

Peng, Y., Leung, H.C., Yiu, S.M. and Chin, F.Y. (2011) Meta-IDBA: a de Novo assembler for metagenomic data. *Bioinformatics*, **27**, i94–i101.

Peura, S., Sinclair, L., Bertilsson, S. and Eiler, A. (2015) Metagenomic insights into strategies of aerobic and anaerobic carbon and nitrogen transformation in boreal lakes. *Science Reports*, **5**, 12102.

Powell, S., Forslund, K., Szklarczyk, D. *et al.* (2014) eggNOG v4.0: nested orthology inference across 3686 organisms. *Nucleic Acids Research*, **42**, D231–D239.

Price, M.N., Dehal, P.S. and Arkin, A.P. (2010) FastTree 2—approximately maximum-likelihood trees for large alignments. *PLoS One*, **5**, e9490.

Quast, C., Pruesse, E., Yilmaz, P. *et al.* (2013) The SILVA ribosomal RNA gene database project: improved data processing and web-based tools. *Nucleic Acids Research*, **41**, D590–D596.

Quince, C., Lanzen, A., Curtis, T.P. *et al.* (2009) Accurate determination of microbial diversity from 454 pyrosequencing data. *Nature Methods*, **6**, 639–641.

Quince, C., Lanzen, A., Davenport, R.J. and Turnbaugh, P.J. (2011) Removing noise from pyrosequenced amplicons. *BMC Bioinformatics*, **12**, 38.

R Development Core Team. (2011). R: A Language and Environment for Statistical Computing. Vienna, Austria: R Foundation for Statistical Computing, http://www.R-project.org (accessed October 28, 2016).

Reeder, J. and Knight, R. (2010) Rapidly denoising pyrosequencing amplicon reads by exploiting rank-abundance distributions. *Nature Methods*, **7**, 668–669.

Rosen, G.L., Reichenberger, E.R. and Rosenfeld, A.M. (2011) NBC: the Naive Bayes Classification tool webserver for taxonomic classification of metagenomic reads. *Bioinformatics*, **27**, 127–129.

Roume, H., Muller, E.E., Cordes, T. *et al.* (2013) A biomolecular isolation framework for ecosystems biology. *ISME Journal*, **7**, 110–121.

Salipante, S.J., Kawashima, T., Rosenthal, C. *et al.* (2014) Performance comparison of Illumina and ion torrent next-generation sequencing platforms for 16S rRNA-based bacterial community profiling. *Applied and Environmental Microbiology*, **80**, 7583–7591.

Salzberg, S.L., Sommer, D.D., Puiu, D. and Lee, V.T. (2008) Gene-boosted assembly of a novel bacterial genome from very short reads. *PLoS Computational Biology*, **4**, e1000186.

Schloss, P.D. (2014) Microbiology: an integrated view of the skin microbiome. *Nature*, **514**, 44–45.

Schloss, P.D., Westcott, S.L., Ryabin, T. *et al.* (2009) Introducing mothur: open-source, platform-independent, community-supported software for describing and comparing microbial communities. *Applied and Environmental Microbiology*, **75**, 7537–7541.

Schloss, P.D., Gevers, D. and Westcott, S.L. (2011) Reducing the effects of PCR amplification and sequencing artifacts on 16S rRNA-based studies. *PLoS One*, **6**, e27310.

Schoch, C.L., Seifert, K.A., Huhndorf, S. *et al.* (2012) Nuclear ribosomal internal transcribed spacer (ITS) region as a universal DNA barcode marker for Fungi. *Proceedings of the National Academy of Sciences of the United States of America*, **109**, 6241–6246.

von der Schulenburg, J.H., Hancock, J.M., Pagnamenta, A. *et al.* (2001) Extreme length and length variation in the first ribosomal internal transcribed spacer of ladybird beetles (Coleoptera: Coccinellidae). *Molecular Biology and Evolution*, **18**, 648–660.

Segata, N., Izard, J., Waldron, L. *et al.* (2011) Metagenomic biomarker discovery and explanation. *Genome Biology*, **12**, 1–18.

Segata, N., Waldron, L., Ballarini, A. *et al.* (2012) Metagenomic microbial community profiling using unique clade-specific marker genes. *Nature Methods*, **9**, 811–814.

Sender, R., Fuchs, S. and Milo, R. (2016) Revised estimates for the number of human and bacteria cells in the body. *PLoS Biology*, **14**, e1002533.

Seshadri, R., Kravitz, S.A., Smarr, L. *et al.* (2007) CAMERA: a community resource for metagenomics. *PLoS Biology*, **5**, 394–397.

Sharpton, T.J. (2014) An introduction to the analysis of shotgun metagenomic data. *Frontiers in Plant Science*, **5**, 209.

Sheneman, L., Evans, J. and Foster, J.A. (2006) Clearcut: a fast implementation of relaxed neighbor joining. *Bioinformatics*, **22**, 2823–2824.

Shoaib, M., Baconnais, S., Mechold, U. *et al.* (2008) Multiple displacement amplification for complex mixtures of DNA fragments. *BMC Genomics*, **9**, 415.

Simister, R.L., Schmitt, S. and Taylor, M.W. (2011) Evaluating methods for the preservation and extraction of DNA and RNA for analysis of microbial communities in marine sponges. *Journal of Experimental Marine Biology and Ecology*, **397**, 38–43.

Singh, B. and Crippen, T.L. (2015) Methodologies in forensic and decomposition microbiology, in *Forensic Entomology: International Dimensions and Frontiers* (eds J.K. Tomberlin and M.E. Benbow), CRC Press, Boca Raton, pp. 263–282.

Singh, B., Crippen, T.L., Sufrin, V., *et al.* (2014a). Utility of soil microbiome in identification of carrion decomposition sites. *Proceedings of the International Association of Forensic Sciences*, Seoul, South Korea, October 15–18, 2014, pp. 431–432.

Singh, B., Crippen, T.L., Tarone, A.M., *et al.* (2014b). Utility of bacteria associated with human cadavers in estimation of postmortem interval (PMI). *Proceedings American Academy of Forensic Science 66th Annual Scientific Meeting*, Seattle, USA, February 17–22, 2014, pp. 360–361.

Singh, B., Crippen, T., Zheng, L. *et al.* (2014c) A metagenomic assessment of the bacteria associated with Lucilia sericata and Lucilia cuprina (Diptera: Calliphoridae). *Applied Microbiology and Biotechnology*, **99**, 1–15.

Singh, B., Crippen, T., Zheng, L. *et al.* (2015) A metagenomic assessment of the bacteria associated with Lucilia sericata and Lucilia cuprina (Diptera: Calliphoridae). *Applied Microbiology and Biotechnology*, **99**, 869–883.

Smith, D.P. and Peay, K.G. (2014) Sequence depth, not PCR replication, improves ecological inference from next generation DNA sequencing. *PLoS One*, **9**, e90234.

Soergel, D.A., Dey, N., Knight, R. and Brenner, S.E. (2012) Selection of primers for optimal taxonomic classification of environmental 16S rRNA gene sequences. *ISME Journal*, **6**, 1440–1444.

Stewart, F.J., Ottesen, E.A. and DeLong, E.F. (2010) Development and quantitative analyses of a universal rRNA-subtraction protocol for microbial metatranscriptomics. *ISME Journal*, **4**, 896–907.

Swofford, D.L. (2003). PAUP*. Phylogenetic Analysis Using Parsimony (*and Other Methods). Version 4. Sunderland: Sinauer Associates.

Tatusov, R.L., Fedorova, N.D., Jackson, J.D. *et al.* (2003) The COG database: an updated version includes eukaryotes. *BMC Bioinformatics*, **4**, 1–14.

Tedersoo, L., Abarenkov, K., Nilsson, R.H. *et al.* (2011) Tidying up international nucleotide sequence databases: ecological, geographical and sequence quality annotation of its sequences of mycorrhizal fungi. *PLoS One*, **6**, e24940.

Tomberlin, J.K., Crippen, T.L., Tarone, A.M. *et al.* (2012) Interkingdom responses of flies to bacteria mediated by fly physiology and bacterial quorum sensing. *Animal Behaviour*, **84**, 1449–1456.

Trapnell, C., Hendrickson, D.G., Sauvageau, M. *et al.* (2013) Differential analysis of gene regulation at transcript resolution with RNA-seq. *Nature Biotechnology*, **31**, 46–53.

Tridico, S.R., Murray, D.C., Addison, J. *et al.* (2014) Metagenomic analyses of bacteria on human hairs: a qualitative assessment for applications in forensic science. *Investigative Genetics*, **5**, 16.

Truong, D.T., Franzosa, E.A., Tickle, T.L. *et al.* (2015) MetaPhlAn2 for enhanced metagenomic taxonomic profiling. *Nature Methods*, **12**, 902–903.

Vazquez-Baeza, Y., Pirrung, M., Gonzalez, A. and Knight, R. (2013) EMPeror: a tool for visualizing high-throughput microbial community data. *GigaScience*, **2**, 16.

Wagner Mackenzie, B., Waite, D.W. and Taylor, M.W. (2015) Evaluating variation in human gut microbiota profiles due to DNA extraction method and inter-subject differences. *Frontiers in Microbiology*, **6**, 130.

Walters, W., Hyde, E.R., Berg-Lyons, D. *et al.* (2016) Improved bacterial 16S rRNA gene (V4 and V4-5) and fungal internal transcribed spacer marker gene primers for microbial community surveys. *mSystems*, **1**, e00009–e00015.

Wang, Q., Garrity, G.M., Tiedje, J.M. and Cole, J.R. (2007) Naive Bayesian classifier for rapid assignment of rRNA sequences into the new bacterial taxonomy. *Applied and Environmental Microbiology*, **73**, 5261–5267.

Wang, Y.F., Ahn, T.H., Li, Z. and Pan, C.L. (2013) Sipros/ProRata: a versatile informatics system for quantitative community proteomics. *Bioinformatics*, **29**, 2064–2065.

Wesolowska-Andersen, A., Bahl, M.I., Carvalho, V. *et al.* (2014) Choice of bacterial DNA extraction method from fecal material influences community structure as evaluated by metagenomic analysis. *Microbiome*, **2**, 19.

White, J.R., Nagarajan, N. and Pop, M. (2009) Statistical methods for detecting differentially abundant features in clinical metagenomic samples. *PLoS Computational Biology*, **5**, e1000352.

Wright, E.S., Yilmaz, L.S. and Noguera, D.R. (2012) DECIPHER, a search-based approach to chimera identification for 16S rRNA sequences. *Applied and Environmental Microbiology*, **78**, 717–725.

Wu, M. and Scott, A.J. (2012) Phylogenomic analysis of bacterial and archaeal sequences with AMPHORA2. *Bioinformatics*, **28**, 1033–1034.

Yarza, P., Yilmaz, P., Pruesse, E. *et al.* (2014) Uniting the classification of cultured and uncultured bacteria and archaea using 16S rRNA gene sequences. *Nature Reviews. Microbiology*, **12**, 635–645.

Ye, Y. and Tang, H. (2009) An ORFome assembly approach to metagenomics sequences analysis. *Journal of Bioinformatics and Computational Biology*, **7**, 455–471.

Yuan, S., Cohen, D.B., Ravel, J. *et al.* (2012) Evaluation of methods for the extraction and purification of DNA from the human microbiome. *PLoS One*, **7**, e33865.

Zheng, L., Crippen, T.L., Singh, B. *et al.* (2013) A survey of bacterial diversity from successive life stages of black soldier fly (Diptera: Stratiomyidae) by using 16S rDNA pyrosequencing. *Journal of Medical Entomology*, **50**, 647–658.

CHAPTER 6

Culture and long-term storage of microorganisms for forensic science

Emily N. Junkins[1,2], Embriette R. Hyde[3] and David O. Carter[2]

[1] *Department of Microbiology and Plant Biology, University of Oklahoma, Norman, OK, USA*

[2] *Laboratory of Forensic Taphonomy, Forensic Sciences Unit, Division of Natural Sciences and Mathematics, Chaminade University of Honolulu, Honolulu, HI, USA*

[3] *Knight Laboratory, Department of Pediatrics, University of California, San Diego, San Diego, CA, USA*

6.1 Introduction

The development of microbiology arguably began in the mid-seventeenth century with the famous observations of fungi and protozoa made by Robert Hooke (Hooke, 1665) and later investigation of microbial diversity by Antonie van Leeuwenhoek (Leewenhoeck, 1684). Even before the study of these organisms, their metabolic properties were already being applied to food and drink production as early as 10,000 BCE in Egyptian and Chinese populations (Slonczewski and Foster, 2011). Despite these organisms probably being the oldest on Earth, mainstream microbiology did not truly flourish until Louis Pasteur disproved spontaneous generation and Robert Koch developed isolation techniques for microbial samples (see Madigan *et al.*, 2012). Pasteur's studies on fermentation and spoilage from airborne microbes not only discredited spontaneous generation of cells but also developed pasteurization methods still used today in the food industry. Pasteur's work was enforced by John Tyndall who demonstrated that particle-free air does not lead to microbial growth and spoilage on solid media and that the heating of media to 100°C could reliably rid samples of unwanted microorganisms (Tyndall, 1877). These early days of microbiology witnessed a great interest in infection and decomposition, primarily that of humans and their food with a focus on understanding microorganisms sufficiently to manage them. A similar spirit exists in forensic microbiology today, where many laboratories maintain collections of culturable microorganisms (see Leighton and Murch, 2011), an approach based on Paul Kirk's (1953) concept of an

Forensic Microbiology, First Edition. Edited by David O. Carter, Jeffery K. Tomberlin, M. Eric Benbow and Jessica L. Metcalf.

evidence "museum" that houses samples and standards that serve as investigative exemplars and research specimens.

Robert Koch recognized the need for a methodical approach when working with microbes, specifically in terms of etiology, so he developed methods for isolation of microbial species and their maintenance on solid media (Koch, 1883). Many of his techniques are still being used today with little to no modification (Atlas and Bartha, 1997; Slonczewski and Foster, 2011). Both Pasteur and Koch ushered in what is known as the pure culture period that established a simple yet profound and fundamental trend in microbiology: isolate a species to obtain a pure culture and see what it does (Atlas and Bartha, 1997).

Today these principles form the foundation for several microbiological tests, many of which are readily available in microbiology laboratory manuals (Cappucino and Sherman, 2011; Brown, 2012). These relatively simple techniques aided in the initial study of the relationships between microorganisms, infection, and death (White, 1899; Richey and Goerhing, 1918). These techniques have continued to develop into rapid and reliable analytical and diagnostic tools (see Chapters 7 and 8). Interestingly, the areas of forensic microbiology research covered in this text do not regularly employ culture-based techniques (Pechal *et al.*, 2013; Chun *et al.*, 2015; Griffin *et al.*, 2015). Most recent research efforts in forensic microbiology have focused on sequencing the microbial communities and microbiomes associated with vertebrate carrion (Hyde *et al.*, 2015; Finley *et al.*, 2016; Metcalf *et al.*, 2016). These efforts have sought to map the temporal and spatial changes in the structure and function of postmortem microbial communities of remains via high-throughput sequencing with culture techniques playing only a minor role. These culture-independent techniques have provided unprecedented insight into microorganisms and their ecology; these techniques should continue to develop and be used regularly in all areas of microbial research. However, there is still great value in culture-dependent techniques. The purpose of this chapter is to provide a brief review on the value of culturing microorganisms as well as protocols that can be used to culture microorganisms for use in research directed for forensic applications.

6.2 The value of culturing microorganisms

There are many ways to culture microorganisms in the laboratory and there are many reasons for doing so. Some laboratories may be interested in metabolic community function of the community as a whole (e.g., Pechal *et al.*, 2013). In these instances, methods such as Biolog plates utilize redox reactions and color changes to detect metabolic activity of certain microbial communities (Stefanowicz, 2006). Though this method helps to determine what processes the microorganisms are likely involved in, it does not identify certain taxa within the community. Additional value can be found in culturing microbes because it allows for the isolation of a specimen from other organisms with its growth uninhibited, something that community culture methods such as Biolog plates are unable to clarify. This single taxon from pure culture can be manipulated via selective assays, differential assays, and/or growth conditions to assess its metabolic capabilities and infer its ecological role(s). For example, Zheng *et al.* (2013) used bacterial cultures to demonstrate that bacteria can mediate the oviposition of the black soldier fly (*Hermetia illucens*), which is a forensically relevant insect (see Chapter 11). In particular, this research highlighted the interaction of *Ignatzschineria* spp. with these flies (Zheng *et al.*, 2013). *Ignatzschineria* spp. have recently become taxa of great interest to forensic microbiology because of their consistent presence in postmortem microbiomes (Hyde *et al.*, 2013; Pechal *et al.*, 2014; Metcalf *et al.*, 2016). Also, Junkins and Carter (2015) recently used both culture methods and sequencing to identify these taxa in the larval mass of decomposing swine (*Sus scrofa domesticus*) (Tables 6.1 and 6.2). These studies provide excellent examples of the insight that can be gained by combining culture-dependent and culture-independent approaches.

It is, however, important to recognize that a debate exists concerning the value of culture-dependent techniques. Ritz (2007), for example, states that using cultured environmental microbes to infer *in situ* microbial community structure is an absurdity; culturable microbial communities are no indication of the representativeness of the entire microbiome associated with that environment because only a small percentage, only thousands of approximately 10^6 microorganisms, can be cultured. Furthermore, cultivability is not related to abundance; it is possible to culture widely distributed as well as rare, fastidious microorganisms (Shade *et al.*, 2012). These points are valid, and it is necessary to accept that identifying the structure of the culturable microbial community from a sample provides limited insight into the structure of the whole microbial community.

However, these limitations should not preclude the use of culture-dependent techniques. Culture-dependent techniques are of great value when viewed as complementary to culture-independent sequencing techniques. This is a perspective consistent with that of Nichols (2007). This complementarity can be achieved in many ways, including biochemical assays to reveal growth and metabolism (e.g., morphology, pH tolerance, salinity tolerance, nutrient specificity, enzyme activity), competition studies to reveal competition and survival tactics (e.g., antibiotic production, motility), and microscopic approaches to understand morphology and motility. These approaches have been used for several decades (Herbert *et al.*, 1956) and provide an effective means to understand the structural and functional properties of microbial taxa. Since one goal of forensic microbiology is to characterize ante- and postmortem microbial communities for use as physical evidence, it is in our best interest to understand as many individual microorganisms as possible. Determining and characterizing the roles of culturable microorganisms can help to interpret the datasets from culture-independent sequencing techniques.

Table 6.1 Subset of postmortem bacteria isolated from the skin of a decomposing swine (*Sus scrofa domesticus*) carcass on Oahu, Hawaii (E. Junkins, Chaminade University of Honolulu, unpublished data), where + indicates the presence of a taxon

				Hours postmortem					
Phylum	**Family**	**Genus**	**Species**	**74**	**80**	**98**	**104**	**122**	**124**
Firmicutes	Bacillaceae	*Bacillus*	*cereus*	+	+	+	+		+
Firmicutes	Bacillaceae	*Bacillus*	*altitudinis*	+					
Firmicutes	Bacillaceae	*Bacillus*	*aryabhattai*	+	+	+	+	+	
Firmicutes	Enterococcaceae	*Enterococcus*	*durans*	+					
Firmicutes	Staphylococcaceae	*Staphylococcus*	*xylosus*	+			+		
Firmicutes	Staphylococcaceae	*Staphylococcus*	*cohnii*	+			+		
Firmicutes	Bacillaceae	*Bacillus*	*badius*	+					+
Firmicutes	Bacillaceae	*Lysinibacillus*	*fusiformis*	+					
Firmicutes	Staphylococcaceae	*Staphylococcus*	*sciuri*		+				
Firmicutes	Bacillaceae	*Bacillus*	*subtilis*		+				
Firmicutes	Bacillaceae	*Bacillus*	*thuringiensis*		+			+	
Firmicutes	Bacillaceae	*Bacillus*	*simplex*		+		+		
Firmicutes	Lactobacillaceae	*Pediococcus*	*acidilactici*		+				
Firmicutes	Bacillaceae	*Bacillus*	*aryabhattai*		+	+			
Firmicutes	—	*Rummeliibacillus*	*stabekisii*		+	+	+		+
Firmicutes	Staphylococcaceae	*Staphylococcus*	*cohnii cohnii*		+				
Firmicutes	Enterococcaceae	*Vagococcus*	*sp.*			+			
Firmicutes	Bacillaceae	*Bacillus*	*megaterium*			+	+		+
Firmicutes	Bacillaceae	*Lysinibacillus*	*boronitolerans*			+		+	+
Firmicutes	Bacillaceae	*Bacillus*	*muralis*			+			
Firmicutes	Enterococcaceae	*Vagococcus*	*lutrae*				+	+	
Firmicutes	Bacillaceae	*Bacillus*	*methylotrophicus*				+		
Firmicutes	Bacillaceae	*Bacillus*	*kochii*				+		
Firmicutes	Staphylococcaceae	*Staphylococcus*	*nepalensis*				+		
Firmicutes	Bacillaceae	*Fictibacillus*	*arsenicus*				+		
Firmicutes	Enterococcaceae	*Enterococcus*	*faecalis*				+		

Table 6.2 Subset of Illumina 16S taxa significantly correlated with oxidation–reduction potential isolated from the larval mass on decomposing swine (*Sus scrofa domesticus*) (Junkins, unpublished data)

Phylum	Class	Order	Family	Genus
Firmicutes	Bacilli	Lactobacillales	Unclassified	Unclassified
Proteobacteria	Gammaproteobacteria	Xanthomonadales	Xanthomonadaceae	*Ignatzschineria*
Firmicutes	Clostridia	Clostridiales	Peptostreptococcaceae	94otu3326
Firmicutes	Clostridia	Clostridiales	Clostridiaceae	*Clostridium*
Firmicutes	Clostridia	Clostridiales	Mogibacteriaceae	*Mogibacterium*
Firmicutes	Bacilli	Lactobacillales	Lactobacillaceae	*Pediococcus*
Firmicutes	Bacilli	Lactobacillales	Lactobacillaceae	*Lactobacillus*
Firmicutes	Clostridia	Clostridiales	Mogibacteriaceae	94otu45033
Firmicutes	Bacilli	Lactobacillales	Lactobacillaceae	*Pediococcus*
Firmicutes	Clostridia	Clostridiales	Clostridiaceae	Unclassified
Firmicutes	Clostridia	Clostridiales	Peptostreptococcaceae	94otu34216
Firmicutes	Clostridia	Clostridiales	Clostridiaceae	Unclassified

6.3 Collection and handling of samples

A forensic investigation should include a well-developed evidence collection and handling plan. And with the development of forensic microbiology, this plan will need to include a strategy for integrating the collection and preservation of microbial samples as well (see Chapter 4). Fortunately, a long history exists on the development of such methods from a forensic science perspective. These methods could easily be integrated into standard operating procedures currently implemented by forensic practitioners to some degree. Three common methods include (i) bulk collection of a complete item, such as in the case of a entire removal of bloodstained carpet or a section of a wall; (ii) collection of a portion of an area of interest, such as cuttings of fabric or vacuuming any liquids; and (iii) swabbing the surface of interest that holds a substance, such as a body fluid or finger print (Smith, 2011). Pechal *et al.* (2014) also provide an effective framework for the management of microbial evidence; handling and storage of these samples will depend on the nature of the sample (e.g., fresh–wet tissue, bone, water, swabs), the distance traveled before storage and analysis, and the capabilities of the collecting agency/institution. Also, sampling for further analysis versus mere detection will alter the methods employed during an investigation.

Figure 6.1 illustrates a general methodology of sample handling. The protocol of each stage will vary depending on the organisms collected and the purpose of their collection (e.g., analysis). Methods of sample collection should also be considered when planning extraction and analysis techniques. Extraction techniques will then determine what further analyses can be done on each sample.

A US Government Accountability Office (GAO) report (2005) concerning the US Postal Service (USPS), the Centers for Disease Control and Prevention (CDC), and the Environmental Protection Agency (EPA) highlighted the lack of validated collection and analysis methods during the 2001 anthrax (*Bacillus anthracis* endospores) scare (Sharma *et al.*, 2011). The GAO found these agencies were severely lacking in most method areas outlined in Figure 6.2. Sampling strategy lacked any statistical confidence in any areas targeted for analysis; GAO suggested incorporating probability sampling based on the likelihood of encountering the target specimen, in this case, anthrax. This issue could potentially come into play in the role of forensic microbiology, specifically dealing with trace analysis of microbes and locating areas of interest based on demonstrated likelihood of detecting representative microbial samples (e.g., Lax *et al.* 2014).

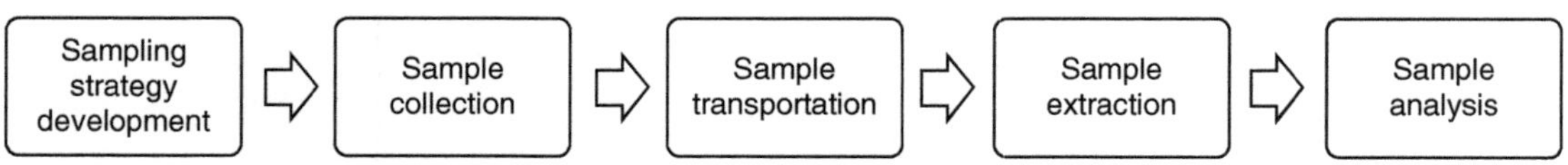

Figure 6.1 Generalized methodology for the collection and analysis of microbial samples. Source: Adapted from United States Government Accountability Office (2005)

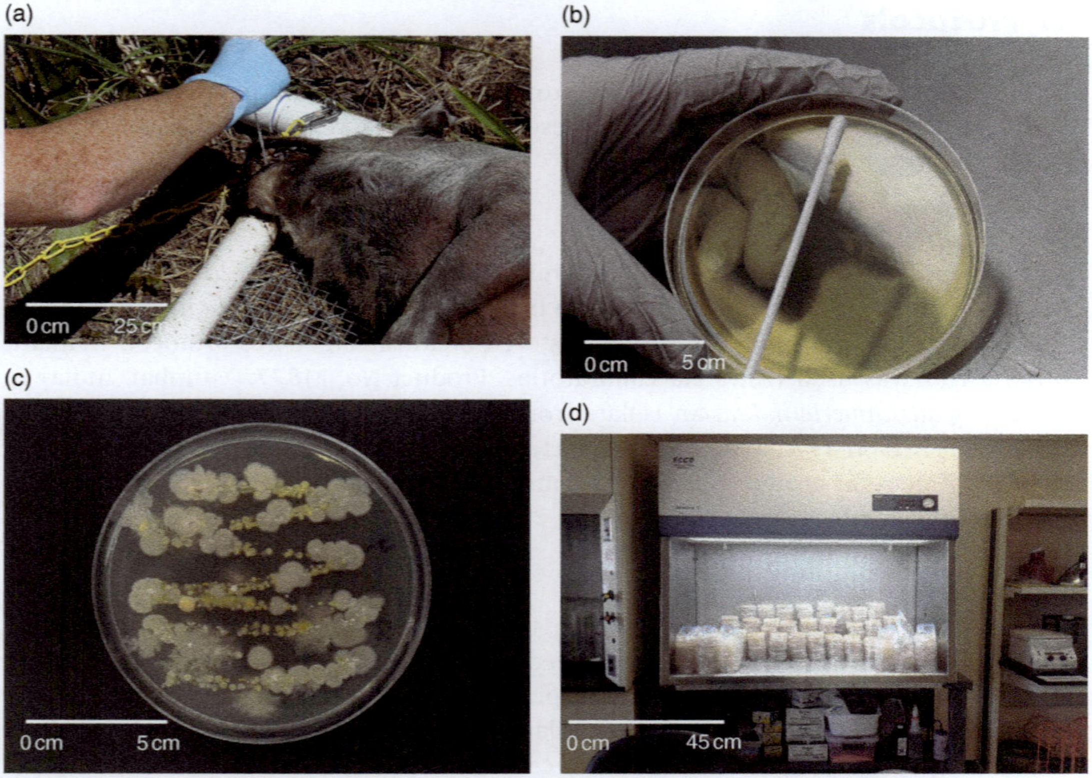

Figure 6.2 Sampling the larval mass in the oral cavity of a decomposing swine (*Sus scrofa domesticus*) carcass (a) and the streaking for growth in a Petri dish containing standard nutrient agar (b) resulting in the growth of several colonies with contrasting morphology (c) at room temperature following incubation in a laminar flow cabinet (d)

Another area that the GAO highlighted for review that would also be applicable to forensic microbiology was the lack of information available to investigators when choosing the methods of extraction and analysis most effective for their investigation. GAO suggested more monetary investment and higher prioritizing of material or technological capabilities be made available. Reports such as this have lead the National Institute of Standards and Technology (NIST) and the National Institute of Justice to form bodies such as the Organization of Scientific Area Committee (OSAC, http://www.nist.gov/forensics/osac.cfm) to develop standards and guidelines to improve quality and consistency of work in the forensic science community. To achieve this, the goal of OSAC is to establish research and measurement standards, documentary standards, and sufficient scientific basis for all forensic disciplines. OSAC consists of subject matter experts across all disciplines of forensic science to ensure equal representation of each area. It is this effort that also provides the stimulus for the current chapter. With the goal in mind to eventually develop validated standards that will endure scrutiny encountered in the court of law, brief culture protocols are outlined in the following sections.

6.4 Protocols

The purpose of this section is to provide protocols for culture and long-term storage of bacteria and fungi. Laboratories can use these protocols as a reliable foundation that can be modified to suit their interests.

6.4.1 Aerobic culture

The protocol for the culturing of aerobic microorganisms and their long-term storage consists of a series of steps that we refer to here as the (i) plate phase, (ii) broth phase, (iii) storage phase, and (iv) reculturing of stored organisms. For this protocol, standard nutrient agar was used (HiMedia Laboratories, Product No. M877, Mumbai, India) to cultivate *Ignatzschineria indica*, an isolate from the postmortem microbiome of a larval mass established on a swine carcass (Chun *et al.*, 2015) (Figure 6.2a). This medium is just one type of general growth medium; a multitude of media exist to select for several microbial taxa (Table 6.3).

6.4.2 Sterile technique

Cultivation of microbes relies heavily on the ability to prevent contamination and maintain sterility. This prevention is achieved through various practices that collectively comprise sterile–aseptic technique. These practices include washing equipment with a biocidal compound (e.g., ethanol, bleach), flaming metal sampling equipment and tubes, and working in laminar flow and biosafety cabinets. Wearing personal protective equipment further decreases the probability of contamination (Chapter 4).

6.4.3 Sample collection, transport, and culture

The methods described here to collect microbial samples for transport to the laboratory should be followed in conjunction with guidelines described in Chapter 4:

- Swab habitat of interest (Figure 6.2a).
- Place the swab in a sterile container for transport.
- In sterile space, streak the swab onto an agar-based medium (Figure 6.2b).
- Incubate in the dark at room temperature for 24–48 hours.

6.4.3.1 The plate phase—isolation

- Select colonies for isolation based on morphology. This step is based on the assumption that all taxa have unique morphology (see Figure 6.2c). This assumption is incorrect so it is imperative to accept that colonies of the same taxon can have different morphologies and colonies of differing taxon can have a similar morphology.
- Culture for isolation onto an agar-based medium and incubate at required conditions (dependent upon taxon growth requirements) until colonies are visible to the naked eye (Figure 6.2c). Each colony should theoretically be a single taxon. Many general

Table 6.3 Overview of agar and broth media that are commonly used to culture microorganisms

Growth medium	Target microorganisms
Standard nutrient agar	General growth
Trypticase soy agar/tryptic soy broth (TSA/TSB)	General growth and storage
Lysogeny broth and agar (LB)	Industry favorite for the cultivation of *Escherichia coli*
Desoxycholate (DOC) agar	Selective (S): Enterobacteriaceae Differential (D): +/– lactose fermentation
Endo agar, eosin methylene blue (EMB) agar, MacConkey agar	S: Gram negative, enteric organisms D: +/– lactose fermentation
Hektoen enteric (HE) agar	S: Gram negative, enteric organisms, specifically *Salmonella* and *Shigella* D: +/– lactose fermentation and *Salmonella* versus *Shigella* spp.; +/– sulfur reduction
Mannitol salt agar (MSA)	S: microbes capable of tolerating high salt concentrations (i.e., halophiles) D: +/– mannitol fermentation
Phenylethyl alcohol (PEA) agar	S: Gram positive
Salmonella–Shigella (*SS*) agar	S: Gram negative, usually *Salmonella* and *Shigella* D: +/– lactose fermentation; +/– thiosulfate reduction
Tellurite glycine agar	S: coagulase (–) staphylococci and Gram negative D: +/– tellurite reduction
Thiosulfate citrate bile salts sucrose (TCBS) agar	S: against Gram positive D: +/– sucrose fermentation; +/– thiosulfate reduction
Xylose lysine deoxycholate (XLD) agar	S: Gram negative D: +/– xylose fermentation; +/– citrate decarboxylation; +/– sulfur reduction
Blood agar	D: hemolytic reactions

Source: Adapted from Leboffe and Pierce (2010).

microbiology textbooks are great resources for the overview of this initial phase (Slonczewski and Foster, 2011; Madigan *et al.*, 2012).

- Prepare multiple replicates, as contamination is always possible. In those cases, the contaminated Petri dishes can be discarded in lieu of those in pure culture.

6.4.3.2 The broth phase—growth of pure culture

- Prepare a sterile broth (type of broth dependent upon taxon growth requirements) and add 3–5 mL to a sterile culture tube.
  - We recommend adding broth to culture tubes prior to autoclaving, but this might not be possible if working with presterilized broth.

  ◦ We preferentially use tryptic soy broth since the isolate pictured in Figure 6.3 was grown in a general growth medium.
- Obtain a loop of organism and transfer into the culture tube by depositing it on the inside of the glass (Figure 6.3).
- Cap the tube and wash the specimen with broth medium by slowly tilting the tube.
- Allow the broth to culture at required conditions in a shaker (Figure 6.3d) and incubate until turbid.

6.4.3.3 The storage phase—storage for future use

- Once the broth is turbid, indicating growth (Figure 6.4a), transfer 1.5 mL of culture into a sterile 2 mL storage tube (Figure 6.4b).

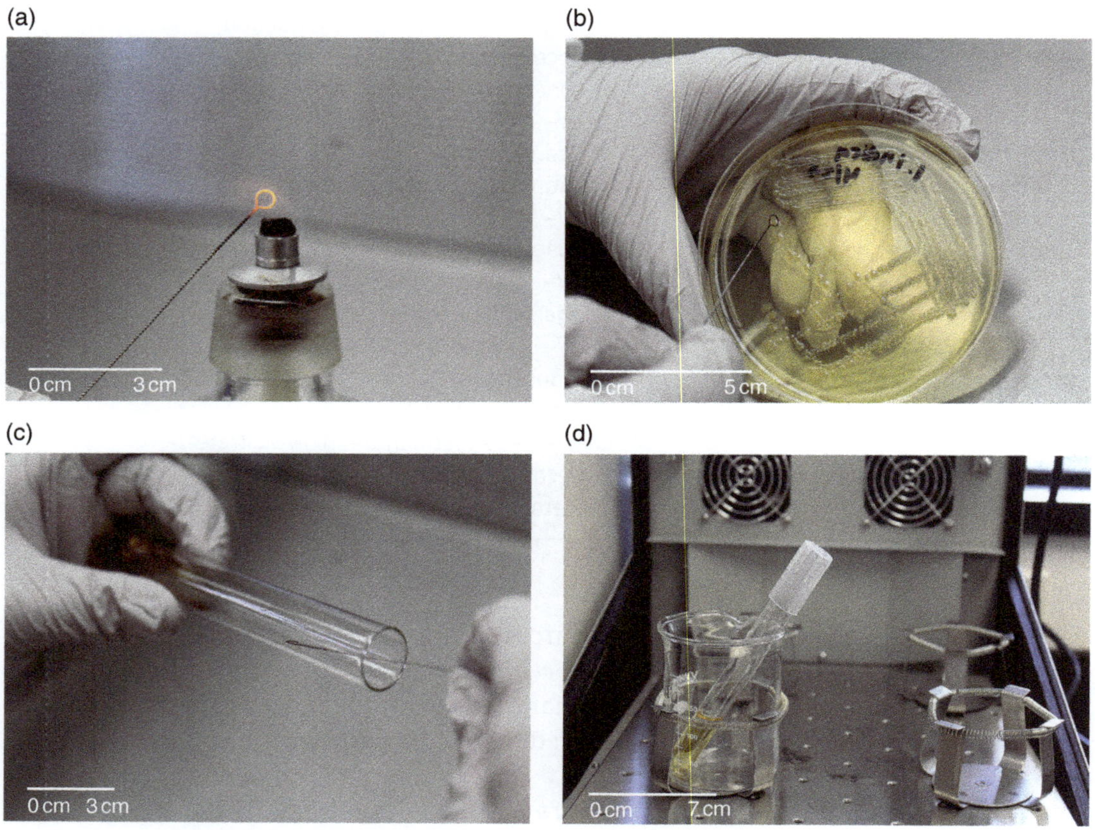

Figure 6.3 Aseptic technique of flaming the collection loop (a) to obtain bacterial cells of *Ignatzschineria indica* for broth culture (b). Cells are scraped on the inside of the culture tube (c) to be washed down into broth medium and placed on incubation shaker (d) until turbid. Turbidity can occur within 24 hours at room temperature

- Transfer 0.5 mL of glycerol into the storage tube (Figure 6.4c). The ratio of culture to glycerol can vary slightly; for this protocol a solution of 25% glycerol was used but 20–80% glycerol can be used.
  - Glycerol acts as a cryoprotectant.
- Vortex the tube for 10 seconds to ensure an even mixture of glycerol and culture (Figure 6.4d).

6.4.3.4 Reculturing stored microorganisms

To bring a taxon from −80°C storage to active pure culture:

- Remove storage tube from cryobox and incubate in water bath (22°C) for 10 minutes.
- Use a sterile loop, or micropipette 50 μL of culture, and place on an agar-based medium for incubation at required conditions.
- Return the storage tube to −80°C storage.

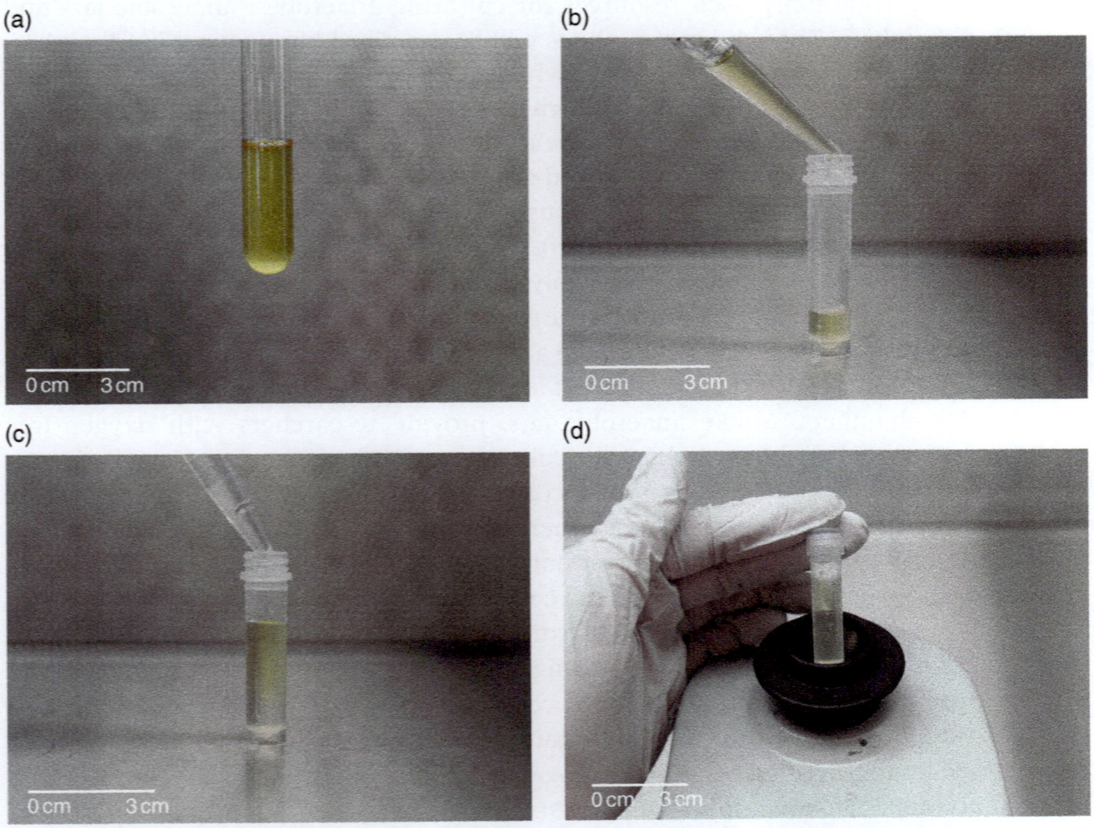

Figure 6.4 Turbidity of broth medium in a culture tube (a) indicates growth and readiness for cryostorage in a 2 mL microfuge tube containing 1.5 mL of broth (b) and 0.5 mL of glycerol (c) mixed thoroughly via vortex (d) and placed in cryobox for storage at −80°C

6.4.4 Anaerobic culture

The protocol for culturing, storing, and reculturing anaerobic organisms follows the same basic steps as the aerobic protocol earlier: (i) the plate phase, (ii) the broth phase, (iii) the storage phase, and (iv) the reculturing of stored organisms. However, the process is not as straightforward with anaerobes as it is with aerobes, for several reasons. Anaerobes (both facultative and obligate) are notoriously fastidious and can require various supplements for growth (Table 6.4). Additionally, the presence of oxygen, by definition, must be inhibited. While facultative anaerobes and even some obligate anaerobes can survive short periods of time in an atmosphere with low oxygen concentrations, each species is different and some true obligate anaerobes cannot be exposed to oxygen. Therefore, it is crucial that when working with anaerobes, all steps, from media preparation to growth of cultures, are conducted anaerobically.

6.4.4.1 The anaerobic growth environment

There are two commonly used resources for culturing anaerobes: anaerobic jars and anaerobic chambers. Though both methods will be described here, anaerobic chambers are recommended over anaerobic jars whenever possible because they allow the investigator to work with cultures in a bench-like fashion and can contain items such as plate readers and liquid handling robots that facilitate growth measurements, high-throughput culturing, colony counting, and other techniques.

The first anaerobic jar, known as a combustion jar, was developed in 1916 (McIntosh *et al.*, 1916) and used heated catalysts to "scrub" oxygen from the jar by adsorbing H_2 and reducing O_2 to H_2O. More recent jars use catalysts active at room temperature, and portable jars, such as the BD GasPak™ (Figure 6.5), are great options for short-term transportation of cultures (e.g., from anaerobic chamber to a core facility or second chamber) and can be used to facilitate the growth of organisms at low oxygen concentrations (2–3%).

Anaerobic chambers, unlike anaerobic jars, provide researchers with oxygen-free enclosures to apply benchtop methods for culturing anaerobes. The anoxic environment is maintained by introducing a gas mix comprised of nitrogen, carbon dioxide, and hydrogen (no more than 5% due to the inherent combustibility of H_2) into the chamber, with catalysts (typically palladium) maintaining the oxygen-free environment. Most anaerobic chambers, such as those produced by Coy Laboratory Products (Grass Lake, MI, USA), are comprised of an enclosed workstation and a vacuum-tight "pass box," a separate but physically connected interchange unit that allows for the passage materials into and from the anaerobic chamber. For vacuum cycle pass boxes, a cycle of vacuum and gas purging replaces the oxygen-containing atmosphere in the pass box with an oxygen-free atmosphere consisting of nitrogen, carbon dioxide, and hydrogen. The pass box that accompanies anaerobic chambers manufactured by Anaerobe Systems (Morgan Hill, CA, USA) does not rely on a vacuum cycle to create the anaerobic environment within the pass box, thus reducing the risk of contamination due to lids or caps flying off of plates, bottles, and containers during the vacuum cycle.

Table 6.4 A non-inclusive list of media designed for the growth of anaerobes

Medium	Notable components/ supplements	Bacterial taxa supported
Brucella blood agar	Sheep blood, vitamin K, and hemin	Nonselective, supports growth of obligate anaerobes
Bacteroides bile esculin agar	Gentamicin, esculin, and hemin	Selective isolation of the *Bacteroides fragilis* group and *Bilophila*
Brain heart infusion agar*	Meat peptones, hemin, and vitamin K	Nonselective, supports the growth of most anaerobes
Campylobacter agar	Trimethoprim, vancomycin, and polymyxin B	Selective for the growth of *Campylobacter jejuni* subsp. *jejuni*; inhibits growth of normal fecal bacteria
Chocolate agar	Cofactor (NAD)	Supports growth of fastidious organisms; cofactor facilitates growth of *Neisseria gonorrhoeae*, *Neisseria meningitidis*, and *Haemophilus influenzae*
Colombia blood agar	Sheep blood, vitamin K, and hemin	Nonselective, supports the growth of many fastidious anaerobes
Fusobacterium selective agar	Josamycin, neomycin, and vancomycin	Selective growth of *Fusobacterium* species
Lactobacillus-MRS agar*	Peptones, yeast extract, glucose, sorbitan monooleate complex, magnesium, sodium acetate, and ammonium citrate	Selective for the growth of *Lactobacillus* found in clinical specimens and dairy and food products
Kanamycin vancomycin laked blood agar	Laked sheep blood, vitamin K, and hemin	Selective isolation of pigmented and other *Prevotella* species and *Bacteroides* species
Phenylethyl alcohol blood agar	Phenylethyl alcohol	Inhibits facultative Gram-negative rods and swarming activity of some Clostridia species
Tryptic soy blood agar	Hemin and vitamin K	Supports growth of many fastidious anaerobes
Chopped meat broth	Chopped meat	Supports growth of most non-spore forming anaerobes; also appropriate as a maintenance medium
Peptone yeast extract broth	Hemin and vitamin K	Facilitates the growth of *Prevotella*, *Porphyromonas*, and the *Bacteroides fragilis* group; serves as a base medium for several additives that can be used to test nutrient requirements of anaerobes
Thioglycolate broth	Vitamin K, hemin, and marble chip (calcium carbonate)	Supports growth of most anaerobes

* Indicates solid medium also available as broth.

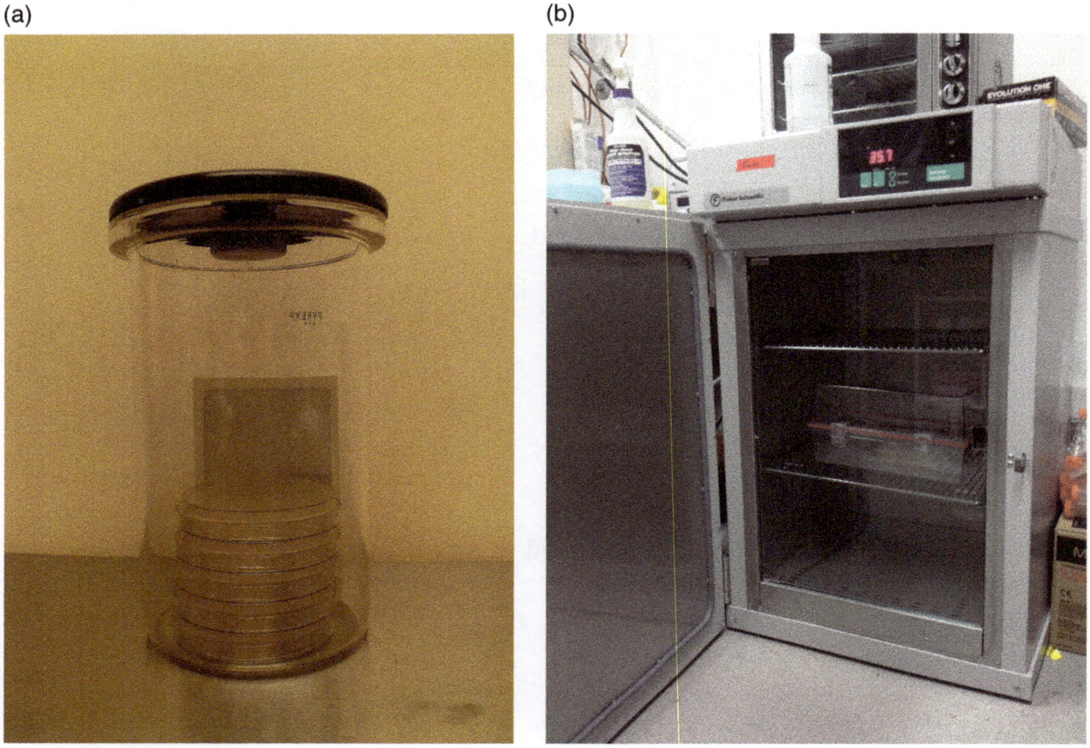

Figure 6.5 Portable anaerobic jars are great options for incubation under anoxic and microaerophilic conditions or for short-term transportation of anaerobes. Pictured are two versions of the BD GasPak™ (Becton, Dickinson and Company, Franklin Lakes, NJ, USA) incubating at room temperature (a) and in a 37°C incubator (b). The anaerobic jar in (a) contains a sachet of activated charcoal to remove oxygen and five Petri dishes (9 cm diameter, 1.5 cm height)

An anaerobic chamber (Figure 6.6) can be soft-sided (constructed of vinyl) or rigid (Plexiglas, steel, or aluminum). Rigid chambers have the advantage over vinyl chambers in that they support gloveless access to the internal workspace, allowing the investigator to use his/her or preferred laboratory gloves in lieu of the bulky black rubber gloves present on most anaerobic chamber models. This also helps reduce contamination by improving dexterity and precision of the investigator. Additionally, rigid chambers are more resistant to puncture than their vinyl counterparts. Nevertheless, leaks are more easily detected in vinyl chambers, and vinyl chambers often allow more range of motion than their rigid counterparts.

6.4.4.2 Anaerobic sterile technique

Because flame cannot be used inside anaerobic chambers, the typical flame sterilization method cannot be used when working with anaerobic organisms. Therefore, disposable items such as sterile plastic loops are recommended for preparing cultures in the

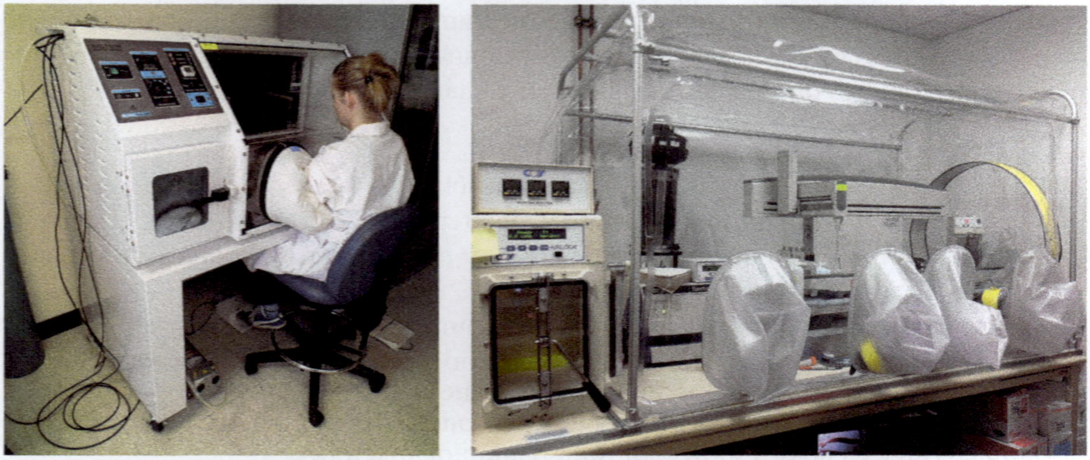

Figure 6.6 Anaerobic chambers come in a range of specifications including rigid (a) and soft-sided (b). These chambers can be enhanced with the inclusion of instrumentation. For example, the large Coy Laboratory Products vinyl chamber (b) contains a plate reader and liquid handling robot to facilitate high-throughput culturing and interrogation of cultures. The pass box is located on the left of the both chambers, which allows for the movement of samples between aerobic and anaerobic conditions

anaerobic chamber. Additionally, sterilized materials such as loops, needles, glass jars, tubes, and pipette tips can be introduced into the chamber through the pass box and used to prepare cultures with little risk of contamination. Anaerobic chambers and the tools inside them should also be regularly wiped down with a cleansing solution such as benzalkonium chloride to ensure the inside environment is clean.

6.4.4.3 Preparation of growth medium

The presence of even small amounts of oxygen can prevent the growth of obligate anaerobes; therefore, it is crucial that media are either prepared in an anaerobic environment or made anaerobic after preparation. Gas impermeable stoppers can be used to close off glass bottles and tubes, and black butyl rubber is commonly used for this purpose.

While some manufacturers, such as Anaerobe Systems, sell media prepared in a completely anaerobic environment, in many cases, the investigator will prepare their own medium. The easiest way to prepare anaerobic liquid medium is to make the medium as usual and then introduce it into the anaerobic chamber with caps loosened. It will take at least 24 hours for the medium to become completely anaerobic. To speed up this process, the headspace of the tube or bottle can be evacuated with nitrogen, and the addition of reducing agents such as cysteine hydrochloride is also recommended. Resazurin, a redox indicator, can be added to the medium to act as an indicator of the presence of oxygen (the resazurin will turn pink if oxygen is present). Like liquid medium, solid medium can also be prepared aseptically outside of the anaerobic chamber and then transferred to the chamber 24 hours before use to ensure the medium is fully anaerobic. Conversely, the medium can be prepared outside of the chamber and then transferred into the chamber where plates can be poured and solidified in the anaerobic environment.

6.4.4.4 The plate phase—anaerobic isolation

The culturing of anaerobes for isolation on plates follows virtually the same steps as culturing aerobically, with the exception that all steps are performed in the anaerobic chamber (or on the bench top, with the plates later moved to an anaerobic jar). The protocol is as follows:

- Culture for isolation onto an agar-based medium (Table 6.4) appropriate for your target taxa and incubate anaerobically until colonies are visible to the naked eye.
  - Most anaerobes are fastidious, and therefore, the growth rate of anaerobes will often be slower than the growth rate of aerobes. The Wadsworth-KTL Anaerobic Bacteriology Manual is an excellent resource for the isolation of anaerobes.
- Prepare multiple replicates, as contamination is always possible.

6.4.4.5 The broth phase—growth of anaerobic pure culture

The culturing of anaerobes in a liquid medium (broth) follows virtually the same steps as culturing aerobically, with the exception that all steps are performed in the anaerobic chamber. The protocol is as follows:

- Prepare anaerobic sterile broth as described earlier (type of broth dependent upon taxon growth requirements), and add 5 mL to a sterile glass culture tube.
- Using a plastic disposable loop, obtain a loop of bacterial colony from an agar plate, and transfer the colony into the culture tube by depositing it on the inside of the glass and tilting the tube so the medium washes over the colony.
- Incubate the liquid cultures anaerobically at the appropriate temperature until the medium is turbid, indicating bacterial growth.

6.4.5 Preparing freezer stocks of pure culture

Before the preparation of freezer stocks, all required materials (pipette tips, culture tubes, sterile glycerol) should be moved into the anaerobic chamber at least 24 hours prior to use.

- Once the broth is turbid, transfer the culture and sterile glycerol to a sterile polystyrene snap cap vial so that the glycerol concentration is at 20%. Polystyrene is used because oxygen is unable to penetrate it, ensuring that the cultures will remain anaerobic once they are removed from the chamber and placed into the freezer. Polystyrene vials come in several sizes, so the amount of culture and glycerol transferred will depend on the size of the vial used.
- Vortex the tube for 10 seconds to ensure an even mixture of glycerol and culture.
- Place the culture in a cryobox and move immediately into a −80°C freezer. If the freezer is located in a long distance from the anaerobic chamber, the culture may be placed on dry ice and transported to the freezer.

6.4.6 Reculturing stored microorganisms

To bring an anaerobic isolate from −80°C storage:

- Remove polystyrene vial from the cryobox and move it into the anaerobic chamber; thaw on ice or at room temperature.
- Using a sterile plastic loop, obtain a loopful of culture, or pipette 50 μL of culture, and place on an appropriate (based on the taxon) agar-based solid medium for growth under anaerobic conditions at the required temperature.
- Anaerobic freezer stocks should never be used more than once! Due to the solubility of oxygen in water, oxygen may concentrate in a culture during thawing and refreezing, thereby ruining the culture.

6.5 Conclusions

The goal of this chapter is to introduce the reader to protocols related to the culture of microorganisms and briefly describe protocols for their long-term storage to facilitate additional analyses in the future. Perhaps most importantly, this chapter attempted to demonstrate the value of culturing techniques as a fundamental practice in not only forensic application but also microbiology as an entire field. We encourage forensic microbiologists to use culture-dependent techniques in conjunction with sequencing techniques when investigating a microbial community of interest. Furthermore, the protocols for long-term storage are included in an attempt to encourage the construction and maintenance of a microbial reference collection. These individual taxa can be brought out of culture and used in competition studies (Zheng *et al.*, 2013) or simply studied in greater detail.

Acknowledgments

The authors kindly thank the editors for helpful comments on the preparation of this manuscript.

References

Atlas, R.M. and Bartha, R. (1997) *Microbial Ecology: Fundamentals and Applications*, 4th edn, Benjamin Cummings, San Francisco, CA.

Brown, A.E. (2012) *Benson's Microbiological Applications: Laboratory Manual in General Microbiology*, 12th edn, McGraw Hill, New York, NY.

Cappucino, J.G. and Sherman, N. (2011) *Microbiology: A Laboratory Manual*, 9th edn, Pearson, Boston, MA.

Chun, L., Miguel, M., Junkins, E.N. *et al.* (2015) An initial investigation in the ecology of culturable postmortem bacteria. *Science and Justice*, **55**, 394–401.

Finley, S.J., Pechal, J.L., Benbow, M.E. *et al.* (2016) Microbial signatures of cadaver gravesoil during decomposition. *Microbial Ecology*, **71**, 524–529.

Griffin, A.R., Junkins, E.N., Carter, D.O. and Jans, M. (2015) *A protocol for the collection and culture of microbes from the external surfaces of bone. Proceedings of 67th Annual Meeting of the American Academy of Forensic Sciences*, Orlando, FL, February 16–21, 89.

Herbert, D., Elsworth, R. and Telling, R.C. (1956) The continuous culture of bacteria: a theoretical and experimental study. *Journal of General Microbiology*, **14**, 601–622.

Hooke, R. (1665) *Micrographia*, Royal Society, London.

Hyde, E.R., Haarmann, D.P., Lynne, A.M. *et al.* (2013) The living dead: bacterial community structure of a cadaver at the onset and end of the bloat stage of decomposition. *PLoS One*, **8**, e77733.

Hyde, E.R., Haarmann, D.P., Petrosino, J.F. *et al.* (2015) Initial insights into bacterial succession during human decomposition. *International Journal of Legal Medicine*, **129**, 661–671.

Junkins, E.N. and Carter, D.O. (2015) *Ecology of decomposition: a chemical and biological profile of the maggot mass. Proceedings of 67th Annual Meeting of the American Academy of Forensic Sciences*, Orlando, FL, February 16–21, 913.

Kirk, P.L. (1953) *Crime Investigation: Physical Evidence and the Police Laboratory*, John Wiley & Sons, Inc, New York, NY.

Koch, R. (1883) The new methods for studying the microcosm of soil, air, and water. *Deutsches Ärzteblatt*, **137**, 244–250.

Lax, S., Smith, D.P., Hampton-Marcell, J. *et al.* (2014) Longitudinal analysis of microbial interaction between humans and the indoor environment. *Science*, **345**, 1048–1052.

Leboffe, M.J. and Pierce, B.E. (2010) *Microbiology: Laboratory Theory and Application*, 3rd edn, Morton Publishing Company, Englewood, CO.

Leewenhoeck, A. (1684) An abstract of a letter from Mr. Anthony Leewenhoeck at Delft, Dated Sep. 17. 1683. Containing some microscopical observations, about animals in the scurf of the teeth, the substance called worms in the nose, the cuticula consisting of scales. *Philosophical Transactions of the Royal Society*, **14**, 568–574.

Leighton, T. and Murch, R. (2011) Biorepositories and their foundations: microbial forensic considerations, in *Microbial Forensics*, 2nd edn (eds B. Budowle, S.E. Schutzer, R.G. Breeze, P.S. Keim and S.A. Morse), Academic Press, Burlington, MA, pp. 581–601.

Madigan, M., Martinko, J., Stahl, D. and Clark, D. (2012) *Brock Biology of Microorganisms*, 13th edn, Pearson, San Francisco, CA.

McIntosh, J., Fildes, P. and Bullock, W. (1916) A new apparatus for the isolation and cultivation of anaerobic microorganisms. *Lancet*, **187**, 768–770.

Metcalf, J.L., Xu, Z.Z., Weiss, S. *et al.* (2016) Microbial community assembly and metabolic function during mammalian corpse decomposition. *Science*, **351**, 158–162.

Nichols, D. (2007) Cultivation gives context to the microbial ecologist. *FEMS Microbiology Ecology*, **60**, 351–357.

Pechal, J.L., Crippen, T.L., Tarone, A.M. *et al.* (2013) Microbial community functional change during vertebrate carrion decomposition. *PLoS One*, **8**, e79035.

Pechal, J.L., Crippen, T.L., Benbow, M.E. *et al.* (2014) The potential use of bacterial community succession in forensics as described by high throughput metagenomic sequencing. *International Journal of Legal Medicine*, **128**, 193–205.

Richey, D.G. and Goerhing, C. (1918) Studies on bacteriemias in the agonal period. *Journal of Medical Research*, **38**, 421–447.

Ritz, K. (2007) The plate debate: cultivable communities have no utility in contemporary environmental microbial ecology. *FEMS Microbiology Ecology*, **60**, 358–362.

Shade, A., Hogan, C.S., Klimowicz, A.K. *et al.* (2012) Culturing captures members of the soil rare biosphere. *Environmental Microbiology*, **14**, 2247–2252.

Sharma, S.K., Bailey, H. and Melling, J. (2011) Sampling for microbial forensic investigations, in *Microbial Forensics*, 2nd edn (eds B. Budowle, S.E. Schutzer, R.G. Breeze, P.S. Keim and S.A. Morse), Academic Press, Burlington, MA, pp. 393–404.

Slonczewski, J.L. and Foster, J.W. (2011) *Microbiology: An Evolving Science*, 2nd edn, W. W. Norton and Company, New York, NY.

Smith, J.A.L. (2011) Collection and preservation of microbial forensic samples, in *Microbial Forensics* (eds B. Budowle, S.E. Schutzer, R.G. Breeze, P.S. Keim and S.A. Morse), Academic Press, Burlington, MA, pp. 379–392.

Stefanowicz, A. (2006) The Biolog plates technique as a tool in ecological studies of microbial communities. *Polish Journal of Environmental Studies*, **15**, 669–676.

Tyndall, J. (1877) Further researches on the deportment and vital persistence of putrefactive and infective organisms from a physical point of view. *Philosophical Transactions of the Royal Society of London*, **167**, 149–206.

United States Government Accountability Office (2005) Anthrax detection: agencies need to validate sampling activities in order to increase confidence in negative results. United States Government Accountability Office, Washington, DC.

White, F.W. (1899) Cultures from the blood in septicaemia, pneumonia, meningitis, and chronic diseases. *Journal of Experimental Medicine*, **4**, 425–450.

Zheng, L., Crippen, T.L., Holmes, L. *et al.* (2013) Bacteria mediate oviposition by the black soldier fly, *Hermetia illucens* (L.), (Diptera: Stratiomyidae). *Scientific Reports*, **3**, 2563.

CHAPTER 7

Clinical microbiology and virology in the context of the autopsy

Elisabeth J. Ridgway, Bala M. Subramanian and Mohammad Raza

Department of Medical Microbiology and Virology, Sheffield Teaching Hospitals NHS Foundation Trust, Sheffield, UK

7.1 Introduction

Microbiological investigation is an important part of the autopsy process. Infection may be present, or suspected, antemortem in which case appropriate investigation can help to establish its role in the cause or manner of death. Alternatively, it may be an unexpected finding, in which case the diagnosis will provide important information to the family of the deceased. It may also have significant implications for public health and may assist in managing individual case contacts as well as the wider community, by documenting patterns of infection, contributing to outbreak control, and identifying novel infections and bioterrorist threats. Unfortunately, partly under the misapprehension that modern imaging and diagnostic tests provide diagnostic precision, autopsy rates have suffered a progressive decline in most countries over recent decades; this decline has important implications for the accuracy of death certification (Sarode *et al.*, 1993; Bonds *et al.*, 2003; Gibson *et al.*, 2004; Roulson *et al.*, 2005; Wilson, 2006; Winters *et al.*, 2012).

In both adults and children, many studies have identified that a considerable proportion of deaths have significant, previously unidentified conditions revealed at autopsy. In one meta-analysis, approximately 50% of autopsies revealed at least one clinically unsuspected finding (range 45–76%) (Roulson *et al.*, 2005). A systematic review of 53 autopsy series between 1966 and 2002 found that major diagnostic errors were clinically undetected in at least 8.4% of cases and about 50% of these patients might have survived to discharge had misdiagnosis not occurred (Shojania *et al.*, 2003). Unexpected infections have been consistently identified at postmortem examination (Stevanovic *et al.*, 1986; Landefeld *et al.*, 1988; Gibson *et al.*, 2004; Newton *et al.*, 2004; Roulson *et al.*, 2005; Spiliopoulou *et al.*, 2005; Wilson, 2006). A review of autopsies performed between 1996

Forensic Microbiology, First Edition. Edited by David O. Carter, Jeffery K. Tomberlin, M. Eric Benbow and Jessica L. Metcalf.

and 2001 found that 75.3% of adults and 48% of fetuses/neonates had an infection identified at autopsy, of which 43% of adults and 58% of infants were unsuspected before death (Bonds *et al.*, 2003). Even common infections such as pneumonia, sepsis, meningitis, peritonitis, and infective endocarditis (IE) fall into the top 10 discrepant diagnoses (Mercer and Talbot, 1985; Gibson *et al.*, 2004; Spiliopoulou *et al.*, 2005). Infection is very commonly overlooked where it coexists with another condition—over 80% of cases in one series (Stevanovic *et al.*, 1986)—and even when it has been diagnosed during life, autopsy reveals that disseminated infection, such as bacterial endocarditis and acute pyelonephritis, is consistently underdiagnosed (Cameron and McGoogan, 1981; Thorn *et al.*, 2010; Tajiri *et al.*, 2012). In developed countries, where infectious deaths are relatively uncommon (Heron, 2013), clinical inexperience may play a part in failure of antemortem diagnosis. However, even in countries such as India, where infectious disease remains a common cause of death (Centers for Disease Control and Prevention, 2013a) and clinical suspicion is higher, infection is frequently only diagnosed at autopsy (Sarode *et al.*, 1993).

There is clearly a need to be suspicious of infection as a contributor to, or cause of, death, and this chapter will review how the pathologist should approach the autopsy in order to obtain the most appropriate and best quality samples for microbiological investigation. Having taken samples, consideration will be given to the diagnostic methods that are available and how the results of investigations should be interpreted. Finally, the clinical conditions where microbiological investigations can play an important role will be discussed.

7.2 The historical view of autopsy microbiology

The relevance and value of postmortem bacteriology has been a contentious subject for over a hundred years. In the early twentieth century, following Fredette's (1916) report of the utility of postmortem blood cultures and, subsequently, Giordano and Barnes (1922) and Hunt *et al.* (1929) studies on tissue cultures, the practice of postmortem bacteriology enjoyed considerable support. However, although postmortem examination could reveal the cause of death, the significance of the microorganisms recovered at autopsy was difficult to clarify and in the following decades, concerns about the reliability of culture results led to a marked decline of interest in this field. Eventually, the prevailing view was that postmortem microbiological examination was unhelpful and, indeed, misleading. This was reinforced by a study by Wilson *et al.* (1993), which compared positive postmortem blood cultures with the results of antemortem blood cultures in 111 patients. The results of antemortem and postmortem blood cultures did agree in 35% of cases, but in the remainder the results were inconsistent, and cultures were often contaminated. The authors concluded that postmortem blood culture results should be regarded with a high index of suspicion, rarely, if ever, provided new, useful information and would only be worthwhile in situations where antemortem cultures were not available.

There are two main hypotheses that may explain the presence of microorganisms in postmortem blood and tissue samples. The first was proposed by Fredette (1916), who suggested that a large proportion of the positive cultures recovered at autopsy were due to "agonal invasion." He hypothesized that, *before* death, cell structure, function, and viability decline allowing endogenous microorganisms to invade the bloodstream as a preterminal event. Recent evidence to support this view comes from a study of tissue donors, which found that the highest positivity rate of blood cultures was seen in the first 5 hours after death and that there was no significant difference in positivity between the first 24 hours of death and thereafter (Saegeman *et al.*, 2009).

The alternative hypothesis was put forward by Carpenter and Wilkins (1964) and proposed that endogenous bacteria multiply and migrate throughout the body only after death, a phenomenon termed "postmortem invasion." They provided evidence to support the theory when they retrospectively studied 2033 autopsies and found that the proportion of positive cardiac blood cultures increased linearly with the interval from death to the time of blood culture collection. Similarly, the positivity rate of postmortem lung cultures also increased. Other clinical observations, *in vitro* models and animal studies have provided support for the theory, (Burn, 1934; Smith *et al.*, 1975; Kellerman *et al.*, 1976) and the evidence for postmortem migration is generally much stronger than for agonal invasion. Nevertheless, the hypothesis has not been easy to substantiate. One would expect that the number of microorganisms per milliliter or milligram of sample would increase with the duration of the postmortem interval (PMI), but culture-based studies have failed to confirm this (Wilson *et al.*, 1993; Hove and Pencil, 1998; Goldwater, 2009; Lobmaier *et al.*, 2009; Prtak *et al.*, 2010; Weber *et al.*, 2010b). It is only with the advent of molecular techniques, and the use of real-time quantitative PCR, that it has become clear that sterility declines with time after death in a range of tissues and that there are changes in the numbers and species of the invading organisms (Tuomisto *et al.*, 2013).

Perhaps more importantly, given that it is largely preventable, is that samples may become contaminated by the pathologist during collection at the postmortem examination. This is particularly important when sampling sites that are relatively protected from bacterial invasion, such as the brain, meninges, and pericardium, where contamination has the potential to generate misleading results. Potentially pathogenic species are commonly mixed with commensals in postmortem blood and tissue culture samples (Wilson *et al.*, 1993; Tuomisto *et al.*, 2013), and, as it can be difficult to attribute causation in this situation, any steps to minimize contaminants are worth the effort.

Several approaches to reducing bacterial contamination have been attempted. Some pathologists have used modified surgical techniques to good effect. At one extreme, using a surgically clean postmortem room, scrubbing the body for 5 minutes with povidone–iodine with a 5-minute rinse with 70% alcohol, routine scrubbing, gowning, gloving, and masking of autopsy staff and collecting portions of each organ using a separate set of instruments for each sample, up to 45% of lung and 75% of kidney specimens may be retrieved sterile (O'Toole *et al.*, 1965; Minckler *et al.*, 1966). However, this complex technique is impractical for routine use and may be unnecessary to achieve reliable microbiological culture results. de Jongh *et al.* (1968) obtained similar results using much less elaborate methods.

Despite the historical controversies, well-taken autopsy specimens may be of immeasurable value both in confirming a presumptive antemortem infective diagnosis and in highlighting an obscure or unsuspected organism as the cause of a patient's illness (Moar and Millar, 1984; du Moulin and Patterson, 1985). With the advent of molecular diagnostic techniques, dependence on culture, with all its limitations, is no longer required. This allows for the possibility of much greater sensitivity and, perhaps, for microbiology to make a greater contribution to the postmortem examination.

7.3 Which samples should you collect and how?

In order to maximize the yield of useful information from autopsy microbiological investigation, a collaborative approach between the pathologist and microbiologist is needed. Microbiologists will ensure that the correct specimens are collected, that they are appropriately processed, and that the clinical and epidemiological significance of subsequent microbiological results is accurately assessed.

To obtain good quality microbiological specimens during routine autopsies, a standardized sequence may be followed (Table 7.1).

There is some debate about the influence of time after death on culture results (Tsokos and Puschel, 2001), but this largely centers around the effect of time on the postmortem invasion of organisms. However, this is only one side of the equation, and prompt sample collection is particularly important if organisms with complex nutritional requirements, or those present in low numbers, are to be retrieved. Studies in guinea pigs indicate that if bacterial counts are low at the time of death, organisms are eliminated from blood and muscle to the point of sterility at 24 hours. With higher counts, rapid multiplication begins within a few hours of death, especially at higher temperatures (Gill and Penney, 1979). This could yield misleading culture results or could compound the difficulty of retrieving a true pathogen present in low numbers as it will be competed out by fast growers. However, there is no doubt that robust pathogens can survive after the death of the host, as transmission of organisms such as *Streptococcus pyogenes* (Hawkey *et al.*, 1980), *Mycobacterium tuberculosis* (Lundgren *et al.*, 1987), and varicella zoster virus (Paul and Jacob, 2006) to postmortem room staff has been well documented. Fastidious organisms such as *Haemophilus influenzae, Neisseria meningitidis,* and *Streptococcus pneumoniae* can also be retrieved—although it should be noted that in one series the mean time from death to autopsy was only 5 hours (Sadler, 1998). The use of molecular tests, which do not rely on the presence of viable organisms for detection, offers the possibility of identifying pathogens well beyond the time when they might be expected to survive.

Table 7.1 Standard procedure to ensure optimal quality in microbiological sampling at autopsy

Procedure	Rationale
The body should be placed in a sealed body bag and refrigerated until the autopsy is performed	Provided the body remains refrigerated at 4–10°C postmortem invasion by microorganisms can be delayed from 15 to 48 hours; perhaps up to 96 hours in the case of anaerobes
Handling and manipulation of the body should be kept to a minimum	Limiting body movement reduces passive recirculation of blood from heavily colonized areas such as the gut vasculature. Consequently the potential for false positive blood cultures is decreased
The autopsy should be performed and samples taken, as soon as possible, ideally within 24 hours of death	Early sampling optimizes the microbial yield
If infection is suspected as a cause of death, the case should be discussed with a microbiologist *before* commencing the autopsy	This ensures that appropriate specimens are collected and the microbiology laboratory is alerted. Guidance exists on samples to be taken in the postmortem investigation of infection in the setting of sudden unexpected death and suspected bioterrorism
The necessary sterile sample containers should be assembled before commencing the autopsy	Sterile screw-topped containers and special transport media such as blood culture sets
Allow sufficient scalpel blades, forceps, and other items so that a separate set of instruments can be used for each sample collected	
Cultures should be taken at the beginning of the postmortem examination, before manipulation of the organs, and evisceration	Once the body has been opened, the fluid on the organ surfaces and in the body cavities becomes heavily contaminated
Blood and fluid samples should be collected first, followed by tissue samples	
If infection is unexpectedly identified, specimens should be taken with as much care as possible	This should be recorded on the specimen request form. It is likely that some degree of contamination will already have occurred and it will need to be considered when interpreting culture results
Collect samples from at least two sampling sites	Multiple samples increase the chances of identifying a significant pathogen, and consistent positive results from multiple, well-taken samples from different sites make them very likely to reflect infection, rather than contamination. It is almost impossible to attribute any significance to a positive culture from a single site

Table 7.1 (Continued)

Procedure	Rationale
The degree of care taken to avoid contamination during sampling is a key factor in obtaining useful results	Contamination may be avoided by careful attention to technique and aseptic practice
Once taken, samples should be sent to the microbiology laboratory with the minimum delay	Overgrowth or death of organisms in transit is undesirable. Refrigeration will kill some fragile organisms such as *Streptococcus pneumoniae* and *Neisseria gonorrhoeae*. For molecular and antigen tests, the timing is less critical and these may be useful where delays are inevitable. However, RNA is relatively labile which may be relevant if it is the target for a molecular test
Record the death to autopsy time (in hours) on the request form and include the sequence and method of sample collection	This information can be helpful when interpreting test results
Full clinical details and a summary of the autopsy finding (e.g., evidence of pneumonia on gross examination) should accompany the specimens	This allows the microbiologist to make the correct interpretation of any positive results, for example, "Did the patient die *of* influenza or *with* influenza?" It allows addition of additional investigations to the standard test battery

See also Moar and Millar (1984); du Moulin and Patterson (1985); and Fernández-Rodríguez *et al.* (2015).

7.3.1 Blood

Ideally, blood cultures should be collected as soon as possible after death and before the body is placed in refrigerated storage, but failing this, they should be collected before the autopsy is commenced. A scrupulous aseptic technique is essential in both taking the sample and inoculating the bottles of blood culture broth. Blood cultures are essentially an enrichment culture that aims to amplify very small numbers of organisms. Consequently, even small numbers of organisms will yield an apparently significant result, and blood cultures are very good at growing contaminants if the sampling technique is substandard. The classical method of autopsy blood culture is to take heart blood during the procedure—either from the right atrium or from the tip of the heart after pointing the apex toward the ceiling. In either case, even after searing the entry point on surface of the heart with a heated spatula, or application of 70% alcohol, contamination is very common.

In contrast, if blood is collected before the autopsy begins, the contamination rate can be reduced dramatically (Hove and Pencil, 1998). One method is to use a sterile needle and syringe to take blood from a large vein. Sampling directly from the heart via the fourth intercostal space is another option, but that could lead to internal lesions, particularly if repeated attempts are made. An aseptic technique should be used and sterile gloves worn. Disinfection of the skin is essential, and the intended puncture site should first be cleaned with a 70% alcohol preparation (with or without povidone–iodine or chlorhexidine), the excess wiped off, and the skin allowed to dry. Blood can be collected from the neck or subclavian veins; the femoral or intra-abdominal veins should not be used as blood collected from sites below the waist is much more likely to be contaminated. The recommended volume of blood is usually 6–10 mL/pair of bottles but will depend on the blood culture system in use, and it should be inoculated directly into the culture bottles (rather than into another sterile container) using the same aseptic technique as for clinical samples. If only a small volume of blood is available, just the aerobic bottle of the culture set should be used. Changing needles before inoculating the blood culture bottles, or between them, does not significantly reduce contamination rates and is inadvisable because of the risk of inoculation injury (Leisure *et al.*, 1990).

Blood for serological investigations, molecular tests, and antigen detection may also be collected before the autopsy, but, as these techniques are much less susceptible to contamination, the site of blood collection is less critical, and samples may be collected later in the procedure. Serological testing is typically performed on serum, which is obtained by spinning clotted whole blood. Sometimes, especially after delayed collection, autopsy "blood" samples consist almost entirely of clot, in which case no serum is obtained on centrifugation—regardless of the blood tube/preservative in which the sample is sent. Although modern analyzers have a degree of compensation, clots and heavily hemolyzed samples generally are not testable. Even if the technical difficulties are overcome, these substandard samples often give low-level false positive results, and confirmation of these is very difficult. If a clot is the only sample available, careful consideration should be given to the need for the test. A false positive blood-borne virus screen, for instance, may unnecessarily invoke contact tracing and public health involvement and create a difficult situation with regard to the family/next of kin.

7.3.2 Cerebrospinal fluid

Cerebrospinal fluid (CSF) should be collected by cisternal or lumbar puncture (L4/5). It is best to obtain lumbar puncture specimen before the postmortem and in the same manner as used in the clinical setting. As for blood cultures, the skin should be cleaned with a 70% alcohol preparation (with or without povidone–iodine or chlorhexidine), the excess wiped off, and the skin allowed to dry. Using an aseptic technique and wearing sterile gloves, CSF should be collected using a sterile needle and syringe and collected in a sterile screw-topped container. In cases of meningitis, samples of pus for Gram stain and culture may be taken directly from the brain surface after opening of the head cavity (Tsokos and Puschel, 2001).

7.3.3 Tissue, pus, and fluids

Depending on the suspected diagnosis or autopsy findings, a variety of different tissues may be collected for microbiological examination (e.g., heart valve, lung, meninges, brain, spleen, abscesses). Samples should be collected immediately after entering the body and may be collected *in situ* or from an excised organ. All possible steps should be taken to minimize contamination, particularly once an area has been identified for sampling. This may entail elevating the organ to prevent contaminated fluids from flowing across it. Organs and tissues deep in the body cavity, such as the paraspinal or retropharyngeal areas, may be a challenge because of the tendency for fluid to accumulate at the site.

Having identified a suitable site, it should be prepared before sampling. A section of the external surface of the organ (~2 cm × 2 cm) should be sterilized by searing to dryness with a flat-faced soldering iron or a heated spatula. Alcohol or iodine cannot be used for this purpose as it may have a persistent antibacterial effect that could interfere with culture. Using a sterile scalpel and forceps, about 1 cm^3 of tissue (if possible) should be removed and placed in a sterile container of appropriate size. Small tissue fragments have a tendency to dry out in transit to the laboratory, and suspending them in a small volume of sterile saline may prevent this. A different set of instruments should be used for each site/culture. Under no circumstances should containers that have held formalin be used, even if the formalin is discarded prior to use. As an alternative to removing a piece of tissue, a sterile swab can be forced through the seared area or fluid aspirated using a sterile needle and syringe. If it is difficult to pierce the seared capsule with the swab, it may help to make a small cut with a sterile scalpel first. The edges of the opening can then be pulled apart by traction on more peripheral sections of the organ, and the swab pushed through the opening without touching the edges. For virological investigations, tissue samples should be submitted in sterile containers in saline, without formalin to maximize the DNA/RNA yield.

Samples of pus may be aspirated with a sterile needle and syringe and placed in a suitably sized airtight sample container without delay. Avoid an oversized container with a large headspace as anaerobic organisms die rapidly on exposure to air. A wide bore needle or irrigation syringe may be required to aspirate thick pus. If an abscess is present, part of the abscess wall should be sent for culture, as well as the contents. Attention should also be given to lymph nodes draining infected areas and samples sent as appropriate. Swabs of pus should only be taken as a last resort as the amount of material they collect is relatively small, and it is difficult to undertake different culture methods (e.g., routine, fungal, mycobacterial, enrichment cultures) on a single swab.

Other fluids (e.g., ascites, joint fluid, pleural fluid) may be collected using an aseptic technique as for blood cultures. An aliquot may be inoculated directly into blood culture bottles, but it is essential that an additional sample is sent in a sterile screw-topped container so that Gram and Ziehl–Neelsen staining can be performed, as well as direct culture, specialized culture (e.g., for mycobacteria), and molecular analysis if required.

7.3.4 Urine and bowel contents/feces

Urine cultures are of limited value. Positive cultures of little clinical significance occur at all ages, but increase as age advances and with comorbidity, and are universal in those with long-term indwelling catheters (Nicolle *et al.*, 2005). If urine samples are taken, the urine should be aspirated directly from the bladder using an aseptic technique and sent promptly to the laboratory in a sterile screw-topped container. Depending on the clinical presentation, urinary antigen tests, for *Legionella*, for example, may be useful.

Bowel contents may be collected from the lumen of the gut with a spatula and placed in a sterile screw-topped container or using a clinical fecal collection container that has a spoon built in to the lid. If the sample is very liquid, a swab may be taken, although this is less satisfactory because of the small volume of sample. An aseptic technique is not required; the bowel contents are not sterile, and culture for enteric pathogens is highly selective, so contamination of the sample is not a concern.

7.4 Which methods are available for the diagnosis of infection?

The range of techniques available for the microbiological investigation of clinical and autopsy specimens is diverse (Table 7.2). Each technique varies in its usefulness depending on the sample type, the conditions under which it was collected, and the clinical condition under investigation. A standardized set of microbiological investigations has recently been proposed for autopsy samples from cases of sudden unexpected death from infancy to adulthood and in suspected bioterrorism (Fernández-Rodríguez *et al.*, 2015).

Staining techniques are likely to be undertaken in both the microbiology and histopathology departments on any samples they receive. The results of stained sections or smears may be so distinctive that a definitive diagnosis is possible, or they may provide corroborating evidence to aid in the interpretation of cultures and other tests. The value and sensitivity of staining techniques is hugely dependent on the stains selected, the expertise of the microscopist, and the time spent on assessing the films and sections (Woods and Walker, 1996).

Traditional culture methods still have an important place in routine investigation. Bacterial and fungal culture is readily available, cheap, and "open" (i.e., asks the question "is infection present?" rather than "is organism X present or not?") and generates an isolate for definitive identification, sensitivity testing, and typing—which may have clinical, epidemiological, and public health significance. Methods have refined over recent years; for example, continuously monitored liquid culture for *M. tuberculosis* is now standard, and highly specific chromogenic agar, with enzyme substrates that release colored dyes on hydrolysis so that pathogens are readily differentiated from commensal flora, is in widespread use (Perry and Freydiere, 2007).

Table 7.2 Techniques routinely available for microbiological analysis of autopsy specimens

Method		Examples and comments on use
Microscopy	Light	*Wet film:* fungal elements, helminth ova, and cysts *Stained film:* for example, Gram stain, Ziehl–Neelsen, calcofluor white
	Fluorescent	*Stained films:* for example, cryptosporidia, mycobacteria *Immunofluorescence:* for example, *Pneumocystis carinii*, respiratory viruses
	Electron	Predominantly virus detection, for example, SRSV, herpesviruses, enteroviruses, adenovirus
Culture	Routine bacterial culture	Standard culture and sensitivity testing methods
	Enrichment culture	Uses a broth to amplify low bacterial counts (including contaminants), for example, blood cultures, pus, fluids, tissue
	Prolonged bacterial culture	*Legionella* spp., *Actinomyces* spp.
	Mycobacterial culture	*Mycobacterium tuberculosis* and nontuberculous mycobacteria
	Fungal culture	Yeasts, molds, dimorphic fungi
	Viral culture—largely replaced by molecular methods	Herpesviruses, adenovirus, enteroviruses
Serology	Enzyme immunoassay (EIA) Complement fixation test (CFT) Particle agglutination Immunofluorescence (IF) Recombinant immunoblot assay (RIBA) Western blot (WB)	Archived sera may be available to allow comparison of titers (e.g., respiratory viruses, *Legionella*, chlamydia, hepatitis viruses, CMV, EBV, measles)
Antigen detection	Particle agglutination	Meningococcus, pneumococcus, cryptococcus, for example, CSF
	EIA	*Legionella pneumophila* (urine), rotavirus, chlamydia, RSV, hepatitis B surface antigen, galactomannan, and beta-D-glucan (fungal)
	IF	RSV, VZV, HSV, respiratory viruses
Molecular tests	Direct hybridization Nucleic acid amplification Post-amplification analysis	Many species including *Mycobacterium tuberculosis*, *Neisseria meningitidis*, *Aspergillus* spp., *Candida* spp., CMV, HSV, and an increasing number of other organism

The contribution of molecular techniques to the diagnosis of infection is expanding rapidly. Once considered research tools, they are now commonplace and can be applied to almost any kind of biological sample (Procop, 2007). As the reliability and format of the technology improves, costs will become less prohibitive, and the technique will open up the possibility of identifying non-cultivable species and defining their role in the pathogenesis of disease. Molecular techniques have proven useful in the identification of a variety of pathogens from autopsy samples, including viruses (Bajanowski *et al.*, 2003; Guarner *et al.*, 2006; Denison *et al.*, 2011; Edler *et al.*, 2011), bacteria (Guarner *et al.*, 2006; Guarner *et al.*, 2007; Luchini *et al.*, 2008), and fungi (Schwarz *et al.*, 2014).

Serological antigen and antibody tests are commonly used in the diagnosis of infection. Antigen detection is preferred, as it provides direct evidence of infection, whereas antibody detection is only an indirect marker—nevertheless, it remains useful because antigen detection assays are not available for all infectious agents. Each diagnostic laboratory will have a range of techniques available, which will depend on the antibody/antigen under investigation and the type of automation in place. The sensitivity of these techniques differs, with enzyme immunoassay being 10-fold more sensitive than immunofluorescence, which, in turn, is 10-fold more sensitive than complement fixation. Although assays are usually performed on serum, some tests can be carried out on CSF and saliva. Advice should be sought from the local microbiology department about the availability of testing and what can additionally be accessed through reference services.

The biochemical markers procalcitonin (PCT) and C-reactive protein (CRP) are readily available from most clinical chemistry departments and, as an adjunct to other postmortem investigations, may provide a useful tool in the investigation of infection. The assays are usually performed on serum, but other sample types, such as pericardial fluid and CSF, have been validated in the autopsy setting.

7.5 How do you put the results into context?

7.5.1 Culture

When conducting a postmortem analysis of a patient with suspected infection of uncertain etiology, it is useful to have as much information as possible from the antemortem clinical history, radiological and microbiology findings to help formulate a differential diagnosis and direct further investigations.

There are many variables that need to be taken in to account when interpreting microbiology results—particularly when assessing sites with a normal flora such as the respiratory, gastrointestinal, and genitourinary tracts. Illness, antimicrobial therapy, and medical intervention can disturb the normal flora, and culture results can be misleading.

In cultures from any site, consideration should be given to whether an organism is:

1. A true pathogen
2. Part of the normal flora that would be expected at that site

3. Present secondary to agonal spread of organisms—that is, antemortem mucosal compromise or possibly arising from prolonged resuscitation
4. Present secondary to postmortem translocation of organisms
5. A contaminant that has been introduced during specimen collection at autopsy

By comparing culture results from different sampling sites and between antemortem and postmortem samples, with the clinical history and gross and microscopic findings, plus an understanding of microbiological methods and the behavior of pathogens at different sites, a collaborative approach between the microbiologist and pathologist will usually allow an assessment of their clinical relevance (Table 7.3). Supporting evidence may also be secured through adjunctive investigations such as molecular tests, serology, or biochemical markers.

Up to about 48 hours, the PMI appears to have a relatively small influence on culture results of postmortem samples, despite the controversy over the potential for postmortem invasion (Goldwater, 2009; Lobmaier *et al.*, 2009; Prtak *et al.*, 2010; Weber *et al.*, 2010b). Rather than overgrowth, Weber *et al.* (2010b) found that positive cultures actually decreased with time, and the proportion of positives fell from 83% for samples taken within 24 hours of death to 67% when taken after 5 or more days. This may reflect the relative insensitivity of routine clinical culture techniques however, as studies on human cadaveric tissue using next-generation sequencing have shown that, with time, an increasing abundance and diversity of microbial species can be detected (Pechal *et al.*, 2014; Hyde *et al.*, 2015). Whether this is clinically relevant is, as yet, unclear, and careful

Table 7.3 Guide to interpretation of culture results from different anatomical sites

	Organism that acts as an unequivocal pathogen (e.g., *Mycobacterium tuberculosis*)	**Organism that acts as both colonizer and pathogen (e.g., *Staphylococcus aureus*, *Aspergillus* spp.)**	**Organism that acts mainly as a colonizer (e.g., coagulase-negative staphylococcus)**
Sterile site (e.g., CSF)	Significant	Significant	Possibly significant. Needs corroboration*
Protected site (e.g., lung)	Significant	Possibly significant. Needs corroboration*	Unlikely to be significant
Site with normal flora (e.g., skin/gut/upper respiratory tract)	Significant	Possibly significant. Needs corroboration*	Unlikely to be significant

* Corroboration = clinical history, results of other microbiological investigations, histology, consideration of how the sample was collected, quantity of growth (heavy growth more significant than scanty), whether in pure or mixed growth (pure more significant), whether isolated on direct culture or only from enrichment (direct more significant).

technique when taking autopsy samples to avoid contamination is probably much more important than theoretical considerations of bacterial invasion.

7.5.1.1 Blood cultures

In life, blood cultures are often contaminated because of poor aseptic technique. Although it is recommended that contamination rates should be below 3% in diagnostic blood culture systems (Self *et al.*, 2013), they are often significantly higher than this (Dawson *et al.*, 2008). Even with the most careful technique, it is inevitable that contamination will also occur in the postmortem setting, and, consequently, when a cultured organism has the capacity to act as both a colonizer and a pathogen, a judgment has to be made as to whether it is a true positive or a contaminant. For example, coagulase-negative staphylococci are very commonly contaminants but may be significant when associated with intravenous and prosthetic devices (Weinstein, 2003).

Generally, the same principles applied in life can be used to interpret postmortem blood cultures, in that a pure growth of a recognized pathogen is usually significant and mixed cultures are typically contaminants. However, some caution is required. Although interpretation of the presence of organisms such as *N. meningitidis*, *H. influenzae*, *S. pneumoniae*, *Salmonella* spp., and *Staphylococcus aureus* is straightforward, the significance of others such as *Enterococcus* spp., Enterobacteriaceae, and other Gram-negative enteric organisms is more difficult to interpret. The latter group arises from the gastrointestinal tract and grows rapidly as contaminants, so supporting evidence, such as isolation of the same organism from different sites, isolation from a "protected" site such as the spleen, histological appearance, or a compatible clinical picture, is required before regarding them as pathogens. As in life, a positive blood culture does not always correlate with symptomatic disease, and therefore it may not indicate the cause of death.

In terms of determining how frequently contamination occurs, autopsy studies have found remarkable consistency—with about 20% of cultures yielding a single pathogen that was compatible with the cause of death, about 10% yielding mixed cultures of doubtful significance, and the remainder sterile (Fredette, 1916; Adelson and Kinney, 1956; Carpenter and Wilkins, 1964; Pryce-Davies and Hurley, 1979; Morris *et al.*, 2006). This would suggest that if contamination rates for autopsy blood cultures are comparable to those in life, agonal spread and postmortem translocation appear to account for only a very small proportion of positive cultures. Further evidence that postmortem invasion is negligible comes from studies where authors have found either no or only a weak association between the yield of cultures and PMI (Carpenter and Wilkins, 1964; Pryce-Davies and Hurley 1979; Morris *et al.*, 2006). The application of next-generation sequencing to postmortem blood cultures and the clinical relevance of the findings remain to be seen.

Historically, mixed blood cultures in the postmortem setting have been deemed to be artifacts arising from contamination or postmortem invasion and reported to occur with a frequency of between 1 and 7% (Carpenter and Wilkins, 1964; Pryce-Davies and

Hurley, 1979; Morris *et al.*, 2007). However, in life, polymicrobial bloodstream infection has been reported with increasing frequency, with rates in hospitalized patients ranging from 6 to 32% of all bloodstream infections. The clinical importance of polymicrobial bacteremia is that the mortality rate is approximately twice that of monomicrobial infection and ranges from 21 to 63%. There is an association with recent hospitalization, malignancy, neutropenia, gastrointestinal disease, intra-abdominal infection, genitourinary disease, recent surgical procedures, and the presence of central venous catheters (Lin *et al.*, 2010). Differentiating contamination from true polymicrobial bloodstream infection is extremely difficult. In life, significance is usually confirmed by repeated isolation from blood cultures taken on different occasions (Lee *et al.*, 2007), but this is clearly not possible at autopsy.

7.5.1.2 Cerebrospinal fluid

Unlike blood cultures where agonal and postmortem spread may account for a small proportion of positive results, this appears to be very uncommon in CSF. Most well-collected autopsy samples are sterile or yield a single, clinically relevant pathogen (Pryce-Davies and Hurley, 1979; Eisenfield *et al.*, 1983). A small number of samples yield mixed cultures, probably arising from contamination during collection, but, with careful attention to aseptic technique, the contamination rate falls well below 10% (Morris *et al.*, 2007). In one study of 38 consecutive SUDI autopsies, none were contaminated (Adelson and Kinney, 1956). This suggests that the blood–brain barrier remains intact in the early period after death, preventing ingress of any bacteria that are present in the bloodstream (by whatever mechanism), and that CSF results at autopsy are reliable as long as contamination during collection is avoided. CSF protein measurements are representative of the antemortem state within the first 24 hours after death (Mangin *et al.*, 1983). This may be a useful adjunctive test when interpreting CSF culture results and supports the premise that the blood–brain barrier remains intact during this period. The use of CSF white cell count as an indicator for inflammation or infection is more problematic, and its value in support of a diagnosis of infection depends greatly on the time after death, and it is imperative that CSF samples are obtained as soon as possible (Morris *et al.*, 2007). The normal CSF cell contains no more than four mononuclear cells per cubic millimeter and no neutrophil polymorphs. After death, the count increases substantially, predominantly due to an increase in mononuclear cells that probably detach from the meninges, but by 12 hours the cells become vacuolated and differentiation of the cell type becomes impossible (Platt *et al.*, 1989; Wyler *et al.*, 1994). Platt *et al.* (1989) found that the cell count in adults ranged from 1 to 108×10^6 cells/L and from 37 to 3250×10^6 cells/L in cases of noninfective sudden infant death. In a body stored at 4°C, the cell count rises significantly more slowly than in one stored at 20°C, but in both cases the counts correlate with the time of death according to a mathematical function (polynomial curve of third order) and can be used to estimate PMI (Wyler *et al.*, 1994). In practical terms, therefore, a raised white cell count does not necessarily indicate the presence of infection or inflammation. Although neutrophils are not seen in the absence of inflammation, they have a short half-life (2–4 hours) and so may be absent if the autopsy is delayed (Morris *et al.*, 2006). However, they do not enter the CSF after death, so their presence does indicate antemortem inflammation. It is important to remember that prior neurosurgery, intraventricular bleeding, or trauma may have a significant impact on the CSF findings and interpretation of culture results.

7.5.1.3 Other samples

In the absence of evidence of local infection, positive cultures from normally sterile tissues can be interpreted in a similar fashion to blood cultures. Careful attention to sample collection technique will minimize the chances of contamination. Although, theoretically, organisms could penetrate into tissue after contamination of the surface of an organ, this is unlikely to be a problem in practice. Studies on muscle have shown that although penetration into the interior can occur, it depends on the bacterial species involved and the density of surface contamination, and it is relatively slow and substantially reduced at refrigeration temperatures (Gill and Penney, 1979). Consequently, it should not confound culture results if there is appropriate storage of the body and timely postmortem examination.

Interpretation of cultures of sites with a normal bacterial flora, such as the upper respiratory tract, is difficult unless an unequivocal pathogen, for example, *M. tuberculosis*, is isolated. The diagnostic contribution of samples from non-sterile sites may, in reality, be limited. In a review of neonatal, infant, and young childhood deaths, Sadler (1998) found that significant pathogens were often demonstrated by bacteriological culture of the CSF, blood, and trachea, whereas culture of swabs from the ears, nose, and throat usually yielded insignificant commensals or contaminants.

7.5.2 Serology and molecular tests

Interpretation of serological tests can be challenging. Investigation of a particular infection may require several serological assays, and interpretation of the results is complex and relies on careful consideration of the laboratory and clinical findings. The usefulness of a result depends on the performance of the assay, the infecting agent under investigation, the timing of the sample in relation to the stage of infection, and the immune status of the patient. Complement fixation tests (CFTs) are particularly challenging to interpret. A significant positive result is usually taken as fourfold increase in titer between an acute and a convalescent sample, but in the autopsy setting, unless a previously archived sample is available for comparison, only a single sample will be available. Basing a diagnosis on a single high titer is unreliable and, even when a result appears to be clearly positive, expert advice should be sought. Because of these difficulties, most laboratories have replaced CFTs with more reliable technology.

Molecular detection assays may help clarify serology or culture results. For postmortem samples, particularly if only very small volumes are obtainable, this may be useful, and these tests offer increased sensitivity and specificity over culture (Tuomisto *et al.*, 2013). However, the enhanced sensitivity is such that a positive result does not necessarily indicate that the organism is significant as a cause of death. The presence of nucleic acid does not always equate to presence of live organisms in the specimen. Even after effective antimicrobial treatment, bacterial DNA may still be detectable, and persistent detection of nucleic acid occurs in chronic and convalescent carriers, not just in an acute episode of infection (Health Protection Agency, 2013). Consequently, results need to be interpreted in conjunction with the clinical history, culture results, and histopathological evidence of infection.

7.5.3 Biochemical markers

In life, two biochemical markers, PCT and CRP, perform consistently well in differentiating infection from a systemic inflammatory response of noninfectious origin (Table 7.4). PCT increases 2–4 hours after bacterial challenge, has levels proportional to the degree of sepsis, and declines rapidly with clinical recovery. Although levels also increase after severe noninfectious inflammatory stimuli, such as major burns, severe trauma, acute multi-organ failure, or major abdominal or cardiothoracic surgery, rises are transient. Studies have shown that an elevated PCT has a sensitivity of 97%, with 78% specificity, for systemic bacterial infection (Harbarth *et al.*, 2001; Tsokos *et al.*, 2001). In the postmortem setting, Tsokos *et al.* (2001) assessed the course of femoral venous serum PCT levels in sepsis-associated deaths and found that levels successfully distinguished infective deaths from non-septic causes. A linear decline in PCT concentration after death allowed the antemortem PCT level to be predicted by taking two timed postmortem samples.

There has been some debate about a suitable PCT cut off for postmortem serum samples. Some authors propose a very low level, such as 0.19 ng/mL, to take into account the postmortem decline in levels (Schmidt *et al.*, 2015), others have suggested 2 ng/mL on the basis that this may be better at discriminating sepsis from other conditions that can lead to elevated levels, and yet others have used 10 ng/mL and found that, at this level, false positives are reliably excluded (Ramsthaler *et al.*, 2008). In general, it would appear that a serum level below 2 ng/mL, taken within a few days of death, almost certainly excludes bacterial sepsis or septic shock as a cause of death (Bode-Janisch *et al.*, 2013) and Tsokos *et al.* (2001) found PCT measurement to be valid until 140 hours after death. Nevertheless, after several days a single low PCT may fail to exclude sepsis as a cause of death (Bode-Janisch *et al.*, 2013).

One problem with serum PCT measurement is its unreliability in the presence of severe hemolysis (free hemoglobin level over 500 mg/dL). Hemolysis is less of a problem in fluids such as aqueous or vitreous humor, pericardial fluid, and CSF, all of which have been compared with serum and are able to reliably identify sepsis-related deaths and provide useful alternatives if serum is unsuitable or unavailable (Schrag *et al.*, 2012a, b; Schmidt *et al.*, 2015). CSF and vitreous humor have particularly high concordance with serum levels (Schrag *et al.*, 2012a; Schmidt *et al.*, 2015). However, a lower

Table 7.4 Characteristics of procalcitonin and C-reactive protein as biochemical markers of sepsis

Marker	Behavior in sepsis	Marker characteristics	Sample types validated at autopsy
Procalcitonin (PCT) Propeptide of calcitonin No hormonal activity	Elevated in bacterial sepsis and, to a lesser extent, in multi-organ dysfunction, resuscitation, inhalation injuries, burns, and cardiogenic shock No elevation in viral infections, autoimmune conditions, or allergy Levels rise rapidly and correlate with the degree of sepsis Marker of choice in postmortem diagnosis of systemic sepsis	Investigation readily available Normal level in life <0.15 ng/mL. Rises over 2 ng/mL in sepsis can exceed 100 ng/mL Half-life 25–30 hours Very stable at 4°C degrees and at room temperature Arterial and venous concentrations comparable Stable on freeze–thaw Linear fall in levels after death	Serum Pericardial fluid CSF Vitreous and aqueous fluid
C-reactive protein (CRP) Acute phase protein made by the liver	Released into the blood within a 6–12 hours after tissue injury, the start of an infection, or other cause of inflammation. Peaks at 24–48 hours Nonspecific response to inflammation rather than infection	Investigation readily available Half-life 19 hours Normal level in life <2 mg/L Very stable at 4°C degrees. Stable on freeze–thaw Postmortem levels approximately 35% reduced but stable up to 6 days	Serum Pericardial fluid CSF Sonicated liver supernatant Vitreous fluid

See also Tsokos *et al.* (2001); Harbarth *et al.* (2001); Ramsthaler *et al.* (2008); Schrag *et al.* (2012a, b); Uhlin-Hansen (2001); Bode-Janisch *et al.* (2013); Schmidt *et al.* (2015); Palmiere *et al.* (2013); and Fujita *et al.* (2002).

cut off is required for these samples; using 0.05 ng/mL for aqueous fluid gave a sensitivity and specificity of 92 and 69%, and using 0.4 ng/mL for CSF gave 81 and 86% (Schmidt *et al.*, 2015).

CRP has been evaluated extensively in a variety of postmortem samples. In nearly all comparisons, CRP has been able to discriminate sepsis from other causes of death but with less sensitivity than PCT (Schrag *et al.*, 2012a, b). The nonspecific response of CRP to inflammation, rather than infection, is acknowledged in the interpretation of levels during life, so it is unsurprising that this should also be the case at autopsy (Fujita *et al.*, 2002). In terms of serum concentration, some authors found that levels after death were comparable to those in life (Astrup and Thomsen, 2007), whereas others found a decline of about 35% (Uhlin-Hansen, 2001). As an alternative to serum, CRP has also been measured in vitreous, CSF, pericardial fluid, and homogenized and sonicated liver (Schrag *et al.*, 2012a, b). Liver and CSF levels correlate well with serum (Astrup and Thomsen, 2007; Schrag *et al.*, 2012a). Although pericardial and serum levels of CRP correlated less well than PCT did, CRP still had a sensitivity and specificity of 80 and 95% for sepsis, and pericardial fluid is a useful alternative if serum is unavailable (Schrag *et al.*, 2012a). In a recent review, 14 of 19 forensic autopsies classified as having an infection-related death had elevated CRP levels (Christofferson, 2015). The author proposed that, where antemortem infection is unsuspected, CRP measurement should be performed at the beginning of each autopsy. If elevated, it would alert the pathologist to the need for microbiological sampling, and this could be performed before significant handling of the body so that the risk of contamination would be minimized.

A number of other cytokines and inflammatory markers have been assessed in the postmortem setting. Compared with PCT and CRP, these have been found to be less useful as markers of sepsis not only because of performance but also because the assays are not as widely available. These include lipopolysaccharide-binding protein (LBP) (Augsburger *et al.*, 2013), soluble triggering receptor expressed on myeloid cells—type 1 (sTREM-1) (Palmiere *et al.*, 2013), interleukins 6 and 8, and tumor necrosis factor (Harbarth *et al.*, 2001; Schrag *et al.*, 2012a).

7.6 What are the risks of transmission of infection in the postmortem room?

The autopsy plays an undisputed role in the context of diagnosing infectious diseases, with studies demonstrating that a substantial number of infections, which are not detected in life, are detected during postmortem examination (Stevanovic *et al.*, 1986; Landefeld *et al.*, 1988; Gibson *et al.*, 2004; Newton *et al.*, 2004; Roulson *et al.*, 2005; Spiliopoulou *et al.*, 2005; Wilson, 2006). There is often a great deal of anxiety among autopsy personnel when dealing with patients with either confirmed or suspected high-risk infections. However, the perceived risk of occupational exposure to infections in these circumstances is probably overstated. In the current climate of stringent infection control precautions and safe working practices, underpinned by governance and legislation, the actual risk to mortuary workers is very low.

A key element in maintaining this low level of risk is an understanding of potential hazards, and this is the responsibility of the pathologist. Risk can be effectively managed by identifying the risk factors for transmission of infection, by performing a risk assessment prior to each autopsy procedure, by following locally agreed protocols, and by seeking advice in unusual circumstances. Risk factors for transmission can be broadly divided into five categories (Table 7.5), and the importance of these is amply illustrated by reports of transmission when they are overlooked. There are many clinical and statutory guidelines on biosafety in the work environment produced by various organizations in different countries. A comprehensive review of this guidance is beyond the scope of this chapter, and, in any case, the information is continuously being updated. A list of useful websites is provided in Table 7.6 for reference.

The vast majority of pathogens encountered at autopsy is BSL-3 or lower (Mazuchowski and Meier, 2005). If postmortem is being conducted where BSL-4 agent is suspected, a specialist high containment level facility must be used, and advice should be sought in all cases. It must be noted, however, that there will be regular occurrences where infectious diseases are only detected postmortem; therefore, suitable standard infection control precautions should be universally adopted.

7.7 How does autopsy microbiology contribute to the diagnosis of specific conditions?

7.7.1 Pneumonia

Despite the "classical" presentations of community-acquired pneumonia, no convincing association has been demonstrated between individual symptoms, physical findings, or laboratory test results and a specific etiology (Bartlett *et al.*, 2000). It is therefore perhaps unsurprising that in about 10% of cases, the diagnosis is missed during life and only made at autopsy (Landefeld *et al.*, 1988). With nosocomial pneumonia, the situation is even worse, and in a necropsy study of 24 patients dying of ARDS, 14 had histological evidence of pneumonia of which only 9 (64%) had been accurately diagnosed in life (Andrews *et al.*, 1981). In contrast, of the 10 non-pneumonia deaths, 8 (80%) were correctly diagnosed antemortem.

Even when apparently straightforward cases of pneumonia come to autopsy, lung samples should be taken. It increases the overall proportion of cases with a definitive diagnosis and will identify where resistant organisms have caused treatment failure and so contributed to the fatal outcome. Resistance in pathogens causing community-acquired pneumonia is increasingly prevalent (Klugman, 2007), and, in ventilated patients, infection with multiresistant pathogens is associated with substantially worse clinical outcome (Parker *et al.*, 2008). A number of autopsy studies have been used to clarify the range of important pathogens causing fatal pneumonia in children (Turner *et al.*, 2012), and many others highlight the role of the autopsy in identifying unexpected fungal and mycobacterial infection in this context (Stevanovic *et al.*, 1986; Sarode *et al.*, 1993; d'Arminio Monforte *et al.*, 1996; Garbino *et al.*, 2011). Care in interpretation of

Table 7.5 Identification of the risk factors for transmission of infection when undertaking a risk assessment

Transmission risk factor	Examples of transmission	Control measures	References
Cadaver-related factors • Viable organisms persist after death	• *Mycobacterium tuberculosis* • Blood-borne viruses • Transmissible spongiform encephalopathies (TSE)	• Classification of infectious diseases according to level of risk to humans • Standard operating protocols for each infectious agent or category	Burton (2003); Mazuchowski and Meier (2005); Centers for Disease Control and Prevention (2009)
Routes of transmission • Inhalation of aerosolized particles • Direct skin or mucosal contact • Accidental percutaneous inoculation • Ingestion • Skin contamination without inoculation	• Cutaneous TB ("prosector's wart") • Blood-borne viruses • *Streptococcus pyogenes*	• Classification of infectious diseases according to level of risk to humans • Appropriate personal protective equipment (PPE) for autopsy personnel for each category, for example, surgical scrubs, surgical cap, impervious gown, and/or apron with full sleeve coverage, eye protection, face mask, N95 mask, shoe covers or waterproof shoes, double surgical gloves, ideally interposed with a cut-resistant synthetic mesh • Autopsy procedures that minimize the risks of aerosolization, contamination, or inoculation • Personnel should be aware of local occupational health protocols in case of exposure or injury	Hawkey *et al.* (1980); Nyberg *et al.* (1990); Ball *et al.* (1991); Collins and Grange (1999); Ganczak *et al.* (2003); Mazuchowski and Meier (2005); Centers for Disease Control and Prevention (2005); Flavin *et al.* (2007); and Centers for Disease Control and Prevention (2009)
Autopsy personnel factors • Immune status	• Blood-borne viruses • Tuberculosis	• Appropriate vaccinations according to local prevalence, type of mortuary, and risk of exposure • As a minimum: hepatitis B and BCG • Others: hepatitis A, *Neisseria meningitidis*, varicella zoster, polio, smallpox, anthrax • Active involvement of the occupational health services • Personnel should be aware of local occupational health protocols in case exposure or injury occurs during autopsy • Participation in a TB screening program	West (1984); Lundgren *et al.* (1987); Rodrigues *et al.* (1993); Burton (2003); Centers for Disease Control and Prevention (2005); Lucas (2010); and Public Health England (2014a)

(*Continued*)

Table 7.5 (Continued)

Transmission risk factor	Examples of transmission	Control measures	References
Design and facilities of the autopsy room		• Design, construction, and ventilation to nationally approved standards • Whole-room ventilation or down-draught tables • Adequate space and distance from other activities • Accessible and safe storage of equipment • Daily environmental cleaning (phenolic disinfectant) • Rigorous cleaning and decontamination of instruments • Safe disposal of waste according to regulations	Healing *et al.* (1995); NHS Estates (2005); Wilson (2006); and Lucas (2010)
Conduct and behavior of personnel	• *Varicella zoster* virus • Tuberculosis	• Adequate supervision of all personnel during the autopsy procedure • The number of staff and trainees should be kept to a minimum	Wilkins *et al.* (1994); Wilson (2006); and Paul and Jacob (2006)

Table 7.6 Clinical and statutory guidelines on biosafety in the work environment

Organization	Website address
UK Advisory Committee on Dangerous Pathogens (ACDP)	www.gov.uk/government/groups/advisory-committee-on-dangerous-pathogens
UK Health and Safety Executive (HSE): Control of Substances Hazardous to Health (COSHH)	www.hse.gov.uk/coshh/
The Centers for Disease Control and Prevention (CDC): Biosafety in Microbiological and Biomedical Laboratories	www.cdc.gov/biosafety/publications/bmbl5/index.htm
CDC: The National Institute for Occupational Safety and Health (NIOSH)	www.cdc.gov/niosh
Royal College of Pathologists: guidelines on autopsy practice 2002	www.rcpath.org www.rcpath.org/NR/rdonlyres/412AEB13-F5B8-4C6B-A087-2833223C7A4D/0/main_document.pdf
Public Health England: infectious diseases A–Z	www.gov.uk/health-protection/infectious-diseases

unexpected findings is necessary as illustrated by a report that identified *Legionella pneumophila* serogroup 8 in cultures of postmortem autopsy lung tissue from two patients. In neither case were the clinical and pathological findings consistent with *Legionella* infection, and investigation revealed that contamination had probably occurred from the hot water outlets in the postmortem room (Lightfoot *et al.*, 1991).

The contribution of the autopsy to understanding the pathogenesis of emerging respiratory pathogens is illustrated by the influenza A H1N1 strain that appeared in 2009. In April of that year, a novel H1N1 influenza A virus was identified in two children in California. Within a month, the World Health Organization declared a "public health emergency of international concern," and within 2 months, sustained human-to-human transmission in several countries on different continents led to the announcement of the highest alert level (phase 6, pandemic). The clinical impact was enormous, and by February 2010, 42–86 million cases had occurred in the United States alone, with between 188,000–389,000 hospital admissions and 8,520–17,620 deaths (Centers for Disease Control and Prevention, 2010). Some groups were particularly at risk, such as pregnant women and those with preexisting illness or immunosuppression, and one study found that a risk factor could be identified in 88% of fatal cases (Edler *et al.*, 2011). The first autopsy findings from the pandemic came from Mauad *et al.* (2010) who reported on 21 Brazilian patients. The cause of death in all cases was extensive involvement of the lungs, and there were no signs of direct virus-induced injury in any other organs. The lungs showed severe diffuse alveolar damage with varying degrees of alveolar hemorrhage, necrotizing bronchiolitis, and tracheobronchitis, and this, combined with an aberrant immune response (marked expression of TLR-3 and IFN-g and a large number of CD81 T cells and granzyme B1 cells), had led to acute respiratory failure. Bacterial coinfection, primarily with *S. pneumoniae*, was found in 8 out of 21 patients (38%). Another autopsy series identified H1N1 virus in the lung tissue of a patient who died more than a month after initial presentation suggesting that PCR on lung tissue may be useful for confirming the diagnosis in patients with a protracted disease course (Harms *et al.*, 2010). Denison *et al.* (2011) examined respiratory tissue from 442 cases of suspected influenza infection by real-time reverse transcription PCR for influenza A and B and the 2009 H1N1 strain. In 84 cases a diagnosis of influenza infection as a contributor to death was only obtained by testing the autopsy tissue—antemortem tests either had not been performed or were negative. The superiority of PCR in detecting small amounts of nucleic acid was illustrated by comparison with IHC. Only 107 (49%) of 218 PCR positives were positive by IHC, and there was a significant inverse relationship between a positive IHC result and duration of illness, particularly beyond 10 days. If IHC had been the only investigation of the autopsy tissue, only 42 cases would have received a diagnosis. Active surveillance and testing is critical in detecting novel infectious diseases early—and testing of autopsy tissue is an essential part of this process.

7.7.2 Mycobacterial infection

Worldwide, tuberculosis (TB) is second only to HIV/AIDS as the greatest killer by a single infectious agent. In 2013, 9 million people fell ill with TB and 1.5 million died from the disease, with over 95% of TB deaths occurring in low- and middle-income countries (World Health Organisation, 2015). The autopsy has an important contribution to make to the epidemiological control of the disease, to the management of the contacts of an individual case, and to the understanding of nontuberculous mycobacterial disease. Many cases of TB are unrecognized during life and only diagnosed at autopsy. The likelihood of this depends very much on the local prevalence of the infection. One meta-analysis of clinical/autopsy discrepancies found that in the 1930s the sensitivity of clinical diagnosis of TB was 91%, but by the 1970s it had fallen to 50%; doctors became less good at recognizing TB as the prevalence declined (Stevanovic *et al.*, 1986; d'Arminio Monforte *et al.*, 1996; Roulson *et al.*, 2005). In addition, where mycobacterial infection occurs in conjunction with some other significant condition, or in AIDS patients with nonspecific symptoms and a rapidly terminal course, it is extremely likely to be overlooked (Stevanovic *et al.*, 1986; d'Arminio Monforte *et al.*, 1996). In developed countries TB is unexpectedly found at autopsy in 0.1–3% of cases (Juul, 1977; Cameron and McGoogan, 1981; Stevanovic *et al.*, 1986). In developing countries, where the prevalence of TB is higher, the situation is different. In India, infectious disease is the most common cause of death (46.8%), and TB accounts for one-third of these cases. High clinical awareness of TB means that over 80% of autopsy cases are identified antemortem (Sarode *et al.*, 1993). However, there is a tendency to overdiagnosis; in a series of 1000 autopsies, 5 of the 13 brain tumors were incorrectly diagnosed as tuberculous meningitis, and 7 of the 30 cases of malignant lymphoma were thought clinically to be disseminated TB (Sarode *et al.*, 1993).

Although TB may be suspected macroscopically at autopsy and mycobacterial infection can be confirmed on later histological examination, it is important to send samples for culture. Because of the public health implications of both the diagnosis and of resistance to antituberculous therapy (in particular multidrug resistance (MDR)), full speciation and sensitivity testing of the infecting organism is required. Resistance has been relatively stable in England and Wales since the mid-1990s, but the proportion of strains showing primary MDR increased from 1.1% in 2004 to 1.6% in 2011 (Public Health England, 2014b). This is not the case in other countries, most notably the former Baltic states, where rates of primary MDR approach 15%. Placed in the context of an estimated incidence of TB of over 250 cases per 100,000 population in that region, this gives an incidence of MDR-TB of 35 cases per 100,000 (Zager and McNerney, 2008). This is reflected in the prevalence in the United Kingdom where the highest proportions of MDR-TB cases were in those born in Ukraine, Latvia, and Lithuania (Public Health England, 2014b). Extensively drug-resistant (XDR) strains (those resistant to most available agents) first emerged as a problem in South Africa in association with HIV infection and with a uniformly fatal course. By 2013, at least one case of XDR-TB had been reported from over 100 countries; an estimated 9% of patients with MDR-TB have XDR-TB. Although XDR-TB is therefore more likely to occur where there is a high prevalence of MDR strains, patients may present unexpectedly in low prevalence countries (Blaas *et al.*, 2008).

Strain typing has now become an integral part of the surveillance system in many countries. In the United Kingdom, a national strain typing service was established in 2010 and prospectively types TB isolates using 24 loci MIRU–VNTR. Whole genome sequencing is likely to replace it as, within days of culture, it has the potential to identify the species, drug resistance, and chain of transmission and, consequently, to inform public health action (Public Health England, 2014b, 2015).

Environmental nontuberculous mycobacteria are ubiquitous organisms with which humans commonly interact. Although they have been recognized as causing human infection since the early 1950s, the epidemiological characteristics of nontuberculous mycobacterial diseases including mortality rate and its associated factors remain largely unknown. Autopsy studies have contributed significantly to the understanding of the role of nontuberculous mycobacteria in disease. O'Connell *et al.* (2012) examined the clinical picture, histopathological findings, and radiology of primary pulmonary or disseminated disease in an autopsy series. They found that patients dying of primary lung disease all had a significant chronic clinical illness at autopsy; there was evidence of extensive structural pathology such as cavity formation and scarring, but no infection outside the chest. In contrast, fatal disseminated infection showed widespread infection, which did not always involve the lungs, but, where it did, there was a distinct histopathology consistent with systemic immune dysfunction. Clearly, in HIV there is immune dysfunction, and early autopsy studies found high bacterial loads of *Mycobacterium avium* and widespread organ involvement, leading to the view that this was typical of nontuberculous mycobacterial infection in AIDS. However, in a subsequent series of autopsies on patients with documented *M. avium* bacteremia, Torriani *et al.* (1994) found that 30% had no histological evidence of infection and the number and distribution of involved sites was very variable. It appeared that the likelihood of developing detectable tissue involvement was directly related to the duration of bacteremia, and they postulated that *M. avium* bacteremia may precede widespread disease. The prevailing clinical practice was to wait for multiple positive blood cultures or for symptoms to develop, before starting treatment, but these findings challenged this and supported both prophylaxis and early treatment, as they should be more effective with a lower organism burden. This approach is now widely accepted as standard in managing patients with HIV (Nelson *et al.*, 2011).

7.7.3 Fungal infections

In terms of infectious causes of death that are identified at autopsy, fungal infection forms a relatively small proportion of the total. In unselected case series, the percentage varies with the population served and the local infectious disease epidemiology but has been found to be 5.9% in India (Sarode *et al.*, 1993) and 16.4% in the United States (Bonds *et al.*, 2003). As a proportion of all deaths, the percentage is generally low (1.4%, Boon *et al.* (1991); 4%, Vogeser *et al.* (1999); 3.4%, Groll *et al.* (1996); and 4.5%, Suzuki *et al.* (2013)), even in patients that have died in intensive care setting (4%, Dimopoulos *et al.* (2003)). However, the vast majority of these patients have a significant predisposing factor such as hematological or other malignancy and AIDS or solid organ transplantation (Sarode *et al.*, 1993; Groll *et al.*, 1996; Vogeser *et al.*, 1999; Suzuki *et al.*, 2013). This is unsurprising, as fungal infections are a significant cause of morbidity and mortality in the immunocompromised and have been found to be one of the most common unexpected autopsy findings in this group of patients (Landefeld *et al.*, 1988). The increasing intensity of treatments for hematological malignancies, particularly allogenic bone marrow and stem cell transplants, has created a group of patients that are at particularly high risk of invasive fungal infection and in this cohort rates of infection at autopsy have ranged between 10.7% (Bonds *et al.*, 2003) and 31% (Chamilos *et al.*, 2006). In the latter study, particularly high rates (42%) were seen in patients with acute myeloid leukemia.

Fungal infections are difficult to diagnose during life, and this is particularly the case in the immunocompromised. Apart from a few well-characterized syndromes, the clinical picture is often unclear with nonspecific symptoms, signs, and radiological changes.

Cultures of readily obtainable specimens are often negative, particularly if the patient is on antifungal therapy, and thrombocytopenia or clotting abnormalities may preclude invasive sampling. Consequently, even in this high-risk group, many infections remain undiagnosed and are only discovered if an autopsy is performed. In one recent autopsy series, only 25% of cases of IFI were diagnosed as proven or probable IFI during life, even though 77% of them were considered to be related significantly to the patient's death (Chamilos *et al.*, 2006). This has also been found by others (Groll *et al.*, 1996; Bonds *et al.*, 2003). Nevertheless, several series indicate that the rate of antemortem diagnosis appears to have progressively improved since the 1970s, possibly reflecting the introduction of more sensitive antigen detection tests, molecular tests, and serum markers such as beta-D-glucan (Groll *et al.*, 1996; Chamilos *et al.*, 2006; Suzuki *et al.*, 2013). In non-immunocompromised patients, antemortem suspicion of IFI is frequently low, and this is reflected in the high frequency of infection only being identified at autopsy—89% in one series (Sarode *et al.*, 1993; Garbino *et al.*, 2011). An autopsy study of intensive care unit deaths found that, of the six unexpected cases of invasive aspergillosis found at autopsy (representing 2.7% of the total deaths), all had been culture positive in life but the *Aspergillus* had been dismissed as a colonizer (Dimopoulos *et al.*, 2003). This is also the case outside intensive care, where *Aspergillus* was cultured within 10 days of death in 63% patients found to have infection at autopsy (Vogeser *et al.*, 1999). These studies clearly illustrate that clinicians should have a high index of clinical suspicion and institute early empirical antifungal therapy and that improved diagnostic techniques are needed. Although serum biomarkers such as galactomannan, 1,3-beta-D-glucan, and *Aspergillus* PCR have started to enter clinical practice, the usefulness of all these assays is still a matter of intense debate and investigation. False positive results are common and serial estimations are required for optimal use (Barton, 2013; Theel and Doern, 2013), and these factors may limit their value in the autopsy setting.

7.7.4 Infective endocarditis

The incidence of IE is estimated to be between 30 and 100 cases per million population per year, amounting to approximately one case per 1000 hospital admissions, so it is not a "rare" condition (Pittet and Harding, 1998; Berlot *et al.*, 2014). In spite of better diagnostic imaging, modern antibiotics, and specialist surgery, it remains a life-threatening illness with a mortality around 20–40%—which has changed little in the last 50 years (Young, 1987; Murdoch *et al.*, 2009). In this time there has been a change in the predisposing factors, with degenerative valve disease or a prosthetic valve featuring prominently, rather than rheumatic heart disease. Nevertheless, an important factor remains the failure to diagnose the condition during life. In the late 1960s Robinson *et al.* (1972) identified endocarditis at autopsy in 47 of 1881 patients (2.5%). Among those patients with native valves, a clinical diagnosis was only made in 39% of cases during life. More recently, Berlot *et al.* (2014) identified IE in 1.6% of intensive care unit deaths of which only 25% were diagnosed in life, and Fernández Guerrero *et al.* (2012) found that, between 1970 and 2008, 38% of IE cases diagnosed at autopsy were not identified antemortem. Even in a specialist cardiology hospital, the diagnosis had been missed in 27% of IE cases identified at postmortem (Saad *et al.*, 2007).

Using a combination of blood culture, valve culture, and examination of sections for the presence of bacteria, Robinson *et al.* (1972) found that the infecting organism could be identified at autopsy in 94% of patients with native valve infections—although in 20% of cases, it was only on the basis of identification of organisms in valve sections. Zeien and Klatt (1990) have described a routine approach for pathologists performing autopsies on patients with valvular implants. The advent of molecular techniques that can be applied to both fresh- and formalin-fixed paraffin-embedded tissue is allowing precise identification of infecting species in situations where samples are culture negative or unavailable for culture and has proven useful in the diagnosis of IE (Herrmann *et al.*, 2014; Heras Cañas *et al.*, 2015; Sax *et al.*, 2015).

It is common for patients with prosthetic valves not to present until many months or years after surgery, and this is illustrated by an episode of nosocomial transmission arising from contaminated heater–cooler units used during cardiac operations. Since 2011, eight cases of cardiovascular infection caused by *Mycobacterium chimera* have been identified in patients who had previously undergone cardiac surgery in Switzerland, The Netherlands, and Germany (European Centre for Disease Prevention and Control, 2015). Three patients were diagnosed with endocarditis and presented between 2.9 and 3.6 years after valve replacement. In these cases the organism was cultured from cardiac tissue, including samples taken at autopsy (Sax *et al.*, 2015).

7.7.5 Gastrointestinal infection

If gastrointestinal infection is considered to be a major cause or contributing factor of death, then attempts must be made to identify the infecting microorganisms. The pattern of infection varies markedly between different countries. In England and Wales, *Campylobacter jejuni* is by far the most common food-borne infection, and, in 2014, nearly 59,000 cases of *Campylobacter* infection were notified to Public Health England, compared with 6,600 cases of Salmonellosis (Public Health England, 2015); norovirus and rotavirus accounted for about 5000 cases each. In the United States, however, a different pattern is seen. Of the 9.4 million episodes of food-borne illness each year, norovirus causes nearly 60% of infections, with nontyphoidal *Salmonella* spp. and *Campylobacter* spp. accounting for about 10% each (Scallan *et al.*, 2011).

Most deaths due to gastroenteritis are due to fluid loss, which affects the young and very old. Viruses such as rotavirus are the leading cause of enteric infection in children, and, although these infections are rarely fatal, a report from Toronto described 21 fatal cases in healthy children over a 5-year period, all dying from severe dehydration with marked hypernatremia (Carlson *et al.*, 1978). Gastroenteritis with an associated septicemia, particularly with *Salmonella*, can lead to death, and this is reflected by figures from the United States that estimate that nontyphoidal *Salmonella* spp. cause nearly

30% of the deaths from food-borne illness each year (Scallan *et al.*, 2011). However, patterns of infection and mortality may alter because of outbreaks, for example, deaths from norovirus increase by 50% in years when epidemics are caused by new strains of the virus (Centers for Disease Control and Prevention, 2012), or secondary to changes in the prevalence of circulating strains.

Bacterial, viral, and parasitic enteric pathogens are readily detected in fecal specimens collected at autopsy. For most clinical laboratories, culture methods are the mainstay of diagnosis, but, as in all fields, molecular diagnostics are becoming more common, either as single target detection for specific pathogens such as *Clostridium difficile* or norovirus or multi-target multiplex and array systems that detect a large number of pathogens. For most agents, PCR has a higher sensitivity than culture, apart from *Salmonella* where traditional selenite enrichment still has higher sensitivity (Trafford *et al.*, 2015). It is imperative that all enteric bacterial isolates have antibiotic susceptibility tests performed because of the increasing numbers of resistant organisms being reported, and all strains should be sent for typing to facilitate outbreak investigation.

7.7.6 Meningitis and central nervous system infections

The development of effective vaccines against the primary pathogens causing meningitis (*N. meningitidis, H. influenzae,* and *S. pneumoniae*) and their adoption into the routine vaccination schedules in many countries have significantly changed the epidemiology of bacterial meningitis in both children and adults over the last few years (Castelblanco *et al.*, 2013). Nevertheless, the management of individual cases requires definitive identification of the causative pathogen wherever possible as, depending on the organism, it may impact on the treatment of close contacts as well having as a wider public health impact. Household contacts of patients with sporadic *N. meningitidis* disease have a secondary attack rate of approximately four cases per 1000, which is 500–800 times greater than the rate for the total population (Centers for Disease Control and Prevention, 2013b). Similarly, the risk of life-threatening secondary disease in children under five who are in contact with a child with *H. influenzae* meningitis is at least 800 times the endemic attack rate (Glode *et al.*, 1980)—although these data predate the adoption of routine childhood Hib vaccination. In childhood tuberculous meningitis, nearly half the cases have been found to be associated with an infected adult family member that is only identified after the diagnosis is made in the child (Doerr *et al.*, 1995).

In all these infections, the risk of secondary cases may be reduced by formal identification and tracing of contacts and by offering appropriate antibiotic prophylaxis or vaccination. As the rate of secondary disease is highest immediately after onset of disease in the index patient, antimicrobial chemoprophylaxis should be administered as soon as possible—ideally within 24 hours (Centers for Disease Control and Prevention, 2013b). Consequently, prompt identification of the causative pathogen is particularly important. Unfortunately, in both adults and children, the diagnosis may be overlooked before a fatal outcome (64% of cases in an autopsy series by Gibson *et al.* (2004)), or a sudden death may occur at home precluding the collection of antemortem cultures. In both these circumstances it falls to the pathologist to take appropriate specimens at autopsy in order to make the microbiological diagnosis.

Although meningitis pathogens can be cultured from autopsy specimens (Ploy *et al.*, 2005; Palmiere *et al.*, 2015), noncultural diagnostic techniques such as PCR lend themselves particularly well to this setting because nonviable organisms are still detectable and the tests are not significantly influenced by the presence of contaminants. Even when there is severe decomposition and there is polymicrobial growth in blood and CSF, so that the possibility of making a microbiological diagnosis by cultural methods is remote, PCR on CSF yields results with a high sensitivity and specificity (Maujean *et al.*, 2013; Palmiere *et al.*, 2015). The sensitivity of PCR and applicability to a wide range of samples means that, by analysis of blood, CSF, and any skin lesions, the causative pathogen in meningitis may even be identified without the need for a full autopsy (Ploy *et al.*, 2005). If, for any reason, samples are not submitted at the time of autopsy, and only formalin-fixed paraffin-embedded tissue sections are available, PCR is still a useful technique (Fernández-Rodríguez *et al.*, 2008).

Although published sampling guidelines advocate taking nasopharyngeal and throat swabs in cases of sudden death in young people, and in those with neurological symptoms (Fernández-Rodríguez *et al.*, 2015), care is needed when interpreting positive results. Meningococcal colonization of the nasopharynx is common, with rates ranging from 10% up to 55% in some UK university students, and is represented by a wide diversity of clonal types. In contrast, disease isolates have a very restricted genetic lineage (Read, 2014). Carriage of *S. pneumoniae* is even more prevalent, but, equally, the serotype distribution is not concordant with that found in invasive infection (Cardozo *et al.*, 2006). Consequently, in meningitis cases, the pathologist should avoid attributing undue significance to an isolated nasopharyngeal isolate. It may be incidental, and corroborating evidence is required before it can be identified as being directly relevant to the cause of death.

7.7.7 Septicemia

Based on a US study, which gave an estimated incidence of 3.0 cases per 1000 population per year, the worldwide burden of sepsis amounts to 20 million cases per year that, with an average mortality of 35%, means around 20,000 deaths per day worldwide (Daniels, 2011). Although sepsis is the most common cause of death in medical and surgical intensive care units, the pathologist needs to be alert to the possibility of overwhelming sepsis as a cause of death even after low-risk procedures (Vongpaisarnsin *et al.*, 2015). It is consistently underdiagnosed in life and, in consequence, is one of the top ten discrepant diagnoses identified at autopsy (Gibson *et al.*, 2004). One study found that 48% of deaths attributable to sepsis were only identified at postmortem examination (Mercer and Talbot, 1985). Nevertheless, it can be difficult to make firm diagnosis at autopsy—antemortem history may be lacking, cultures may be contaminated, and the autopsy macroscopic and microscopic findings are insensitive and nonspecific. Serological markers such as CRP and PCT may help in this respect by providing corroborating evidence.

The commonest cause of septic shock is bacterial infection, although it is also seen in fungal, viral, and protozoal infection, with most infections arising from the lungs, abdomen, and urinary tract. Bacteremia occurs in 40–60% of patients, with the proportion of Gram positive and negative infections being approximately equal. A causative organism may not be identified from blood cultures in 10–30% of sepsis cases, probably because of prior exposure to antibiotics. Postmortem yield of blood cultures is inversely proportional to the length of antemortem antibiotic therapy and is only 35% for patients who have received antibiotic treatment for more than 4 days leading to death (Roberts, 1998). The autopsy has an important role to play in the monitoring of the efficacy of empirical antibiotic treatment in patients dying of sepsis, and it is likely to become critically important in an era of increasing antimicrobial resistance (World Health Organization, 2014). MDR strains are associated with increased mortality, which primarily arises from failure of empirical antibiotic therapy (Walsh, 2010). Where cultures are negative in life, or if samples are unavailable, autopsy cultures may allow infection with resistant strains to be identified and allow differentiation between treatment failure arising from resistance and treatment failure due to an uncontrolled focus.

7.7.8 Neonates and sudden unexplained death in infancy

Sudden unexplained death in infancy (SUDI/SUID) is defined as the sudden and unexpected death of an infant less than a year old where the cause of death is not immediately obvious before investigation. In the United States it is the third leading cause of infant deaths, the leading cause of death in infants 1–12 months old (Centers for Disease Control and Prevention, 2015), and caused about 3500 deaths in 2014. Twenty-five percent of these arose from accidental suffocation or strangulation in bed, forty-four percent were categorized as Sudden Infant Death Syndrome (SIDS), where the death could not be explained after a thorough investigation, including a complete autopsy, and 31% as "unknown cause" because a thorough investigation was not conducted and cause of death could not be determined.

Neonatal autopsies, or those in cases of SUDI/SUID, require comprehensive sampling in order to identify both congenital and acquired bacterial and viral infections that may be implicated as a cause of death. In many countries, these cases are investigated by governmental agencies following standardized protocols for specialized autopsy, and there are a number of published guidelines, all of which include recommendations for microbiological sampling (The Royal College of Pathologists and The Royal College of Paediatrics and Child Health, 2004; Howatson, 2006; Centers for Disease Control and Prevention, 2013c; Fernández-Rodríguez *et al.*, 2015). Nevertheless, interpretation of the results of the microbiology samples is not straightforward.

Despite intensive study the pathogenesis of SUDI remains unclear. The possibility that underlying infection may be a cause of SUDI, based on autopsy findings, histology, and the detection of organisms in autopsy samples, has been proposed for many years (Morris *et al.*, 2006; Moon *et al.*, 2007; Highet, 2008). A central role for infection is supported by the age distribution of cases, which is reciprocal to the concentration of serum immunoglobulin that protects against infection by bacteria and bacterial toxins (Morris and Harrison, 2008). However, the exact contribution it makes has remained unclear—not least because of the uncertainty of the role of potential pathogens in the absence of a tissue reaction. A number of autopsy studies have attempted to ascertain the role of infection in determining the cause of death.

One study reviewed the results of samples taken in 470 SUDI autopsies (Weber *et al.*, 2008). Although the majority of samples yielded organisms, most appeared unrelated to the cause of death, with 68% of cultures mixed and only 32% yielding pure growth. The overall rates of positive cultures were the same between patient groups having death attributed to different causes, that is, bacterial infection, unexplained, or noninfective causes, indicating that the mere presence of organisms was not diagnostic of the cause of death. However, significantly more pathogens with the capacity to cause septicemia without an obvious focus were isolated from infants whose death was explained by bacterial infection or was unexplained, than in those with a noninfective cause. In particular, *S. aureus* and *Escherichia coli* were found significantly more often in infants with unexplained death than in with those with a noninfective cause, especially in samples from lungs or spleen. A preponderance of *S. aureus* has also been found by others: in one study it was twice as common in unexplained SUDI (40%) as in noninfective SUDI (21%) (Weber *et al.*, 2011), and, in another, which focused on sterile site samples (heart blood, spleen, and CSF), *S. aureus* was found in sterile site samples from 11% of SIDS and 19% of infection-related deaths, but not in samples from accidental deaths. In contrast, there was no difference in the isolation of coliforms between the different groups (Goldwater, 2009).

Many authors have queried the role that *S. aureus* plays in SUDI and, in particular, whether the production of toxins, both by *S. aureus* and other organisms, may be important (Highet, 2008; Weber *et al.*, 2008; Goldwater, 2009). *S. aureus* toxins are superantigens that can induce significant pro-inflammatory responses that are lethal if uncontrolled, for example, in toxic shock syndrome. *S. aureus* toxins have been demonstrated in the tissues of a large proportion of babies dying of SIDS—59 of 105 cases in one study (Blackwell, 2008), and, as they are only produced between 37 and 40°C, detection implies that it occurred antemortem (Goldwater, 2009). The production of toxin does not appear to be a simple explanation, however. Weber *et al.* (2011) compared the prevalence of toxigenic *S. aureus* (as identified by the presence of a range of staphylococcal toxin genes) in explained and unexplained SUDI cases. Although toxin gene-carrying *S. aureus* was common, accounting for 78% of all *S. aureus* strains in this study, and the prevalence in unexplained SUDI was higher than in the noninfection group

(81% vs. 66%), it failed to reach statistical significance. The response of the host immune system to *S. aureus* and its toxins may be critical. Highet (2008) and Highet *et al.* (2009) have proposed that SUDI occurs as a consequence of a transient bacteremia with a toxigenic species in an infant with a defect in pathogen recognition, which triggers a lethal, immune response with little tissue response.

In terms of the most useful samples to take from SUDI cases for diagnosing infection, and the best time to take them, autopsy studies have generated useful data. Results from sterile site samples (heart blood, spleen, CSF, lung) appear to reliably predict sepsis and are less confounded by normal flora or incidental carriage of potential pathogens (Goldwater, 2009; Lobmaier *et al.*, 2009). Prtak *et al.* (2010) also found that 27% of cases had a middle ear exudate, of which nearly 50% yielded a potential pathogen, but emphasized the importance of taking samples from multiple sites and including investigations for virology. The role that viral infection plays in SUDI remains a matter of debate. Although there is evidence of a viral etiology in some cases, Bajanowski *et al.* (2003) postulated that they act simply as a trigger. This theory is supported by a virological analysis of 490 SUDI autopsies, which identified virus in 4% of cases, but in only half of these was death attributed to a viral infection. In the remainder, death was considered to be due to other infective and noninfective causes (Weber *et al.*, 2010a). The authors concluded that routine virological analysis was of only limited benefit, and, although more sensitive molecular techniques may identify more cases, the contribution to the final cause of death may still be difficult to establish.

Although many SUDI investigation protocols emphasize the need to take microbiology samples as soon as possible after death, usually on arrival in the accident and emergency department (AED), a delay may not be detrimental to the inquiry. Pryce *et al.* (2011a) evaluated 109 SUDI cases with blood cultures samples taken in both AED and at autopsy. A total of 43 cases had positive AED blood cultures, of which 79% were also positive at autopsy, and 37 of 66 cases with negative blood cultures from AED had subsequent positive autopsy blood cultures. Although there were two cases where *S. aureus* was isolated only in AED, its significance was unclear and the authors concluded that in no case was a definite pathogen identified in an AED sample that was not also found at autopsy (Pryce *et al.*, 2011a). Interestingly, the study also found that, despite documentation to the contrary, in 40% of SUDI cases, no samples were actually submitted from AED so that, in these, the autopsy played an important role in identifying the presence of significant pathogens that may have contributed to death. If microbiology samples are overlooked, or there is no facility for them to be taken immediately after death, culture results appear to be stable as long as the autopsy is performed within 48 hours (Lobmaier *et al.*, 2009). Although cultures taken after 48 hours yield fewer isolates, routine SUDI sampling should still take place even with a PMI of several days (Weber *et al.*, 2010b).

Perhaps the biggest challenge in investigation of SUDI is to distinguish true infection from simple postmortem contaminants, postmortem translocation, and incidental antemortem colonization. Since most SUDI samples are associated with some growth, it can be extremely difficult to definitively rule out infection, and finding a pathogenic organism does not necessarily indicate it as significant with regard to death (Morris *et al.*, 2006; Prtak *et al.*, 2010). As there are no objective criteria by which to judge significance, interpretation is usually based on opinion and this can lead to dissent. Pryce *et al.* (2011b) distributed a number of SUDI scenarios, differing only in their microbiology results, to 63 specialists in the field. Although there was general agreement in scenarios where infection was highly likely or highly unlikely, this was not the case when attributing significance to the cause of death in others, for example, when *S. aureus* was present in mixed culture. Given the recommendation for an expert, multidisciplinary review of autopsy and other findings before death certification (The Royal College of Pathologists and The Royal College of Paediatrics and Child Health, 2004), differences in interpretation will need to be overcome, and a consensus reached.

7.7.9 Emerging infectious diseases and bioterrorism agents

Over the last few decades, there has been a steep increase in the numbers of emerging and reemerging infectious diseases threats—many of which are zoonoses (Khabbaz *et al.*, 2010). International travel, growing populations, urban development, social disruption, global environmental changes, and overuse of antibiotics are key factors in the epidemiology of these diseases as they provide ideal conditions for microorganisms to evolve, adapt, and survive (Schwartz and Herman, 1996; Procop and Wilson, 2001). These new and reemerging infectious diseases, along with increased concerns of bioterrorism, have created new challenges in diagnosis, surveillance, and public health response efforts. Due to the potential for large numbers of casualties, pathologists are often among the first healthcare professionals to encounter infectious diseases outbreaks, new infectious agents, or reemerging infections. Expertise in infectious diseases pathology is therefore crucial, not only in understanding the diagnosis but also in informing public health strategies on outbreak investigation and management.

In the event of a request to undertake an autopsy on an individual who may have died from a bioterrorism attack, the pathologist should immediately seek expert advice from national Microbiology and Public Health departments, such as Public Health England or the US Centers for Disease Control and Prevention (CDC). Close liaison with local services, including the microbiology laboratory, will be required. CDC has developed a classification system for bioterrorism agents based on how well they align with the properties of an ideal biological agent, namely, (i) highly lethal, (ii) easily dispersed, (iii) easily produced in vast quantities, (iv) stable, (v) preferably transmissible via the aerosol route, or from person-to-person, (vi) resistant to standard treatment, and (vii) not vaccine preventable (Nulens and Voss, 2002; Centers for Disease Control and Prevention, 2006) (Table 7.7). In the United Kingdom, the Advisory Committee on Dangerous Pathogens (ACDP) and the Control of Substances Hazardous to Health (COSHH) Regulations categorize biological agents by hazard

Table 7.7 Agents with bioterrorism potential as categorized by US Centers for Disease Control and Prevention (Centers for Disease Control and Prevention, 2006)

Category	Features	Agent
A **High priority agents**	Easily disseminated Result in high mortality Potential for public panic and social disruption Require special action for public health preparedness	Anthrax Botulism Plague Smallpox Tularemia Viral hemorrhagic fever
B **Second highest priority agents**	Moderately easy to disseminate Result in moderate morbidity and low mortality rates Require specific enhancements of CDC's diagnostic capacity and enhanced disease surveillance	Brucellosis Epsilon toxin of *Clostridium perfringens* Food safety threats (e.g., Salmonellosis, *Escherichia coli 0157:H7*, shigellosis) Glanders Melioidosis Psittacosis Q fever Ricin toxin Staphylococcal enterotoxin B Typhus fever Viral encephalitis Water safety threats (e.g., cholera, cryptosporidiosis)
C **Third highest priority agents**	Emerging pathogens that are readily available, easy to produce/disseminate, and have potential for high morbidity and mortality	Examples include: Nipah virus Hantavirus

group (ACDP, 2013). These classifications are critical in identifying the risk to autopsy personnel and the safety precautions required.

Documentation of the full range of samples and investigations that may be required is beyond the scope of this chapter. It is extensive and depends on the antemortem clinical presentation and pathological findings during autopsy (Nulens and Voss, 2002; Nolte *et al.*, 2010; Fernández-Rodríguez *et al.*, 2015) but would include samples of the brain, heart, lung, liver, spleen, kidney, and other significant diseased or potentially diseased organs for routine histopathology, bacterial and viral culture or PCR, whole blood for bacterial culture, serum for serologic assays or PCR, and nasopharyngeal swab for viral culture/PCR. Samples of all tissues and serum should be frozen for further analysis if required. An outline of the samples required for category A bioterrorism agents is given in Table 7.8.

Table 7.8 Diagnostic samples and microbiology tests for category A bioterrorism agents

Bioterrorism agent (disease)	Presentation	Diagnostic microbiology specimens	Tests available in Microbiology/Public Health Reference Laboratories
Bacillus anthracis (Anthrax)	Cutaneous	Skin biopsy from both the center and periphery of an eschar, blood cultures	Culture
	Inhalational	Hilar lung with regional lymph nodes, bronchi, and trachea Peripheral pulmonary parenchyma Any major organs showing significant pathology Blood cultures	PCR
	Gastrointestinal	Blood cultures, gastric aspirate, ascites, stool	Serology Antigen detection with electrochemiluminescence Toxin detection Phage type (penicillin sensitivity)
Francisella tularensis (Tularemia)	Ulceroglandular	Culture potential port of entry (skin, throat, conjunctiva)	Culture
	Typhoidal/ pulmonary	Bronchial/tracheal wash Gastric aspirate Biopsy of multiple organs and necrotic lymph nodes	Serology PCR
Yersinia pestis (Plague)	Pneumonic Bubonic Septicemic	Blood cultures Bronchial/tracheal wash Sputum Throat Hemorrhagic lymph nodes	Culture F1 antigen detection PCR

Table 7.8 (Continued)

Bioterrorism agent (disease)	Presentation	Diagnostic microbiology specimens	Tests available in Microbiology/Public Health Reference Laboratories
Clostridium botulinum toxin (Botulism)	Descending paralysis	Tissue for anaerobic cultures from suspected entry sites (wound, GI, respiratory)	Detection and identification of *C. botulinum* by PCR and culture
		Serum Food Environmental samples	Detection of botulinum neurotoxins by mouse bioassay
Ebola, Lassa, Marburg viruses (viral hemorrhagic fever)	Viral hemorrhagic fever syndrome	Blood Serum Throat swab/washings Liver biopsy CSF	Culture PCR Serology EM
Variola virus (smallpox)	Rash	Fluid from vesicle Biopsy skin lesion Saliva Serum	EM Giemsa stain Cell culture Serology PCR

Source: Nulens and Voss (2002). Reproduced with permission of Elsevier.

Historically, postmortem analysis has played an integral part in delineating the pathogenesis of new and reemerging infectious diseases—including the identification of Lassa virus as the causative pathogen in hemorrhagic fever cases in Nigeria in 1969 and *L. pneumophila* as the agent causing Legionnaires' disease during the initial outbreak in Philadelphia in 1977 (Schwartz and Herman, 1996). During the 1990s, several unexplained outbreaks occurred in the United States, and autopsy data was instrumental in identifying the causative organisms; in 1993 autopsy data helped characterize the Hantavirus syndrome and virus, and, in 1999, a large cluster of cases of meningoencephalitis associated with muscle weakness in New York City, NY, USA was recognized as West Nile virus encephalitis, which was the first time it had been seen in the Western Hemisphere (Nash *et al.*, 2001).

Healthcare workers are traditionally taught that "common things are common." However, in the face of several recent bioterrorism events, it is important to have an enhanced index of suspicion, especially when the clinical picture does not entirely fit with a natural infectious disease outbreak. There are a number of epidemiological indicators that can help differentiate between a deliberate and natural infectious disease outbreak. A highly unusual event with large numbers of casualties, or associated with higher morbidity or mortality than is expected, would be suspicious for a bioterrorism attack. If a disease is uncommon for a particular environment or host, or if the vector required for its spread is not normally present, this should also arouse suspicion. Other clues include a point source outbreak, multiple epidemics occurring simultaneously, unnatural spread of a zoonosis in which human disease precedes animal disease or unusual disease manifestations, or antibiotic resistance patterns (Dembek *et al.*, 2007). Due to the high mortality rates associated with category A bioterrorism agents, it is likely that pathologists will be at the forefront of initial investigations. The hazards of dealing with unsuspected cases were highlighted in a report where an autopsy was performed on a patient with tularemia without adequate precautions because the clinical service failed to communicate their clinical suspicion. Two autopsy personnel required prophylactic doxycycline but neither developed signs nor symptoms of tularemia (Shapiro and Schwartz, 2002).

7.8 Conclusion

Despite a decline in the numbers that are being performed, the autopsy still has an important part to play in quality assuring clinical diagnostic processes and developing our understanding of disease, not least when infection is involved. By ensuring that the collection of samples for microbiology is an integral part of the autopsy process, that optimal use is made of standard and molecular techniques, and that open channels of communication exist between pathologists, microbiologists, and public health experts, the information that is generated is central to establishing the cause of death, identifying emerging pathogens, and directing epidemiological investigations in outbreak settings.

References

d'Arminio Monforte, A., Vago, L., Gori, A. *et al.* (1996) Clinical diagnosis of mycobacterial diseases versus autopsy findings in 350 patients with AIDS. *European Journal of Clinical Microbiology and Infectious Diseases*, **15**, 453–458.

Adelson, L. and Kinney, E.R. (1956) Sudden and unexpected death in infancy and childhood. *Pediatrics*, **17**, 663–699.

Advisory Committee on Dangerous Pathogens (ACDP) (2013) *The Approved List of Biological Agents*, 3rd edn. Health and Safety Executive, London, www.hse.gov.uk/pubns/misc208.htm (accessed March 10, 2016).

Andrews, C.P., Coalson, J.J., Smith, J.D. and Johanson, W.G. (1981) Diagnosis of nosocomial bacterial pneumonia in acute, diffuse lung injury. *Chest*, **80**, 254–258.

Astrup, B.S. and Thomsen, J.L. (2007) The routine use of C-reactive protein in forensic investigations. *Forensic Science International*, **172**, 49–55.

Augsburger, M., Iglesias, K., Bardy, D. *et al.* (2013) Diagnostic value of lipopolysaccharide-binding protein and procalcitonin for sepsis diagnosis in forensic pathology. *International Journal of Legal Medicine*, **127**, 427–435.

Bajanowski, T., Rolf, B., Jorch, G. and Brinkmann, B. (2003) Detection of RNA viruses in sudden infant death (SID). *International Journal of Legal Medicine*, **117**, 237–240.

Ball, J., Desselberger, U. and Whitwell, H. (1991) Long-lasting viability of HIV after patient's death. *Lancet*, **338**, 63.

Bartlett, J.G., Dowell, S.F., Mandell, L.A. *et al.* (2000) Practice guidelines for the management of community-acquired pneumonia in adults. Infectious Diseases Society of America. *Clinical Infectious Diseases*, **31**, 347–382.

Barton, R.C. (2013) Laboratory diagnosis of invasive aspergillosis: from diagnosis to prediction of outcome. *Scientifica*, **2013**, 459405.

Berlot, G., Calderan, C., Fiorenza, C. *et al.* (2014) Infective and non-infective endocarditis in critically ill patients: a clinical–pathological study. *Internal and Emergency Medicine*, **9**, 773–778.

Blaas, S.H., Mutterlein, R., Weig, J. *et al.* (2008) Extensively drug resistant tuberculosis in a high income country: a report of four unrelated cases. *BMC Infectious Diseases*, **8**, 60.

Blackwell, C. (2008) Bacterial toxins and sudden unexpected death in infancy. *Lancet*, **372**, 714.

Bode-Janisch, S., Schults, S., Schmidt, A. *et al.* (2013) Serum procalcitonin levels in the postmortem diagnosis of sepsis. *Forensic Science International*, **226**, 266–272.

Bonds, L.A., Gaido, L., Woods, J.E. *et al.* (2003) Infectious diseases detected at autopsy at an urban public hospital, 1996–2001. *American Journal of Clinical Pathology*, **119**, 866–872.

Boon, A.P., O'Brien, D. and Adams, D.H. (1991) 10 Year review of invasive aspergillosis detected at necropsy. *Journal of Clinical Pathology*, **44**, 452–454.

Burn, C.G. (1934) Experimental studies of postmortem bacterial invasion in animals. *Journal of Infectious Diseases*, **54**, 388–394.

Burton, J.L. (2003) Health and safety at necropsy. *Journal of Clinical Pathology*, **56**, 254–260.

Cameron, H.M. and McGoogan, E. (1981) A prospective study of 1152 hospital autopsies. II. Analysis of inaccuracies in clinical diagnosis and their significance. *Journal of Pathology*, **133**, 285–301.

Cardozo, D.M., Nascimento-Carvalho, C.M.C., Souza, F.R. and Silva, N.M.S. (2006) Nasopharyngeal colonization and penicillin resistance among pneumococcal strains: a worldwide 2004 update. *Brazilian Journal of Infectious Diseases*, **10**, 293–303.

Carlson, J.A., Middleton, P.J., Szymanski, M.T. *et al.* (1978) Fatal rotavirus gastroenteritis: an analysis of 21 cases. *American Journal of Diseases of Children*, **132**, 477–479.

Carpenter, H.M. and Wilkins, R.M. (1964) Autopsy bacteriology: review of 2033 cases. *Archives of Pathology*, **77**, 73–81.

Castelblanco, R.L., Lee, M. and Hasbun, R. (2013) Epidemiology of bacterial meningitis in the USA from 1997 to 2010: a population-based observational study. *Lancet Infectious Diseases*, **14**, 813–819.

Centers for Disease Control and Prevention (2005) Guidelines for preventing the transmission of *M. tuberculosis* in healthcare settings. Centers for Disease Control and Prevention, Atlanta, GA.

Centers for Disease Control and Prevention (2006) Emergency preparedness and response. Bioterrorism overview. Centers for Disease Control and Prevention, Atlanta, GA.

Centers for Disease Control and Prevention (2009) *Biosafety in Microbiological and Biomedical Laboratories*, 5th edn, Centers for Disease Control and Prevention, Atlanta, GA.

Centers for Disease Control and Prevention (2010) Estimates of 2009 H1N1 influenza cases, hospitalizations and deaths in the United States, April 2009–February 13, 2010, http://www.cdc.gov/H1N1flu/estimates/April_February_13.htm (accessed November 19, 2016).

Centers for Disease Control and Prevention (2012) Press release: deaths from gastroenteritis double. http://www.cdc.gov/media/releases/2012/p0314_gastroenteritis.html (accessed June 6, 2015).

Centers for Disease Control and Prevention (2013a) CDC in India factsheet. http://www.cdc.gov/globalhealth/countries/india/pdf/india.pdf (accessed June 8, 2015).

Centers for Disease Control and Prevention (2013b) Prevention and control of meningococcal disease. *Morbidity and Mortality Weekly Report, Recommendations and Reports*, **62** (RR02), 1–22.

Centers for Disease Control and Prevention (2013c) Specimen submission guidelines for pathologic evaluation of sudden unexplained infant death with pathologic or clinical suspicion of infection, http://www.cdc.gov/ncezid/dhcpp/idpb/specimen-submission/suid.html (accessed May 24, 2015).

Centers for Disease Control and Prevention (2016). Sudden unexpected infant death and sudden infant death syndrome, http://www.cdc.gov/sids/aboutsuidandsids.htm (accessed November 19, 2016).

Chamilos, G., Luna, M., Lewis, R.E. *et al.* (2006) Invasive fungal infections in patients with hematologic malignancies in a tertiary care cancer center: an autopsy study over a 15-year period (1989–2003). *Haematologica*, **91**, 986–989.

Christofferson, S. (2015) The importance of microbiological testing for establishing cause of death in 42 forensic autopsies. *Forensic Science International*, **250**, 27–32.

Collins, C.H. and Grange, J.M. (1999) Tuberculosis acquired in laboratories and necropsy rooms. *Communicable Disease and Public Health*, **2**, 161–167.

Daniels, R. (2011) Surviving the first hours in sepsis: getting the basics right (an intensivist's perspective). *Journal of Antimicrobial Chemotherapy*, **66** (Suppl 2), ii11–ii23.

Dawson, S.J., Wilson, M.P., White, D.G. *et al.* (2008) Audit of blood culture contamination rate. *Bulletin of the Royal College of Pathologists*, **144**, 290–291.

Dembek, Z.F., Kortepeter, M.G. and Pavlin, J.A. (2007) Discernment between deliberate and natural infectious diseases outbreaks. *Epidemiology and Infection*, **135**, 353–371.

Denison, A.M., Blau, D.M., Jost, H.A. *et al.* (2011) Diagnosis of influenza from respiratory autopsy tissues detection of virus by real-time reverse transcription-PCR in 222 cases. *Journal of Molecular Diagnostics*, **13**, 123–128.

Dimopoulos, G., Piagnerelli, M., Berré, J. *et al.* (2003) Disseminated aspergillosis in intensive care unit patients: an autopsy study. *Journal of Chemotherapy*, **15**, 71–75.

Doerr, C.A., Starke, J.R. and Ong, L.T. (1995) Clinical and public health aspects of tuberculous meningitis in children. *Journal of Pediatrics*, **127**, 27–33.

Edler, C., Klein, A., Gehl, A. *et al.* (2011) The new influenza A (H1N1/09): symptoms, diagnostics, and autopsy results. *International Journal of Legal Medicine*, **125**, 157–161.

Eisenfield, L., Ermocilla, R., Wirtschafter, D. and Cassady, G. (1983) Systemic bacterial infections in neonatal deaths. *American Journal of Diseases of Children*, **137**, 645–649.

European Centre for Disease Prevention and Control (2015) Invasive cardiovascular infection by *Mycobacterium chimaera*—30 April 2015. European Centre for Disease Prevention and Control, Stockholm.

Fernández Guerrero, M.L., Álvarez, B., Manzarbeitia, F. and Renedo, G. (2012) Infective endocarditis at autopsy: a review of pathologic manifestations and clinical correlates. *Medicine*, **91**, 152–164.

Fernández-Rodríguez, A., Alcalá, B. and Álvarez-Lafuente, R. (2008) Real-time polymerase chain reaction detection of *Neisseria meningitidis* in formalin-fixed tissues from sudden deaths. *Diagnostic Microbiology and Infectious Disease*, **60**, 339–346.

Fernández-Rodríguez, A., Cohen, M.C., Lucena, J. *et al.* (2015) How to optimise the yield of forensic and clinical post-mortem microbiology with an adequate sampling: a proposal for standardisation. *European Journal of Clinical Microbiology and Infectious Diseases*, **34**, 1045–1057.

Flavin, R.J., Gibbon, N. and O'Briain, D.S. (2007) Mycobacterium tuberculosis at autopsy—exposure and protection: an old adversary revisited. *Journal of Clinical Pathology*, **60**, 487–491.

Fredette, J.W. (1916) Bacteremias in the agonal period. *Journal of Laboratory and Clinical Medicine*, **2**, 180–188.

Fujita, M.Q., Zhu, B.L., Ishida, K. *et al.* (2002) Serum C-reactive protein levels in postmortem blood—an analysis with special reference to the cause of death and survival time. *Forensic Science International*, **130**, 160–166.

Ganczak, M., Boron-Kaczmarska, A. and Dziuba, I. (2003) Pathologist and HIV—are safe autopsies possible? *Polish Journal of Pathology*, **54**, 143–146.

Garbino, J., Fluckiger, U., Elzi, L. *et al.* (2011) Survey of aspergillosis in non-neutropenic patients in Swiss teaching hospitals. *Clinical Microbiology and Infection*, **17**, 1366–1371.

Gibson, T.N., Shirley, S.E., Escoffery, C.T. and Reid, M. (2004) Discrepancies between clinical and postmortem diagnoses in Jamaica: a study from the University Hospital of the West Indies. *Journal of Clinical Pathology*, **57**, 980–985.

Gill, C.O. and Penney, N. (1979) Survival of bacteria in carcasses. *Applied and Environmental Microbiology*, **37**, 667–669.

Giordano, A.S. and Barnes, A.R. (1922) Studies in postmortem bacteriology: value and importance of cultures made postmortem. *Journal of Laboratory and Clinical Medicine*, **7**, 538–546.

Glode, M.P., Daum, R.S., Goldmann, D.A. *et al.* (1980) *Haemophilus influenzae* type B meningitis: a contagious disease of children. *British Medical Journal*, **280**, 899–901.

Goldwater, P.N. (2009) Sterile site infection at autopsy in sudden unexpected deaths in infancy. *Archives of Disease in Childhood*, **94**, 303–307.

Groll, A.H., Shah, P.M., Mentzel, C. *et al.* (1996) Trends in the postmortem epidemiology of invasive fungal infections at a university hospital. *Journal of Infection*, **33**, 23–32.

Guarner, J., Paddock, C.D., Shieh, W.J. *et al.* (2006) Histopathologic and immunohistochemical features of fatal influenza virus infection in children during the 2003–2004 season. *Clinical Infectious Diseases*, **43**, 132–140.

Guarner, J., Packard, M.M., Nolte, K.B. *et al.* (2007) Usefulness of immunohistochemical diagnosis of *Streptococcus pneumoniae* in formalin-fixed, paraffin-embedded specimens compared with culture and Gram stain techniques. *American Journal of Clinical Pathology*, **127**, 612–618.

Harbarth, S., Holeckova, K., Froidevaux, C. *et al.* (2001) Diagnostic value of procalcitonin, interleukin-6, and interleukin-8 in critically ill patients admitted with suspected sepsis. *American Journal of Respiratory and Critical Care Medicine*, **164**, 396–402.

Harms, P.W., Schmidt, L.A., Smith, L.B. *et al.* (2010) Autopsy findings in eight patients with fatal H1N1 influenza. *American Journal of Clinical Pathology*, **134**, 27–35.

Hawkey, P.M., Pedler, S.J. and Southall, P.J. (1980) *Streptococcus pyogenes*: a forgotten occupational hazard in the mortuary. *British Medical Journal*, **281**, 1058.

Healing, T.D., Hoffman, P.N. and Young, S.E.J. (1995) The infection hazards of human cadavers. *Communicable Disease Report. CDR Review*, **5**, R61–R68.

Health Protection Agency (2013) Guidance for the interpretation of PCR assays for gastrointestinal pathogens, https://www.gov.uk/government/uploads/system/uploads/attachment_data/file/340692/Guidance_for_the_interpretation_of_PCR_assays_for_Gastrointestinal_Pathogens.pdf (accessed May 21, 2015).

Heras Cañas, V., Pérez Ramirez, M.D., Bermudez Jiménez, F. *et al.* (2015) *Lactococcus garvieae* endocarditis in a native valve identified by MALDITOF MS and PCR-based 16s rRNA in Spain: a case report. *New Microbes and New Infections*, **5**, 13–15.

Heron, M. (2013) Deaths: leading causes for 2010. *National Vital Statistics Reports*, **62** (6), 1–96.

Herrmann, M.D., Neumayr, A., Essig, A. *et al.* (2014) Isolated Whipple's endocarditis: an underestimated diagnosis that requires molecular analysis of surgical material. *Annals of Thoracic Surgery*, **98**, e1–e3.

Highet, A.R. (2008) An infectious aetiology of sudden infant death syndrome. *Journal of Applied Microbiology*, **105**, 625–635.

Highet, A.R., Berry, A.M. and Goldwater, P.N. (2009) Novel hypothesis for unexplained sudden unexpected death in infancy (SUDI). *Archives of Disease in Childhood*, **94**, 841–843.

Hove, M. and Pencil, S.D. (1998) Effect of postmortem sampling technique on the clinical significance of autopsy blood cultures. *Human Pathology*, **29**, 137–139.

Howatson, A.G. (2006) The autopsy for sudden unexpected death in infancy. *Current Diagnostic Pathology*, **12**, 173–183.

Hunt, H.F., Barrow, E., Thompson, L. and Waldron, G.W. (1929) A bacteriological study of 567 postmortem examinations. *Journal of Laboratory and Clinical Medicine*, **14**, 907–912.

Hyde, E.R., Haarmann, D.P., Petrosino, J.F. *et al.* (2015) Initial insights into bacterial succession during human decomposition. *International Journal of Legal Medicine*, **129**, 661–671.

de Jongh, D.S., Loftis, J.W., Green, G.S. *et al.* (1968) Postmortem bacteriology: a practical method for routine use. *American Journal of Clinical Pathology*, **49**, 424–428.

Juul, A. (1977) Clinical underdiagnosed active tuberculosis. Experience from autopsy material. *Acta Medica Scandinavica*, **202**, 225–229.

Kellerman, G.D., Waterman, N.G. and Scharefenberger, L.F. (1976) Demonstration *in vitro* of postmortem bacterial transmigration. *American Journal of Clinical Pathology*, **66**, 911–915.

Khabbaz, R.F., Ostroff, S.M., LeDuc, J.W. *et al.* (2010) Emerging and reemerging infectious diseases threats, in *Principles and Practice of Infectious Diseases*, 7th edn (eds G.L. Mandell, J.E. Bennett and R. Dolin), Churchill Livingstone Elsevier, Philadelphia, PA, pp. 199–219.

Klugman, K.P. (2007) Clinical impact of antibiotic resistance in respiratory tract infections. *International Journal of Antimicrobial Agents*, **29** (Suppl. 1), S6–S10.

Landefeld, C.S., Chren, M.M., Myers, A. *et al.* (1988) Diagnostic yield of the autopsy in a university hospital and a community hospital. *New England Journal of Medicine*, **318**, 1249–1254.

Lee, A., Mirrett, S., Reller, B. and Weinstein, M.P. (2007) Detection of bloodstream infections in adults: how many blood cultures are needed? *Journal of Clinical Microbiology*, **45**, 3546–3548.

Leisure, M.K., Moore, D.M., Schwartzman, J.D. *et al.* (1990) Changing the needle when inoculating blood cultures. A no-benefit and high-risk procedure. *Journal of the American Medical Association*, **264**, 2111–2112.

Lightfoot, N.F., Richardson, I.R., Shrimanker, J. and Farrell, D.J. (1991) Post-mortem isolation of *Legionella* due to contamination. *Lancet*, **337**, 376.

Lin, J., Lai, C., Chen, Y. *et al.* (2010) Characteristics and outcomes of polymicrobial bloodstream infections in the emergency department: a matched case–control study. *Academic Emergency Medicine*, **17**, 1072–1079.

Lobmaier, V.K., Vege, A., Gaustad, P. and Rognum, T.O. (2009) Bacteriological investigation—significance of time lapse after death. *European Journal of Clinical Microbiology and Infectious Diseases*, **28**, 1191–1198.

Lucas, S. (2010) Autopsies of people with high-risk infection, in *The Hospital Autopsy: A Manual of Fundamental Autopsy Practice*, 3rd edn (eds J.L. Burton and G.N. Rutty), Hodder Arnold, London, pp. 71–89.

Luchini, D., Meacci, F., Oggioni, M.R. *et al.* (2008) Molecular detection of *Leptospira interrogans* in human tissues and environmental samples in a lethal case of leptospirosis. *International Journal of Legal Medicine*, **122**, 229–233.

Lundgren, R., Norrman, E. and Asberg, I. (1987) Tuberculosis infection transmitted at autopsy. *Tubercle*, **68**, 147–150.

Mangin, P., Lugnier, A.A., Chaumont, A.J. *et al.* (1983) Forensic significance of postmortem estimation of the blood cerebrospinal fluid barrier permeability. *Forensic Science International*, **2**, 143–149.

Mauad, T., Hajjar, L.A., Callegari, G.D. *et al.* (2010) Lung pathology in fatal novel human influenza A (H1N1) infection. *American Journal of Respiratory and Critical Care Medicine*, **181**, 72–79.

Maujean, G., Guinet, T., Fanton, L. and Malicier, D. (2013) The interest of postmortem bacteriology in putrefied bodies. *Journal of Forensic Sciences*, **58**, 1069–1070.

Mazuchowski, E.L. and Meier, P.A. (2005) The modern autopsy: what to do if infection is suspected. *Archives of Medical Research*, **36**, 713–723.

Mercer, J. and Talbot, I.C. (1985) Clinical diagnosis: a post-mortem assessment of accuracy in the 1980s. *Postgraduate Medical Journal*, **61**, 713–716.

Minckler, T.M., Newell, G.R., O'Toole, W.F. *et al.* (1966) Microbiology experience in collection of human tissue. *American Journal of Clinical Pathology*, **45**, 85–92.

Moar, J.J. and Millar, S.D. (1984) The value of autopsy bacteriology. *South African Medical Journal*, **66**, 192–193.

Moon, R.Y., Horne, R.S.C. and Hauck, F.R. (2007) Sudden infant death syndrome. *Lancet*, **370**, 1578–1587.

Morris, J.A. and Harrison, L.M. (2008) Sudden unexpected death in infancy: evidence of infection. *Lancet*, **371**, 1815–1816.

Morris, J.A., Harrison, L.M. and Partridge, S.M. (2006) Postmortem bacteriology: a re-evaluation. *Journal of Clinical Pathology*, **59**, 1–9.

Morris, J.A., Harrison, L.M. and Partridge, S.M. (2007) Practical and theoretical aspects of postmortem bacteriology. *Current Diagnostic Pathology*, **13**, 65–74.

du Moulin, G.C. and Patterson, D.G. (1985) Clinical relevance of postmortem microbiologic examination: a review. *Human Pathology*, **16**, 539–548.

Murdoch, D.R., Corey, G.R., Hoen, B. *et al.* (2009) Clinical presentation, etiology and outcome of infective endocarditis in the 21st century: the International Collaboration on Endocarditis-Prospective Cohort Study. *Archives of Internal Medicine*, **169**, 463–473.

Nash, D., Mostashari, F., Fine, A. *et al.* (2001) The outbreak of West Nile virus infection in the New York city area in 1999. *New England Journal of Medicine*, **344**, 1807–1814.

Nelson, M., Dockrell, D.H., Edwards, S. and on behalf of the BHIVA Guidelines Subcommittee (2011) British HIV Association and British Infection Association guidelines for the treatment of opportunistic infection in HIV-seropositive individuals 2011. *HIV Medicine*, **12** (Suppl. 2), 1–5.

Newton, D., Coffin, C.M., Clark, E.B. and Lowichik, A. (2004) How the pediatric autopsy yields valuable information in a vertically integrated health care system. *Archives of Pathology and Laboratory Medicine*, **128**, 1239–1246.

NHS Estates (2005) *HBN 20: Facilities for Mortuary and Post Mortem Room Services*, The Stationery Office, London.

Nicolle, L.E., Bradley, S., Colgan, R. *et al.* (2005) Infectious Diseases Society of America guidelines for the diagnosis and treatment of asymptomatic bacteriuria in adults. *Clinical Infectious Diseases*, **40**, 643–654.

Nolte, K.B., Fischer, M., Reagan, S. and Lynfield, R. (2010) Guidelines to implement medical examiner/coroner-based surveillance for fatal infectious diseases and bioterrorism ("Med-X"). *American Journal of Forensic Medicine and Pathology*, **31**, 308–312.

Nulens, E. and Voss, A. (2002) Laboratory diagnosis and biosafety issues of biological warfare agents. *Clinical Microbiology and Infection*, **8**, 455–466.

Nyberg, M., Suni, J. and Haltia, M. (1990) Isolation of human immunodeficiency virus (HIV) at autopsy one to six days post-mortem. *American Journal of Clinical Pathology*, **94**, 422–425.

O'Connell, M.L., Birkenkamp, K.E., Kleiner, D.E. *et al.* (2012) Lung manifestations in an autopsy-based series of pulmonary or disseminated nontuberculous mycobacterial disease. *Chest*, **141**, 1203–1209.

O'Toole, W.F., Saxena, H.M., Golden, A. and Ritts, R.E. (1965) Studies of postmortem microbiology using sterile autopsy technique. *Archives of Pathology and Laboratory Medicine*, **80**, 540–547.

Palmiere, C., Bardy, D., Mangin, P. and Ausburger, M. (2013) Value of sTREM-1, procalcitonin and CRP as laboratory parameters for post-mortem diagnosis of sepsis. *Journal of Infection*, **67**, 545–555.

Palmiere, C., Vanhaebost, J., Ventura, F. *et al.* (2015) Cerebrospinal fluid PCR analysis and biochemistry in bodies with severe decomposition. *Journal of Forensic and Legal Medicine*, **30**, 21–24.

Parker, C.M., Kutsogiannis, J., Muscedere, J. *et al.* (2008) Ventilator-associated pneumonia caused by multidrug-resistant organisms or *Pseudomonas aeruginosa*: prevalence, incidence, risk factors, and outcomes. *Journal of Critical Care*, **23**, 18–26.

Paul, N. and Jacob, M.E. (2006) An outbreak of cadaver-acquired chickenpox in a health care setting. *Clinical Infectious Diseases*, **43**, 599–601.

Pechal, J.L., Crippen, T.L., Benbow, M.E. *et al.* (2014) The potential use of bacterial community succession in forensics as described by high throughput metagenomic sequencing. *International Journal of Legal Medicine*, **128**, 193–205.

Perry, J.D. and Freydiere, A.M. (2007) The application of chromogenic media in clinical microbiology. *Journal of Applied Microbiology*, **103**, 2046–2055.

Pittet, D. and Harding, I. (1998) Infective endocarditis and glycopeptides. *Journal of Infection*, **37**, 127–135.

Platt, M.S., McClure, S., Clarke, R. *et al.* (1989) Postmortem cerebrospinal fluid pleocytosis. *American Journal of Forensic Medicine and Pathology*, **10**, 209–212.

Ploy, M.C., Garnier, F., Languepin, J. *et al.* (2005) Interest of postmortem-collected specimens in the diagnosis of fulminant meningococcal sepsis. *Diagnostic Microbiology and Infectious Disease*, **52**, 65–66.

Procop, G.W. (2007) Molecular diagnostics for the detections and characterisation of microbial pathogens. *Clinical Infectious Diseases*, **45**, S99–S111.

Procop, G.W. and Wilson, M. (2001) Infectious disease pathology. *Clinical Infectious Diseases*, **32**, 1589–1601.

Prtak, L., Al-Adnani, M., Fenton, P. *et al.* (2010) Contribution of bacteriology and virology in sudden unexpected death in infancy. *Archives of Disease in Childhood*, **95**, 371–376.

Pryce, J.W., Roberts, S.E., Weber, M.A. *et al.* (2011a) Microbiological findings in sudden unexpected death in infancy: comparison of immediate postmortem sampling in casualty departments and at autopsy. *Journal of Clinical Pathology*, **64**, 421–425.

Pryce, J.W., Weber, M.A., Hartley, J.C. *et al.* (2011b) Difficulties in interpretation of post-mortem microbiology results in unexpected infant death: evidence from a multidisciplinary survey. *Journal of Clinical Pathology*, **64**, 706–710.

Pryce-Davies, J. and Hurley, R. (1979) Infections and perinatal mortality. *Journal of Antimicrobial Chemotherapy*, **5** (Suppl. A), 59–70.

Public Health England (2014a) Chapter 12: Immunisation of healthcare and laboratory staff. In: *The Green Book*, Public Health England, London, https://www.gov.uk/government/uploads/system/uploads/attachment_data/file/147882/Green-Book-Chapter-12.pdf (accessed June 14, 2015).

Public Health England (2014b) Tuberculosis in the UK: 2014 report. Public Health England, London.

Public Health England (2015) Collaborative Tuberculosis Strategy for England 2015 to 2020. Public Health England, London.

Ramsthaler, F., Kettner, M., Mall, G. and Bratzke, H. (2008) The use of rapid diagnostic test of procalcitonin serum levels for the postmortem diagnosis of sepsis. *Forensic Science International*, **178**, 139–145.

Read, R.C. (2014) *Neisseria meningitidis*; clones, carriage, and disease. *Clinical Microbiology and Infection*, **20**, 391–395.

Roberts, F.J. (1998) Procurement, interpretation, and value of postmortem cultures. *European Journal of Clinical Microbiology and Infectious Diseases*, **17**, 821–827.

Robinson, M.J., Greenberg, J.J., Korn, M. and Rywlin, A.M. (1972) Infective endocarditis at autopsy: 1965–1969. *American Journal of Medicine*, **52**, 492–498.

Rodrigues, L.C., Diwan, V.K. and Wheeler, J.G. (1993) Protective effect of BCG against tuberculous meningitis and military tuberculosis: a meta-analysis. *International Journal of Epidemiology*, **22**, 1154–1158.

Roulson, J., Benbow, E.W. and Hasleton, P.S. (2005) Discrepancies between clinical and autopsy diagnosis and the value of post mortem histology; a meta-analysis and review. *Histopathology*, **47**, 551–559.

Saad, R., Yamada, A.T., Pereira da Rosa, F.H. *et al.* (2007) Comparison between clinical and autopsy diagnoses in a cardiology hospital. *Heart*, **93**, 1414–1419.

Sadler, D.W. (1998) The value of a thorough protocol in the investigation of sudden infant deaths. *Journal of Clinical Pathology*, **51**, 689–694.

Saegeman, V., Verhaegen, J., Lismont, D. *et al.* (2009) Influence of postmortem time on the outcome of blood cultures among cadaveric tissue donors. *European Journal of Clinical Microbiology and Infectious Diseases*, **28**, 161–168.

Sarode, V.R., Datta, B.N., Banerjee, A.K. *et al.* (1993) Autopsy findings and clinical diagnoses: a review of 1,000 cases. *Human Pathology*, **24**, 194–198.

Sax, H., Bloemberg, G., Hasse, B. *et al.* (2015) Prolonged outbreak of *Mycobacterium chimaera* infection after open-chest heart surgery. *Clinical Infectious Diseases*, **61**, 67–75.

Scallan, E., Hoekstra, R.M., Angulo, F.J. *et al.* (2011) Foodborne illness acquired in the United States—major pathogens. *Emerging Infectious Diseases*, **17**, 7–15.

Schmidt, S., Windgassen, M., Nofer, J.R. *et al.* (2015) Procalcitonin as a postmortem sepsis marker. A comparison of the validity of results obtained from blood serum, aqueous humour and cerebrospinal fluid. *International Journal of Legal Medicine*, **129**, 117–123.

Schrag, B., Iglesias, K., Mangin, P. and Palmiere, C. (2012a) Procalcitonin and C-reactive protein in pericardial fluid for postmortem diagnosis of sepsis. *International Journal of Legal Medicine*, **126**, 567–572.

Schrag, B., Roux-Lombard, A., Schneiter, D. *et al.* (2012b) Evaluation of C-reactive protein, procalcitonin, tumour necrosis factor alpha, interleukin-6 and interleukin-8 as diagnostic parameters in sepsis-related fatalities. *International Journal of Legal Medicine*, **126**, 505–512.

Schwartz, D.A. and Herman, C.J. (1996) Editorial response: the importance of the autopsy in emerging and reemerging infectious diseases. *Clinical Infectious Diseases*, **23**, 248–254.

Schwarz, P., Dannaoui, E., Gehl, A. *et al.* (2014) Molecular identification of fungi found on decomposed human bodies in forensic autopsy cases. *International Journal of Legal Medicine*, **129**, 785–791.

Self, W.H., Speroff, T., Grijalva, C.G. *et al.* (2013) Reducing blood culture contamination in the emergency department: an interrupted time series quality improvement study. *Academic Emergency Medicine*, **20**, 89–97.

Shapiro, D.S. and Schwartz, D.R. (2002) Exposure of laboratory workers to *Francisella tularensis* despite a bioterrorism procedure. *Journal of Clinical Microbiology*, **40**, 2278–2281.

Shojania, K.G., Burton, E.C., McDonald, K.M. and Goldman, L. (2003) Changes in rates of autopsy-detected diagnostic errors over time. *Journal of the American Medical Association*, **289**, 2849–2856.

Smith, R.F., Linares, H.A. and Jorgensen, J.H. (1975) Bacteremia and postmortem microbiology in burned children. *American Journal of Clinical Pathology*, **63**, 502–508.

Spiliopoulou, C., Papadodima, S., Kotakidis, N. and Koutselinis, A. (2005) Clinical diagnoses and autopsy findings—a retrospective analysis of 252 cases in Greece. *Archives of Pathology and Laboratory Medicine*, **129**, 210–214.

Stevanovic, G., Tucakovic, G., Dotlic, R. and Kanjuh, V. (1986) Correlation of clinical diagnoses with autopsy findings: a retrospective study of 2,145 consecutive autopsies. *Human Pathology*, **17**, 1225–1230.

Suzuki, Y., Kume, H., Togano, T. *et al.* (2013) Epidemiology of visceral mycoses in autopsy cases in Japan: the data from 1989 to 2009 in the Annual of Pathological Autopsy Cases in Japan. *Medical Mycology*, **51**, 522–526.

Tajiri, T., Tate, G., Masunaga, A. *et al.* (2012) Autopsy cases of fulminant bacterial infection in adults: clinical onset depends on the virulence of bacteria and patient immune status. *Journal of Infection and Chemotherapy*, **18**, 637–645.

The Royal College of Pathologists and The Royal College of Paediatrics and Child Health (2004) Sudden unexpected death in infancy—a multi-agency protocol for care and investigation. www.rcpath.org/NR/rdonlyres/30213EB6-451B-4830-A7FD-4EEFF0420260/0/SUDIreportforweb.pdf (accessed May 24, 2015).

Theel, E.S. and Doern, C.D. (2013) β-d-Glucan testing is important for diagnosis of invasive fungal infections. *Journal of Clinical Microbiology*, **51**, 3478–3483.

Thorn, J.L., Gilchrist, K.B., Sobonya, R.E. *et al.* (2010) Postmortem candidemia: marker of disseminated disease. *Journal of Clinical Pathology*, **63**, 337–340.

Torriani, F.J., McCutchan, J.A., Bozzette, S.A. *et al.* (1994) Autopsy findings in AIDS patients with *Mycobacterium avium* complex bacteremia. *Journal of Infectious Diseases*, **170**, 1601–1605.

Trafford, G., Ratnaraja, N. and Wickramasinghe, N. (2015) Molecular diagnostic testing for common stool pathogens. *Journal of Hospital Infection*, **90**, 196–198.

Tsokos, M. and Puschel, K. (2001) Postmortem bacteriology in forensic pathology: diagnostic value and interpretation. *Legal Medicine*, **3**, 15–22.

Tsokos, M., Reichelt, U., Nierhaus, A. and Püschel, K. (2001) Serum procalcitonin (PCT): a valuable biochemical parameter for the post-mortem diagnosis of sepsis. *International Journal of Legal Medicine*, **114**, 237–243.

Tuomisto, S., Karhunen, P.J., Vuento, R. *et al.* (2013) Evaluation of postmortem bacterial migration using culturing and real-time quantitative PCR. *Journal of Forensic Science*, **58**, 910–916.

Turner, G.D., Bunthi, C., Wonodi, C.B. *et al.* (2012) The role of postmortem studies in pneumonia etiology research. *Clinical Infectious Diseases*, **54** (Suppl 2), S165–S171.

Uhlin-Hansen, L. (2001) C-reactive protein (CRP), a comparison of pre- and post-mortem blood levels. *Forensic Science International*, **124**, 32–35.

Vogeser, M., Wanders, A., Haas, A. and Ruckdeschel, G. (1999) A four-year review of fatal aspergillosis. *European Journal of Clinical Microbiology and Infectious Diseases*, **18**, 42–45.

Vongpaisarnsin, K., Tansrisawad, N., Hoonwijit, U. and Jongsakul, T. (2015) *Pseudomonas aeruginosa* septicemia causes death following liposuction with allogenic fat transfer and gluteal augmentation. *International Journal of Legal Medicine*, **129**, 815–818.

Walsh, T.R. (2010) Emerging carbapenemases: a global perspective. *International Journal of Antimicrobial Agents*, **36** (Suppl 3), S8–S14.
Weber, M.A., Klein, N.J., Hartley, J.C. *et al.* (2008) Infection and sudden unexpected death in infancy: a systematic retrospective case review. *Lancet*, **371**, 1848–1853.
Weber, M.A., Hartley, J.C., Ashworth, M.T. *et al.* (2010a) Virological investigations in sudden unexpected deaths in infancy (SUDI). *Forensic Science, Medicine, and Pathology*, **6**, 261–267.
Weber, M.A., Hartley, J.C., Brooke, I. *et al.* (2010b) Post-mortem interval and bacteriological culture yield in sudden unexpected death in infancy (SUDI). *Forensic Science International*, **198**, 121–125.
Weber, M.A., Hartley, J.C., Klein, N.J. *et al.* (2011) Staphylococcal toxins in sudden unexpected death in infancy: experience from a single specialist centre. *Forensic Science, Medicine, and Pathology*, **7**, 141–147.
Weinstein, M.P. (2003) Blood culture contamination: persisting problems and partial progress. *Journal of Clinical Microbiology*, **41**, 2275–2278.
West, D.J. (1984) The risk of hepatitis B infection among health professionals in the United States: a review. *American Journal of the Medical Sciences*, **287**, 26–33.
Wilkins, D., Woolcock, A.J. and Cossart, Y.E. (1994) Tuberculosis: medical students at risk. *Medical Journal of Australia*, **160**, 395–397.
Wilson, M.L. (2006) Infectious diseases and the autopsy. *Clinical Infectious Diseases*, **43**, 602–603.
Wilson, S.J., Wilson, M.L. and Reller, L.B. (1993) Diagnostic utility of post-mortem blood cultures. *Archives of Pathology and Laboratory Medicine*, **117**, 986–988.
Winters, B., Custer, J., Galvagno, S.M. *et al.* (2012) Diagnostic errors in the intensive care unit: a systematic review of autopsy studies. *BMJ Quality and Safety*, **21**, 894–902.
Woods, G.L. and Walker, D.H. (1996) Detection of infection or infectious agents by use of cytologic and histologic stains. *Clinical Microbiology Reviews*, **9**, 382–404.
World Health Organization (2015) Tuberculosis. Fact sheet no. 104. http://www.who.int/mediacentre/factsheets/fs104/en/ (accessed June 4, 2015).
World Health Organization (2014) Antimicrobial resistance: global report on surveillance 2014. WHO Press, Geneva.
Wyler, D., Marty, W. and Bär, W. (1994) Correlation between the post-mortem cell content of cerebrospinal fluid and time of death. *International Journal of Legal Medicine*, **106**, 194–199.
Young, S.E.J. (1987) Aetiology and epidemiology of infective endocarditis in England and Wales. *Journal of Antimicrobial Chemotherapy*, **20** (Suppl. A), 7–14.
Zager, E.M. and McNerney, R. (2008) Multidrug-resistant tuberculosis. *BMC Infectious Diseases*, **8**, 10.
Zeien, L.B. and Klatt, E.C. (1990) Cardiac valve prostheses at autopsy. *Archives of Pathology and Laboratory Medicine*, **114**, 933–937.

CHAPTER 8

Postmortem bacterial translocation

Vadim Mesli[1], Christel Neut[2] and Valery Hedouin[1]
[1] *Lille University School of Medicine, CHU Lille, Forensic Taphonomy Unit, UTML, Lille, France*
[2] *Clinical Bacteriological Laboratory, INSERM U995 LIRIC, University of Lille, Lille, France*

8.1 Introduction

Microorganisms participate actively in the decomposition of a human cadaver and are the core of many postmortem changes. The cadaver microbiota shifts temporally after the death of the subject, and temperature, as well as anaerobic conditions, seems to be the preponderant factors driving the decomposition process (Morris *et al.*, 2006; Beauthier, 2011). The cadaver microbiota can have practical consequences in forensic practice, when trying to interpret the postmortem interval or bacteriological samples. Bacterial translocation is a recently developed concept referring to the passage of microorganisms from the gastrointestinal tract to extraintestinal sites, such as the blood (Plantefève and Bleichner, 2001; Balzan *et al.*, 2007). Unfortunately, the timing of bacterial translocation after death is not fully understood. This chapter will examine the different factors that influence postmortem bacterial translocation and its relevance to the forensic practice.

8.1.1 The intestinal microbiota in health

The human intestinal microbiota, also called gut flora, are defined as the microorganisms (mostly bacteria) colonizing the intestinal tract. This microbial community contains from 10^{12} to 10^{14} bacteria, which represents more than 10 times the number of cells in the human body (Marteau, 2013). Recent progress made in molecular biology and genetics is allowing for a better understanding of the physiology and composition of the human intestinal microbiome after death in comparison to culture-based methods (Goulet, 2009; Arumugam *et al.*, 2011; Lozupone *et al.*, 2012; Yatsunenko *et al.*, 2012).

Forensic Microbiology, First Edition. Edited by David O. Carter, Jeffery K. Tomberlin, M. Eric Benbow and Jessica L. Metcalf.

The intestinal tract of a newborn is sterile at birth; however, microorganisms encountered at the earliest stages of life rapidly colonize. The first contact with the mother's microbiota is initiated at birth and then during cutaneous contacts and breastfeeding, referred to as horizontal transmission. Microbial colonization also occurs through interaction with the external environment. The human microbiota is considered not to be matured before the age of two (Grönlund *et al.*, 1999; Rougé *et al.*, 2010). Early in life (and still increased in breast-fed infants), *Bifidobacterium* is predominant, and later the Gram-negative anaerobes like *Bacteroides* and *Prevotella* increase. The composition of the microbiome will evolve during life, and in the elderly (due to the diminution of immune functions with aging, drug therapies, nutritional behavior, etc.) the proportion of endogenous opportunistic bacteria with a higher pathogenic potential will increase, implicated in an increased risk of infection (Cherbuy *et al.*, 2013).

Bacteria normally present in the intestinal tract are called commensal, as opposed to pathogenic bacteria. They exist as an ecosystem that includes the microorganisms, the human host, and the abiotic characteristics (e.g., pH) of the tract. The host consumes food that allows the reproduction of bacteria, which themselves have an impact on the host metabolism (Ducluzeau, 1998). The intestinal microbiota can have positive effects on human health and is now considered an organ as related to its metabolic abilities (referred as symbiosis). On the other hand, alterations of the microbiota (or dysbiosis) can have negative effects, such as in small intestinal bacterial overgrowth, inflammatory bowel diseases, or outgrowth of resistant bacteria after antibiotic administration (for instance, *Clostridium difficile*) (Marteau, 2013).

The use of molecular markers at the phylum (second level of the living species classification) or genus level shows similarities in the core microbiome between individuals, with variations depending on age, native countries, some diets, or diseases. On the contrary, fecal studies that utilized specific molecular markers indicated significant differences between bacterial composition from one individual to another, making the composition of the colonic microflora as specific for each individual as the fingerprint (Costello *et al.*, 2009; Doré and Corthier, 2010; Arumugam *et al.*, 2011; Consortium THMP, 2012; Yatsunenko *et al.*, 2012).

The dominant bacteria of the intestinal microbiome, such as *Bacteroides* and *Bifidobacterium*, digest food residues in anaerobic conditions (also called saccharolytic fermentation). But proteolytic bacteria are still present, such as Enterobacteriaceae, *Enterococcus*, *Clostridium*, or even *Fusobacterium* (Orrhage and Nord, 2000) even if proteins are not really abundant at this level. Symbiosis is still possible between the human host and potential pathogens due to homeostasis between the host and the microorganisms (Steinberg, 2003).

All along the intestinal tract, quantity and diversity of bacterial species are increasing and reaching a maximum of biodiversity in the colon, with more than a thousand different bacterial species in each individual (Goulet, 2009). This evolution is synthesized in Figure 8.1.

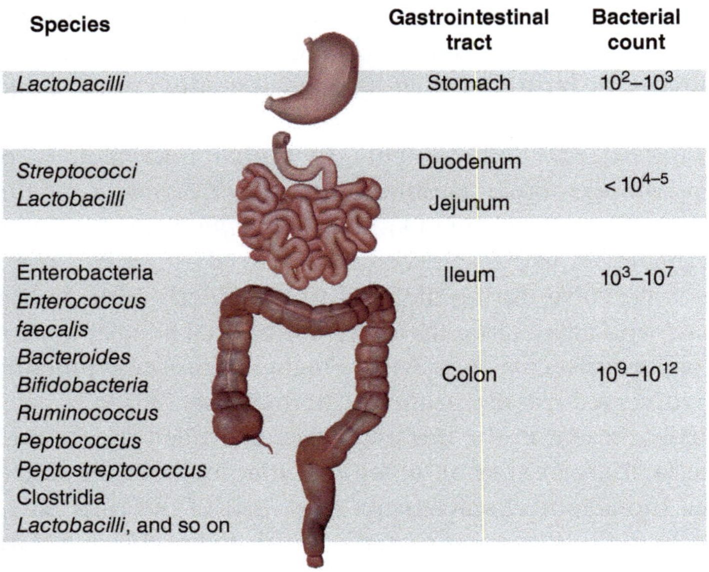

Figure 8.1 Bacterial species and quantification depending on the location on the gastrointestinal tract

8.1.1.1 Microbiome of the stomach

The inside of the stomach is very acidic, with hydrogen potential (pH) around two (Beasley *et al.*, 2015). Those physicochemical conditions select for only a few endogenous bacteria, such as *Helicobacter*, *Streptococcus*, and *Lactobacillus*, which will survive in a healthy host without consequent multiplication. Those bacteria have been implicated in disease pathologies, such as hypochloridia (Ducluzeau, 1998), or may also have a beneficial effect with potentially probiotic bacteria, such as *Lactobacillus*.

8.1.1.2 Microbiome of the duodenum and jejunum

This part of the gastrointestinal tract is characterized by the presence of bile acids. The number of bacteria is about 10^2–10^5 colony-forming units per gram (CFU/g). Facultative anaerobic bacteria are the dominant species here, particularly *Streptococci* and *Lactobacillus* (Girard-Pipau *et al.*, 1981). For optimal health, bacterial levels should remain as low as possible as nutriments should be absorbed by the epithelium and not be degraded by bacteria (which occurs in the syndrome of small intestinal overgrowth, leading to malnutrition) (Donowitz and Petri, 2015).

8.1.1.3 Microbiome of the ileum

Bacterial species are more numerous in the ileum than upstream, continuously increasing from 10^3 to 10^7 CFU/g, with anaerobes becoming more dominant. *Bacteroides*, Enterobacteriaceae, and *Streptococci* are classically identified in this part of the gastrointestinal tract (Ghnassia, 1979; Ducluzeau, 1998). Luminal bacteria have only few interactions with the intestinal mucosa as the passage of the bacteria is rapid due to the important peristalsis at this site. However, interactions may occur with bacteria that are able to adhere to and sometimes invade the epithelial layer such as with pathogenic species of *Salmonella, Shigella, Yersinia,* or *Listeria.*

8.1.1.4 Microbiome of the colon

After the ileocaecal valve, anaerobic species are the dominant taxa, and their number is a 100-fold higher than upstream. The transit is slow, allowing bacteria to have enough time to interact with their environment and to develop, and bacterial amounts are reaching a peak concentration up to 10^{12} CFU/g (Marteau, 2013). Differences are observed inside the colon, between cecal or ascending, transverse or descending colonic locations. This ecosystem is closer and closer to the one described in the feces (Marteau, 1996; Ducluzeau, 1998). Both endogenous and exogenous substrates can be used for bacterial fermentation. Exogenous substrates concern mainly complex polysaccharides not digested in the small intestine (resistant starch, pectin, cellulose), which underlines the importance of diet (and mainly the consumption of fruits and vegetables) in maintaining a homeostasis between the human body and the microbiome. Bacteria degrading these polysaccharides (*Bifidobacterium, Bacteroides, Ruminococcus*) have a low pathogenic potential. An important endogenous substrate is the mucus secreted continuously by the epithelium. This layer is stiffer and larger in the colon than in the small intestine and will protect the epithelium from contact with bacteria. The inner layer is usually sterile despite the high luminal amount of bacteria. It is one of the most important factors of protection assuring the sterility of the inner organs of the human body (McGuckin *et al.*, 2011). Here also only bacteria with a very low pathogenic potential are able to adhere and to degrade slowly the mucins in health. They occupy ligands of the mucins, which are no longer available for pathogenic bacteria. Mucin structure comprises large amounts of sugar residues linked to a protein core. Typically mucin-degrading bacteria are *Akkermansia, Ruminococcus, Bifidobacterium,* and *Bacteroides* (McGuckin *et al.*, 2011).

8.2 Bacterial translocation in health and disease

The migration of bacteria through the intestinal mucosa has been observed since 1950 (Schweinburg *et al.*, 1950). With numerous studies characterizing this phenomenon, the definition of the intestinal bacterial translocation has changed several times over the two last decades. It is currently acknowledged that bacterial translocation is defined by the migration of viable bacteria or bacterial fragments through the gastrointestinal tract toward the mesenteric lymph nodes (where bacteria are most likely eliminated), to reach the systemic circulation (Plantefève and Bleichner, 2001; Balzan *et al.*, 2007).

Anaerobic and aerobic bacteria are both capable of bacterial translocation. Some bacteria have a higher potential of translocation, such as Enterobacteriaceae, *Enterococcus*, and *Clostridium* in particular (Laffineur *et al.*, 1992). Bacterial translocation can cause systemic inflammation, or even a sepsis, in the living human host (Roy and Gatt, 2012). However, bacterial translocation is not always observed in pathological situations. It was indeed revealed in healthy individuals, without any negative consequence detected (Sedman *et al.*, 1994; O'Boyle *et al.*, 1998) if the immune defenses of the human body are able to eliminate them in the lymph nodes or at least in the blood. Bacterial translocation is thought to occur mainly after eating (Kelly *et al.*, 2012). In food industry, animals are slaughtered after weaning to reduce the presence of bacteria in the tissues used for consumption.

8.2.1 Pathophysiological mechanisms

Bacteria are present in the intestinal lumen and have to go through several "obstacles" until they reach the lamina propria (connective tissue under the epithelium). Those main obstacles are:

- The commensal bacteria, present in large numbers, act as a barrier to access to the intestinal epithelium (Marteau, 2013).
- The mucus, a viscous substance produced by specialized epithelial cells, covers the surface of the gastrointestinal tract epithelium. It constitutes a physical barrier and also inhibits the microorganisms' adhesion to the epithelium. Its thickness is variable and decreases from the antrum of the stomach to the jejunum and then increases to reach a maximum thickness in the colon (McGuckin *et al.*, 2011).
- The migration through the enterocytes, which may be carried out by two options: a paracellular migration (through tight junctions) or a transcellular migration (through the cell itself). The transcellular migration is the process most observed in the living patient, through undamaged enterocytes (Wells and Erlandsen, 1996).

After going through the lamina propria, bacteria cells can access the systemic circulation by two means: (i) by the venous system till the portal vein and (ii) by the lymphatic drainage channels. The lymphatic system seems to be preferred, the positive cultures in mesenteric lymph nodes being in greater number than in the blood (Moore *et al.*, 1991; Plantefève and Bleichner, 2001; Deitch, 2002).

Simultaneously, the digestive immune system of the host is also involved; it is mainly constituted of Peyer's patches (aggregated lymphoid nodules) and B-type lymphoid follicles. Those patches are localized between the mucosa and the submucosa and are topped by a dome rich in B and T lymphocytes and macrophages. This dome also contains M cells, able to perform endocytosis of bacterial agents. After passing by the basal pole of the epithelial cells, bacteria are in contact with the other cells that constitute the gut-associated lymphoid tissue (GALT). All those interactions are allowing the development of a local immunity that could also be transferred to the whole host organism (Neutra, 1998; Plantefève and Bleichner, 2001).

8.2.2 Factors responsible for an increase in the bacterial translocation

Bacterial translocation is higher when the intestinal tract, acting here as a barrier, is altered. Three main types of damage could allow for translocation to occur (Plantefève and Bleichner, 2001; Balzan *et al.*, 2007): (i) intestinal mucosa alteration, (ii) modification of the intestinal microbiota, and (iii) immunodeficiency. The main factors involved in bacterial translocation are synthesized in Figures 8.2 and 8.3. Each will be discussed in the following text in more detail.

8.2.2.1 An intestinal mucosa alteration

Mucus obstructs bacterial adhesion to the epithelium wall. Bacterial translocation facilitation is thus observed during alterations of mucus composition or secretion, such as cases of inflammatory bowel diseases (Katayama *et al.*, 1997). Enteral and parenteral nutrition, inducing atrophied mucosa, are also associated with cases of bacterial translocation (Spaeth *et al.*, 1994). Intestinal ischemia, such as the ones that can be encountered in large cutaneous burns, may facilitate bacterial translocations (Baron *et al.*, 1994). Additionally, decreasing the intestinal blood perfusion induced by intra-abdominal hyper pressure in the mouse model resulted in bacterial translocations (Eleftheriadis *et al.*, 1996).

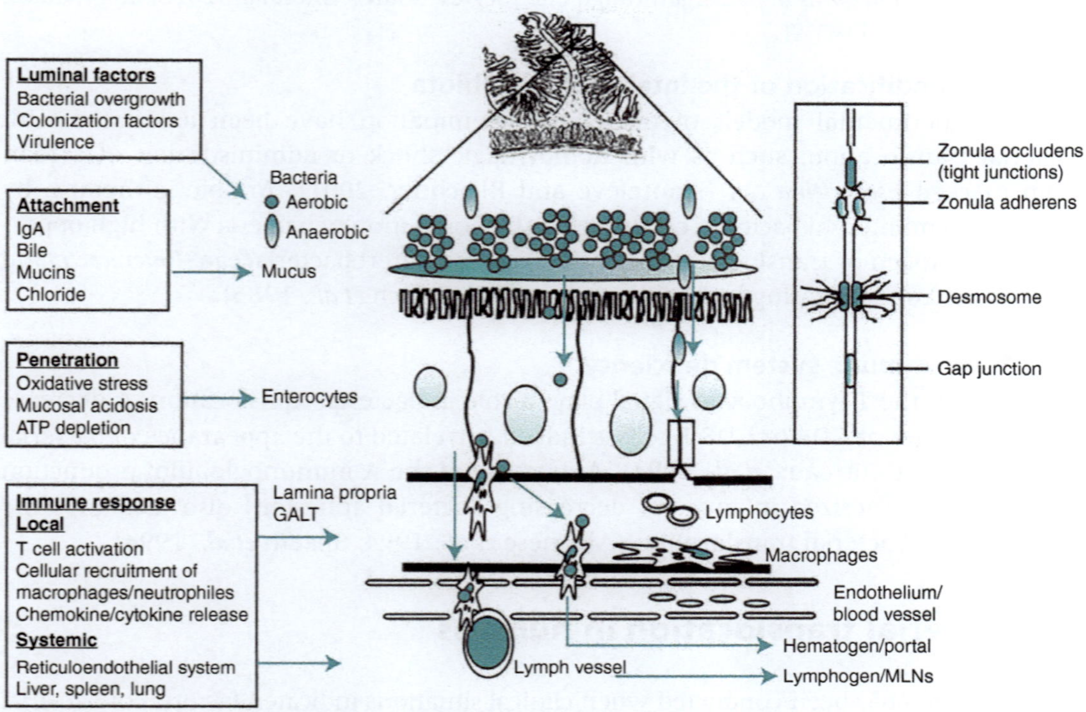

Figure 8.2 Mechanisms and factors involved in bacterial translocation. Source: Wiest and Rath (2003). Reproduced with permission of Elsevier

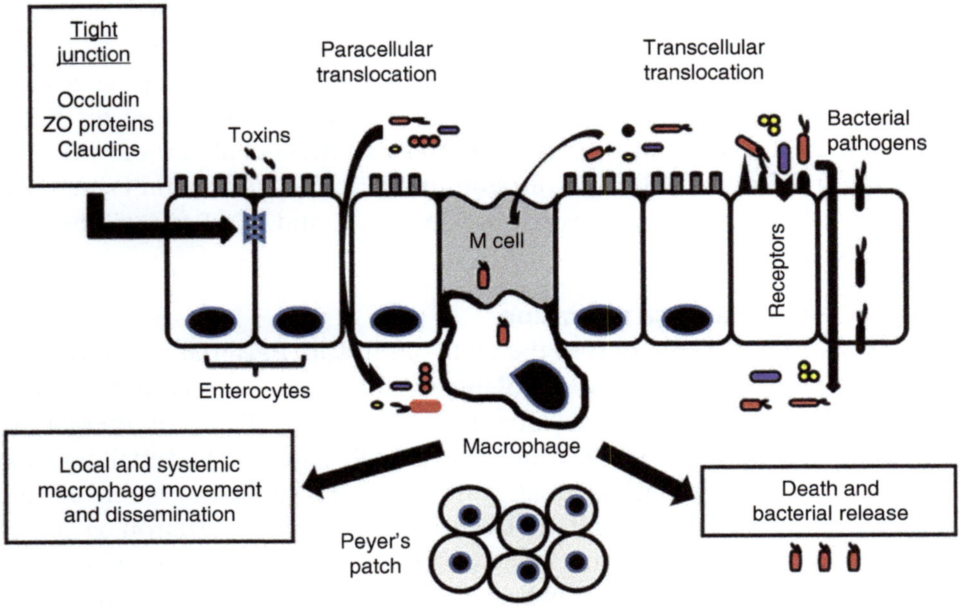

Figure 8.3 Running paths of bacteria through enterocytes. Source: Sherman (2010). Reproduced with permission of Elsevier

8.2.2.2 A modification of the intestinal microbiota

Several experimental models of microbial dissemination have been associated with bacterial translocation, such as with hemorrhagic shock or administration of carbon atom marked *Escherichia coli* (Plantefève and Bleichner, 2001). Antibiotic therapy, by destroying commensal bacteria, can increase the population of bacteria with high opportunistic pathogen or translocation potential (such as Enterobacteriaceae, *Enterococci*, and *Clostridium*) and facilitating bacterial translocation (Deitch *et al.*, 1985).

8.2.2.3 An immune system deficiency

Alteration of the T lymphocytes signal plays a role in bacterial translocation. A decrease in macrophages or CD4 or CD8 T cells is indeed correlated to the appearance of bacterial translocation (Gautreaux *et al.*, 1994). A decrease of the A immunoglobulin production (antibodies synthesized in mucosa, decreasing bacterial adhesion) also facilitates the occurrence of bacterial translocation (Albanese *et al.*, 1994; Spaeth *et al.*, 1994).

8.3 Bacterial translocation in humans

Much research has been conducted when clinical situations indicated factor(s) of bacterial translocation. Thus, bacterial translocation has been identified in several pathological states in humans: hemorrhagic shock (Nadler and Ford, 2000), inflammatory bowel diseases such as Crohn's disease (Laffineur *et al.*, 1992), type 1 or 2 viral immunodeficiency virus infections (Brenchley *et al.*, 2006; Nowroozalizadeh *et al.*, 2010), hepatitis B or C virus infection (Sandler *et al.*, 2011), graft versus host disease (Langrehr *et al.*, 2000), intestinal occlusion (Deitch *et al.*, 1990), diabetes (Amar *et al.*, 2011), or colorectal cancer (Lescut *et al.*, 1990).

8.3.1 Bacterial translocation after death

Viable bacteria originating from the intestinal microbiota have been found in blood samples taken from human cadavers (Table 8.1), providing evidence that bacterial translocation occurs after death (Mesli, 2013). The three main factors resulting in bacterial translocation in the living human (previously described) also occur after death. But there are no precise data concerning the timing or the determining factors of the postmortem bacterial translocation, which could be facilitated by an increased permeability of gastrointestinal wall due to a long agonal phase, the absence of blood supply responsible for ischemia, or the absence of mucus secretion. A medical history could also play a role, such as an intestinal bowel disease.

Table 8.1 Bacteria isolated from blood samples on different remains, in relation to the postmortem interval (Mesli, 2013)

Postmortem interval	Bacteria (CFU/mL)
Case 1 (27 hours)	*Streptococcus oralis* (10)
	Enterobacter agglomerans (10)
	Escherichia coli (10)
	Klebsiella pneumoniae (10)
	Clostridium perfringens (10)
	Clostridium sordellii (1)
	Enterococcus sp. (0,2)
	Veillonella dispar (0,2)
	Bacteroides thetaiotaomicron (1)
	Bacteroides fragilis (0,2)
	Bacteroides ovatus (0,2)
Case 2 (28 hours)	*Staphylococcus* sp. (10)
	E. coli (10)
	Bifidobacterium longum (10)
	Streptococcus anginosus (1)
	Bacteroides vulgatus (0,2)
Case 3 (63 hours)	*Enterobacter agglomerans* (10)
	Streptococcus sp. (10)
	Enterococcus faecium (0,2)
	C. perfringens (0,2)
	Lactobacillus curvatus (0,2)

8.3.2 Identification of bacterial metabolites around the corpse

Analysis of volatile organic compounds in association with vertebrate remains (in the soil and air) has identified several specific metabolites of anaerobic bacteria, providing an indirect evidence of a previous bacterial translocation. Methane and volatile fatty acids were found, such as butyric acid, isobutyric acid, propionic acid, valeric acid, and isovaleric acid. Products of the anaerobic degradation of amino acids were also identified, such as phenylacetic and phenylpropionic acids (Vass *et al.*, 2004, 2008; Swann *et al.*, 2010a; Vass, 2012). Those two acids have also been isolated in analysis of decomposition fluids of pigs' cadavers, when they were disposed on a matrix without soil (Swann *et al.*, 2010a).

8.4 Physiological changes after death influencing the selection of commensal bacteria

Death is often characterized by a final and irreversible cessation of vital functions (cardiovascular, respiratory, and encephalic). The term "death" implies blood circulation has ceased along with oxygen supply to tissues at a macroscopic level and in cells at a microscopic level (Beauthier, 2011). The agonal period preceding death is characterized by hypoxia, which induces multiple organ failure and progressive alteration of vital organismal functions, until death is occurring. This period is variable depending on the cause of death (Morris *et al.*, 2006). Though metabolism in living human cells is aerobic, human remains following death shift from an aerobic environment to anaerobic, as oxygen supply by blood is no longer ensured. The different changes that occur after death are called the postmortem changes of the cadaver. Those modifications have a relatively standardized chronology, even though they can be affected in speed and intensity, mainly depending on individual and environmental factors (Campana and Fornes, 2010; Beauthier, 2011). Human enzymes are not well adapted after death, and anaerobic bacteria can act more rapidly with a higher energetic yield (Beauthier, 2011).

8.4.1 Variations of available substrates for bacterial proliferation

The postmortem degradation of the cadaver is characterized by a succession of different stages described in the forensic literature (Campana and Fornes, 2010; Beauthier, 2011):

- The cadaver is initially "fresh," without any macroscopic sign of decomposition.
- An initial stage decomposition.
- An advanced stage of decomposition.
- A stage of dehydration/desiccation and skeletonization.

During the cadaver decay, the common body modifications classically described are (Durigon, 2007; Beauthier, 2011):

- The release of a characteristic odor
- A modification of the skin color, initially in the abdominal area, with a green color

- A presence of gas in tissues, which can be manifested by a subcutaneous venous dilatation or a distension of the tissues, commonly in the abdomen

Two other phenomena can also be observed depending on environmental conditions, mummification, and adipocerous transformation:

- Mummification is a body transformation by early and accelerated desiccation, which can be observed at the same time with the other decomposition stages previously described. It is favored by a dry environment and gives the skin a brownish and hardened aspect. The body, or some parts of the body, can thus dehydrate without putrefaction or can dehydrate after a putrefactive stage.
- The adipocerous transformation of the cadaver is the transformation of the fat tissues in the presence of water, also called saponification. This phenomenon is more prevalent when the environment is warmer and wetter or when fats are abundant or even when the environment is anaerobic. Some metabolites of aerobic and anaerobic bacteria play an important role in this process, such as some fatty acids (e.g., 10-hydroxystearic) (Takatori, 2001).

8.4.2 Temperature

The body temperature of the living human, approximately 37.2°C under ideal conditions, gradually decreases after death, to meet equilibrium with the ambient temperature. This heat loss is described as about 1°C/hour, with an initial plateau, followed by a rapid phase of heat loss, and then a final plateau (Beauthier, 2011). The skin cools relatively fast compared with internal organs, but the core temperature depends on the exchanges between the skin and the deep tissues (muscles, organs). The rate of the decrease in temperature depends on multiple individual and environmental factors (Henssge, 1988; Muggenthaler *et al.*, 2012). The main factors include weight of the subject (and the thickness of the fat tissues), presence of clothes, initial temperatures of the body, environment (exposure to wind, humidity), and position of the body (Beauthier, 2011). Thus, many thermal factors are recognized as accelerating the speed of the corpse decomposition (Vass, 2001; Zhou and Byard, 2011):

- Exogenous factors:
  - A high ambient temperature, whether it is a heated home or having several layers of clothes, in a vehicle, outside when exposed to high temperatures (in summer), or when the remains have been exposed to the flames of a fire
  - The absence of refrigeration of the remains
- Endogenous factors:
  - A high initial corporal temperature, for example, during hyperthermia, some drug abuse (e.g., cocaine, amphetamine, neuroleptics), or an infection.
  - Obesity, with the thermic isolation provided by the fat tissues.
  - Diabetes and hyperglycemia: this could be explained by a bacterial growth facilitated by glucose fermentation (Zhou and Byard, 2011).

On the contrary, situations with a lower temperature (immersion, refrigeration, or freezing) are known to delay the cadaver decomposition (Campana and Fornes, 2010).

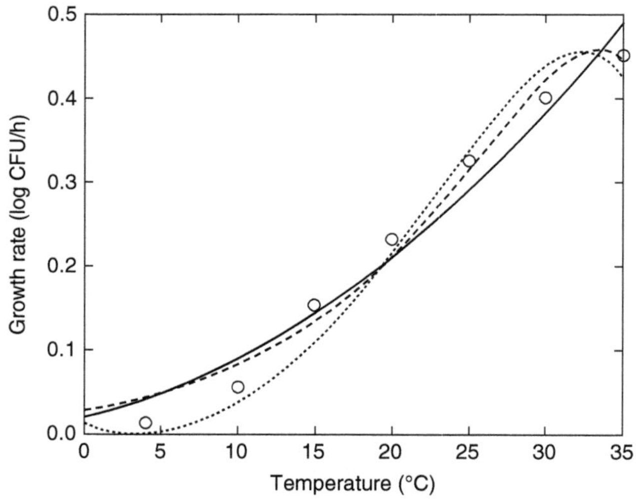

Figure 8.4 Bacterial proliferation of *Escherichia coli* in relation to temperature. *Square root model (full line), Ratkowsky model (slight dotted line), and exponential sum model (heavy dotted line) obtained from nonlinear regression analysis.* Source: Oh *et al.* (2012). Reproduced with permission of Springer

When the temperature falls below 30°C, human enzymes are no longer able to transform substrates and autolysis will be stopped. Many bacteria, in particular those from the intestinal tract, are able to metabolize in those conditions. As seen in the following text, many bacteria are still able to transform substrates.

In forensic practice, when confronted to short postmortem intervals, the thermal decay is also a key factor in the estimation of the moment of death, especially when coupled with the observation of macroscopic postmortem changes of the body. Mathematical models are used in order to evaluate an interval of time, which includes the time of death. Those models use corrective factors, taking into account the variability induced by the factors previously described (Henssge *et al.*, 1988; Knight 1988; Vass, 2011).

The heat loss of the corpse is selecting for specific bacterial taxa or functional groups instead of others. Indeed, bacterial proliferation depends on temperature, and each species has ideal growth temperatures (Juneja *et al.*, 2011; Oh *et al.*, 2012). Before death, except for specific situations previously described, *Bacteroides* and *Bifidobacterium* are the dominant bacteria in the intestinal microbiota. Such species can be observed in a cadaver's blood soon after death (Mesli, 2013). The heat loss is then selecting bacteria that are able to develop at a lower temperature. For example, Figures 8.4 and 8.5 show growth rate curves of *E. coli* and *Clostridium perfringens* in relation to temperature. Such bacteria are able to proliferate even at temperatures between 15 and 20°C.

8.4.3 Anaerobic conditions

Metabolism is defined as all the biochemical reactions of the cells of an organism. The energetic production of the human organism is mainly based on the oxygen and glucose consumption (e.g., glycolysis). It is an aerobic metabolism, which produces adenosine triphosphate. The biggest actor of this metabolism is the mitochondrion, which

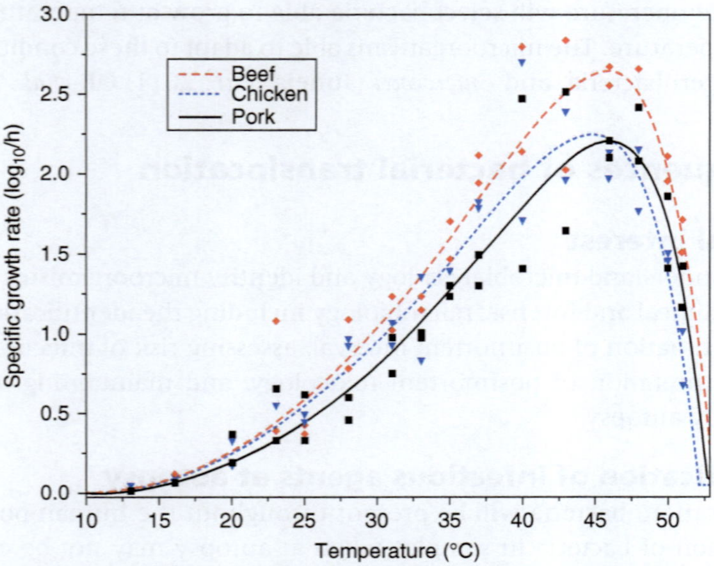

Figure 8.5 Bacterial proliferation of *Clostridium perfringens* in relation to temperature on three different substrates. Source: Juneja *et al.* (2011). Reproduced with permission of Elsevier

synthesizes adenosine triphosphate by a set of chemical reactions called oxidative phosphorylation. After death, oxygen can no longer reach the tissues, and the metabolism thus becomes anaerobic and oxidative phosphorylation is interrupted. Intermediate products accumulate, which results in an intracellular acidification. This acidification releases proteolytic enzymes, which result in cell lysis (Oh *et al.*, 2012). Nevertheless, anaerobic bacteria are able to produce energy from those substrates in the absence of oxygen. Those bacteria then produce specific metabolites, such as methane, propionic acid, or butyric acid, in a soluble or gaseous form (Duerden and Drasar, 1990; Matarazzo *et al.*, 2012). Human enzymes, having a maximum activity around 37°C, become increasingly inefficient as the body temperature decreases. On the other hand, bacterial enzymes are able to perform on a wider range of temperatures (Voet and Voet, 2005). Thus, after death, aerobic cellular metabolism gives way to anaerobic metabolism and temperature conditions unfavorable to human enzymes but favorable to bacterial growth. This is macroscopically confirmed by the presence of gas in the tissues and by the evidence of specific substances of an anaerobic bacterial metabolism around the cadaver (Swann *et al.*, 2010b; Vass, 2012).

In conclusion, after death all factors implicated in the control of the intestinal microflora (blood flow, mucus secretion, immune defenses, etc.) will be lost, and a massive translocation will take place. Among the great variety of bacteria present in the bowel, the selection of bacteria involved in putrefaction will take place by different physicochemical changes:

- Absence of oxygen will favor organisms able of anaerobic metabolism.
- Substrates available are mainly proteins, which will select proteolytic bacteria.

- Decrease in temperature will select bacteria able to grow at temperature approaching ambient temperature. The microorganisms able to adapt to these conditions are mainly *Clostridia*, Enterobacteria, and *Enterococci* (Juneja *et al.*, 2011; Oh *et al.*, 2012).

8.5 Consequences of bacterial translocation

8.5.1 Clinical interest

The ability to understand microbial ecology and identify microorganisms has many consequences for clinical and forensic microbiology including the identification of infectious diseases, the estimation of postmortem interval, assessing risk of infection during transplant, the interpretation of postmortem toxicology, and maintaining adequate safety standards during autopsy.

8.5.2 Identification of infectious agents at autopsy

Due to translocation, bacteria will be present throughout the human body. This means that identification of bacteria in samples taken at autopsy may not be easily related to the identification of an infectious agent linked to death. Bacterial samples at autopsy can thus be often difficult to interpret for forensic application. Indeed, the identified potential pathogens could come from translocation or from a contamination of the sample, inducing a false positive. Generally speaking, a pure bacterial culture (only one species isolated) has more clinical value than a polymicrobial sample. As in samples collected from a living human, they have to be interpreted in relation to the sample site, the clinical history, the characteristics of the bacteria isolated, and the presence of an inflammation (Tsokos and Püschel, 2001; Morris *et al.*, 2006, 2007).

The use of sterile technique while collecting and analyzing the sample as soon as possible are crucial in order to limit the impact of bacterial contamination, whether endogenous or exogenous. Modern techniques of molecular biology, such as polymerase chain reaction, could have a complementary interest when suspecting a specific pathogenic agent. All those precautions being taken, positive bacterial diagnosis can still be possible even several days after death and even if the body is showing signs of decomposition (Maujean *et al.*, 2013). See Chapter 7 for detailed discussion on microbiology as an adjunct to autopsy.

8.5.3 Postmortem interval estimation

As it was developed beforehand, bacteria are influencing the macroscopic postmortem changes of the remains, which are used to estimate the postmortem interval in forensic practice. Different studies have found correlation between the postmortem interval and the bacterial communities:

- Research has been conducted to evaluate the value of the remains' microbial communities succession as a tool for estimating the postmortem interval, due to the recent progress in sequencing (Metcalf *et al.*, 2013; Pechal *et al.*, 2014); those issues and methods are described in more details in other chapters.

- The detection of metabolites of the encephalic decomposition on sheep remains by *in situ* proton magnetic resonance spectroscopy was also used with the purpose of estimating postmortem intervals. Some of those metabolites were products of the anaerobic metabolism and proteolysis, and a correlation between estimated postmortem intervals and real ones has been found. Those findings are in favor of an influence of the intestinal microbiota on the encephalic decomposition (Ith *et al.*, 2011; Musshoff *et al.*, 2011).
- In a study on 33 human cases, relative amounts of intestinal bacterial DNA (such as *Bifidobacteria, Bacteroides, Enterobacter,* and *Clostridia*) increased with time after death on several sample sites, including the blood or the liver (Tuomisto *et al.*, 2013).

The bacterial translocation (and all the factors of its occurrence) has to be taken into account while interpreting those results. Thus, much research is needed in order to appreciate the chronological, quantitative, and qualitative aspects of the postmortem bacterial translocation, in order to implement them in those new methods of postmortem interval estimation.

8.5.4 Infectious risk for postmortem organ transplantation

Infectious risk exists concerning patients receiving an allograft from a donor. Indeed, several cases of complications due to infection are described in the literature, even though donors were healthy and without any infectious disease (Eastlund, 2006). Many authors describe infection rates of *Clostridium sordellii* at 8.1% on 795 musculoskeletal donors (Malinin *et al.*, 2003), of *C. difficile* colitis at 12.4% on 1331 solid organ donors (Boutros *et al.*, 2012), or rare *C. perfringens* infections after liver transplant (Doblecki-Lewis *et al.*, 2008). Those findings make sense that it is actually recommended to graft as soon as possible (principally in order to decrease the cold ischemia time), with strict asepsis rules.

8.5.5 Postmortem toxicological analysis

The putrefying human remains involves anaerobic bacterial metabolism. Anaerobes are able to form a large range of substances normally not present in the living. One of these is ethanol, whose levels are measured in cases where alcohol intoxication is suspected as playing a role in the death (e.g., a car accident). However, it should be noted that anaerobic bacteria are also capable of synthesizing ethanol through alcohol fermentation. This phenomenon is well known and has to be taken into account when interpreting postmortem ethanol levels (Corry, 1978). Recent research in this area is focusing on ethanol metabolites, such as ethyl glucuronide, phosphatidylethanol, and fatty acid ethyl esters, which are more sensitive and more specific (Kugelberg and Jones, 2007).

Gamma-hydroxybutyric acid (GHB) is a central nervous system depressant that has to be considered in forensics investigations, because it can be used as a tracer in cases of drug abuse or drug-facilitated crimes. Even though GHB is present in the human body in tiny concentrations (nanomolar), it is also naturally produced during the decomposition of the body and by some bacteria (Elliott, 2004; Alan and Sarah, 2012). The use of ethylene diamine tetra-acetic acid (EDTA) tubes is thus recommended, in order to limit the bacterial proliferation. Those aspects also have to be taken into account when interpreting GHB postmortem dosing (Beránková *et al.*, 2006). See Chapter 9 for more detail on the relationships between microorganisms and toxicology.

8.5.6 Prevention of biological risk at autopsy

Strict sanitary principles should be followed when handling human remains due to the presence and growth of associated bacteria (Werbrouck *et al.*, 2007). Pathogenic agents can be transmitted from the remains to others through respiratory, digestive, or cutaneous routes. Thus, personal protective equipment must be worn. This equipment includes, but is not limited to, professional clothing with surgical gown, plastic apron, surgical cap, protective glasses, gloves, cut-resistant gloves, dedicated closed shoes, and shoe covers. Bacteria such as *C. perfringens* have been isolated from forensic samples; they have a high pathogenicity (generating gas gangrene) and are very resistant (due to their endospores). The presence of such pathogenic agents in association with human remains demonstrates the importance of specific preventive measures; in cases of accidental blood exposure, an antiseptic effective on spores should be used (e.g., Dakin solutions); disinfectants effective on spores should also be used when cleaning the equipment and the floor (e.g., bleach) (Werbrouck *et al.*, 2007; Malik and Singh, 2011).

8.5.7 Environmental consequences

As bacteria are spreading outside of the gut by translocation, they contaminate the environment close to the remains through time after death. As mentioned before, bacterial metabolites have indeed been found in the air or in the soil around remains. Therefore, it has to be taken into account in forensic investigations or, for example, in civil engineering. These issues are described in more detail in Chapter 10.

8.6 Conclusion

Bacterial translocation is a well-described phenomenon in the living subjects, but remains not fully understood in the cadaver. Several specific metabolites of anaerobic bacteria have been found around human corpses, as well as bacteria of the intestinal microbiota in blood samples taken from human cadavers. On the other hand, the main factors resulting in bacterial translocation in the living human are also occurring after death. All those elements are providing evidence for a postmortem bacterial translocation, whose determining factors and timing still remain unknown. A better knowledge of this phenomenon could have several applications, such as in the interpretation of false positives in bacterial samples at autopsy, the appreciation of infectious risks at autopsy, the estimation of postmortem interval based on microbiological approaches, or a better prevention of allograft contamination.

References

Alan, G. and Sarah, J.P. (2012) Microbes as forensic indicators. *Tropical Biomedicine*, **29**, 311–330.

Albanese, C.T., Smith, S.D., Watkins, S. *et al.* (1994) Effect of secretory IgA on transepithelial passage of bacteria across the intact ileum *in vitro*. *Journal of the American College of Surgeons*, **179**, 679–688.

Amar, J., Chabo, C., Waget, A. *et al.* (2011) Intestinal mucosal adherence and translocation of commensal bacteria at the early onset of type 2 diabetes: molecular mechanisms and probiotic treatment. *EMBO Molecular Medicine*, **3**, 559–572.

Arumugam, M., Raes, J., Pelletier, E. *et al.* (2011) Enterotypes of the human gut microbiome. *Nature*, **473**, 174–180.

Balzan, S., de Almeida Quadros, C., de Cleva, R. *et al.* (2007) Bacterial translocation: overview of mechanisms and clinical impact. *Journal of Gastroenterology and Hepatology*, **22**, 464–471.

Baron, P., Traber, L.D., Traber, D.L. *et al.* (1994) Gut failure and translocation following burn and sepsis. *Journal of Surgical Research*, **57**, 197–204.

Beasley, D.E., Koltz, A.M., Lambert, J.E. *et al.* (2015) The evolution of stomach acidity and its relevance to the human microbiome. *PLoS One*, **10** (7), e0134116.

Beauthier, J.P. (2011) *Traité de Médecine Légale*, 2e édn edn, De Boeck, Brussels, Belgium.

Beránková, K., Mutňanská, K. and Balíková, M. (2006) Gamma-hydroxybutyric acid stability and formation in blood and urine. *Forensic Science International*, **161**, 158–162.

Boutros, M., Al-Shaibi, M., Chan, G. *et al.* (2012) *Clostridium difficile* colitis: increasing incidence, risk factors, and outcomes in solid organ transplant recipients. *Transplantation*, **93**, 1051–1057.

Brenchley, J.M., Price, D.A., Schacker, T.W. *et al.* (2006) Microbial translocation is a cause of systemic immune activation in chronic HIV infection. *Nature Medicine*, **12**, 1365–1371.

Campana, J.P. and Fornes, P. (2010) *Principes de Médecine Légale*, 2e édn edn, Arnette Blackwell, Paris, France.

Cherbuy, C., Thomas, M. and Langella, P. (2013) Le microbiote intestinal: une composante santé qui évolue avec l'âge. *Innovations Agronomiques*, **33**, 37–46.

Consortium THMP (2012) Structure, function and diversity of the healthy human microbiome. *Nature*, **486**, 207–214.

Corry, J.E.L. (1978) Possible sources of ethanol ante- and post-mortem: its relationship to the biochemistry and microbiology of decomposition. *Journal of Applied Bacteriology*, **44**, 1–56.

Costello, E.K., Lauber, C.L., Hamady, M. *et al.* (2009) Bacterial community variation in human body habitats across space and time. *Science*, **326**, 1694–1697.

Deitch, E.A. (2002) Bacterial translocation or lymphatic drainage of toxic products from the gut: what is important in human beings? *Surgery*, **131**, 241–244.

Deitch, E.A., Maejima, K. and Berg, R. (1985) Effect of oral antibiotics and bacterial overgrowth on the translocation of the GI tract microflora in burned rats. *Journal of Trauma and Acute Care Surgery*, **25**, 385–392.

Deitch, E.A., Bridges, W.M., Ma, J.W. *et al.* (1990) Obstructed intestine as a reservoir for systemic infection. *American Journal of Surgery*, **159**, 394–401.

Doblecki-Lewis, S., Palaios, E., Bejarano, P.A. *et al.* (2008) Hepatic gas gangrene following orthotopic liver transplantation: three cases treated with re-transplantation and a review of the literature. *Transplant Infectious Disease*, **10**, 280–285.

Donowitz, J.R. and Petri, W.A. (2015) Pediatric small intestine bacterial overgrowth in low-income countries. *Trends in Molecular Medicine*, **21** (1), 6–15.

Doré, J. and Corthier, G. (2010) Le microbiote intestinal humain. *Gastroentérologie Clinique et Biologique*, **34**, 7–16.

Ducluzeau, R. (1998) Ecosystème microbien du tube digestif, in *Encyclopédie Médico-Chirurgicale—Gastro-Enterologie*, Elsevier, Paris.

Duerden, B.I. and Drasar, B.S. (1990) *Anaerobes and Human Disease*, Hodder Arnold, London.

Durigon, M. (2007) *Pratique Médico-Légale*, 2e édn edn, Masson, Paris.

Eastlund, T. (2006) Bacterial infection transmitted by human tissue allograft transplantation. *Cell and Tissue Banking*, **7**, 147–166.

Eleftheriadis, E., Kotzampassi, K., Papanotas, K. *et al.* (1996) Gut ischemia, oxidative stress, and bacterial translocation in elevated abdominal pressure in rats. *World Journal of Surgery*, **20**, 11–16.

Elliott, S.P. (2004) Further evidence for the presence of GHB in postmortem biological fluid: implications for the interpretation of findings. *Journal of Analytical Toxicology*, **28**, 20–26.

Gautreaux, M.D., Deitch, E.A. and Berg, R.D. (1994) T lymphocytes in host defense against bacterial translocation from the gastrointestinal tract. *Infection and Immunity*, **62**, 2874–2884.

Ghnassia, J.C. (1979) Etude quantitative de la Flore bactérienne intestinale. *Médecine et Maladies Infectieuses*, **9**, 507–510.

Girard-Pipau, F.R., Rambaud, J.C. and Perol, Y. (1981) Ecologie microbienne du jéjunum supérieur normal et pathologique de l'homme. *Médecine et Maladies Infectieuses*, **11**, 238–248.

Goulet, O. (2009) La flore intestinale: un monde vivant à préserver. *Journal de Pediatrie et de Puericulture*, **22**, 102–106.

Grönlund, M.M., Lehtonen, O.P., Eerola, E. and Kero, P. (1999) Fecal microflora in healthy infants born by different methods of delivery: permanent changes in intestinal flora after cesarean delivery. *Journal of Pediatric Gastroenterology and Nutrition*, **28**, 19–25.

Henssge, C. (1988) Death time estimation in case work. I. The rectal temperature time of death nomogram. *Forensic Science International*, **38**, 209–236.

Henssge, C., Madea, B. and Gallenkemper, E. (1988) Death time estimation in case work. II. Integration of different methods. *Forensic Science International*, **39**, 77–87.

Ith, M., Scheurer, E., Kreis, R. *et al.* (2011) Estimation of the postmortem interval by means of 1H MRS of decomposing brain tissue: influence of ambient temperature. *NMR in Biomedicine*, **24**, 791–798.

Juneja, V.K., Marks, H., Huang, L. and Thippareddi, H. (2011) Predictive model for growth of *Clostridium perfringens* during cooling of cooked uncured meat and poultry. *Food Microbiology*, **28**, 791–795.

Katayama, M., Xu, D., Specian, R.D. and Deitch, E.A. (1997) Role of bacterial adherence and the mucus barrier on bacterial translocation: effects of protein malnutrition and endotoxin in rats. *Annals of Surgery*, **225**, 317–326.

Kelly, C.J., Colgan, S.P. and Frank, D.N. (2012) Of microbes and meals: the health consequences of dietary endotoxemia. *Nutrition in Clinical Practice*, **27** (2), 215–225.

Knight, B. (1988) The evolution of methods for estimating the time of death from body temperature. *Forensic Science International*, **36**, 47–55.

Kugelberg, F.C. and Jones, A.W. (2007) Interpreting results of ethanol analysis in postmortem specimens: a review of the literature. *Forensic Science International*, **165**, 10–29.

Laffineur, G., Lescut, D., Vincent, P. *et al.* (1992) Bacterial translocation in Crohn disease. *Gastroentérologie Clinique et Biologique*, **16**, 777–781.

Langrehr, J.M., Machens, C., Zill, E. *et al.* (2000) Bacterial translocation during graft-versus-host disease after small bowel transplantation is reduced following inhibition of inducible nitric oxide synthesis. *Transplantation*, **69**, 2415–2421.

Lescut, D., Colombel, J.F., Vincent, P. *et al.* (1990) Bacterial translocation in colorectal cancers. *Gastroentérologie Clinique et Biologique*, **14** (11), 811–814.

Lozupone, C.A., Stombaugh, J.I., Gordon, J.I. *et al.* (2012) Diversity, stability and resilience of the human gut microbiota. *Nature*, **489**, 220–230.

Malik, P. and Singh, G. (2011) Health considerations for forensic professionals: a review. *Forensic Science Policy and Management*, **2**, 1–4.

Malinin, T.I., Buck, B.E., Temple, H.T. *et al.* (2003) Incidence of clostridial contamination in donors' musculoskeletal tissue. *Journal of Bone and Joint Surgery*, **85**, 1051–1054.

Marteau, P. (1996) Composition of the caecal flora in healthy humans and comparison with the fecal flora. *Gut*, **39**, A135.

Marteau, P. (2013) Microbiote intestinal, in *Encyclopédie Médico-Chirurgicale—Gastro-Enterologie*, Elsevier, Paris.

Matarazzo, F., Ribeiro, A.C., Faveri, M. *et al.* (2012) The domain Archaea in human mucosal surfaces. *Clinical Microbiology and Infection*, **18**, 834–840.

Maujean, G., Guinet, T., Fanton, L. and Malicier, D. (2013) The interest of postmortem bacteriology in putrefied bodies. *Journal of Forensic Sciences*, **58**, 1069–1070.

McGuckin, M.A., Lindén, S.K., Sutton, P. and Florin, T.H. (2011) Mucin dynamics and enteric pathogens. *Nature Reviews. Microbiology*, **9**, 265–278.

Mesli V (2013) Post-mortem bacterial translocation. Unpublished MD thesis, Lille 2 University, France.

Metcalf, J.L., Wegener Parfrey, L., Gonzalez, A. *et al.* (2013) A microbial clock provides an accurate estimate of the postmortem interval in a mouse model system. *eLife*, **2**, e01104.

Moore, F.A., Moore, E.E., Poggetti, R. *et al.* (1991) Gut bacterial translocation via the portal vein: a clinical perspective with major torso trauma. *Journal of Trauma*, **31**, 629–638.

Morris, J.A., Harrison, L.M. and Partridge, S.M. (2006) Postmortem bacteriology: a re-evaluation. *Journal of Clinical Pathology*, **59**, 1–9.

Morris, J.A., Harrison, L.M. and Partridge, S.M. (2007) Practical and theoretical aspects of postmortem bacteriology. *Current Diagnostic Pathology*, **13**, 65–74.

Muggenthaler, H., Sinicina, I., Hubig, M. and Mall, G. (2012) Database of post-mortem rectal cooling cases under strictly controlled conditions: a useful tool in death time estimation. *International Journal of Legal Medicine*, **126**, 79–87.

Musshoff, F., Klotzbach, H., Block, W. *et al.* (2011) Comparison of post-mortem metabolic changes in sheep brain tissue in isolated heads and whole animals using 1H-MR spectroscopy—preliminary results. *International Journal of Legal Medicine*, **125**, 741–744.

Nadler, E.P. and Ford, H.R. (2000) Regulation of bacterial translocation by nitric oxide. *Pediatric Surgery International*, **16**, 165–168.

Neutra, M.R. (1998) Current concepts in mucosal immunity. V. Role of M cells in transepithelial transport of antigens and pathogens to the mucosal immune system. *American Journal of Physiology*, **274**, G785–G791.

Nowroozalizadeh, S., Månsson, F., da Silva, Z. *et al.* (2010) Microbial translocation correlates with the severity of both HIV-1 and HIV-2 infections. *Journal of Infectious Diseases*, **201**, 1150–1154.

O'Boyle, C., MacFie, J., Mitchell, C. *et al.* (1998) Microbiology of bacterial translocation in humans. *Gut*, **42**, 29–35.

Oh, D.-H., Ding, T. and Jin, Y.-G. (2012) A new secondary model developed for the growth rate of *Escherichia coli* O157:H7 in broth. *Indian Journal of Microbiology*, **52**, 99–101.

Orrhage, K. and Nord, C.E. (2000) Bifidobacteria and lactobacilli in human health. *Drugs under Experimental and Clinical Research*, **26**, 95–111.

Pechal, J.L., Crippen, T.L., Benbow, M.E. *et al.* (2014) The potential use of bacterial community succession in forensics as described by high throughput metagenomic sequencing. *International Journal of Legal Medicine*, **128**, 193–205.

Plantefève, G. and Bleichner, G. (2001) Translocation bactérienne: mythe ou réalité? *Reanimation*, **10**, 550–561.

Rougé, C., Goldenberg, O., Ferraris, L. *et al.* (2010) Investigation of the intestinal microbiota in preterm infants using different methods. *Anaerobe*, **16**, 362–370.

Roy, J. and Gatt, M. (2012) Causes of surgical sepsis including bacterial translocation. *Surgery*, **30**, 645–650.

Sandler, N.G., Koh, C., Roque, A. *et al.* (2011) Host response to translocated microbial products predicts outcomes of patients with HBV or HCV infection. *Gastroenterology*, **141**, 1220–1230.

Schweinburg, F.B., Seligman, A.M. and Fine, J. (1950) Transmural migration of intestinal bacteria; a study based on the use of radioactive *Escherichia coli*. *New England Journal of Medicine*, **242**, 747–751.

Sedman, P.C., Macfie, J., Sagar, P. *et al.* (1994) The prevalence of gut translocation in humans. *Gastroenterology*, **107**, 643–649.

Sherman, M.P. (2010) New concepts of microbial translocation in the neonatal intestine: mechanisms and prevention. *Clinics in Perinatology*, **37**, 565–579.

Spaeth, G., Gottwald, T., Specian, R.D. *et al.* (1994) Secretory immunoglobulin A, intestinal mucin, and mucosal permeability in nutritionally induced bacterial translocation in rats. *Annals of Surgery*, **220**, 798–808.

Steinberg, S.M. (2003) Bacterial translocation: what it is and what it is not. *American Journal of Surgery*, **186**, 301–305.

Swann, L., Chidlow, G.E., Forbes, S. and Lewis, S.W. (2010a) Preliminary studies into the characterization of chemical markers of decomposition for geoforensics. *Journal of Forensic Sciences*, **55**, 308–314.

Swann, L.M., Forbes, S.L. and Lewis, S.W. (2010b) Analytical separations of mammalian decomposition products for forensic science: a review. *Analytica Chimica Acta*, **682**, 9–22.

Takatori, T. (2001) The mechanism of human adipocere formation. *Legal Medicine*, **3**, 193–204.

Tsokos, M. and Püschel, K. (2001) Postmortem bacteriology in forensic pathology: diagnostic value and interpretation. *Legal Medicine*, **3**, 15–22.

Tuomisto, S., Karhunen, P.J., Vuento, R. *et al.* (2013) Evaluation of postmortem bacterial migration using culturing and real-time quantitative PCR. *Journal of Forensic Sciences*, **58**, 910–916.

Vass, A.A. (2001) Beyond the grave—understanding human decomposition. *Microbiology Today*, **28**, 190–192.

Vass, A.A. (2011) The elusive universal post-mortem interval formula. *Forensic Science International*, **204**, 34–40.

Vass, A.A. (2012) Odor mortis. *Forensic Science International*, **222**, 234–241.

Vass, A.A., Smith, R.R., Thompson, C.V. *et al.* (2004) Decompositional odor analysis database. *Journal of Forensic Sciences*, **49**, 760–769.

Vass, A.A., Smith, R.R., Thompson, C.V. *et al.* (2008) Odor analysis of decomposing buried human remains. *Journal of Forensic Sciences*, **53**, 384–391.

Voet, D. and Voet, J.G. (2005) *Biochimie*, De Boeck Supérieur, Brussels.

Wells, C.L. and Erlandsen, S.L. (1996) Bacterial translocation: intestinal epithelial permeability, in *Gut Dysfunction in Critical Illness* (eds P.J.L. Rombeau and P.J. Takala), Springer, Berlin , pp. 131–149.

Werbrouck, A., Manaouil, C., Gignon, M. *et al.* (2007) Le risque infectieux en salle d'autopsie. *Journal de Medecine Legale Droit Medical*, **50**, 426–439.

Wiest, R. and Rath, H.C. (2003) Bacterial translocation in the gut. *Best Practice & Research. Clinical Gastroenterology*, **17**, 397–425.

Yatsunenko, T., Rey, F.E., Manary, M.J. *et al.* (2012) Human gut microbiome viewed across age and geography. *Nature*, **486**, 222–227.

Zhou, C. and Byard, R.W. (2011) Factors and processes causing accelerated decomposition in human cadavers—an overview. *Journal of Forensic and Legal Medicine*, **18**, 6–9.

CHAPTER 9

Microbial impacts in postmortem toxicology

Jared W. Castle[1], Danielle M. Butzbach[2], G. Stewart Walker[1], Claire E. Lenehan[1], Frank Reith[3,4] and K. Paul Kirkbride[1]

[1] *School of Chemical and Physical Sciences, Flinders University, Adelaide, South Australia, Australia*

[2] *Toxicology Section, Forensic Science SA, Adelaide, South Australia, Australia*

[3] *School of Biological Sciences, Sprigg Geobiology Centre, University of Adelaide, Adelaide, South Australia, Australia*

[4] *CSIRO Land and Water, Environmental Contaminant Mitigation and Technologies, PMB2, Adelaide, South Australia, Australia*

9.1 Introduction

Forensic toxicologists are required to determine the drugs and poisons present in an individual in order to assist law enforcement, the coroner or criminal courts in establishing cause of death, manner of death, or potential impairment of an individual. Screening methods in postmortem toxicology rely on the identification of drugs and known metabolites. Therefore, any phenomenon that leads to the inaccurate identification or quantification of drugs or poisons in postmortem specimens is of great concern. In particular, the collection and analysis of postmortem specimens is often hindered by the (i) decomposition of the human body and (ii) sometimes rapid degradation of the very substances that have caused the death of the individual—both processes that are largely mediated by microorganisms. Partial or complete conversion of a drug or metabolite into another compound could result in that drug not being detected, and/or create challenges in interpretation, particularly if potentially toxic concentrations or drug combinations were present. In the worst-case scenario, an incorrect determination of what substances were present at death, and their concentrations, may lead to erroneous "cause/manner of death" findings. This could have further repercussions in legal proceedings, for example, in court trials and insurance claims. Furthermore, the presence or absence of drugs

Forensic Microbiology, First Edition. Edited by David O. Carter, Jeffery K. Tomberlin, M. Eric Benbow and Jessica L. Metcalf.

causing impairment of an individual may not be able to be established, or incorrect medication adherence determined.

Thus, knowing the stability of any drugs or poisons under investigation is imperative to ensure the following: (i) appropriate degradation products are screened; (ii) appropriate storage conditions are employed; and (iii) results obtained can be analyzed, presented, and interpreted with confidence. In addition, toxicology laboratories are required by law to store case specimens for an extended length of time to enable reanalysis if requested, and so knowledge of the stability of drugs over this time period is important. This chapter reviews the literature that covers the impact of microbial activity on the amount of drug, poison, or their metabolites present in postmortem tissues and fluids.

9.2 Microbial factors complicating postmortem toxicological analyses

9.2.1 Cadaver decomposition and specimen contamination

Decomposition predominantly occurs through the action of heterotrophic microorganisms present in an individual after their death (i.e., putrefaction) (Evans, 1963). Cadaver decomposition, and the microbial community shifts that occur throughout, is discussed in Chapters 2, 8, and 10. In many cases, particularly those of sudden or unexpected death, only postmortem specimens are available for analysis in postmortem toxicology. Thus, specimen selection is partially dependent upon how far putrefaction has progressed. Each specimen type has its own distinct advantages and disadvantages; in some cases, a combination thereof can be valuable in elucidating the role of a drug or poison in the death of an individual.

The postmortem migration of gastrointestinal microorganisms (see Chapter 8) into the lymphatic and circulatory system decreases the likelihood of obtaining sterile specimens commonly collected at autopsy, such as blood, liver, and urine (Evans, 1963; Kellerman *et al.*, 1976; Melvin *et al.*, 1984). The pH of postmortem blood becomes mildly acidic during the autolytic phase of early decomposition before becoming mildly alkaline during putrefaction (Saukko and Knight, 2004). Thus, microorganisms that thrive in such environments are expected to dominate in blood, provided they are capable of fermentation or anaerobic respiration. Despite microbial contamination, the use of blood is usually most convenient because pharmacological data concerning drug concentrations found during antemortem therapeutic drug monitoring and potentially fatal and nonfatal drug concentrations found in postmortem specimens are available (Druid and Holmgren, 1997; Musshoff *et al.*, 2004; Reis *et al.*, 2007; Jones and Holmgren, 2009; Schulz *et al.*, 2012; Launiainen and Ojanperä, 2014; Skov *et al.*, 2015a, b).

Putrefactive amines such as tyramine, tryptamine, and 2-phenylethylamine are produced from the degradation of proteins (Stevens, 1984). These amines could interfere with the detection of some drugs (i.e., amphetamines) in decomposing specimens analyzed by some immunoassays and gas chromatography with nitrogen–phosphorus detection. However, the use of modern and more selective analytical techniques has reduced the extent of such interference.

9.2.2 Postmortem drug and metabolite degradation

Along with the body, some drugs and metabolites may microbially degrade during the time between death and specimen collection at autopsy (the postmortem interval, PMI). This is of particular concern if the body has been exposed to warm enough temperatures to favor microbial growth, as this increases the likelihood of degradation prior to autopsy. Similarly, drugs and metabolites may continue to microbially degrade in collected specimens before analysis.

Gastrointestinal microorganisms that translocate into the lymphatic and circulatory system after death have the ability to carry out a diverse array of xenobiotic metabolic reactions (see Chapters 8 and 9). Numerous reviews describe the nature of these reactions and the substrates affected (Scheline, 1968; Sousa *et al.*, 2008; Kang *et al.*, 2013). Microbial and mammalian metabolic pathways can sometimes be similar, for example fungi, particularly *Cunninghamella* spp., have been used as model organisms in pharmacological studies instead of laboratory animals and *in vitro* models (Abourashed *et al.*, 1999; Asha and Vidyavathi, 2009).

Chemical functional groups prone to decomposition, although not necessarily due to microbial activity, were established in a landmark study published by Stevens (1984). In these experiments, human liver macerates spiked with drugs were placed outside, with protection from rain but not necrophagous flies. Nitro groups, N-oxides, oximes, thiono sulfur groups, heterocyclic sulfur atoms, and amine and hydroxyl functionality on an aromatic ring were determined to be the chemical functionalities most prone to decomposition. Metabolism by bacteria carried by blow flies was suggested and direct inoculation using fecal bacteria reportedly increased degradation rates (however, no data were provided).

Data from studies investigating the microbial metabolism of drugs in the gastrointestinal tract support the results of Stevens (1984), showing that many reactions in which functional groups are interconverted, added, or removed may take place (Scheline, 1968; Soleim and Scheline, 1972; Sousa *et al.*, 2008). Some functionalities that are found susceptible to microbial metabolism are shown in Table 9.1.

9.3 Precautions taken to limit microbial impacts

Chemical preservatives may be added to postmortem specimens in order to limit further microbial action, with sodium fluoride (2% w/v) most commonly used. Its preservative action is primarily achieved by fluoride ions inhibiting enolase and H^+/ATPase (Hamilton, 1990). Dual inhibition of these enzymes both prevents the catabolism of carbohydrates

Table 9.1 Some functionalities found susceptible to microbial metabolism

Reaction	Example	References
Hydrolysis of		
Glucuronide conjugates	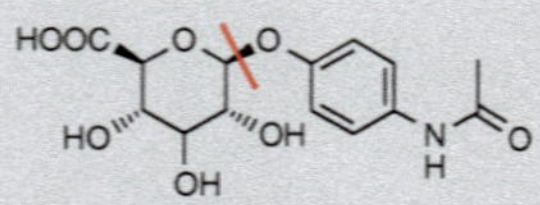 paracetamol glucuronide	Thai *et al.* (2014) and Scott *et al.* (2015)
Glycoside conjugates	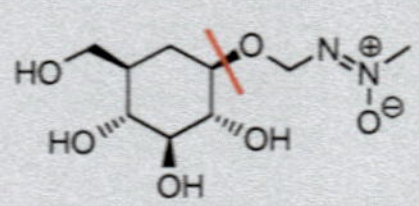 cycasin → methylazoxymethanol	Spatz *et al.* (1967)
Glycine conjugates	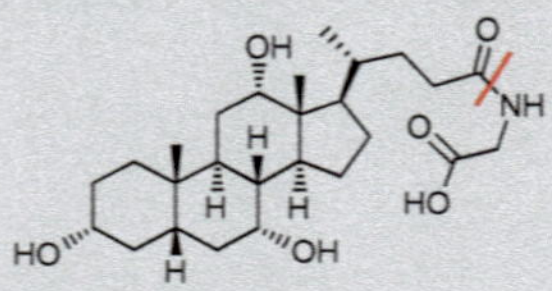 glycocholic acid	Norman and Grubb (1955)
Amides	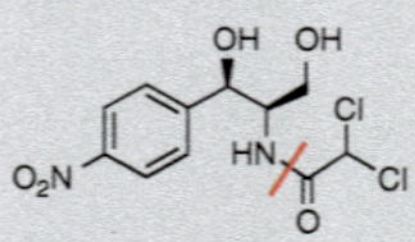 chloramphenicol	Holt (1967)
Esters	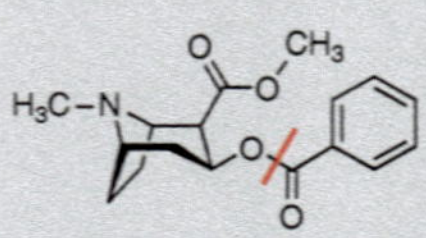 cocaine → ecgonine methyl ester	Grassi *et al.* (2006)
Reduction of		
Alkenes	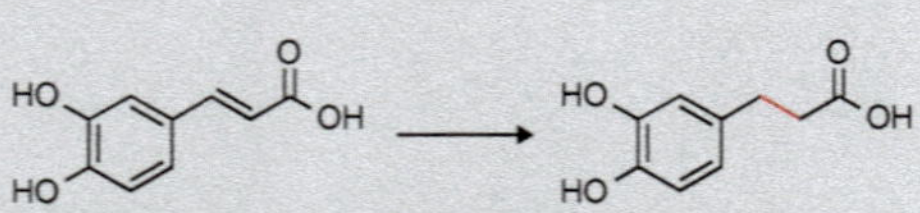 caffeic acid → dihydrocaffeic acid	Peppercorn and Goldman (1971)
Azo groups	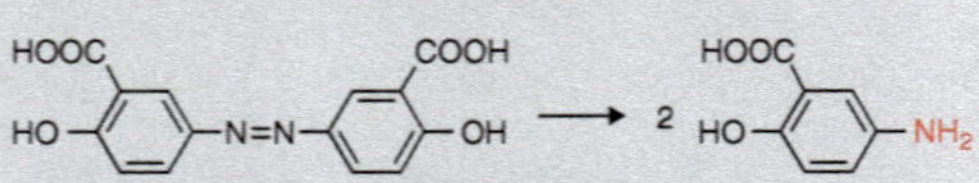 olsalazine → 5-aminosalicylic acid	Wadworth and Fitton (1991)
Nitro groups	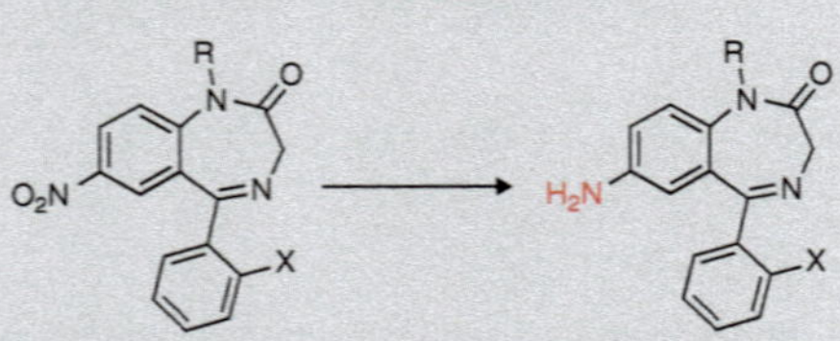 X = F (flunitrazepam), Cl (clonazepam), H (nitrazepam)	Robertson and Drummer (1995)

(*Continued*)

Table 9.1 (Continued)

Reaction	Example	References
Nitroalkenes	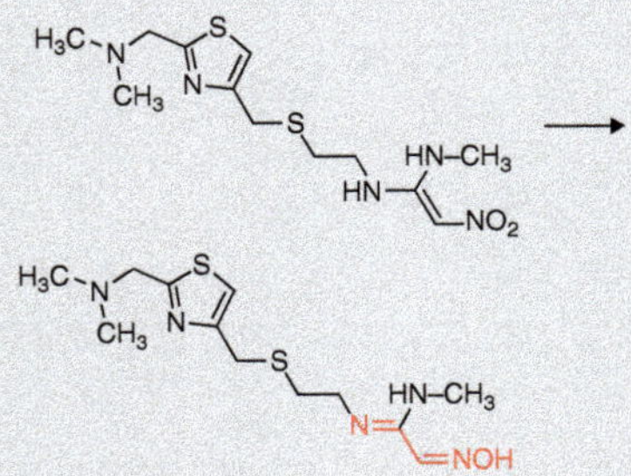 nizatidine	Basit *et al.* (2002)
Sulfoxides	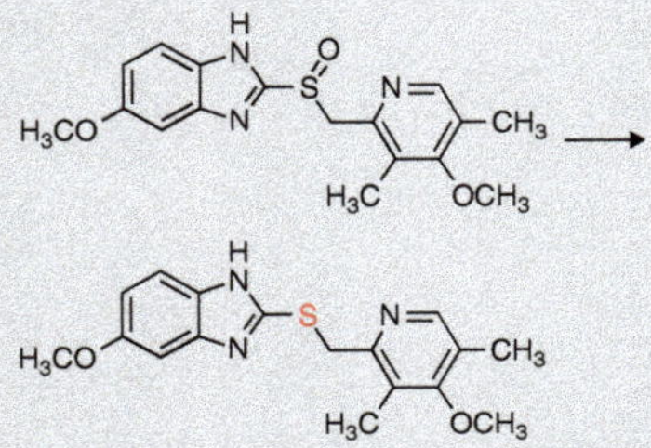 omeprazole → omeprazole sulfide	Watanabe *et al.* (1995)
Functionalization		
Aromatization	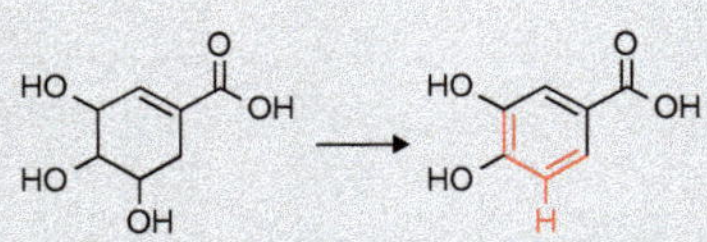 shikimic acid → protocatechuic acid	Scheline (1968)
Esterification	Esterification of coprostanone	Rosenfeld *et al.* (1967)
Hydroxylation	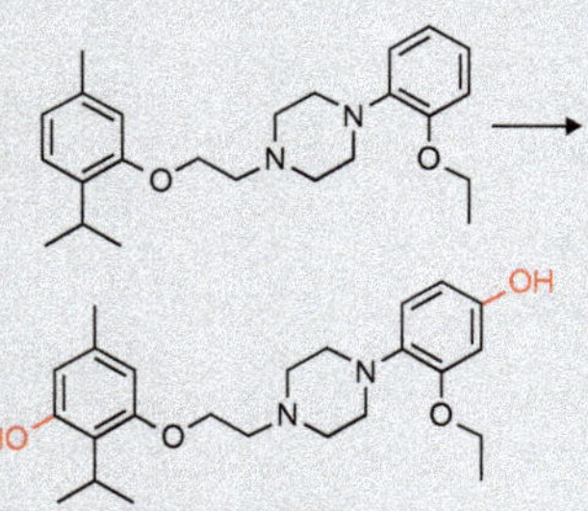 B1178 dihydroxylation	Abourashed *et al.* (1999)
N-Acetylation	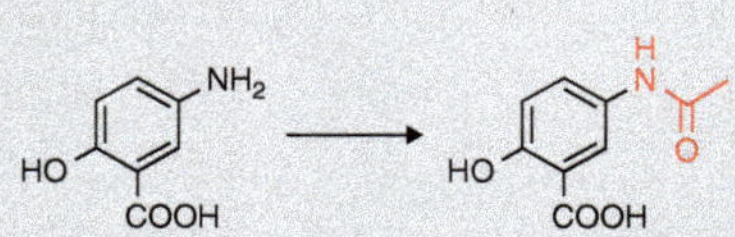 5-aminosalicylic acid → *N*-acetyl-5-aminosalicylic acid	Dull *et al.* (1987)

Table 9.1 (Continued)

Reaction	Example	References
Other		
Deamination	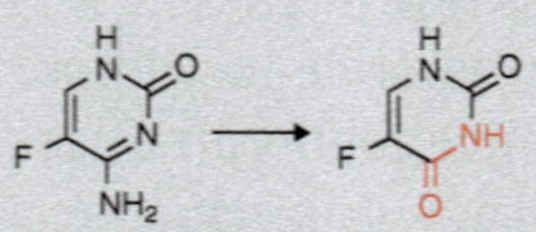 flucytosine → 5-fluorouracil	Harris *et al.* (1986)
Decarboxylation	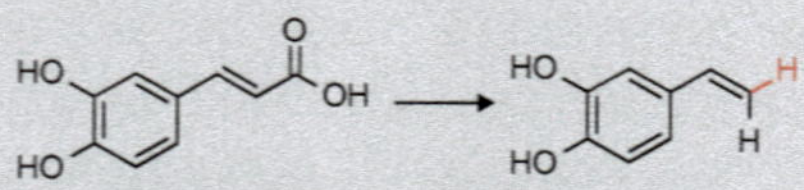 caffeic acid → 4-vinylcatechol	Indahl and Scheline (1968)
Dehalogenation	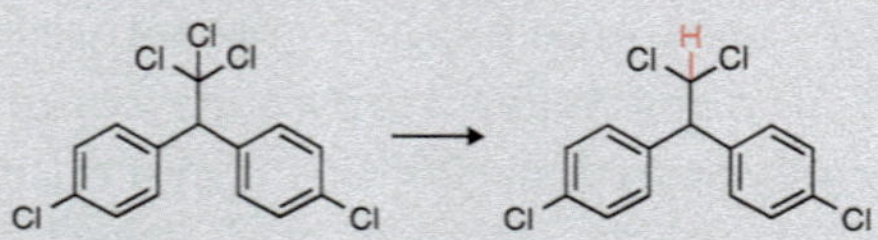 DDT → DDD	Mendel and Walton (1966)
Dehydroxylation	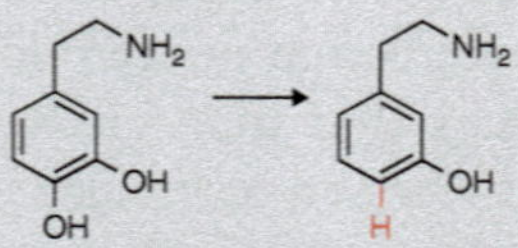 dopamine → m-tyramine	Goldin *et al.* (1973)
Denitration	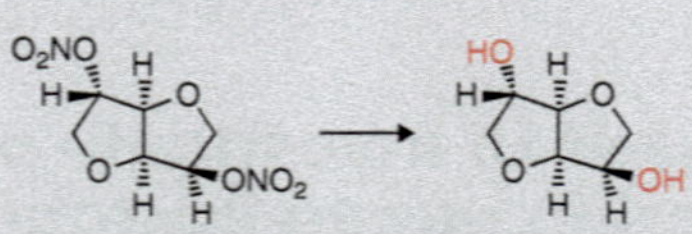 isosorbide dinitrate → isosorbide	Shamat (1993)
Heterocyclic ring scission	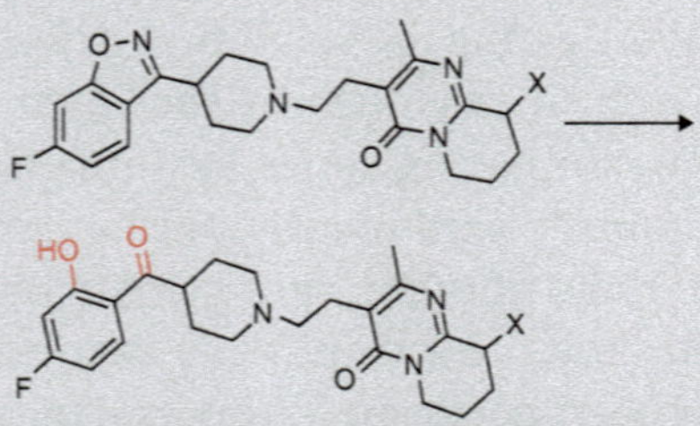 X = H (risperidone), OH (paliperidone)	Butzbach (2011) and Butzbach *et al.* (2013a)
N-Deacetylation	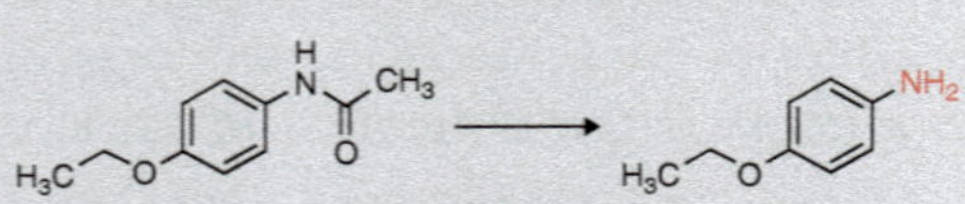 phenacetin → p-phenetidine	Smith and Griffiths (1974)
N-Demethylation	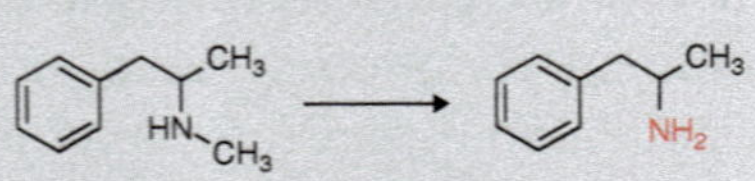 methamphetamine → amphetamine	Caldwell and Hawksworth (1973)

for adenosine triphosphate (ATP) production and leads to acidification in the cytoplasm, further reducing enzyme activity (Hamilton, 1990). Other enzymes also inhibited by fluoride include blood cholinesterase (which is only partially inhibited), whose activity otherwise remains in postmortem specimens containing only citrate and facilitates the hydrolysis of cocaine to ecgonine methyl ester (EME) (Baselt *et al.*, 1985; Isenschmid *et al.*, 1989). However, the efficiency of sodium fluoride in minimizing the degradation of different drugs is variable and, in the case of organophosphorus pesticides, its use may lead to accelerated degradation (Moriya *et al.*, 1999; Pannell *et al.*, 2001; Rositano *et al.*, 2014). Because sodium fluoride may both stabilize and degrade drugs, the collection of multiple aliquots of blood is therefore preferred to enable the storage of both preserved and unpreserved specimens. Other preservatives, such as sodium azide, have been used; however, its high reactivity with carbonyl compounds prevents its practical use (Zaitsu *et al.*, 2008).

Postmortem toxicology specimens are stored at low temperatures in order to decrease the rate of microbial metabolism. In a mortuary, corpses are commonly stored at 4°C. Autopsy specimens are then generally stored at −20°C, a temperature that is considered to inhibit the growth of most microorganisms (Megyesi *et al.*, 2005). Yet the relationships between temperature, microorganisms, and enzyme activity are complex and merit more research. One particularly complicated factor is that many microorganisms release enzymes extracellularly to break down substrates that then cross the cell membrane. This means that while a microbe might cease activity at a particular temperature regime, the extracellular enzymes might remain active.

9.4 Experimental protocols used to investigate postmortem drug and metabolite degradation due to microbial activity

There have been numerous studies investigating the stability of drugs during postmortem decomposition that emphasize the role of microbes in specimen degradation. Some experiments have been carried out by monitoring concentration changes within decaying cadavers; however, in these situations it is inherently unclear whether changes may be attributed to drug decomposition, redistribution, or a combination thereof (Maskell *et al.*, 2016). Therefore, *in vitro* studies are most commonly carried out. In addition to human specimens, research has also been performed using surrogate tissue and blood derived from pigs, rats, or other mammals. This is becoming more commonplace following an investigation in the late 1990s in the United Kingdom regarding organs and other specimens being retained after autopsy without the next-of-kin's consent or knowledge (Bennett, 2001).

Where local legislation and ethical constraints prevent the use of postmortem specimens for research purposes, a common compromise is to inoculate sterile specimens with cultured species in order to simulate postmortem conditions. This species-specific approach, regardless of the blood matrix chosen, has proven invaluable in predicting analyte stability in real postmortem cases, and even in discovering important microbial degradation products whose significance may have otherwise remained unknown (Butzbach *et al.*, 2013a). While effective, this approach does not account for the presence of microbial communities in postmortem specimens, where there will be complex interspecies interactions (Janaway *et al.*, 2009; Paczkowski and Schütz, 2011; Madigan *et al.*, 2014). Some microorganisms may also be unable to carry out all of the enzymatic steps required to degrade an analyte to the extent that could be observed in postmortem blood, or may produce metabolites that other species do not (Peppercorn and Goldman, 1971; Abourashed *et al.*, 1999; Butzbach, 2011; Martínez-Ramírez *et al.*, 2015). Collecting the postmortem blood of animals for use as a model may also be limited due to matrix and microbiome variations from human specimens (Li *et al.*, 2005; Ley *et al.*, 2008; Furet *et al.*, 2009). Studies using only one source of blood matrix, such as a blood donation from an individual, have been found to be subject to interindividual variation (Saar *et al.*, 2012).

Research investigating postmortem fungal metabolism has seldom ventured beyond exploring the postmortem production of ethanol by fermentation (Corry, 1978; Lough and Fehn, 1993; Yajima *et al.*, 2006). Of note is the work carried out by Martínez-Ramírez and colleagues, wherein the fungal metabolism of the four psychoactive drugs amitriptyline, mirtazapine, promethazine, zolpidem, and the beta-blocker metoprolol was assessed using fungi found to colonize cadavers (Martínez-Ramírez *et al.*, 2012, 2015, 2016). In the case of four of the five drugs studied, metabolites previously unreported in mammals formed; of the 28 fungal strains tested, all produced metabolites found in mammals and 18 produced novel metabolites in growth medium (Martínez-Ramírez *et al.*, 2015). One of these novel metabolites, α hydroxy zolpidem, was subsequently detected in 2 of 33 reanalyzed postmortem case samples, confirming that postmortem fungal degradation of drugs is possible in postmortem specimens (Martínez-Ramírez *et al.*, 2016).

Assessment of drug stability in sterilized specimens or antemortem whole blood provides information that is necessary to determine whether degradation observed in postmortem specimens, simulated or not, is a result of microbial metabolism, enzymatic degradation, or instability in the matrix toward chemical processes such as hydrolysis. However, appropriate abiotic controls have not been included in many studies where casework specimens have been analyzed over time due to the nature of the experiments.

9.5 Examples of microbially mediated drug degradation

In the following discussion, the current literature on the stability of specific drugs and their metabolites in postmortem specimens is reviewed, highlighting potential interpretative issues that may be of consequence following drug or metabolite degradation and production. The significance of potential microbial contributions is highlighted by comparison to stability studies in sterile specimens and antemortem whole blood, where possible.

9.5.1 Drugs

9.5.1.1 Alcohol (ethanol)

Alcohol is the dominant drug encountered in many countries and is routinely determined in postmortem toxicology. Ethanol can be produced postmortem by the fermentation of sugars by bacteria and yeasts (Corry, 1978; Kugelberg and Jones, 2007; Paczkowski and Schütz, 2011). It is also possible for some microorganisms, for example, *Pseudomonas* spp. and *Serratia marcescens*, to consume ethanol, thus lowering the concentration found in blood at autopsy (Dick and Stone, 1987). Postmortem blood concentrations averaging 0.36 g/100 mL (0.36% w/v) are usually present if acute intoxication was the cause of death (Kugelberg and Jones, 2007). In cases where ethanol is produced postmortem, concentrations are usually less than <0.03 g/100 mL (Kugelberg and Jones, 2007). However, there are case studies where concentrations as high as 0.19 g/100 mL (Kugelberg and Jones, 2007) and 0.22 g/100 mL have been attributed to postmortem production (Hunsaker and Hunsaker, 2005). This means that determining whether blood alcohol concentrations were from antemortem ingestion or from postmortem production can prove very difficult, especially if decomposition is advanced. In some jurisdictions, any alcohol in the blood is proscribed, and thus ethanol synthesis by microbes postmortem could be problematic for any investigation.

To gauge the extent of the potential problem, Gilliland and Bost (1993) showed that in a retrospective survey of 286 autopsies, no alcohol was detected in 39 cases (14%), consumption was attributed to 130 cases (45%), endogenous alcohol production was indicated in 55 cases (19%), and the source of alcohol was unable to be determined in 62 cases (22%). Therefore, of the 247 cases that tested positive to alcohol, up to 117 (or 41%) could have had the source of ethanol misidentified from blood alcohol concentrations alone (Gilliland and Bost, 1993).

The greater likelihood of vitreous humor and urine remaining sterile has led to many laboratories testing for ethanol in these specimens. Different alcohol levels in these fluids can be indicative of postmortem ethanol production (Gilliland and Bost, 1993; Kugelberg and Jones, 2007). For example, the absence, or a lower concentration, of ethanol in the vitreous humor compared to the blood indicates postmortem production of alcohol in the blood (Kugelberg and Jones, 2007). The concentration of ethanol in vitreous humor averages about 1.2 times that of the blood concentration (Kugelberg and Jones, 2007). For cases involving severe trauma, such as fatal aviation accidents, blood or urine may be unavailable, thus, the concentration of ethanol in alternative tissue specimens becomes particularly relevant (Lewis *et al.*, 2004). A fermentation study by Lewis *et al.* (2004) showed that ethanol formed between 22 and 75 mg/hg (0.022–0.075 g/100 mg), after 96 hours at 4°C, and 19 and 84 mg/hg (0.19–0.084 g/100 mg), after 48 hours at 25°C, in unpreserved macerated muscle and kidney specimens. In fluoride-preserved specimens, no ethanol was formed (Lewis *et al.*, 2004).

Because of the challenges surrounding the interpretation of postmortem ethanol levels, the metabolites ethyl glucuronide (EtG) and ethyl sulfate (EtS) have been explored as potential marker chemicals for the ingestion of ethanol. However, both false-negative and false-positive results for EtG are possible: given sufficient decomposition, complete bacterial hydrolysis of EtG can take place (due to the action of β-glucuronidase-possessing bacteria, such as *Escherichia coli*) (Høiseth *et al.*, 2008), or EtG may be produced from ethanol that has been synthesized postmortem (Helander *et al.*, 2007). In antemortem unpreserved urine, it was found that 5 days of storage at 22°C resulted in EtG either becoming undetectable, increasing in concentration, or being synthesized and then degraded (Helander *et al.*, 2007). EtG was found to completely degrade within a day in some bacterially contaminated unpreserved urine specimens (Helander and Dahl, 2005); however in another study, Schloegl *et al.* (2006) found no degradation over 5 weeks in urine when stored at 4°C. In unpreserved postmortem blood, Høiseth *et al.* (2008) showed that EtG completely degraded in some of the specimens as early as 11 days at 30°C and 3 days at 40°C. EtG was found to be stable in fluoride-preserved specimens compared to corresponding experiments in unpreserved specimens, with no concentration changes reported after 5 days at 22°C in antemortem urine or after 18 days at room temperature in postmortem blood (Helander and Dahl, 2005; Høiseth *et al.*, 2008). Schloegl *et al.* (2006) also examined EtG stability in postmortem liver and muscle tissue (not specified if preserved or unpreserved) and found that room temperature storage for 4 weeks resulted in decreases in concentration of 28% on average. As with ethanol, the use of vitreous humor or urine as complimentary specimens to blood may be beneficial, as EtG has been found to not always be present in all specimens (Thierauf *et al.*, 2011).

Studies on EtS are less frequent—it appears that only one study (Halter *et al.*, 2009) has shown its degradation in microbial growth medium, but not in human tissues (Helander and Dahl, 2005; Baranowski *et al.*, 2008; Thierauf *et al.*, 2011). Concentrations of EtS have not been found to change in studies performed using urine in which uropathogens were present (Helander and Dahl, 2005; Hernández Redondo *et al.*, 2012). Thus, the presence of both EtG and EtS or EtS alone is considered to be a strong indicator of perimortem ingestion of ethanol (Høiseth *et al.*, 2010). However, the slower excretion of the glucuronide and sulfate conjugates means they can be present days after ethanol ingestion, and conjugate concentrations do not correlate with ethanol concentrations; thus results must be cautiously interpreted (Helander *et al.*, 2009; Thierauf *et al.*, 2011).

9.5.1.2 Amphetamines

Methamphetamine can be metabolized by gastrointestinal microbes (sourced from guinea-pig cecal and fecal contents) to yield amphetamine, with species of the genera *Enterobacterium*, *Enterococcus*, *Lactobacillus*, *Clostridium*, and *Bacteroides* capable of performing the N-demethylation (Table 9.1) (Caldwell and Hawksworth, 1973). Therefore, microbial degradation of methamphetamine is plausible in postmortem specimens. However, in rabbit unpreserved postmortem blood, liver, and skeletal muscle stored for 2 years at 25°C, methamphetamine concentrations did not decrease significantly, indicating stability (Nagata *et al.*, 1990). Amphetamine concentrations in the same specimens increased in liver, however, presumably from degradation of another methamphetamine metabolite (Nagata *et al.*, 1990). In human antemortem preserved blood stored at ambient temperature, these amphetamines decreased in concentration; however, there was no clear trend of degradation over a 5-year period (Giorgi and Meeker, 1995). Holmgren *et al.* (2004) found no significant amphetamine concentration changes in preserved postmortem blood specimens stored for a year at −20°C. A longer study by Karinen *et al.* (2014) under the same conditions reanalyzed specimens after 16–18 years, finding that amphetamine concentrations remained within 30% of their initial concentration in 11 out of 16 specimens. In a study by Zaitsu *et al.* (2008), antemortem urine specimens were collected and all found to be contaminated with bacteria to differing degrees. The urines were spiked with methamphetamine and amphetamine, then stored at 4 or 25°C, with either no treatment, addition of sodium azide, or filter-sterilized using a 0.2 µm membrane (Zaitsu *et al.*, 2008). Only specimens stored at 25°C without treatment resulted in losses of 44% and 32% for amphetamine and methamphetamine, respectively, after 150 days (Zaitsu *et al.*, 2008).

One study has examined 3,4-methylenedioxymethamphetamine (MDMA), 3,4-methylenedioxyethylamphetamine (MDEA), and 3,4-methylenedioxyamphetamine (MDA) in serum, whole blood, water, and urine at −20, 4, and 20°C, with no observable decrease in concentrations after 21 weeks (Clauwaert *et al.*, 2001). However, the presence or absence of microbes was not mentioned. Another study found no concentration changes for methamphetamine, amphetamine, MDMA, MDA, ephedrine, and pseudoephedrine in urine specimens stored at both −20 and 4°C after 24 months (for filtration-sterilized specimens) and 6 months (nonsterilized specimens). Filtration-sterilized specimens of the aforementioned drugs, excluding pseudoephedrine, stored for 7 days at 37°C gave statistically insignificant losses, apart from MDMA where losses were practically insignificant (1.8%). All specimens contained sodium azide (0.1% w/v) (Jiménez *et al.*, 2006). Wenholz *et al.* (2016) found MDMA to be moderately stable in putrefying porcine liver, with only some losses in liver at room temperature compared to aqueous controls. No significant change to MDMA concentrations were observed in two postmortem blood specimens stored for 16–18 years at −20°C (Karinen *et al.*, 2014). However, in one specimen from the same study, MDA disappeared and there was a 56% increase in the concentration of amphetamine, suggesting MDA degrades to amphetamine.

9.5.1.3 Antidepressants

The tricyclic antidepressant dothiepin has been reported to completely degrade in putrefying human liver macerates within 3 days (Stevens, 1984). Subsequent investigations where bacteria were inoculated into unpreserved blood (both antemortem and postmortem) have found that the decomposition of dothiepin is not mediated by either *Clostridium perfringens* or *Proteus mirabilis*, bacteria that are observed in postmortem blood (Batziris *et al.*, 1999). Another study found that no degradation occurred in postmortem blood inoculated with human fecal bacteria and incubated at 23°C for 7 days, yet rabbit fecal bacteria degraded dothiepin by 24% in a paired experiment (Pounder *et al.*, 1994). It is therefore possible that the degradation in the liver macerates reported by Stevens (1984) resulted from instability in the liver matrix or potential degradation by microorganisms introduced by blow flies or other contamination. Other tricyclic antidepressants that have been studied and determined to be stable during putrefaction include doxepin, amitriptyline, and imipramine (Stevens, 1984). However, Lutfi (1998) found that amitriptyline and imipramine exhibited losses of 84 and 69% in postmortem blood stored at 25°C for a year, with decreasing losses at lower temperatures. It was not specified whether specimens were preserved or not. Holmgren *et al.* (2004) found amitriptyline and clomipramine to be stable in preserved postmortem blood specimens stored for a year at −20°C.

The monoamine oxidase inhibitor tranylcypromine was observed to decrease in concentration in blood in a case study with obvious signs of putrefaction, indicating microbial activity. The researchers further investigated that finding by spiking tranylcypromine into antemortem blood under nonsterile conditions and incubating it at 37°C for up to 48 hours, by which time only 42% of the drug remained (Yonemitsu and Pounder, 1993).

The serotonin-selective reuptake inhibitor antidepressants citalopram, paroxetine, sertraline, fluoxetine, and venlafaxine have been found to be stable for 57 days in putrefying porcine liver (Butzbach, 2011; Butzbach *et al.*, 2013b). In this same study, fluvoxamine was found to significantly decrease in concentration; however, this may be at least partially due to chemically mediated hydrolysis (Butzbach, 2011; Butzbach *et al.*, 2013b). In a different study (Sykutera *et al.*, 2002), fluoxetine decreased in concentration in blood, urine, and liver stored at 4 and 25°C. The maximum loss was in liver at 25°C with a 36% decrease in concentration over 90 days. However, no information about the presence of microbes or evidence of putrefaction was reported. In a study by Holmgren *et al.* (2004), losses of 28%, although not statistically significant, were observed for fluoxetine in preserved postmortem blood specimens stored for a year at −20°C. However, another study by Karinen *et al.* (2014) found losses of fluoxetine under the same conditions to reach 61% after 16–18 years of storage. Holmgren *et al.* (2004) found citalopram and sertraline to be stable in preserved postmortem blood specimens stored for a year at −20°C.

The microbial contribution to the degradation of trazodone is unknown. McIntyre *et al.* (2015) found degradation up to 20% after 8 months of storage at 4°C in preserved postmortem whole blood. Blood in this study was collected from deceased who exhibited no obvious signs of decomposition.

9.5.1.4 Antipsychotics

Butzbach *et al.* (2013a) found that the microbial degradation of both paliperidone and risperidone to their 2-hydroxybenzoyl degradation products is possible in unpreserved porcine blood inoculated with bacterial species cultured from porcine cecum contents

(Table 9.1). They also found that within 2 days at 37°C, degradation was complete and that losses were >40% after almost 4 days at 20°C, whereas the drugs were stable in sterile controls under the same conditions and in the inoculated specimens at 7°C (Butzbach *et al.*, 2013a). The addition of sodium fluoride was successful in minimizing the formation of the 2-hydroxybenzoyl microbial degradation product of risperidone to trace levels and preventing further degradation *in vitro* (Butzbach, 2011). In postmortem cases, the presence of these degradation products in the absence of the parent drugs has been reported (Butzbach, 2011; Butzbach, 2013a; Taylor and Elliott, 2013). Based on mass spectral data Taylor and Elliott (2013) also suggested that a dihydroxyrisperidone ring scission degradation product was present in the urine, presumably from microbial degradation of the dihydroxy metabolite of risperidone (Taylor and Elliott, 2013). This compound had previously only been reported as present in feces (Mannens *et al.*, 1993; Vermeir *et al.*, 2008). Reference standards were not available to enable any quantification of the degradation products in toxicological analyses of postmortem blood; consequently, for the Taylor and Elliot case, it was only possible to state that the decedent did ingest risperidone some time before death (Taylor and Elliott, 2013). In comparison to the decomposing blood studies, risperidone losses in preserved antemortem whole blood were only about 30% in a 20-week study by Saar *et al.* (2012) at temperatures up to 20°C, further implicating microorganisms in the aforementioned cases as the dominant factor leading to the degradation observed. In the same study, paliperidone losses did not exceed 29% after 10 weeks at 20°C.

Thioridazine concentrations have been reported to both increase and decrease in postmortem blood. Batziris *et al.* (1999) reported that thioridazine degrades more rapidly in putrefying blood in comparison to antemortem blood. It was found that *P. mirabilis* and *C. perfringens* were not responsible for drug losses in unpreserved blood after 8 hours of storage at 37°C. Previously, Stevens reported that thioridazine completely degraded within 5 days when spiked into liver left to putrefy under environmental conditions (Stevens, 1984). Holmgren *et al.* (2004) reported about 40% increase in thioridazine concentrations ($n = 5$) when postmortem femoral blood specimens were stored at −20°C for a year, suggesting possible conversion from a sulfoxide metabolite, even with potassium fluoride (1% w/v) used as a preservative. No experiments have been performed yet to confirm that this conversion occurs and whether or not microbial activity is the cause.

There are contradicting results in the literature regarding the stability of chlorpromazine in both ante- and postmortem specimens, making it unclear whether microorganisms are pivotal to its decomposition or chemical instability may be responsible. Saar *et al.* (2012) found that after 20 weeks in preserved antemortem blood 65–70%, 35%, and about 50% losses of chlorpromazine occurred for storage at −20°C, 4°C, and 20°C, respectively. On the other hand, McKay *et al.* (1984) found chlorpromazine to be stable in unpreserved antemortem blood after 12 weeks at −20°C. Chlorpromazine has been found to degrade at room temperature in unpreserved postmortem blood, urine, liver, kidney, and brain specimens, with refrigeration considerably reducing losses (in blood at room temperature: 75% losses after 3 months, trace levels, or undetectable after 6 months) (Coutselinis *et al.*, 1974). Lutfi (1998) still observed chlorpromazine after 1 year in postmortem blood, although only at 30% of the initial concentration. Chlorpromazine losses in refrigerated unpreserved postmortem blood did not exceed 25% after 6 months, which would suggest, in light of the prior discussed antemortem study by Saar *et al.* (2012), that factors other than microorganism activity may play a role in the decrease in chlorpromazine concentrations (Coutselinis *et al.*, 1974). However, in the putrefying liver study by Stevens (1984), chlorpromazine was found to degrade completely in two separate specimens after 17 and 36 days. Batziris *et al.* (1999) also found chlorpromazine degraded more rapidly in postmortem blood than in antemortem whole blood. Duffort *et al.* (2005) found no degradation of chlorpromazine; however, there was no difference in the extent of degradation in preserved and unpreserved blood specimens. These authors also reported on the stability of 15 other antipsychotics and 21 antidepressants. Unfortunately, whether the matrix used was ante- or postmortem blood was not clearly specified for any drug, making conclusions regarding microbe involvement difficult to draw.

The involvement of microorganisms in olanzapine degradation is unknown. Experimental evidence suggests that chemical degradation is the primary mechanism by which losses occur. Saar *et al.* (2012) observed 100% losses in preserved antemortem blood specimens (20 weeks at −60°C) and 100% losses in stock solutions of olanzapine have been observed for both room temperature and refrigerated (4–6°C) storage (Heller *et al.*, 2004; Karinen *et al.*, 2011).

9.5.1.5 Benzodiazepines

The conversion of nitrobenzodiazepines to their 7-aminometabolites (Table 9.1) occurs in both ante- and postmortem blood, although the rate appears to increase significantly in the latter due to microbial activity. In unpreserved postmortem blood, flunitrazepam and clonazepam decreased in concentration by 96% and 89%, respectively, within 8 hours at 22°C. In comparison, flunitrazepam and clonazepam decreased by 50% and 15%, respectively, after 10 days in unpreserved sterile whole blood at 22°C (Robertson and Drummer, 1998). In an earlier study using bacterially contaminated postmortem blood, complete degradation of flunitrazepam, nitrazepam, and clonazepam occurred within 2, 3, and 4 hours, respectively (Robertson and Drummer, 1995). The addition of 1% sodium fluoride reduced the rate of conversion, with 24%, 46%, and 60% of each drug remaining, respectively, after 6 hours. Addition of 1% sodium fluoride to bacterially contaminated whole fresh blood reduced the rate of bacterial conversion of flunitrazepam by 95%, 92%, and 65% for *C. perfringens*, *Staphylococcus aureus*, and *Bacillus cereus*, respectively (Robertson and Drummer, 1995). Skopp *et al.* (1997) found it took up to 240 days for flunitrazepam to completely degrade in unpreserved antemortem whole blood stored at 4°C. In a separate study by El Mahjoub and Staub (2000), complete degradation was observed within 168 days in unpreserved antemortem whole blood stored at room temperature. Work by Pépin *et al.* (1998) later found flunitrazepam, nitrazepam, and clonazepam to degrade more rapidly at −20 and 4°C in unpreserved postmortem blood, in contrast to the Robertson and Drummer studies under these conditions. The large variation between the results of both sets of experiments suggests that there could be some difference in the postmortem blood matrix or microbial communities present. A long-term study by Karinen *et al.* (2014) found that flunitrazepam and clonazepam concentrations after 16–18 years of storage at −20°C did not significantly decrease compared to concentrations initially determined in preserved postmortem blood.

The 7-aminometabolites, which may be present due to human or microbial metabolism, may also subsequently degrade. In bacterially-contaminated unpreserved postmortem blood, losses of between 10 and 20% were reported after 45 hours of storage at 22°C for the 7-aminometabolites of flunitrazepam, nitrazepam, and clonazepam (Robertson and Drummer, 1998). As with the parent benzodiazepines, degradation of the 7-aminometabolites also occurred in the study by Pépin *et al.* (1998). In preserved postmortem blood stored for a year at −20°C, an average loss of about 50% ($n = 10$) was reported for 7-aminonitrazepam (Holmgren *et al.*, 2004). For other benzodiazepines and their metabolites (diazepam, nordiazepam, flunitrazepam, and 7-aminoflunitrazepam), concentrations changes were statistically insignificant (Holmgren *et al.*, 2004).

Temazepam also degrades more readily in postmortem rather than antemortem specimens. Al-Hadidi and Oliver (1995) reported up to about 50% degradation of the drug when stored at 5 and 25°C for a year in unpreserved antemortem blood, whereas Lutfi (1998) reported 62 and 70% losses for storage of the drug at the same temperatures and for the same time but in postmortem blood (addition of a preservative was unspecified).

The instability of the *N*-oxide benzodiazepines, chlordiazepoxide, and demoxepam in postmortem specimens has been addressed in various studies. For chlordiazepoxide, sodium fluoride has proven effective in significantly slowing degradation rates at both 4 and 25°C in postmortem blood (Levine *et al.*, 1983), while in another study, the same amount of loss at 4°C in unpreserved postmortem blood took approximately six times as long to be achieved in preserved postmortem blood (Melo *et al.*, 2012). At lower temperatures of −80 and −20°C degradation was inhibited, with stability observed after 24 weeks in postmortem bile, blood, and vitreous humor, in both preserved and unpreserved specimens. Degradation products were identified as demoxepam, nordiazepam, and desoxychlordiazepoxide (Levine *et al.*, 1983; Stevens, 1984; Melo *et al.*, 2012). Desmethyldesoxychlordiazepoxide has also been observed in sudden deaths involving chlordiazepoxide, presumably a breakdown product of the metabolite norchlordiazepoxide (Entwistle *et al.*, 1986). Regarding demoxepam, Pépin *et al.* (1998) reported its disappearance in unpreserved postmortem blood after 1 month of storage at 25°C. Losses of 97% were reported after 6 months at 4°C. Stevens determined in experiments using putrefying human liver macerate that both chlordiazepoxide and demoxepam degraded completely within 3 days (Stevens, 1984). For nordiazepam, the following losses were reported: 27% and 88% (75 days in unpreserved postmortem blood at 4°C and 25°C, respectively); 30% and 60% losses (50 days in unpreserved postmortem liver at 4°C and 25°C, respectively) (Levine *et al.*, 1983). Concentration changes of 30% for nordiazepam were reported by Karinen *et al.* (2014) for 20 preserved postmortem blood specimens stored for 16–18 years at −20°C, although statistically insignificant.

Degradation of the triazolo benzodiazepines appears less enhanced in postmortem blood in comparison to the nitro and *N*-oxide benzodiazepines. Alprazolam has been found to degrade by 15% and 30% in unpreserved postmortem blood stored at −20°C and 25°C, respectively, after 2 months (Pépin *et al.*, 1998). Comparatively, alprazolam remained stable after a year of storage at 25°C in packed red blood cells reconstituted in isotonic saline (Alfazil, 2009). In the same experiment, estazolam was also found to be stable, with no apparent change in stability for specimens with sodium fluoride present. In unpreserved postmortem blood, a loss of 30% after 8 weeks at 25°C was reported by Pépin *et al.* (1998). Results from Zaitsu *et al.* (2008) indicate that the enhanced degradation observed may not be due to microbial action, as they found no notable difference in estazolam stability in antemortem urine specimens that underwent filtration sterilisation compared to those that did not. The effectiveness of this sterilization technique was demonstrated by its ability to preserve nitrazepam, which, as previously discussed, degrades microbially within hours at room temperature (Robertson and Drummer, 1998; Zaitsu *et al.*, 2008).

In a recent investigation into the stability of benzodiazepines in Reinforced Clostridial Medium, Martindale *et al.* (2015) suggested that *E. coli* and *Bacteroides fragilis* may have metabolized flunitrazepam and diazepam to minor human metabolites. It was also suggested that *C. perfringens* may secrete compounds that prevent the degradation of diazepam, as it is degraded in the matrix without *C. perfringens* present. This apparent stabilization was observed even when a mixed culture of all three bacterial species was used. These experiments highlight the advantage of performing experiments with microbial communities rather than isolated species.

Diazepam has been observed to be stable in unpreserved postmortem blood, brain, and liver stored at 4 and 25°C for 3–5 months (Levine *et al.*, 1983). Stevens (1984) also found diazepam to be stable in putrefying human liver macerates. However, diazepam has been reported as unstable in unpreserved blood. Skopp *et al.* (1997) reported losses of about 45% for antemortem specimens stored for 240 days at 4°C; Pépin *et al.* (1998) found 25% and 50% losses for postmortem specimens stored for 2 months at 4°C and 25°C, respectively; and Lutfi (1998) reported 68% losses for postmortem specimens stored for a year at 25°C. A long-term study found diazepam concentrations to predominantly remain within 30% of their initial concentration after 16–18 years in preserved postmortem blood stored at −20°C (Karinen *et al.*, 2014).

For other benzodiazepines, such as bromazepam, clobazam, flurazepam, lorazepam, and midazolam, degradation does not appear to be enhanced in postmortem specimens compared to antemortem specimens (Levine *et al.*, 1983; Stevens, 1984; Skopp *et al.*, 1997; Pépin *et al.*, 1998; El Mahjoub and Staub, 2000; Melo *et al.*, 2012).

9.5.1.6 Cannabis

In fresh antemortem blood and plasma, Johnson *et al.* (1984) found THC (Δ^9-tetrahydrocannabinol) and THC–COOH (11-nor-9-carboxy-Δ^9-tetrahydrocannabinol) to be stable at room temperature for 32 days, with significant decreases for THC after that point. No information regarding sterility was provided. Holmgren *et al.* (2004) reported a significant decrease in THC concentrations (about 66%) after storage at −20°C for 1 year in preserved postmortem femoral blood. Skopp and Pötsch (2004) performed a study examining THC–COO–glucuronide in the urine of cannabis users. The glucuronide was hydrolyzed to form THC–COOH at 20 and 40°C without preservatives and it was noted that after 10 days, signs of microbial growth were observed in half of the urine specimens. While this is likely partially due to microorganisms hydrolyzing the glucuronide, the pH of the urine was also a factor (Skopp and Pötsch, 2004). Decreases in THC–COO–glucuronide concentrations were observed by Scheidweiler *et al.* (2016) in preserved antemortem blood stored at −20°C, 4°C, and room temperature for 12, 12, and 1 week, respectively. No concurrent significant increases in THC–COOH concentrations were reported. Holmgren *et al.* (2004) reported a significant decrease in THC concentrations (about 66%) after storage at −20°C for 1 year in preserved postmortem femoral blood (Holmgren *et al.*, 2004). To our knowledge, no studies have been performed to examine whether bacteria or fungi found in human remains can specifically degrade THC or THC–COOH.

9.5.1.7 Cocaine

Some fungi and bacteria have been found to be able to hydrolyze cocaine to ecgonine methyl ester (EME). Notably, this includes *Aspergillus niger*, which is a fungus that is found in the gastrointestinal tract (Table 9.1) (Larsen *et al.*, 2002; Grassi *et al.*, 2006; Rajilić-Stojanović and de Vos, 2014). However, nonmicrobial hydrolysis is also possible; the methyl ester moiety of cocaine may chemically hydrolyze *in vitro* in neutral and alkaline conditions to benzoylecgonine (BE) or the phenyl ester may, by the action of blood cholinesterase, hydrolyze to produce EME (Stewart *et al.*, 1977; Isenschmid *et al.*, 1989). Even with sodium fluoride (1% w/v), which inhibits cholinesterase, complete cocaine degradation was observed by Baselt *et al.* (1993) in preserved antemortem sheep blood after 1 year at 4°C. Cocaine in antemortem whole blood adjusted to a pH of 5 was found to be considerably more stable than at pH 10 (Isenschmid *et al.*, 1989). BE or EME may also degrade to produce ecgonine.

When cocaine and ethanol are concurrently consumed, cocaethylene is formed as a metabolite. There are indications that cocaethylene might be formed by microbes; however, this has not been observed to occur *in vitro* (Moriya and Hashimoto, 1996). In the absence of other cocaine metabolites, analyzing for both ecgonine and cocaethylene is advised (Giorgi and Meeker, 1995; Moriya and Hashimoto, 1996; Logan, 2001).

9.5.1.8 γ-Hydroxybutyrate

γ-Hydroxybutyrate (GHB) is produced in the body at low levels as an endogenous metabolite of γ-aminobutyric acid (GABA), and this necessitates the use of interpretive cutoff values to discriminate between endogenous levels and those arising from ingestion (Castro *et al.*, 2014). Endogenous GHB levels are not expected to exceed 5 mg/L and 10 mg/L in antemortem blood and urine, respectively. For postmortem specimens, greater cutoff values have been advised -30 mg/L and 10 mg/L for femoral blood and urine, respectively (Moriya and Hashimoto, 2005). However, values above these cutoffs are possible and may be as high as 193 (mean = 30 mg/L in that study) and 217 mg/L (mean = 56 mg/L in that study) in postmortem femoral blood and urine, respectively, for specimens collected from deceased where GHB use was not suspected (Elliott, 2001; Kintz *et al.*, 2004; Korb and Cooper, 2014). On the other hand, blood concentrations of GHB in GHB-related deaths can be as low as 18 and 9 mg/L for single drug and poly-drug overdose deaths, respectively (Zvosec *et al.*, 2011). As a consequence, elevated levels in other specimens such as urine, vitreous humor, or brain, or evidence from the scene, are necessary to identify the cause of death as GHB-related (Zvosec *et al.*, 2011).

Decreases in GHB concentrations following initial increases have been observed in unpreserved postmortem blood (Beránková *et al.*, 2006). Addition of fluoride has been shown to result in specimens that are stable for at least 6 months at 4°C and confidently reanalyzed up to 7 years later if stored at −20°C without detrimentally affecting the interpretation of GHB concentrations (Fjeld *et al.*, 2012; Jones *et al.*, 2015). Postmortem specimens have been found to be stable when stored at −20°C for at least 2 years; however, there was no mention whether any preservative was used or not (Andresen-Streichert *et al.*, 2015). Though the precise cause of the rising and then falling concentrations is not clear, microbes are implicated as it has been observed that greater concentration changes take place in postmortem blood rather than antemortem serum (Beránková *et al.*, 2006). One untested hypothesis is that the metabolite GHB-glucuronide, which has a longer elimination half-life than GHB, is slowly hydrolyzed back into GHB (Petersen *et al.*, 2013). This may be mediated by microbial enzymes. Another hypothesis includes the degradation of putrescine, both an endogenous chemical and one that is produced from putrefactive amino acid degradation, to GABA and its subsequent degradation to GHB (Marinetti, 2001; Paczkowski and Schütz, 2011).

Interpretation of GHB concentrations is complex; low levels are expected due to the drug's rapid elimination, and at these levels it is difficult to establish whether concentrations are endogenous, result from consumption, or from postmortem production. As GHB is labile in the absence of preservatives at room temperature (i.e., during the PMI or incorrectly stored antemortem specimens), an alternative biomarker for GHB consumption that can both afford greater sensitivity and behave more predictably is sought after. Recently, GHB sulfate has been discovered as a metabolite which, analogous to EtS in ethanol analysis, may be resistant to microbial degradation and thus more valuable than GHB or GHB glucuronide alone in inferring endogenous or administered GHB for postmortem specimens (Hanisch *et al.*, 2016). Further work is necessary to establish endogenous levels of GHB sulfate, as well as whether microbes may have an impact upon conjugate concentrations.

9.5.1.9 Opiates

Concentrations of 6-monoacetylmorphine (6-MAM) — the key metabolite that indicates heroin consumption — have been found to decrease in femoral blood between body receipt at the mortuary and autopsy, with an average time interval of 64 hours (Gerostamoulos *et al.*, 2012). Høiseth *et al.* (2014) found that in both ante- and postmortem blood, 6-MAM was detected in only 4 of 19 cases in a subsequent reanalysis (between 4.2 and 9.4 years later) despite storage with sodium fluoride and at −20°C. Papoutsis *et al.* (2014) found that losses of 6-MAM were significant after storage for 3 months, with about 35% losses reported in preserved antemortem whole blood. 6-MAM may be found in preserved vitreous humor even when it is no longer detectable in preserved postmortem blood, similar to ethanol (Holmgren *et al.*, 2004; Thierauf *et al.*, 2011). A significant microbiological component to the instability of 6-MAM is implied by the results of Zaitsu *et al.* (2008), in which sodium azide and filtration sterilization decreased and prevented, respectively, 6-MAM degradation in antemortem urine.

Compared to 6-MAM, morphine and codeine are more stable. Insignificant changes to their concentrations were found in preserved postmortem blood stored at −20°C after 1 year (Holmgren *et al.*, 2004), in ante- and postmortem blood at −20°C after 9 years (Høiseth *et al.*, 2014), and less than 30% change for most preserved postmortem specimens reanalyzed after 16–18 years at −20°C (Karinen *et al.*, 2014). Another year-long study found that morphine concentrations in postmortem blood exhibited losses of 15–24% when stored between −20 and 25°C — no mention was made as to the use of preservatives or anticoagulants in this study (Hadidi and Oliver, 1998). Stevens (1984) observed no degradation of morphine in putrefying human liver macerates after 29 days.

The situation relating to overall change in morphine concentrations over time appears dependent upon hydrolysis of its glucuronides (Giorgi and Meeker, 1995; Lin *et al.*, 1995). A study by Moriya and Hashimoto (1997) found that postmortem blood and urine specimens, but not liver, maintained morphine glucuronide concentrations when stored at 4°C, 18–22°C (room temperature), and 37°C over a period of 10 days (no use of preservatives mentioned by the authors). Longer-term studies in four specimens of postmortem blood have shown that morphine and its glucuronides are stable if stored at −20°C (no use of preservatives mentioned by the authors) (Skopp *et al.*, 2001). At 4°C significant changes in concentration were observed for morphine-3-glucuronide (M3G) in postmortem blood, which averaged about 20% losses after 124 days of storage (Skopp *et al.*, 2001). Morphine-6-glucuronide (M6G) and morphine concentrations were more notably affected by storage at 20°C in postmortem blood, wherein both were ultimately reduced to undetectable levels in two of four cases (Skopp *et al.*, 2001). M3G concentrations became undetectable by the 124th day also, although no specimens were analyzed between the 70th and 124th day (Skopp *et al.*, 2001). Giorgi and Meeker (1995) reported that in preserved antemortem blood stored at ambient temperature, morphine concentrations decreased in specimens stored for 2 years, increased in specimens stored for 3 years, and decreased in specimens stored for 5 years. Skopp *et al.* (2001) in their study found that in the postmortem blood specimens, concentrations increased during the first 70 days for two of four cases and then became undetectable in two of the four cases; however, they did not mention whether preservatives were used.

9.5.1.10 New psychoactive substances

New psychoactive substances (NPS) encompass a variety of drugs that may be further classified by their chemical structure (e.g., the cathinones, phenethylamines, tryptamines, and piperazines) or psychotropic effects (e.g., stimulants, hallucinogens, entactogens, synthetic opioids, sedatives, and cannabimimetics) (Munro and Wilkin, 2014; Goggin *et al.*, 2015). Despite the proliferation of NPS and their involvement in fatalities, there is a lack of information regarding their stability toward microbial degradation; as of early 2016, only mephedrone (4-methyl methcathinone) and *N*-benzylpiperazine (BZP) have been studied.

Studies on mephedrone show that microbial activity could be contributing toward its degradation. Johnson and Botch-Jones (2013) reported 30% losses and >50% losses of mephedrone after 7 and 14 days of storage, respectively, in whole blood at 4°C. At room temperature, mephedrone was undetectable after 7 days — the presence or absence of sodium fluoride was not mentioned. Degradation rates were slower in a subsequent study that indicated mephedrone was more stable in antemortem blood compared to postmortem blood. The use of sodium fluoride with potassium oxalate (1.67%/0.2%) as a preservative was also found to be preferred to no preservative at all (Busardò *et al.*, 2015). After 185 days, specimens stored at −20°C in sodium fluoride preserved antemortem blood resulted in losses of 6.6%, whereas preserved postmortem blood specimens showed losses of 9.1% (Busardò *et al.*, 2015). Complete mephedrone degradation was observed within 94 and 59 days at temperatures of 4°C and 20°C, respectively, for both ante- and postmortem specimens (Busardò *et al.*, 2015). Greater than 30% losses were observed for mephedrone in putrefying porcine liver after 30 days and 14 days at 4°C and room temperature, respectively (Wenholz *et al.*, 2016). Degradation in the putrefying porcine liver was significantly greater than in aqueous controls.

Wenholz *et al.* (2016) investigated BZP stability in putrefying porcine liver and water at 4 and 20°C after 14, 30, and 90 days, finding significantly greater degradation of BZP in the putrefying samples after 14 days and 90 days for the samples stored at 4°C and 20°C, respectively. Degradation differences between the two matrices in all other analyzed samples were insignificant, indicating that microbial activity is likely not crucial for the degradation of BZP.

9.5.1.11 Other drugs

Paracetamol concentrations have been found to increase in preserved and unpreserved postmortem blood specimens, with corresponding decreases in concentration of the glucuronide metabolite. Storage of casework specimens below −15°C is effective in inhibiting the conversion of paracetamol glucuronide to paracetamol, regardless of whether a preservative is used (Thai *et al.*, 2014). In unpreserved postmortem blood, this hydrolysis was complete within 3 days at room temperature, whereas in postmortem blood preserved with sodium fluoride (2% w/v), the rate is reduced such that an approximate 50% reduction in instrument response was observed after 5 days (Thai *et al.*, 2014; Scott *et al.*, 2015). Thus, sodium fluoride is efficient in reducing the rate of paracetamol-glucuronide degradation but does not completely prevent degradation from occurring. In comparison, unpreserved sterile antemortem blood showed no conversion of the glucuronide to the parent drug after 48 hours of incubation at 37°C, although some degradation of the glucuronide was observed after 8 weeks storage at 20 and 30°C in preserved antemortem blood (Thai *et al.*, 2014; Scott *et al.*, 2015).

Thai *et al.* (2014) added 20 bacterial species to sterile antemortem unpreserved blood incubated at 37°C and found that paracetamol glucuronide completely degraded within 10 hours, with a corresponding increase in response for paracetamol. Paracetamol glucuronide was then spiked into sterile antemortem unpreserved blood and portions inoculated with individual bacterial species with incubation at 37°C, and it was found that 10 of the species produced paracetamol within 24 hours and at 48 hours paracetamol was detected for another two species. An abiotic control with aqueous solutions of paracetamol glucuronide confirmed that bacteria are crucially involved in the hydrolysis of the glucuronide to the parent drug (Table 9.1) (Thai *et al.*, 2014; Scott *et al.*, 2015). No degradation of the sulfate or mercapturate metabolites was reported (Thai *et al.*, 2014; Scott *et al.*, 2015).

The stability of anti-asthmatic β_2-agonists was investigated by Couper and Drummer (1999) in preserved postmortem blood. While salbutamol and terbutaline were found to be stable after 1 week at 23°C, 6 months at 4°C, and 1–2 years at −20°C, fenoterol degradation was significant under all three conditions (losses of 83%, 93%, and 66%, respectively). Structurally, fenoterol differs from salbutamol and terbutaline by the presence of an additional phenolic group, with its degradation potentially attributed to oxidation at this site. Furthermore, as there were no observed increases in salbutamol concentration over 2 years, the authors also hypothesized that the sulfate metabolite of salbutamol might not degrade back into the parent drug.

Trihexyphenidyl was found to degrade slowly in postmortem blood and urine over a 6-month period at 25°C giving maximal losses of 12.8% (Battah and Hadidi, 1998). Postmortem blood and urine specimens analyzed over the same period and stored at 4 and −20°C were stable. Antemortem blood and urine specimens were collected from four people receiving trihexyphenidyl, with no change in concentrations after storage at 4°C for 6 months.

Valproic acid has been found to degrade in postmortem blood stored at 4°C and 20°C, resulting in losses of about 25% and 85%, respectively, over 28 days (Kiencke *et al.*, 2013).

It appears that chemical hydrolysis is the dominant mechanism of zopiclone degradation with the rate of formation of 2-amino-5-chloropyridine (ACP), the only known degradation product of zopiclone and its metabolites, occurring most rapidly in alkaline conditions (Nilsson *et al.*, 2014). Pounder and Davies (1994) observed no degradation of zopiclone in unpreserved postmortem blood inoculated with fecal bacteria after 10 days at 23°C. Other studies, in both ante- and postmortem blood have found that zopiclone does degrade. After 8 days of storage at 20°C, zopiclone losses were 75%, and hydrolysis was complete after 1 week of storage at 40°C in unpreserved antemortem blood (Jantos *et al.*, 2013). In preserved antemortem blood, complete degradation was reported after ~25 days storage at 20°C, and ~50% decreases after almost 1 year of storage at −20°C (Nilsson *et al.*, 2010). In unpreserved postmortem blood, the degradation of zopiclone is similar to its degradation in preserved antemortem blood (Pépin *et al.*, 1998). Furthermore, in preserved postmortem blood, losses >20% have been observed after 12 months of storage at −20°C by Holmgren *et al.* (2004), whereas no significant degradation was observed by Nilsson *et al.* (2011) in preserved antemortem blood.

9.5.2 Poisons

9.5.2.1 Cyanide

Obtaining meaningful quantitative results for cyanide is complicated by the possibility of increasing or decreasing postmortem concentrations. Lokan *et al.* (1987) suggested that bacterial species such as *Pseudomonas aeruginosa* could utilize glycine as a substrate to form cyanide. A more recent review highlighted that these changes may also be attributed to the formation of cyanide from endogenous thiocyanate, or the degradation of cyanide to thiocyanate or ammonium formate, or the reaction between cyanide with aldehydes and polysulfides (McAllister *et al.*, 2008). Later, McAllister *et al.* (2011) found that changes in cyanide concentrations could be prevented using sodium fluoride (2% w/v), whereas average increases of 35% were reported in unpreserved specimens after 25–30 days at 4°C.

9.5.2.2 Ethylene glycol

Although ethylene glycol is not an endogenous compound, its presence is not uncommon in human blood and urine due to environmental exposure (Gomes *et al.*, 2002; Wurita *et al.*, 2013, 2014b). Threefold increases in ethylene glycol concentrations are possible after 7 days of storage in unpreserved whole blood spiked with putrefying blood (10% v/v) at 26–28°C (Wurita *et al.*, 2014a). However, the microbial production of ethylene glycol is unlikely to have significant consequences on determining ethylene glycol poisoning as the cause of death. Ethylene glycol poisoning results in characteristic findings at autopsy and blood concentrations greater than those reported in the study by Wurita et al (Leikin and Watson, 2003; Rosano *et al.*, 2009; Toth-Manikowski *et al.*, 2014; Wurita *et al.*, 2014a; Viinamäki *et al.*, 2015).

9.5.2.3 Formic acid

Formic acid is a toxic metabolite of formaldehyde and methanol, and is responsible for decreased vision and blindness associated with methanol poisoning. As with ethylene glycol and GHB, background levels of formic acid are present in both ante- and postmortem specimens, even when methanol or formaldehyde ingestion antemortem has been definitively excluded (Viinamäki *et al.*, 2011). This occurs due to consuming foods and drinks that contain trace quantities of methanol and formaldehyde (Magnuson *et al.*, 2007). The increase in formic acid concentrations in putrefying specimens is well established, having been explored as a marker for the estimation of PMI (Donaldson and Lamont, 2013). Cases have been reported where formic acid concentrations have exceeded the likely fatal concentration of 0.5 g/L despite the absence of methanol or formaldehyde ingestion ante mortem, presumably due to protein and amino acid degradation by microbes (Jones *et al.*, 2007; Wallage and Watterson, 2008; Viinamäki *et al.*, 2011; Donaldson and Lamont, 2013). In urine specimens separately inoculated with *Enterococci* and *E. coli*, formic acid concentrations have been observed to both increase, presumably from citric acid decomposition, and decrease when stored at room temperature and 37°C, at a rate dependent on pH (Schiwara *et al.*, 1992). Storage with sodium fluoride as a preservative was sufficient in preventing further significant formic acid synthesis in putrefying blood and urine stored at 4°C for 3–4 months (Viinamäki *et al.*, 2011). Thus, the postmortem production of formic acid may only be significant during the PMI although further work needs to be performed in order to determine this.

9.6 Concluding remarks

Many drugs, poisons, and their metabolites are unstable in postmortem specimens. Having an understanding of the microbial contribution to this instability due to the near-pervasive presence of microorganisms in postmortem specimens is crucial for the interpretation of casework results in forensic toxicology. Despite the use of preservative agents and frozen storage conditions after specimen collection to inhibit both enzymatic and microbial activity, changes in drug concentrations are still observed. This may be because either the microbes or enzymatic processes are not inhibited by the preservation methods or because abiotic processes are responsible, such as chemical hydrolysis. The use of preservatives and frozen storage also cannot counteract the degradation of drugs that occurs during the PMI prior to specimen collection at autopsy.

Although microbial degradation often results in the loss of drugs and/or their metabolites, for those drugs that have conjugate metabolites, microbial activity may result in the formation of the parent drug, thus increasing its concentration over time. Several instances are reported where both microbial production and degradation of drugs and metabolites take place concurrently (Beránková *et al.*, 2006; Helander *et al.*, 2007). Thus, the overall drug concentration outcome over time may be complex and depends upon the kinetics of competing processes.

Whereas a good body of knowledge is available in regard to drug decomposition kinetics, there is less information available concerning drug production kinetics. It therefore follows that a more extensive study of the conversion of glucuronides and other conjugates into their parent drug in the presence of microbes is a research priority. Such investigations are encouraged by prior studies in the literature where the hydrolysis rate of glucuronide metabolites has been found to increase in postmortem blood compared to antemortem blood (Carroll *et al.*, 2000; Thai, 2013; Thai *et al.*, 2014; Scott *et al.*, 2015). Furthermore, in some laboratories, enzymatic or chemical hydrolysis of glucuronide conjugates is deliberately carried out to hydrolyze the conjugates back into the parent drug in order to achieve greater analytical sensitivity or longer detection windows. If conjugates are lost as a result of microbial action, then that is important information for the forensic toxicologist.

Another avenue for further research would be to determine whether there are undiscovered sulfate metabolites of those drugs that are known to undergo glucuronidation. Both EtS and paracetamol sulfate have been observed to be stable in putrefying environments (Thierauf *et al.*, 2011; Thai *et al.*, 2014; Scott *et al.*, 2015). Therefore, other sulfate metabolites may also be resistant to microbial activity and thus useful markers of antemortem drug use. This would particularly be the case if concentrations of the parent drug or other metabolites fluctuate.

Experimental approaches that investigate how the entire microbial community present in postmortem specimens impact drug and metabolite concentrations are required to ensure meaningful toxicology result interpretations. Postmortem specimens are ecological systems, and forensic toxicology would benefit greatly if these microbial habitats were understood in more detail. Further clarification regarding which microorganisms are responsible for degradation may be achieved utilizing high-throughput sequencing to investigate microbial community structure in conjunction with culture-dependent methods such as matrix-assisted laser desorption/ionization time-of-flight mass spectrometry (culturomics).

References

Abourashed, E.A., Clark, A.M. and Hufford, C.D. (1999) Microbial models of mammalian metabolism of xenobiotics: an updated review. *Current Medicinal Chemistry*, **6**, 359–374.

Alfazil, A.A. (2009) Stability of drugs and pesticides of forensic toxicological interest and their metabolites in biological samples. PhD thesis, University of Glasgow.

Al-Hadidi, K.A. and Oliver, J.S. (1995) Stability of temazepam in blood. *Science & Justice*, **35**, 105–108.

Andresen-Streichert, H., Jensen, P., Kietzerow, J. *et al.* (2015) Endogenous gamma-hydroxybutyric acid (GHB) concentrations in post-mortem specimens and further recommendation for interpretative cut-offs. *International Journal of Legal Medicine*, **129**, 57–68.

Asha, S. and Vidyavathi, M. (2009) Cunninghamella—a microbial model for drug metabolism studies—a review. *Biotechnology Advances*, **27**, 16–29.

Baranowski, S., Serr, A., Thierauf, A. *et al.* (2008) In vitro study of bacterial degradation of ethyl glucuronide and ethyl sulphate. *International Journal of Legal Medicine*, **122**, 389–393.

Baselt, R.C., Shaw, R.F. and McEvilly, R. (1985) Effect of sodium fluoride on cholinesterase activity in postmortem blood. *Journal of Forensic Sciences*, **30**, 1206–1209.

Baselt, R.C., Yoshikawa, D., Chang, J. and Li, J. (1993) Improved long-term stability of blood cocaine in evacuated collection tubes. *Journal of Forensic Sciences*, **38**, 935–937.

Basit, A.W., Newton, J.M. and Lacey, L.F. (2002) Susceptibility of the H2-receptor antagonists cimetidine, famotidine and nizatidine, to metabolism by the gastrointestinal microflora. *International Journal of Pharmaceutics*, **237**, 23–33.

Battah, A.H. and Hadidi, K.A. (1998) Stability of trihexyphenidyl in stored blood and urine specimens. *International Journal of Legal Medicine*, **111**, 111–114.

Batziris, H.P., McIntyre, I.M. and Drummer, O.H. (1999) The effect of sulfur-metabolising bacteria on sulfur-containing psychotropic drugs. *International Biodeterioration and Biodegradation*, **44**, 111–116.

Bennett, J.R. (2001) The organ retention furore: the need for consent. *Clinical Medicine (London, England)*, **1**, 167.

Beránková, K., Mutnanská, K. and Balíková, M. (2006) Gamma-hydroxybutyric acid stability and formation in blood and urine. *Forensic Science International (Online)*, **161**, 158–162.

Busardò, F.P., Kyriakou, C., Tittarelli, R. *et al.* (2015) Assessment of the stability of mephedrone in ante-mortem and post-mortem blood specimens. *Forensic Science International*, **256**, 28–37.

Butzbach, D.M. (2011) The stability of selected psychiatric medications during simulated post-mortem decomposition. PhD thesis, Flinders University.

Butzbach, D.M., Stockham, P.C., Kobus, H.J. *et al.* (2013a) Bacterial degradation of risperidone and paliperidone in decomposing blood. *Journal of Forensic Sciences*, **58**, 90–100.

Butzbach, D.M., Stockham, P.C., Kobus, H.J. *et al.* (2013b) Stability of serotonin-selective antidepressants in sterile and decomposing liver tissue. *Journal of Forensic Sciences*, **58**, S117–S125.

Caldwell, J. and Hawksworth, G.M. (1973) The demethylation of methamphetamine by intestinal microflora. *Journal of Pharmacy and Pharmacology*, **25**, 422–424.

Carroll, F.T., Marraccini, J.V., Lewis, S. and Wright, W. (2000) Morphine-3-D glucuronide stability in postmortem specimens exposed to bacterial enzymatic hydrolysis. *The American Journal of Forensic Medicine and Pathology*, **21**, 323–329.

Castro, A.L., Dias, M., Reis, F. and Teixeira, H.M. (2014) Gamma-hydroxybutyric acid endogenous production and post-mortem behaviour—the importance of different biological matrices, cut-off reference values, sample collection and storage conditions. *Journal of Forensic and Legal Medicine*, **27**, 17–24.

Clauwaert, K.M., Van Bocxlaer, J.F. and De Leenheer, A.P. (2001) Stability study of the designer drugs "MDA, MDMA and MDEA" in water, serum, whole blood, and urine under various storage temperatures. *Forensic Science International*, **124**, 36–42.

Corry, J.E. (1978) Possible sources of ethanol ante- and post-mortem: its relationship to the biochemistry and microbiology of decomposition. *Journal of Applied Bacteriology*, **44**, 1–56.

Couper, F.J. and Drummer, O.H. (1999) Postmortem stability and interpretation of beta 2-agonist concentrations. *Journal of Forensic Sciences*, **44**, 523–526.

Coutselinis, A., Dimopoulos, G. and Dritsas, C. (1974) Fatal intoxication with chlorpromazine with special regard to the influence of putrefaction on its toxicological analysis. *Forensic Science*, **4**, 191–194.

Dick, G.L. and Stone, H.M. (1987) Alcohol loss arising from microbial contamination of drivers' blood specimens. *Forensic Science International*, **34**, 17–27.

Donaldson, A.E. and Lamont, I.L. (2013) Biochemistry changes that occur after death: potential markers for determining post-mortem interval. *PLoS One*, **8**, e82011.

Druid, H. and Holmgren, P. (1997) A compilation of fatal and control concentrations of drugs in postmortem femoral blood. *Journal of Forensic Sciences*, **42**, 79–87.

Duffort, G., Deveaux, M., Perrin, M. *et al.* (2005) Étude de la stabilité in vitro de 21 antidépresseurs et de 16 neuroleptiques dans le sang total. *Annales de Toxicologie Analytique*, **17**, 195–202.

Dull, B.J., Salata, K. and Goldman, P. (1987) Role of the intestinal flora in the acetylation of sulfasalazine metabolites. *Biochemical Pharmacology*, **36**, 3772–3774.

El Mahjoub, A. and Staub, C. (2000) Stability of benzodiazepines in whole blood samples stored at varying temperatures. *Journal of Pharmaceutical and Biomedical Analysis*, **23**, 1057–1063.

Elliott, S. (2001) The presence of gamma-hydroxybutyric acid (GHB) in postmortem biological fluids. *Journal of Analytical Toxicology*, **25**, 152.

Entwistle, N., Owen, P., Patterson, D.A. *et al.* (1986) The occurrence of chlordiazepoxide degradation products in sudden deaths associated with chlordiazepoxide overdosage. *Journal of the Forensic Science Society*, **26**, 45–54.

Evans, W.E.D. (1963) *The Chemistry of Death*, Charles C Thomas Pub Ltd, Springfield, IL.

Fjeld, B., Burns, M.L., Karinen, R. *et al.* (2012) Long-term stability of GHB in post-mortem samples and samples from living persons, stored at −20°C, using fluoride preservatives. *Forensic Science International (Online)*, **222**, 47–51.

Furet, J.-P., Firmesse, O., Gourmelon, M. *et al.* (2009) Comparative assessment of human and farm animal faecal microbiota using real-time quantitative PCR. *FEMS Microbiology Ecology*, **68**, 351–362.

Gerostamoulos, D., Beyer, J., Staikos, V. *et al.* (2012) The effect of the postmortem interval on the redistribution of drugs: a comparison of mortuary admission and autopsy blood specimens. *Forensic Science, Medicine, and Pathology*, **8**, 373–379.

Gilliland, M. and Bost, R. (1993) Alcohol in decomposed bodies: postmortem synthesis and distribution. *Journal of Forensic Sciences*, **38**, 1266–1274.

Giorgi, S.N. and Meeker, J.E. (1995) A 5-year stability study of common illicit drugs in blood. *Journal of Analytical Toxicology*, **19**, 392–398.

Goggin, L.S., Gately, N. and Bridle, R.I. (2015) Novel psychoactive substance and other drug use by young adults in Western Australia. *Journal of Psychoactive Drugs*, **47**, 140–148.

Goldin, B.R., Peppercorn, M.A. and Goldman, P. (1973) Contributions of host and intestinal microflora in the metabolism of L-dopa by the rat. *Journal of Pharmacology and Experimental Therapeutics*, **186**, 160–166.

Gomes, R., Liteplo, R. and Meek, M. (2002) *Ethylene Glycol: Human Health Aspects*, World Health Organization, Geneva.

Grassi, H.C., Andrades, E.D.J. and Marquina, F. (2006) Preliminary report on isolation and characterization of cocaine metabolizing fungi and their involvement in human mycoses. *Journal of Medical Mycology*, **16**, 1–8.

Hadidi, K. and Oliver, J. (1998) Stability of morphine and buprenorphine in whole blood. *International Journal of Legal Medicine*, **111**, 165–167.

Halter, C.C., Laengin, A., Al-Ahmad, A. *et al.* (2009) Assessment of the stability of the ethanol metabolite ethyl sulfate in standardised degradation tests. *Forensic Science International*, **186**, 52–55.

Hamilton, I.R. (1990) Biochemical effects of fluoride on oral bacteria. *Journal of Dental Research*, **69**, 660–667.

Hanisch, S., Stachel, N. and Skopp, G. (2016) A potential new metabolite of gamma-hydroxybutyrate: sulfonated gamma-hydroxybutyric acid. *International Journal of Legal Medicine*, **130**, 411–414.

Harris, B.E., Manning, B.W., Federle, T. and Diasio, R.B. (1986) Conversion of 5-fluorocytosine to 5-fluorouracil by human intestinal microflora. *Antimicrobial Agents and Chemotherapy*, **29**, 44–48.

Helander, A. and Dahl, H. (2005) Urinary tract infection: a risk factor for false-negative urinary ethyl glucuronide but not ethyl sulfate in the detection of recent alcohol consumption. *Clinical Chemistry*, **51**, 1728–1730.

Helander, A., Olsson, I. and Dahl, H. (2007) Postcollection synthesis of ethyl glucuronide by bacteria in urine may cause false identification of alcohol consumption. *Clinical Chemistry*, **53**, 1855–1857.

Helander, A., Böttcher, M., Fehr, C. *et al.* (2009) Detection times for urinary ethyl glucuronide and ethyl sulfate in heavy drinkers during alcohol detoxification. *Alcohol and Alcoholism*, **44**, 55–61.

Heller, S., Hiemke, C., Stroba, G. *et al.* (2004) Assessment of storage and transport stability of new antidepressant and antipsychotic drugs for a nationwide TDM service. *Therapeutic Drug Monitoring*, **26**, 459–461.

Hernández Redondo, A., Körber, C., König, S. *et al.* (2012) Inhibition of bacterial degradation of EtG by collection as dried urine spots (DUS). *Analytical and Bioanalytical Chemistry*, **402**, 2417–2424.

Høiseth, G., Karinen, R., Johnsen, L. *et al.* (2008) Disappearance of ethyl glucuronide during heavy putrefaction. *Forensic Science International*, **176**, 147–151.

Høiseth, G., Karinen, R., Christophersen, A. and Mørland, J. (2010) Practical use of ethyl glucuronide and ethyl sulfate in postmortem cases as markers of antemortem alcohol ingestion. *International Journal of Legal Medicine*, **124**, 143–148.

Høiseth, G., Fjeld, B., Burns, M.L. *et al.* (2014) Long-term stability of morphine, codeine, and 6-acetylmorphine in real-life whole blood samples, stored at −20°C. *Forensic Science International*, **239**, 6–10.

Holmgren, P., Druid, H., Holmgren, A. and Ahlner, J. (2004) Stability of drugs in stored postmortem femoral blood and vitreous humor. *Journal of Forensic Sciences*, **49**, 820–825.

Holt, R. (1967) The bacterial degradation of chloramphenicol. *The Lancet*, **289**, 1259–1260.

Hunsaker, D. and Hunsaker, J., III (2005) Alcohol: blood and body fluid analysis, in *Encyclopedia of Forensic and Legal Medicine* (eds J. Payne-James and R.W. Byard), Elsevier Science, Amsterdam/Boston, pp. 29–38.

Indahl, S. and Scheline, R. (1968) Decarboxylation of 4-hydroxycinnamic acids by Bacillus strains isolated from rat intestine. *Applied Microbiology*, **16**, 667.

Isenschmid, D.S., Levine, B.S. and Caplan, Y.H. (1989) A comprehensive study of the stability of cocaine and its metabolites. *Journal of Analytical Toxicology*, **13**, 250–256.

Janaway, R., Percival, S., and Wilson, A. (2009) Decomposition of human remains, in *Microbiology and Aging* (ed. S. Percival), Humana Press, Towata, pp. 313–334, http://www.springer.com/gp/book/9781588296405?wt_mc=ThirdParty.SpringerLink.3.EPR653.About_eBook (accessed December 5, 2016).

Jantos, R., Vermeeren, A., Sabljic, D. *et al.* (2013) Degradation of zopiclone during storage of spiked and authentic whole blood and matching dried blood spots. *International Journal of Legal Medicine*, **127**, 69–76.

Jiménez, C., de la Torre, R., Ventura, M. *et al.* (2006) Stability studies of amphetamine and ephedrine derivatives in urine. *Journal of Chromatography B*, **843**, 84–93.

Johnson, R.D. and Botch-Jones, S.R. (2013) The stability of four designer drugs: MDPV, mephedrone, BZP and TFMPP in three biological matrices under various storage conditions. *Journal of Analytical Toxicology*, **37**, 51–55.

Johnson, J.R., Jennison, T.A., Peat, M.A. and Foltz, R.L. (1984) Stability of Δ9-tetrahydrocannabinol (THC), 11-hydroxy-THC, and 11-nor-9-carboxy-THC in blood and plasma. *Journal of Analytical Toxicology*, **8**, 202–204.

Jones, A.W. and Holmgren, A. (2009) Concentration distributions of the drugs most frequently identified in post-mortem femoral blood representing all causes of death. *Medicine, Science and the Law*, **49**, 257–273.

Jones, G.R., Singer, P.P. and Rittenbach, K. (2007) The relationship of methanol and formate concentrations in fatalities where methanol is detected. *Journal of Forensic Sciences*, **52**, 1376–1382.

Jones, A.W., Gladh, S.-Å., Windberg, C.N. and Johansen, S.S. (2015) Stability of γ-hydroxybutyrate in blood samples from impaired drivers after storage at 4°C and comparison of GC–FID–GBL and LC–MS-MS methods of analysis. *Journal of Analytical Toxicology*, **39**, 294–299.

Kang, M.J., Kim, H.G., Kim, J.S. *et al.* (2013) The effect of gut microbiota on drug metabolism. *Expert Opinion on Drug Metabolism and Toxicology*, **9**, 1295–1308.

Karinen, R., Øiestad, E.L., Andresen, W. *et al.* (2011) Comparison of the stability of stock solutions of drugs of abuse and other drugs stored in a freezer, refrigerator, and at ambient temperature for up to one year. *Journal of Analytical Toxicology*, **35**, 583–590.

Karinen, R., Andresen, W., Smith-Kielland, A. and Mørland, J. (2014) Long-term storage of authentic postmortem forensic blood samples at −20°C: measured concentrations of benzodiazepines, central stimulants, opioids and certain medicinal drugs before and after storage for 16–18 years. *Journal of Analytical Toxicology*, **38**, 686–695.

Kellerman, G., Waterman, N. and Scharefenberger, L. (1976) Demonstration in vitro of postmortem bacterial transmigration. *American Journal of Clinical Pathology*, **66**, 911–915.

Kiencke, V., Andresen-Streichert, H., Müller, A. and Iwersen-Bergmann, S. (2013) Quantitative determination of valproic acid in postmortem blood samples—evidence of strong matrix dependency and instability. *International Journal of Legal Medicine*, **127**, 1101–1107.

Kintz, P., Villain, M., Cirimele, V. and Ludes, B. (2004) GHB in postmortem toxicology: discrimination between endogenous production from exposure using multiple specimens. *Forensic Science International*, **143**, 177–181.

Korb, A.-S. and Cooper, G. (2014) Endogenous concentrations of GHB in postmortem blood from deaths unrelated to GHB use. *Journal of Analytical Toxicology*, **38**, 582–588.

Kugelberg, F.C. and Jones, A.W. (2007) Interpreting results of ethanol analysis in postmortem specimens: a review of the literature. *Forensic Science International*, **165**, 10–29.

Larsen, N.A., Turner, J.M., Stevens, J. *et al.* (2002) Crystal structure of a bacterial cocaine esterase. *Nature Structural and Molecular Biology*, **9**, 17–21.

Launiainen, T. and Ojanperä, I. (2014) Drug concentrations in post-mortem femoral blood compared with therapeutic concentrations in plasma. *Drug Testing and Analysis*, **6**, 308–316.

Leikin, J.B. and Watson, W.A. (2003) Post-mortem toxicology: what the dead can and cannot tell us. *Journal of Toxicology: Clinical Toxicology*, **41**, 47–56.

Levine, B., Blanke, R. and Valentour, J. (1983) Postmortem stability of benzodiazepines in blood and tissues. *Journal of Forensic Sciences*, **28**, 102–115.

Lewis, R.J., Johnson, R.D., Angier, M.K. and Vu, N.T. (2004) Ethanol formation in unadulterated postmortem tissues. *Forensic Science International (Online)*, **146**, 17–24.

Ley, R.E., Hamady, M., Lozupone, C. *et al.* (2008) Evolution of mammals and their gut microbes. *Science*, **320**, 1647–1651.

Li, B., Sedlacek, M., Manoharan, I. *et al.* (2005) Butyrylcholinesterase, paraoxonase, and albumin esterase, but not carboxylesterase, are present in human plasma. *Biochemical Pharmacology*, **70**, 1673–1684.

Lin, D.-L., Liu, H. and Chen, C.-Y. (1995) Storage temperature effect on the stability of morphine and codeine in urine. *Journal of Analytical Toxicology*, **19**, 275–280.

Logan, B.K. (2001) Ecgonine is an important marker for cocaine use in inadequately preserved specimens. *Journal of Analytical Toxicology*, **25**, 219–220.

Lokan, R.J., James, R.A. and Dymock, R.B. (1987) Apparent post-mortem production of high levels of cyanide in blood. *Journal of the Forensic Science Society*, **27**, 253–259.

Lough, P.S. and Fehn, R. (1993) Efficacy of 1% sodium fluoride as a preservative in urine samples containing glucose and Candida albicans. *Journal of Forensic Sciences*, **38**, 266–271.

Lutfi, L.A. (1998) Stability of drugs of forensic interest in post mortem blood. PhD thesis, University of Glasgow.

Madigan, M.T., Bender, K.S., Martinko, J.M. *et al.* (2014) *Brock Biology of Microorganisms*, Pearson, London.

Magnuson, B., Burdock, G., Doull, J. *et al.* (2007) Aspartame: a safety evaluation based on current use levels, regulations, and toxicological and epidemiological studies. *Critical Reviews in Toxicology*, **37**, 629–727.

Mannens, G., Huang, M.L., Meuldermans, W. *et al.* (1993) Absorption, metabolism, and excretion of risperidone in humans. *Drug Metabolism and Disposition*, **21**, 1134–1141.

Marinetti, L. (2001) γ-Hydroxybutyric acid and its analogs, γ-butyrolactone and 1,4-butanediol, in *Benzodiazepines and GHB* (ed. S. Salamone), Humana Press, Totowa, pp. 95–126, http://link.springer.com/chapter/10.1007/978-1-59259-109-1_6 (accessed December 5, 2016).

Martindale, S.M., Powers, R.H. and Bell, S.C. (2015) Production of human metabolites by gastrointestinal bacteria as a potential source of post-mortem alteration of antemortem drug/metabolite concentrations. *Drug Testing and Analysis*, **7**, 75–82.

Martínez-Ramírez, J., Voigt, K. and Peters, F. (2012) Studies on the metabolism of five model drugs by fungi colonizing cadavers using LC-ESI-MS/MS and GC-MS analysis. *Analytical and Bioanalytical Chemistry*, **404**, 1339–1359.

Martínez-Ramírez, J.A., Walther, G. and Peters, F.T. (2015) Studies on drug metabolism by fungi colonizing decomposing human cadavers. Part II: biotransformation of five model drugs by fungi isolated from post-mortem material. *Drug Testing and Analysis*, **7**, 265–279.

Martínez-Ramírez, J.A., Strien, J., Walther, G. and Peters, F.T. (2016) Search for fungi-specific metabolites of four model drugs in postmortem blood as potential indicators of postmortem fungal metabolism. *Forensic Science International*, **262**, 173–178.

Maskell, P., Albeishy, M., De Paoli, G. *et al.* (2016) Postmortem redistribution of the heroin metabolites morphine and morphine-3-glucuronide in rabbits over 24 h. *International Journal of Legal Medicine*, **130**, 519–531.

McAllister, J.L., Roby, R.J., Levine, B. and Purser, D. (2008) Stability of cyanide in cadavers and in postmortem stored tissue specimens: a review. *Journal of Analytical Toxicology*, **32**, 612–620.

McAllister, J.L., Roby, R.J., Levine, B. and Purser, D. (2011) The effect of sodium fluoride on the stability of cyanide in postmortem blood samples from fire victims. *Forensic Science International*, **209**, 29–33.

McIntyre, I.M., Mallett, P. and Stabley, R. (2015) Postmortem distribution of trazodone concentrations. *Forensic Science International*, **251**, 195–201.

McKay, G., Cooper, J.K. and Midha, K.K. (1984) The effect of sample storage on the stability of chlorpromazine in plasma and whole blood. *Journal of Pharmacological Methods*, **12**, 221–223.

Megyesi, M.S., Nawrocki, S.P. and Haskell, N.H. (2005) Using accumulated degree-days to estimate the postmortem interval from decomposed human remains. *Journal of Forensic Sciences*, **50**, 618–626.

Melo, P., Bastos, M.L. and Teixeira, H.M. (2012) Benzodiazepine stability in postmortem samples stored at different temperatures. *Journal of Analytical Toxicology*, **36**, 52–60.

Melvin, J., JR, Cronholm, L.S., Simson, L., Jr and Isaacs, A. (1984) Bacterial transmigration as an indicator of time of death. *Journal of Forensic Sciences*, **29**, 412–417.

Mendel, J.L. and Walton, M.S. (1966) Conversion of p, p-DDT to p, p-DDD by intestinal flora of the rat. *Science*, **151**, 1527–1528.

Moriya, F. and Hashimoto, Y. (1996) Postmortem stability of cocaine and cocaethylene in blood and tissues of humans and rabbits. *Journal of Forensic Sciences*, **41**, 612–616.

Moriya, F. and Hashimoto, Y. (1997) Distribution of free and conjugated morphine in body fluids and tissues in a fatal heroin overdose: is conjugated morphine stable in postmortem specimens? *Journal of Forensic Sciences*, **42**, 736–740.

Moriya, F. and Hashimoto, Y. (2005) Site-dependent production of [gamma]-hydroxybutyric acid in the early postmortem period. *Forensic Science International (Online)*, **148**, 139–142.

Moriya, F., Hashimoto, Y. and Kuo, T.-t. (1999) Pitfalls when determining tissue distributions of organophosphorus chemicals: sodium fluoride accelerates chemical degradation. *Journal of Analytical Toxicology*, **23**, 210–215.

Munro, G. and Wilkin, C. (2014) *New Psychoactive Drugs: No Easy Answer*, Australian Drug Foundation, Melbourne.

Musshoff, F., Padosch, S., Steinborn, S. and Madea, B. (2004) Fatal blood and tissue concentrations of more than 200 drugs. *Forensic Science International (Online)*, **142**, 161–210.

Nagata, T., Kimura, K., Hara, K. and Kudo, K. (1990) Methamphetamine and amphetamine concentrations in postmortem rabbit tissues. *Forensic Science International*, **48**, 39–47.

Nilsson, G.H., Kugelberg, F.C., Kronstrand, R. and Ahlner, J. (2010) Stability tests of zopiclone in whole blood. *Forensic Science International*, **200**, 130–135.

Nilsson, G.H., Kugelberg, F.C., Ahlner, J. and Kronstrand, R. (2011) Influence of pre-analytical conditions on the interpretation of zopiclone concentrations in whole blood. *Forensic Science International*, **207**, 35–39.

Nilsson, G.H., Kugelberg, F.C., Ahlner, J. and Kronstrand, R. (2014) Quantitative analysis of zopiclone, N-desmethylzopiclone, zopiclone N-oxide and 2-amino-5-chloropyridine in urine using LC–MS-MS. *Journal of Analytical Toxicology*, **38**, 327–334.

Norman, A. and Grubb, R. (1955) Hydrolysis of conjugated bile acids by *clostridia* and *enterococci*. *Acta Pathologica Microbiologica Scandinavica*, **36**, 537–547.

Paczkowski, S. and Schütz, S. (2011) Post-mortem volatiles of vertebrate tissue. *Applied Microbiology and Biotechnology*, **91**, 917–935.

Pannell, M., Gilbert, J., Gardiner, J. and Byard, R. (2001) Death due to malathion poisoning. *Journal of Clinical Forensic Medicine*, **8**, 156–159.

Papoutsis, I., Nikolaou, P., Pistos, C. *et al.* (2014) Stability of morphine, codeine, and 6-acetylmorphine in blood at different sampling and storage conditions. *Journal of Forensic Sciences*, **59**, 550–554.

Pépin, G., Dubourvieux, N. and Gaillard, Y. (1998) Difficulty of interpretation of rates concerning benzodiazepines and other anxiolytic molecules in post mortem blood taken during autopsy: study of their degradation in vitro after 6-months storage at different temperatures. *Journal de Medecine Legale Droit Medical*, **41**, 341–353.

Peppercorn, M.A. and Goldman, P. (1971) Caffeic acid metabolism by bacteria of the human gastrointestinal tract. *Journal of Bacteriology*, **108**, 996–1000.

Petersen, I.N., Tortzen, C., Kristensen, J.L. *et al.* (2013) Identification of a new metabolite of GHB: gamma-hydroxybutyric acid glucuronide. *Journal of Analytical Toxicology*, **37**, 291–297.

Pounder, D.J. and Davies, J.I. (1994) Zopiclone poisoning: tissue distribution and potential for postmortem diffusion. *Forensic Science International*, **65**, 177–183.

Pounder, D.J., Hartley, A.K. and Watmough, P.J. (1994) Postmortem redistribution and degradation of dothiepin human case studies and an animal model. *The American Journal of Forensic Medicine and Pathology*, **15**, 231–235.

Rajilić-Stojanović, M. and de Vos, W.M. (2014) The first 1000 cultured species of the human gastrointestinal microbiota. *FEMS Microbiology Reviews*, **38**, 996–1047.

Reis, M., Aamo, T., Ahlner, J. and Druid, H. (2007) Reference concentrations of antidepressants. A compilation of postmortem and therapeutic levels. *Journal of Analytical Toxicology*, **31**, 254–264.

Robertson, M.D. and Drummer, O.H. (1995) Postmortem drug metabolism by bacteria. *Journal of Forensic Sciences*, **40**, 382–386.

Robertson, M.D. and Drummer, O.H. (1998) Stability of nitrobenzodiazepines in postmortem blood. *Journal of Forensic Sciences*, **43**, 5–8.

Rosano, T.G., Swift, T.A., Kranick, C.J. and Sikirica, M. (2009) Ethylene glycol and glycolic acid in postmortem blood from fatal poisonings. *Journal of Analytical Toxicology*, **33**, 508–513.

Rosenfeld, R.S., Paul, I. and Yamauchi, T. (1967) Sterol esterification in feces. *Archives of Biochemistry and Biophysics*, **122**, 653–657.

Rositano, J., Kostakis, P., Gilbert, J., and Kostakis, C. (2014) *Two suicides by mevinphos presented at the 22nd International Symposium on the Forensic Sciences, Australian and New Zealand Forensic Science Society*, August 31–September 4, 2014, Adelaide, Australia.

Saar, E., Gerostamoulos, D., Drummer, O.H. and Beyer, J. (2012) Assessment of the stability of 30 antipsychotic drugs in stored blood specimens. *Forensic Science International*, **215**, 152–158.

Saukko, P. and Knight, B. (2004) *Knight's Forensic Pathology*, Arnold Publishers, London.

Scheidweiler, K., Himes, S., Desrosiers, N. and Huestis, M. (2016) In vitro stability of free and glucuronidated cannabinoids in blood and plasma collected in plastic gray-top sodium fluoride tubes following controlled smoked cannabis. *Forensic Toxicology*, **34**, 179–185.

Scheline, R.R. (1968) Drug metabolism by intestinal microorganisms. *Journal of Pharmaceutical Sciences*, **57**, 2021–2037.

Schiwara, H., Siegel, H. and Goebel, A. (1992) Increase and decrease in formic acid concentration in urine samples stored at room temperature. *European Journal of Clinical Chemistry and Clinical Biochemistry*, **30**, 75–79.

Schloegl, H., Dresen, S., Spaczynski, K. *et al.* (2006) Stability of ethyl glucuronide in urine, postmortem tissue and blood samples. *International Journal of Legal Medicine*, **120**, 83–88.

Schulz, M., Iwersen-Bergmann, S., Andresen, H. and Schmoldt, A. (2012) Therapeutic and toxic blood concentrations of nearly 1,000 drugs and other xenobiotics. *Critical Care*, **16**, R136.

Scott, T., Thai, K., Partridge, E., *et al.* 2015) *Post-mortem production of paracetamol via degradation of its metabolites presented at The 53rd TIAFT Meeting*, August 30–September 4, 2015, Firenze, Italy.

Shamat, M.A. (1993) The role of the gastrointestinal microflora in the metabolism of drugs. *International Journal of Pharmaceutics*, **97**, 1–13.

Skopp, G. and Pötsch, L. (2004) An investigation of the stability of free and glucuronidated 11-nor-Δ9-tetrahydrocannabinol-9-carboxylic acid in authentic urine samples. *Journal of Analytical Toxicology*, **28**, 35–40.

Skopp, G., Pötsch, L., König, I. and Mattern, R. (1997) A preliminary study on the stability of benzodiazepines in blood and plasma stored at 4°C. *International Journal of Legal Medicine*, **111**, 1–5.

Skopp, G., Pötsch, L., Klingmann, A. and Mattern, R. (2001) Stability of morphine, morphine-3-glucuronide, and morphine-6-glucuronide in fresh blood and plasma and postmortem blood samples. *Journal of Analytical Toxicology*, **25**, 2–7.

Skov, L., Johansen, S.S. and Linnet, K. (2015a) Postmortem femoral blood reference concentrations of aripiprazole, chlorprothixene, and quetiapine. *Journal of Analytical Toxicology*, **39**, 41–44.

Skov, L., Johansen, S.S. and Linnet, K. (2015b) Postmortem quetiapine reference concentrations in brain and blood. *Journal of Analytical Toxicology*, **39**, 557–561.

Smith, G. and Griffiths, L. (1974) Metabolism of N-acylated and O-alkylated drugs by the intestinal microflora during anaerobic incubation in vitro. *Xenobiotica*, **4**, 477–487.

Soleim, H.A. and Scheline, R.R. (1972) Metabolism of xenobiotics by strains of intestinal bacteria. *Acta Pharmacologica et Toxicologica*, **31**, 471–480.

Sousa, T., Paterson, R., Moore, V. *et al.* (2008) The gastrointestinal microbiota as a site for the biotransformation of drugs. *International Journal of Pharmaceutics*, **363**, 1–25.

Spatz, M., Smith, D., McDaniel, E. and Laqueur, G. (1967) Role of intestinal microorganisms in determining cycasin toxicity. *Experimental Biology and Medicine*, **124**, 691–697.

Stevens, H.M. (1984) The stability of some drugs and poisons in putrefying human liver tissues. *Journal of the Forensic Science Society*, **24**, 577–589.

Stewart, D.J., Inaba, T., Tang, B.K. and Kalow, W. (1977) Hydrolysis of cocaine in human plasma by cholinesterase. *Life Sciences*, **20**, 1557–1563.

Sykutera, M., Pufal, E. and Sliwka, K. (2002) The influence of temperature and time of storage on the stability of fluoxetine in biological material. *Z Zagadnien Nauk Sadowych*, **50**, 5–16.

Taylor, K. and Elliott, S. (2013) An unusual case of risperidone instability in a fatality presenting an analytical and interpretative challenge. *Drug Testing and Analysis*, **5**, 748–752.

Thai, K. (2013) Investigation of paracetamol production in-vitro in biological samples via degradation of its three main metabolites. Honours thesis, Flinders University.

Thai, K., Lenehan, C.E., Kirkbride, K.P. *et al.* 2014) *Post-mortem production of paracetamol via degradation of its metabolites presented at the 22nd International Symposium on the Forensic Sciences, Australian and New Zealand Forensic Science Society*, August 31–September 4, 2014, Adelaide, Australia.

Thierauf, A., Kempf, J., Perdekamp, M.G. *et al.* (2011) Ethyl sulphate and ethyl glucuronide in vitreous humor as postmortem evidence marker for ethanol consumption prior to death. *Forensic Science International (Online)*, **210**, 63–68.

Toth-Manikowski, S.M., Menn-Josephy, H. and Bhatia, J. (2014) A case of chronic ethylene glycol intoxication presenting without classic metabolic derangements. *Case Reports in Nephrology*, **2014**, 128145.

Vermeir, M., Naessens, I., Remmerie, B. *et al.* (2008) Absorption, metabolism, and excretion of paliperidone, a new monoaminergic antagonist, in humans. *Drug Metabolism and Disposition*, **36**, 769–779.

Viinamäki, J., Rasanen, I., Vuori, E. and Ojanperä, I. (2011) Elevated formic acid concentrations in putrefied post-mortem blood and urine samples. *Forensic Science International (Online)*, **208**, 42–46.

Viinamäki, J., Sajantila, A. and Ojanperä, I. (2015) Ethylene glycol and metabolite concentrations in fatal ethylene glycol poisonings. *Journal of Analytical Toxicology*, **39**, 481–485.

Wadworth, A. and Fitton, A. (1991) Olsalazine. a review of its pharmacodynamic and pharmacokinetic properties, and therapeutic potential in inflammatory bowel disease. *Drugs*, **41**, 647–664.

Wallage, H.R. and Watterson, J.H. (2008) Formic acid and methanol concentrations in death investigations. *Journal of Analytical Toxicology*, **32**, 241–247.

Watanabe, K., Yamashita, S., Furuno, K. *et al.* (1995) Metabolism of omeprazole by gut flora in rats. *Journal of Pharmaceutical Sciences*, **84**, 516–517.

Wenholz, D.S., Luong, S., Philp, M. *et al.* (2016) A study to model the post-mortem stability of 4-MMC, MDMA and BZP in putrefying remains. *Forensic Science International*, **265**, 54–60.

Wurita, A., Suzuki, O., Hasegawa, K. *et al.* (2013) Sensitive determination of ethylene glycol, propylene glycol and diethylene glycol in human whole blood by isotope dilution gas chromatography–mass spectrometry, and the presence of appreciable amounts of the glycols in blood of healthy subjects. *Forensic Toxicology*, **31**, 272–280.

Wurita, A., Suzuki, O., Hasegawa, K. *et al.* (2014a) Occurrence of postmortem production of ethylene glycol and propylene glycol in human specimens. *Forensic Toxicology*, **32**, 162–168.

Wurita, A., Suzuki, O., Hasegawa, K. *et al.* (2014b) Presence of appreciable amounts of ethylene glycol, propylene glycol, and diethylene glycol in human urine of healthy subjects. *Forensic Toxicology*, **32**, 39–44.

Yajima, D., Motani, H., Kamei, K. *et al.* (2006) Ethanol production by Candida albicans in postmortem human blood samples: effects of blood glucose level and dilution. *Forensic Science International*, **164**, 116–121.

Yonemitsu, K. and Pounder, D.J. (1993) Postmortem changes in blood tranylcypromine concentration: competing redistribution and degradation effects. *Forensic Science International*, **59**, 177–184.

Zaitsu, K., Miki, A., Katagi, M. and Tsuchihashi, H. (2008) Long-term stability of various drugs and metabolites in urine, and preventive measures against their decomposition with special attention to filtration sterilization. *Forensic Science International*, **174**, 189–196.

Zvosec, D.L., Smith, S.W., Porrata, T. *et al.* (2011) Case series of 226 γ-hydroxybutyrate–associated deaths: lethal toxicity and trauma. *The American Journal of Emergency Medicine*, **29**, 319–332.

CHAPTER 10

Microbial communities associated with decomposing corpses

Embriette R. Hyde[1], Jessica L. Metcalf[2], Sibyl R. Bucheli[3], Aaron M. Lynne[3] and Rob Knight[1,4]

[1] *Department of Pediatrics, University of California, San Diego, San Diego, CA, USA*
[2] *Department of Animal Sciences, Colorado State University, Fort Collins, CO, USA*
[3] *Department of Biological Sciences, Sam Houston State University, Huntsville, TX, USA*
[4] *Department of Computer Science and Engineering, University of California, San Diego, San Diego, CA, USA*

10.1 Introduction

The breakdown of organic remains, or decomposition, is a mosaic community facilitated by the interaction of biotic and abiotic components. Conditions of weather, as well as scavenger, and microbial activity (among others) combine to generate unique scenarios of decomposition. This process may be further influenced by geography, season, and antemortem individuality or postmortem manipulation of the corpse. For example, a typical outdoor decomposition scenario (e.g., hunting accident death, a body left in the woods after a homicide) involves scavenger access to the cadaver and exposure to the elements, whereas a typical indoor decomposition scenario (e.g., an elderly person passes away at home or an indoor homicide) does not include these factors. To understand decomposition in a way that is helpful to law enforcement personnel, forensic researchers need to define how abiotic and biotic factors influence the ecology of decomposition. This knowledge is important in estimating the time since death (i.e., postmortem interval (PMI)) as well as enabling forensics personnel to detect events such as movement of a body at some point during decomposition or to answer a variety of other relevant questions. Importantly, these questions can only be answered via well-designed studies grounded in the scientific method. Indeed, the National Research Council's Committee on Identifying the Needs of the Forensic Sciences Community (2009) underscored the importance of focusing forensic techniques on the principles of the scientific method (National Research Council Committee on Identifying the Needs of the Forensic Sciences Community, 2009). A number of investigators have answered this call, and although

Forensic Microbiology, First Edition. Edited by David O. Carter, Jeffery K. Tomberlin, M. Eric Benbow and Jessica L. Metcalf.

several areas are being explored, the microorganisms associated with decomposition have been receiving an increasing amount of research attention (Parkinson *et al.*, 2009; Hyde *et al.*, 2013, 2015; Pechal *et al.*, 2013, 2014; Metcalf *et al.*, 2013; Lauber *et al.*, 2014; Carter *et al.*, 2015).

While each corpse undergoes its own individual decomposition trajectory based on specific ecological factors, discernible patterns to the decomposition process exist. Generally, decomposition is characterized by five stages, attributed primarily to microbial and necrophagous arthropod activity: fresh decay, putrefaction, black putrefaction, butyric fermentation, and dry decay (Pinheiro, 2006). During fresh decay, bacteria on the inside of the corpse begin digesting away surrounding tissues (Janaway *et al.*, 2009). During putrefaction, bacteria inside the corpse undergo anaerobic respiration, which leads to the buildup of gaseous by-products (Vass, 2001) that inflate the cadaver (bloat) and eventually force fluids out of the body (purge; Janaway *et al.*, 2009). The corpse is then opened to the environment, facilitating wet tissue decomposition and leading to the dry stages of decomposition.

10.1.1 Overview of the importance of bacteria in decomposition and Arpad Vass' original efforts to catalogue this diversity

Although bacteria are recognized as a major driving force behind decomposition, few studies describing the microbial ecology of decomposition have been published (Schoenen and Schoenen, 2013; Hyde *et al.*, 2013, 2015; Metcalf *et al.*, 2013; Pechal *et al.*, 2013; Carter *et al.*, 2015; and Weiss *et al.*, 2015), particularly in comparison with the extensive work done in the field of forensic entomology. Bacteria not only are an intrinsic factor of the corpse, both internally and externally, but may also be introduced from a variety of sources including vertebrate scavengers, arthropods, and soil. Only a few investigations into the potential sources of corpse-colonizing bacteria have been reported. Using 454 pyrosequencing, Zheng *et al.* (2013) characterized the microbial communities associated with successive life stages of the black soldier fly, with the major phyla being Bacteroidetes and Proteobacteria, while Singh *et al.* (2015) reported the presence of the bacterial genera *Providencia, Ignatzschineria, Lactobacillus, Lactococcus, Vagococcus, Morganella,* and *Myroides* associated with blow flies (*Lucilia sericata* and *Lucilia cuprina*). Recently, Roggenbuck *et al.* (2014) utilized 454 pyrosequencing to describe the bacterial communities present on the faces and in the stool of 50 vultures (black and turkey vultures), finding that microbial communities were dominated by Clostridia and Fusobacteria. Additionally, we also reported differential microbial communities associated with carcasses placed on sterile or untreated soils (Lauber *et al.*, 2014); nevertheless, whether soil or carcass was the source of these decomposer bacterial communities was unclear. Arpad Vass abandoned early attempts to catalogue the bacteria involved in human decomposition due to the sheer number of species that he isolated from decomposing cadavers, concluding that "with the exception of microorganisms living in deep sea vents, every microorganism known is involved in some aspect of the human decompositional cycle from *Acetobacter* to *Zooglea*" (Vass, 2001). The microbial ecology of decomposition can truly be appreciated taking into account that in some environments, less than 1% of bacterial diversity is cultivable (Pace *et al.*, 1986). If Vass was overwhelmed by the bacterial diversity that he was able to cultivate from decomposing cadavers, and this overwhelming diversity may represent only 1% of the decomposer community, imagine what truly astounding level of microbial diversity must be associated with decomposition!

Though the sheer number of bacteria present on human corpses deterred Vass' attempts to establish a comprehensive catalogue of the bacterial diversity of decomposition, others have not been so easily swayed, and there are a number of studies in the primary literature detailing the microbial ecology of various decomposition situations. Much literature focuses on investigations of microbial communities in gravesoils, mainly using T-RFLP analysis, which will be discussed further in the chapter. Several reports have also demonstrated that bacteria are useful biomarkers in forensic science (Evans, 1963; Melvin *et al.*, 1984; Carter and Tibbett, 2008; Kakizaki *et al.*, 2008; Meyers and Foran, 2008; Janaway *et al.*, 2009; Howard *et al.*, 2010; Dickson *et al.*, 2011; Schoenen and Schoenen 2013). Specifically, Evans (1963), Janaway *et al.* (2009), Melvin *et al.* (1984), and Vass (2001), among others, list a number of microorganisms that may be significant during the bloat stage of decomposition, and these studies will be discussed in more detail in Section 10.4, while Metcalf *et al.* (2013) and Pechal *et al.* (2014) describe microbial signatures that can be used to estimate the PMI (described in detail in Section 10.7).

10.1.2 Marker gene and metagenomics methods for facilitating studies of the microbial ecology of decomposition

With the advent of marker gene and metagenomics technologies to assess bacterial DNA obtained directly from environmental samples, Vass's impossible problem is becoming solvable. Building on the studies by Janaway, Vass, Melvin, Carter, and others, a few groups have begun to use microbial marker gene sequencing and analysis (described in detail in Chapter 5) to survey the microbial diversity associated with decomposition, providing a higher resolution view of the microbial ecology of decomposition than what has ever been available. For example, Metcalf *et al.* (2013) and Pechal *et al.* (2014) utilized next-generation bacterial and eukaryotic marker gene sequencing and analysis to survey the bacterial and eukaryotic signatures associated with decomposition and bacterial signatures associated with decomposition in mice and pigs, respectively, while others have begun to apply next-generation sequencing techniques toward the characterization of microbial communities associated with human decomposition (Hyde *et al.*, 2013, 2015; Cobaugh *et al.*, 2015; Metcalf *et al.*, 2016a).

This chapter will provide an overview of the current knowledge of the microbial ecology of vertebrate remains decomposition, obtained by numerous culture-dependent and culture-independent studies. First, we will explore the microbiology of gravesoils

(Section 10.2), focusing on how microbial communities in soil are not only affected by decomposition but also promote decomposition. We will then discuss decomposition in water (Section 10.3), not only exploring the decomposition of marine and freshwater organisms such as whales and fish but also discussing human decomposition in water. We will next focus on two important research foci: nonhuman (Section 10.4) and human (Section 10.5) models of decomposition. Here we will provide an overview of experiments using nonhuman models and human cadavers that have advanced forensic microbiology. We will also discuss whether animal models can be used as a proxy for studies of human decomposition, by discussing whether or not evidence suggests the presence of a universal decomposition signature. Finally, we will present evidence for the power of using microbial community trends in estimating the PMI.

10.2 The soil microbiology of decomposition

Although corpses are intrinsically a microbial ecosystem, the soil associated with a corpse during terrestrial decomposition scenarios also represents a complex microbial community during decomposition and even after the remains have disintegrated. Soil microbial biomass (Child, 1995; Hopkins *et al.*, 2000; Carter and Tibbett, 2006) and activity (Putman, 1978; Hopkins *et al.*, 2000; Carter and Tibbett, 2006), as well as nematode abundance (Todd *et al.*, 2006; Metcalf *et al.*, 2013), increase during decomposition. Moreno *et al.* (2011) demonstrated that the microbial communities characterizing gravesoil were over 50% dissimilar within themselves over a time period of weeks when compared with those characterizing control soil. During putrefaction, the majority of the microbes within a corpse are expected to become part of the soil as the body breaks open and decomposition fluids and materials are transferred to the ground; indeed, human intestine, mouth, and skin commensal microorganisms not normally present in soil have been reported in gravesoils (Moreno *et al.*, 2011). The transfer of corpse microbes and metabolic by-products to the soil is likely affected by scavenger access to the corpse, however. For example, Carter *et al.* (2007) reported that significant amounts of cadaveric material (including microbes) might enter the soil only when the decomposition process is dominated by microbes and insects rather than scavengers. This phenomenon is because scavengers can very quickly remove flesh from a carcass, reducing the carcass to a skeleton before any cadaveric material—microbes included—has time to penetrate the soil.

10.2.1 Microbial diversity of gravesoil and the rate of decomposition

Carter *et al.* (2015) described bacterial and microbial eukaryotic communities associated with gravesoils and how gravesoil's microbial ecology differs across seasons. Using a pig model of decomposition, Carter *et al.* (2015) observed significantly different ($p<0.001$) microbial community structures in gravesoils collected from carcasses decomposing in the winter compared with those decomposing in the summer; specifically, the bacterial family Chitinophagaceae (Bacteroidetes: Sphingobacteriales) and rhabditid nematodes were significantly more abundant in summer gravesoils than in winter gravesoils (Carter *et al.*, 2015). Metcalf *et al.* (2013) also reported increased abundances of nematodes during decomposition, suggesting that nematode population spikes may be characteristic signatures of gravesoils. Carter *et al.* (2015) also observed that after carcass rupture, bacterial and microbial eukaryotic gravesoil communities were different from control soil microbial communities; however, this difference was more apparent in summer soils than in winter soils and was only significant in both bacterial and eukaryotic communities in the summer (Carter *et al.*, 2015). Significantly greater abundances of the bacterial genus *Sphingobacterium*, nematodes in the family Rhabditidae, the slime mold *F. alba*, the amoeba Euamoebida, and the fungi Eurotiomycetes and Tremellomycetes were observed in summer gravesoils compared with those in control soils.

Soil microbial communities become conditioned by previous episodes of decomposition; for example, Carter and Tibbett (2008) demonstrated that soil previously containing decomposing muscle tissue promoted the decomposition of fresh tissue much faster than did virgin soil that had never contained decomposing muscle tissue. The improved decomposition rate associated with the "conditioned soil" was attributed to the presence of zymogenous bacteria, which was able to utilize cadaveric material better compared to native soil microbes. Lauber *et al.* (2014) performed a similar experiment, placing dead mice on sterile soil or on soil with intact microbial communities, and observed that corpses placed on nonsterile soil reached the advanced stage of decomposition two to three times faster than corpses placed on sterile soil. Frund and Schoenen (2009) similarly demonstrated that the decomposition of adipocere, a grayish white, crumbly lipid bulk originating from fat (and that may persist for decades), was increased when adipocere was buried in biologically active field soil.

Burial of vertebrate remains affects the decomposition process by inhibiting vertebrate scavenger and arthropod access, leaving the decomposition process up to microbes, soil-dwelling animals such as nematodes, and soil-dwelling insects such as mites, resulting in slower decomposition (VanLaerhoven and Anderson, 1999; Fiedler and Graw, 2003; Carter *et al.*, 2007). Several factors affect microbe-mediated decomposition in soil, and these factors represent a delicate balance. We know that the influence of moisture in the decomposition of corpses in soil is usually due to the effect of moisture on microbial activity. The matric potential of the soil can be used as a standard reference point for microbial activity and most clearly illustrates the delicate balance that exists; at a matric potential (the suction with which water is held between soil particles) greater than −0.01 megapascals (MPa) (drier soil), microbial motility is limited and hydrolytic extracellular enzymes are inhibited (Carter *et al.*, 2010), slowing or inhibiting decomposition. For example, sandy soil, with its low moisture content, promotes desiccation (Fiedler and Graw, 2003), which can result in natural preservation of a corpse for thousands of years (Micozzi, 1991). In contrast, at a matric potential less than −0.1 MPa (wetter soil), gas diffusion through the soil is slowed and microbial activity declines (Carter *et al.*, 2010; Schoenen and Schoenen, 2013). Moisture also lowers the rate of oxygen exchange with CO_2 to a rate insufficient to meet the aerobic microbial demand (Carter *et al.*, 2007). In these cases, adipoceres form. Aerating the soil can improve microbe-mediated decomposition, particularly in the context of only partially decomposed remains when adipocere is present (Fründ and Schoenen, 2009). Decomposition of adipocere upon soil aeration or movement of the adipocere to the ground surface is typically achieved by *Bacillus* spp., *Cellulomonas* spp., and *Nocardia* spp. (Pfeiffer *et al.*, 1998). Temperature also affects microbial activity during decomposition; Carter and Tibbett (2006) reported an increase in microbial activity at higher temperatures, with an approximately two-fold increase with a temperature increase from 12 to 22°C. Paul and Clark (1996) had also previously reported similar increases in microbial activity with higher temperature. Acidic soils may inhibit the activity of gravesoil microbes; in acidic soils, plants produce high levels of tannins (Swift *et al.*, 1979), which inhibit microbial activity by combining with proteins and carbohydrates in the organic matter present in the soil.

10.2.2 Detecting decomposition signatures in soil and clandestine graves

Knowledge of soil microbiology during and after decomposition is critical in forensic science. Further, knowing how long a decomposition "signature" may remain in the soil after a corpse has decomposed, or how detectable a microbial signature is on surface soils above a buried corpse, may be key in locating clandestine graves. Studies have demonstrated that gravesoils contain higher microbial biomass carbon and increased rates of respiration and microbial mineralization of nitrogen, even after 430 days of decomposition (Hopkins *et al.*, 2000). And, the decomposition-related changes in bacterial nitrogen cycling genes (due to the high influx of nitrogenous compounds from the cadaver during decomposition) not only impacted critical processes in the intrinsic soil nitrogen cycle but could also discriminate between gravesoil and control soil (Moreno *et al.*, 2011). Increased levels of carbon and nitrogen in cadaver decomposition islands (CDIs), which are associated with increased soil microbial biomass, mineralization of carbon by microbes, and increase in nematode abundance, have been detected for up to 1 year after decomposition in elephants (Coe, 1978).

The plant communities under and around decomposing organisms are greatly affected, sometimes for years (Towne, 2000; Carter *et al.*, 2007), and this likely affects both the soil microbial communities (Johnson *et al.*, 2003) and the nematodes that feed on them (Carter *et al.*, 2007). Identification of these microbial community changes could be useful in identifying past decomposition events or those ongoing (e.g., clandestine graves). Clearly, specific microbe-mediated soil signatures exist, sometimes for many years, and it may one day be possible to use microbial community composition and metabolic activity to identify the location of clandestine graves.

In a laboratory mouse model experiment with three contrasting soil types, Metcalf *et al.* (2016a) demonstrated signatures of microbial decomposer communities that remained for at least 30 days after corpse removal. Using 16S rRNA, 18S rRNA, and ITS region markers, Metcalf *et al.* (2016a) showed that the distinct microbial diversity associated with decomposition is detectable in deserts, shortgrass prairies, and forest soils at 10, 20, and 30 days after the mouse cadaver was removed from the soil. This shift suggests that soil microbial ecology is not only potentially useful for locating clandestine graves but also for detecting whether a decomposing body was moved from a location. As part of the same study, 16S rRNA data generated from each of the three soil types were combined with data from gravesoils associated with human decomposition as well as the 16S rRNA dataset of soils from diverse habitats from around the world that were not associated with decomposition. A supervised learning method was implemented to show that gravesoils can be discerned from nondecomposition soils with high probabilities (Metcalf *et al.*, 2016a). This suggests that the gravesoils' microbial signature is very distinct and easily identifiable regardless of soil type, which makes it an attractive forensic tool. Future studies further characterizing the microbial ecology of gravesoils will be key to furthering this arm of forensics.

10.2.3 Plant litter

Although this chapter focuses on vertebrate decomposition, it would not be complete without a discussion of plant decomposition, as it is estimated that roughly 90% of organic resources decomposing terrestrially are in fact plant derived (i.e., leaf litters, root exudates, and stems; Swift *et al.*, 1979; Carter *et al.*, 2007), and the presence of plant litters around decomposing animals or humans can also affect the decomposition trajectory of the carcass. Mant (1950) reported that surrounding a buried corpse with plant materials such as straw and pine branches increased the rate of decomposition. The addition of plant material not only aerates the space between the corpse and the soil, increasing microbial activity, but also widens the carbon to nitrogen ratio, which also promotes microbial activity (Tibbett and Carter, 2008). Additionally, because plant litter is primarily decomposed by microorganisms, the placement of dead plant material around the corpse may introduce a decomposer community already primed for decomposition (see discussion in Section 9.2.1).

Plant litter, comprised of dead leaves, barks, needles, fruits, seeds, and twigs that have fallen to the ground, plays a key role in maintaining "healthy" soil by releasing nutrients into the ground. Microorganisms are the primary decomposers of leaf litters, although nematodes also play an important role (Bjornlund and Christensen, 2005). The nutrient content of the plant litter as well as climate can significantly affect both the rate of litter decomposition and the decomposer microbial communities (Bjornlund and Christensen, 2005). For example, Bjornlund and Christensen reported a more extensive decomposition and a 4–15-fold higher density of bacteria, active fungal mycelia, and nematodes in ash leaves than in beech leaves, which are of lower nutritional quality (Bjornlund and Christensen, 2005). More recently, Mora-gómez *et al.* (2016) reported season-specific microbial decomposer communities on plant litters: fungi played a more important role in spring, whereas in summer, bacteria were the more important litter decomposers. Nevertheless, irrespective of season, leaf litter decomposition followed a two-stage process: the initial stages of decomposition dominated by bacterial biomass and related to disaccharide and lignin degradation and the later stages of decomposition dominated by fungal biomass and related to lignin decomposition. Duarte *et al.* (2016) also observed seasonal variation in leaf litter decomposer communities and noted that seasonal changes in stream water had a more significant impact on both the activity and diversity of microbial decomposers than did a stream water–warming designed to predict a global warming scenario. Both studies seem to support the hypothesis posited by Mora-gómez *et al.* (2016): different microbial assemblages associated with leaf litter decomposition may differ in their sensitivity to local environmental changes, a potentially very relevant phenomenon given that intermittent streams are highly seasonal ecosystems.

10.3 Freshwater and marine decomposition

Decomposition of vertebrate carrion in water is generally accepted to be different from terrestrial environs. This difference is due to the unique conditions a corpse is presented with in the water, including temperature, water salinity, water depth, currents or tides, chemical content of the water, scavengers including fish, and the ability of the body to move in three dimensions. Marine bacterial communities are also inherently different from terrestrial communities, and therefore, the microbial ecology of marine decomposition is likely to be considerably different from that of terrestrial decomposition. Several early studies reported the presence of bacteria (Hobischak and Anderson, 2002), fungi (Hobischak and Anderson, 2002), and algae (Casamatta and Verb, 2000; Haefner *et al.*, 2004) on corpses decomposing in freshwater, and their potential forensic value is recognized (Haefner *et al.*, 2004; Zimmerman and Wallace, 2008); however, only a few studies have attempted to describe specific microbial communities associated with decomposition in water and whether and how these communities change throughout decomposition.

10.3.1 Freshwater decomposition: Fish

Salmon provide a wonderful model for studying freshwater decomposition because a large number of carcasses naturally occur after the annual salmon run, when salmon travel from the ocean upstream into rivers to spawn. It has long been recognized that nitrogen released from the carcasses of salmons that have died after spawning can enhance microbial and algal growth (Richey *et al.*, 1975; Mathisen *et al.*, 1988; Parmenter and Lamarra, 1991). Several groups have shown that salmon carcasses release significant amounts of nitrogen into the water (Brickell and Goering, 1970; Richey *et al.*, 1975;

Durbin *et al.*, 1979), and Parmenter and Lamarra (1991) reported that up to 95% of nitrogen originating from the carcass was leached into the water column in less than a year. It is easy to imagine how these levels of nitrogen could affect freshwater microbial communities. Indeed, Wold and Hershey (1999) found that microbial biofilms and algal growth were increased in areas where salmon carcasses were located and reported that carcass-derived nitrogen was actively taken up by the biofilms—and this nitrogen was incorporated into wood biomass by those microorganisms actively decomposing nearby wood. Additionally, chlorophyll concentrations also increase with carcass biomass (Wold and Hershey, 1999; Johnston *et al.*, 2004). Richey *et al.* (1975) reported that peak microbial activities and nitrate concentrations were associated with the height of salmon decomposition. Decomposing fish have additionally been associated with increased ammonia concentrations, which in turn can affect the ecosystem in general by increasing phytoplankton biomass and leaf litter respiration rates (Brickell and Goering, 1970).

Recently, a salmon model of decomposition was utilized to characterize potential interactions between the decomposer microbiomes and decomposer insects.

10.3.2 Freshwater decomposition: Swine

Swine have also been used as a model to study freshwater decomposition. Using stillborn pig carcasses, Benbow *et al.* (2015) described bacterial community succession associated with carcasses placed in a freshwater stream during two seasons: summer and winter. Consistent with decomposition studies in other models, Proteobacteria and Firmicutes were the most abundant bacterial phyla present on the carcasses, during both seasons. During decomposition, Proteobacteria decreased, while Firmicutes increased. During both seasons, four differing dominant bacterial communities were described on the carcasses. For example, summer carcasses were dominated by *Pseudomonas*, *Psychrobacter*, and *Ewingella* on day 0; *Klebsiella*, *Psychrobacter*, and *Citrobacter* on day 7; *Zoogloea*, *Clostridium*, *Dechloromonas*, and *Desulfosporomusa* on day 14; and *Clostridium*, *Enterococcus*, and *Lactobacillus* on day 21. A similar pattern was observed in winter carcasses, with *Psychrobacter*, *Pseudomonas*, and *Enterococcus* dominating on day 0; *Lactococcus* dominating on day 14; *Proteocatella*, *Clostridium*, and Veillonellaceae dominating on day 28; and *Veillonellaceae* and *Clostridium* dominating on day 42. Consistent with other studies assessing seasonal differences in the decompositional microbiome, bacterial communities were found to be significantly different between the summer and winter seasons.

10.3.3 Marine decomposition: Whale falls

Marine ecosystems represent an inherently different setting for decomposition due to the high salinity levels of marine waters compared to freshwater. Additionally, conditions in the deep sea are inherently different from those in freshwater in a number of ways, from temperature to oxygen levels to pressure (Goffredi and Orphan, 2010). Whales are particularly useful organisms for studies of marine decomposition; their large size provides a nutrient-rich environment for a variety of organisms—including microbes—in the relatively nutrient-poor deep sea (Goffredi and Orphan, 2010). The first whale fall ecosystem was described in 1989 (Smith *et al.*, 1989), with the importance of microbes in this ecosystem recognized shortly thereafter (Allison *et al.*, 1991; Smith, 1992).

The activity of mobile scavengers feeding on whale corpses spreads the whale biomass out on the ocean floor, which is important for eliciting microbial activity (Goffredi and Orphan, 2010; Smith *et al.*, 2015), with decreased microbial diversity and increased microbial biomass (Goffredi *et al.*, 2008; Treude *et al.*, 2009) associated with decomposing whales. The decomposition islands formed around whale falls are extremely organic and sulfide rich (Smith *et al.*, 2015), and high levels of sulfide and methane have been correlated with high levels of microbial methanogenesis and sulfate reduction (Naganuma *et al.*, 1996; Goffredi *et al.*, 2008; Treude *et al.*, 2009). These processes represent an unusual coexistence of methanogenic archaea and sulfate-reducing bacteria (Goffredi and Orphan, 2010), highlighting the unique, complex nature of whale fall ecosystems.

In 2010, Goffredi and Orphan used 16S rDNA sequencing to characterize the bacterial communities in sediments associated with two whale carcasses in Monterey Bay, California, compared with those in control sediments with no decomposition event (Goffredi and Orphan, 2010). While higher abundances of Bacteroidetes, Firmicutes, and Epsilonproteobacteria were found in whale fall-associated sediments (comprising 51% of bacterial diversity in decomposition sediments but only 14% of bacterial diversity in control sediments), higher abundances of Planctomycetes and Gammaproteobacteria were found in control sediments (32% of bacterial diversity in control sediments vs. 7% of bacterial diversity in whale fall sediments). Additionally, although Deltaproteobacteria was the most abundant bacterial class detected in both decomposition and control sediments, the abundances of bacterial families within this class differed, with Desulfobacteraceae and Desulfobulbaceae detected primarily in sediments underneath the whale falls.

Due to their high lipid content and tough matrix, whale bones also provide a long-lasting, rich energy source for microbes (Smith *et al.*, 2015). Some bacteria found in whale bones, such as collagen and cholesterol-hydrolyzing bacteria of the order Oceanospirallales, act as symbionts for other whale fall resident organisms, such as the bone worm (*Osedax*) (Smith *et al.*, 2015), further highlighting the complex and diverse ecosystem characterizing whale falls.

10.3.4 Marine decomposition: Swine

Although studies of salmons and whale falls have provided important information regarding how microbial communities may be involved in decomposition in the water, the applicability of this work to human decomposition is limited due to the inherent differences between organisms living in water and humans, which are terrestrial. Some groups have used swine models of decomposition as a proxy for human decomposition in water because in general, swine are understood to be similar to humans in a number of ways, including size, hair cover, diet, and gastrointestinal (GI) bacteria (Schoenly *et al.*, 2006). In 2011, Dickson and colleagues reported on the colonizing bacteria detected on decomposing pig heads submerged in an open-water system (Otago Harbor, New Zealand) in autumn and in winter (Dickson *et al.*, 2011). They reported a rapid bacterial colonization of the pig heads and observed a succession-like pattern in bacterial colonization. Several genera, including *Fusobacterium* and *Nevskia*, were detected on the first sampling event (3 days after submersion) but were never detected again. Others (*Shewanella*, *Aeromonas*, *Carnobacterium*, and *Marimonas* for autumn carcasses and *Polaribacter* and *Tenacibaculum* for winter carcasses) were detected only after 107 accumulated degree days (ADDs) (7 days after submission) for autumn carcasses or 60 ADDs (10 days after submission) for winter carcasses. Notably, a visible microbial film covering the pig head was observed in the advanced putrefaction stage of decomposition. Variation in microbial communities between body sites—cheek, snout, or neck wound—was also observed. Bacteria of the order Flavobacteriales were more abundant on the cheek and snout than in the neck wound and were most abundant on the cheek. Bacteria genera specific to a body site were also observed; *Flavobacterium*, *Photobacterium*, and *Lutibacter* were detected only on the cheek, *Fusobacterium* and a novel genus of family Pseudomonadaceae were detected only on the snout, and *Vagococcus* was detected only on the neck wound. Not surprisingly, seasonal differences in the rate of decomposition and the colonizing bacteria present were also observed. Among the genera detected only on swine decomposing in autumn were *Carnobacterium*, *Marimonas*, *Aeromonas*, and two unclassified genera of the order Bacteroidales, while *Polaribacter* and an unclassified genus of the order Bacteroidales were found only on winter swine. This study has been incredibly valuable, as it very clearly demonstrated a successive and season-specific pattern of bacterial colonization of a terrestrial animal decomposing in water, which, if applied to humans, can have significant forensic applications.

10.4 The microbiology of nonhuman models of terrestrial decomposition

Nonhuman models are essential for establishing statistically significant changes in microbial community, taxonomy, and diversity and isolating variables in controlled lab settings to test their effects on these trends (see Chapter 3). Swine have been used extensively to study terrestrial decomposition, and rat and mouse models also provide an excellent opportunity to test the effects of variables such as temperature (Carter *et al.*, 2008), the presence of soil (Lauber *et al.*, 2014), soil moisture (Carter *et al.*, 2010), and soil type (Metcalf *et al.*, 2016a).

10.4.1 Terrestrial decomposition: Rats

By 1986, several studies on the various processes of postmortem change had been performed using animal models that had been previously frozen and then thawed, although none had actually determined the specific effects of freezing and thawing on the decomposition processes, including the microbial component, to determine whether these models could act as a proxy for freshwater decomposition. In 1986, Dr. Marc Micozzi reported an elegant study addressing this important knowledge gap in the field. Micozzi rightly pointed out that these studies may not have provided an accurate picture of decomposition, with serious consequences in the context of estimating the PMI, and added that studies on frozen–thawed animals could be valuable in the context of human deaths occurring during the frost season (Micozzi, 1986).

Using Wistar rats that were laid in the field either fresh (within 2 hours of death) or after a freeze–thaw cycle, Micozzi (1986) reported differences in gross morphology, insect invasion, and microbial succession in the two groups of animals. After just 2 days in the field, bacterial overgrowth was observed in the GI tract of fresh animals but in the upper airways of frozen animals. Although over a period of 6 days, a microbiologic succession sequence from enteric to soil organisms was observed in both fresh and frozen animals, the growth rate of enteric organisms was higher in the fresh animals. Consistent with this observation, frozen animals seemed more susceptible to invasion by microorganisms from the outside. While putrefaction (anaerobic decomposition mediated by the microorganisms intrinsic to the corpse) characterized the decay process of the fresh animals, aerobic decomposition characterized the decay process of the previously frozen animals, proceeding faster in these animals than the previously frozen ones. Micozzi (1986) therefore concluded that the freeze–thaw cycle not only diminished the ability of enteric organisms to grow and participate in decomposition but also disrupted the tissues of the corpses, facilitating aerobic decomposition and the entrance of environmental organisms. Therefore, this study, although it is already nearly 30 years old, serves as an important reminder of how various factors can influence the decomposition cycle, including the microbial ecology of that cycle, and investigators would be dutiful to keep this knowledge in the back of their minds when designing and performing decomposition studies.

10.4.2 Terrestrial decomposition: Mice

Although Micozzi's work was certainly important and yielded untimely information about the microbial ecology of decomposition, it did not focus solely on the microbes—which ones are involved or where do they come from? Though Micozzi did report a succession pattern in the microorganisms of decomposing rats, the detail of the study ends there. Others have been more direct in their efforts to describe and define the microbial ecology of decomposition, and much work has been done using mice as models. Though some believe that swine make a better model for human decomposition (discussed more in Section 10.4.3), mice are inherently easier to work with—they are smaller and cost less and it is much easier to work with them in larger numbers as they take up less space and produce offspring quickly.

One of the earliest microbial decomposition studies using mice reported bacterial transmigration (see also Chapter 8) through mouse intestines during decomposition (Melvin *et al.*, 1984) and described a succession-like pattern, with *Staphylococcal* species transmigrating out of the intestine first, followed by coliform-type organisms, *Candida* species, and, finally, a variety of anaerobic species. Transmigration also occurred quicker at higher temperatures—within 3 hours at 37°C but not until 72 hours at 4°C. This study, although extremely informative, was culture based, yielded poor taxonomic resolution, and was performed on intestinal sections rather than in intact organisms.

With the recent development of next-generation sequencing technologies and tools for analyzing sequencing data, more recent work has been able to describe specific microbial communities associated with decomposition using mice as a model. For example, Metcalf *et al.* (2013) surveyed the bacterial and microbial eukaryotic communities associated with murine decomposition on soil, taking advantage of a controlled laboratory setting and a large number of replicates. They found that the decomposition of a microbiome was different across body sites, with the soil microbial communities also different from those of the corpse, and that the communities changed in a similarly successive manner in each replicate as decomposition progressed. During the bloat stage in the abdominal cavity, common gut bacteria such as Lactobacillaceae (e.g., *Lactobacillus*) and Bacteroidaceae (e.g., *Bacteroides*) increased in relative abundance. Once the abdominal cavity is exposed to oxygen, however, these taxa decreased in relative abundance and we see an increase in oxygen-tolerant bacteria such as taxa in the order Rhizobiales, as well as recognized opportunistic pathogens in the family Enterobacteriaceae (e.g., *Serratia, Escherichia, Klebsiella, Proteus*). In both skin and associated soils, bacterial taxa in the families Xanthomonadaceae and Sphingobacteriaceae increased in abundance during decomposition. In soils we also detect a decrease in the relative abundance of bacterial taxa in the phylum Acidobacteria, which is likely a response to increases in soil pH after rupture of the corpse and release of ammonia-rich fluids. Interestingly, they noted an increase in the bacterivorous nematode *O. tipulae*, an observation that had never before been reported. Several recent studies have since confirmed that a nematode bloom is commonly associated with decomposition of different host organisms in outdoor scenarios (Carter *et al.*, 2015; Weiss *et al.*, 2015; Metcalf *et al.*, 2016a).

10.4.3 Terrestrial decomposition: Swine

As mentioned in Section 9.2, swine are often viewed as good proxies for studying human decomposition. They are of similar size to humans, have sparse hair covering, and harbor similar bacteria in their GI tracts (Schoenly *et al.*, 2006). A number of researchers have utilized swine models of decomposition to study several forensic questions, including the effects of season (Pechal *et al.*, 2013; Carter *et al.*, 2015) and carcass mass (Weiss *et al.*, 2015) on the structure or function of microbial communities associated with decomposition.

Using three terrestrially decomposing swine and pyrosequencing of the bacterial 16S rRNA gene, Pechal *et al.* (2014) described significant changes in the relative abundances of bacterial phyla and families as decomposition progressed and also reported a decrease in microbial community diversity in the later stages of decomposition.

Proteobacteria was the dominant phylum at the start of decomposition but was replaced by Firmicutes as the dominant phylum in the later stages of decomposition. Early decomposition time points were characterized by the presence of Moraxellaceae, Aerococcaceae, and Enterobacteriaceae, with Planococcaceae and Clostridiaceae as the bacterial families dominating at the late decomposition stage.

In another similar study, Pechal *et al.* (2013) assessed changes in bacterial metabolic community profiles throughout decomposition, at different seasons, and with or without insect access to the carcasses. Although Hopkins *et al.* (2000) determined total C, total N, and microbial biomass C as a proxy for microbial activity throughout decomposition, Pechal *et al.* (2013) are the first (and to date, only) to *directly* assess the functional activity of microbial communities associated with decomposition by using a series of plates to measure microbial community metabolic activity. Pechal *et al.* (2013) reported that bacterial community carbon usage varied across seasons, being highest in spring, as well as throughout single decomposition events, being higher at the later stages of decomposition (with the exception of fall decomposition events, which were characterized by decreased bacterial carbon usage as decomposition progressed). Significant differences in overall bacterial community functional profiles were also observed across seasons and across decomposition phases in spring and summer but not in fall or winter. Although the mean microbial functional activity decreased on carcasses not exposed to insects and increased on carcasses with insect exposure, this difference was not significant.

In the same study, bacterial community composition was also assessed. An early decomposition microbial community dominated by Proteobacteria, followed by Firmicutes, was observed in carcasses with and without insect exposure; however, in carcasses with insect exposure (and as observed in their other studies, Pechal *et al.*, 2014), the Proteobacteria decreased and the Firmicutes increased as decomposition progressed, while in carcasses without insect exposure, the Firmicutes decreased and the Proteobacteria remained a dominating member of the bacterial community as decomposition progressed. The difference in both bacterial community structure (including succession pattern) and function observed between carcasses with or without insect exposure is intriguing and suggests that important insect–microbe interactions occur during decomposition (see Chapter 11). These results also show that decomposition scenarios excluding insects will likely differ microbially than those that do not exclude insects, a fact that should inform future studies into the microbial ecology of decomposition.

10.5 The microbiology of terrestrial human decomposition

Controlled studies of human decomposition are notoriously difficult to conduct due to inherent challenges associated with cadaver research including legal and social aspects of "gift of body" donations, locating tracts of appropriate land, and navigating public perception of such research facilities (i.e., commonly referred to as "body farms"; see literature by

Dr. William M. Bass discussing these challenges further). Very few facilities equipped for human decomposition studies exist; these facilities are extensively described in Chapter 3 and therefore will only be briefly addressed in this section. The first facility, the University of Tennessee Forensic Anthropological Research Facility built by Dr. William M. Bass, forged the way for human decomposition studies. Since the creation of Dr. Bass's facility, we have seen the establishment of four additional US facilities: the Forensic Osteology Research Station (Western Carolina University), the Forensic Anthropology Center at Texas State (San Marcos University), the Southeast Texas Applied Forensic Science (STAFS) Facility (Sam Houston State University), and, most recently, the Forensic Investigation Research Station (Colorado Mesa University), each exploring unique geographic influences on decomposition while trying to generate universal models useful to law enforcement agencies.

10.5.1 Initial insights into the microbial ecology of human decomposition

The first bacterial marker gene study on the microbial communities associated with human decomposition was published in 2013. Using 454 pyrosequencing of variable regions 3–5 of the bacterial 16S rRNA marker gene, Hyde *et al.* (2013) described the pre- and end-bloat microbial communities associated with two human cadavers placed at the STAFS Facility in Huntsville, TX, and allowed to decompose under natural, terrestrial conditions. As Metcalf *et al.* (2013) observed in mice, the bacterial communities of different body sites (mouth, stomach, small intestine, colon, and rectum) differed, both in terms of diversity and composition. For example, rectal/stool samples contained the most diverse bacterial community (400 operational taxonomic units (OTUs)), while the stomach contained the least diverse bacterial community (26 OTUs). The microbial community of the stomach was dominated by *Morganella* (Proteobacteria), which is a bacterial genus that has shown acid tolerance (Young *et al.*, 1996). The oral microbiome of one cadaver was similar to that of healthy Human Microbiome Project participants, dominated by *Streptococcus*, *Prevotella*, and *Veillonella*; the other cadaver has a less typical oral microbiome, dominated by *Staphylococcus*, *Peptoniphilus*, and *Clostridium*. In both cadavers, the GI tract microbiome contained typical GI tract organisms such as *Bacteroides*, *Clostridium*, *Lactobacillus*, *Escherichia*, and *Faecalibacterium*.

Some of the bacterial taxa that (Hyde *et al.*, 2013) detected on these two cadavers have previously been described in decomposition literature (i.e., *Staphylococcus*, previously described by Melvin *et al.*, 1984; *Clostridium*, previously described by Janaway *et al.* (2009)), though other bacterial taxa, such as *Bacteroides*, *Eggerthella*, *Erwinia*, *Eubacterium*, and some unclassified bacterial families were also detected. Consistent with the shift from aerobic bacteria (*Staphylococcus* and Enterobacteriaceae) to anaerobic bacteria (*Clostridium* and *Bacteroides*) reported by Evans (1963), Janaway *et al.* (2009), Vass (2001), Melvin *et al.* (1984), and Carter *et al.* (2008), who used animal models and culture-based techniques, Hyde *et al.* also reported a shift from aerobic to anaerobic bacteria.

Pre- and end-bloat samples were collected from the oral cavity of each cadaver (other pre-bloat samples, i.e., stomach, small intestine, and rectum, could not be collected as they would have been destructive), and shifts in the microbial community composition between these two decompositional stages were also observed. In one cadaver, the pre-bloat oral community was composed mainly of the phylum Proteobacteria (*Pseudomonas*), while in the same cadaver post-bloat, the oral community was mainly composed of Firmicutes (*Peptoniphilus* and *Clostridium*). In the second cadaver, the pre-bloat oral community was dominated by Actinobacteria (*Microbacterium*) and Firmicutes (*Streptococcus* and *Veillonella*) while the post-bloat oral community was dominated by Firmicutes (*Clostridiaceae* and *Planococcaceae*) and Proteobacteria (*Ignatzschineria*). The presence of *Clostridium* was also observed in end-bloat samples from the GI tract, and this observation is not surprising given that *Clostridium* hydrolytic enzymes convert carbohydrates to organic acids and alcohols and that *Clostridium* lipases may significantly contribute to the hydrolysis of body fat (Janaway *et al.*, 2009).

10.5.2 Identification of microbial signatures associated with decomposition

The Hyde *et al.* (2013) study was important in that it was the first to use culture-independent techniques to characterize microbial communities associated with two different stages of human decomposition. However, due to inherent weaknesses in the study, only two time points were tested and limited body sites were sampled due to the destructive nature of some samples; a new study was designed in which several skin sites, as well as the mouth and the rectum, were sampled from two cadavers every 2 days for 2 weeks followed by every 4 days for another 2 weeks (Hyde *et al.*, 2015). This dataset provided more consistent, in-depth, and precise data than did the previous dataset, and a clear succession pattern was observed in both cadavers. Both cadavers were characterized by a similar successive trend at the phylum level, with oral and skin samples dominated initially by Proteobacteria with Firmicutes and, to a lesser extent, the Actinobacteria, dominating at the later stages of decomposition. Fecal/rectal samples also followed a similar trend, being dominated by Firmicutes and Actinobacteria at the later stages of decomposition, although the starting community had higher relative abundance of Bacteroidetes than did the skin or the mouth. Most apparent was the stark post-purge increase in the genus *Ignatzschineria*, which is associated with myiasis by fly larvae in the Sarcophagidae family (Gupta *et al.*, 2011), a significant observation given that insect activity drastically increases after purge.

As with the prior dataset, body site differences in microbial communities were observed, though these differences became less apparent as decomposition progressed, with microbial communities becoming more similar particularly in the last 2 weeks of the sampling period. There were also differences between the cadavers: while one remained dominated by *Ignatzschineria* until the very end of the sampling period, at which point *Acinetobacter* and *Corynebacterium* became the dominant genera, the other cadaver lost the *Ignatzschineria* population after only 10 days, with *Clostridium* as the dominant genus from that point on.

10.5.3 Microbial eukaryotic decomposers

Although less data exist for microscopic eukaryotes than for bacteria, nematodes and fungi are clearly important in the decomposition process. Metcalf *et al.* (2013) have characterized the full diversity of microbes across the tree of life, including bacteria, archaea, and microbial eukaryotes, by sequencing multiple taxonomic marker genes to describe each microbial branch of the tree of life. To characterize the microbial eukaryotic community, the 18S rRNA maker was used, and to focus on the fungal communities, the ITS gene region was sequenced. Using the 18S rRNA marker gene, nematodes of class Chromadorea, for example, Rhabditidae and Aphelenchidae, fungi in the classes Saccharomycetes (phylum Ascomycota), Microbotryomycetes (phylum Basidiomycota), and Mycotypha (subdivision Mucoromycotina), as well as the slime mold *F. alba* were detected at various stages during decomposition. The ITS region marker data provided a higher resolution of the fungal taxonomy. Fungal species that become abundant during decomposition included *Yarrowia* (Saccharomycetes), *Stromatonectria* (Ascomycota), *Lysurus* (Basidiomycota), and particularly in gravesoils *Rodentomyces* (Ascomycota) and *Mortierella* (Zygomycota). Common bacterial and eukaryotic microbial decomposers associated with human decomposition are illustrated in Figure 10.1.

10.5.4 Linking cadaver and soil microbial communities

In both life and death, vertebrate skin communities are intimately tied to their surrounding environment. We have already discussed gravesoil microbial communities (Section 10.2) and carcass/corpse microbial communities (Sections 10.4 and 10.5) separately; however, these communities likely interact and affect each other during the process of decomposition. It may be possible, for example, that a certain proportion of the decomposer community comes from the soil surrounding the carcass, particularly since gravesoil promotes decomposition much quicker than does soil upon which decomposing material has never been placed (Carter and Tibbett, 2008; see Section 10.2). To obtain a complete picture of terrestrial decomposition, cataloguing the microbial communities present on the carcass and in the surrounding soil is essential.

During decomposition similar microbial taxa become abundant in both cadaver skin and associated soils. For example, Metcalf *et al.* (2016a) show that bacterial taxa in the family *Rhizobiaceae* significantly increased in relative abundance in both cadaver skin and gravesoil communities, which is similar to trends uncovered in controlled mouse model experiments. Furthermore, nematodes of the class Chromadorea (e.g., Rhabditidae and Aphelenchidae) as well as the cellular slime mold *F. alba* is also commonly associated with both skin and soils associated with decomposition. Additionally, Lauber *et al.* (2014) demonstrated that the presence of a soil microbial community affects the skin microbial

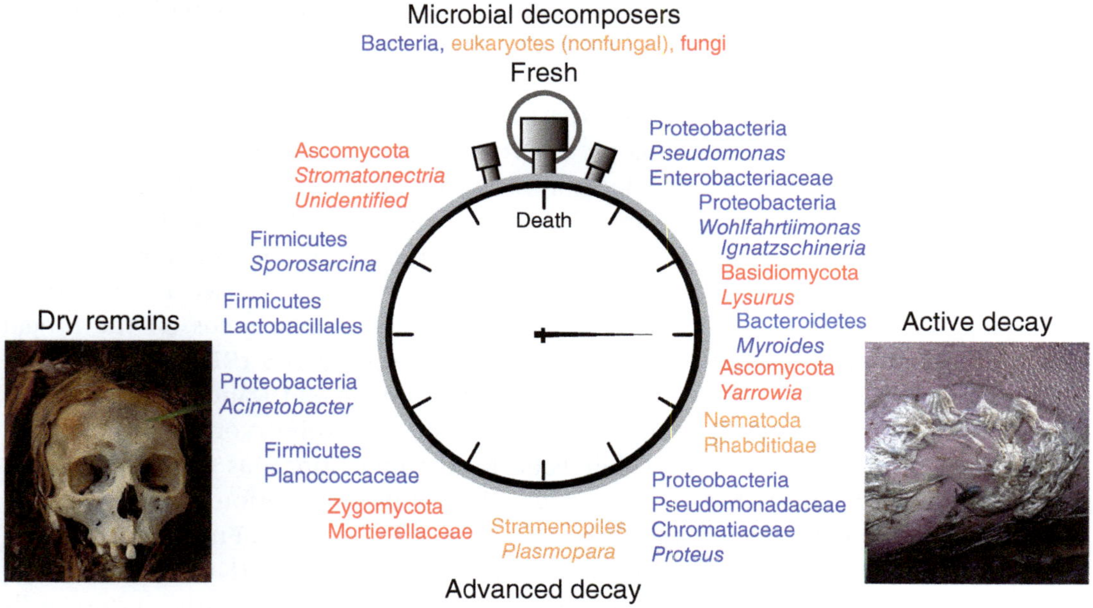

Figure 10.1 Succession of microbial decomposers during human decomposition. The microbial (bacterial, eukaryotic, and fungal) communities associated with two human cadavers at Sam Houston State University's Southeast Texas Applied Forensic Science Facility were determined using deep marker gene sequencing and analysis of samples collected from the cadavers throughout decomposition. A small subset of taxa associated with specific decomposition stages are highlighted here. Shown here is a photo of a fly larval mass under the skin during active decay and a skull during dry remains. Source: Photo credit: laboratory of Dr. Sibyl Bucheli. Reprinted with permission from Metcalf *et al.* (2016b)

community during decomposition by placing mice on untreated and sterilized soils and comparing the microbial communities for eight time points during decomposition. A striking difference in the microbial eukaryotic communities was evident on the skin. For example, mice decomposing on sterilized soils had skin colonized primarily by fungi, whereas those decomposing on untreated soils had skin colonized by a more diverse set of microbial eukaryotes including Metamonada, Metazoa (primarily nematodes), Apicomplexa, and Stramenopiles. Notably, although Lauber *et al.* (2014) were unable to demonstrate the soil as a source for decomposer microbial communities, Metcalf *et al.* (2016a) used dynamic Bayesian inference neural information flow networks to show that soil was the primary source of decomposer microbes.

10.5.5 Linking cadaver and insect microbial communities

Much as there appears to be a connection between soils and cadaver microbial communities, there is likely an interaction between insects and cadaver microbial communities (Chapter 11). Work on this topic is scarce but likely to increase over the coming months and years. Using salmon carcasses in freshwater streams, Pechal and Benbow (2016) demonstrated that the internal microbial communities were significantly different in three mayfly species collected from streams containing salmon carcasses compared with mayflies collected from carcass-free streams. In two mayfly species, the bacterial phylum Proteobacteria increased in flies collected from carcass-containing streams, while the Tenericutes increased in all three fly species; however, these organisms, while they may have been present due to fish corpse decomposition, they may also be sourced from fish excrements, sheddings of the mucosal lining, or environmental biofilms (Pechal and Benbow, 2016).

After placing salmon carcasses on the stream bank, the authors collected blow fly larvae and adults associated with salmon carcasses, as well as samples from the individual carcasses. Although some bacterial taxa observed on all carcasses belong to genera containing fish pathogens, for example, *Bacteroides* in salmonid gill disease (Bowman and Nowak, 2004) and *Myroides* as a general salmon pathogen (Loch and Faisal, 2015), only 22.4–26.2% of OTUs were shared between salmon carcasses and the blow fly larvae feeding on those carcasses (Pechal and Benbow, 2016). This indicates that while some of the internal blow fly larvae's microbial community may indeed come from the carcass, a majority of the community appears to be independently sourced from what has typically been considered the larvae's main food source. Additionally, there was little evidence that carcasses had any effect on the adult blow fly microbiome.

Using human corpses and adult blow flies collected from those corpses, Metcalf *et al.* (2016a) also assessed the potential connection between decomposing corpses and insects. Though the results suggested that blow flies may represent a source for decomposer microbes on corpses, particularly in spring (as compared with winter), soil was a larger source of microbial decomposers (see Section 10.5.4).

10.6 Is there a universal decomposition signature?

The question of whether or not there is a universal signature in the microbial communities associated with decomposition—among humans or between humans and various animal models of decomposition—is an important one, for a number of reasons. The presence of universal decomposition signature in humans, pigs, and mice (current models that have been typically investigated) would lend further support to the hypothesis that these animal models serve as good proxies for human decomposition research and can provide forensic science with accurate, useful information that can be extrapolated to human decomposition. Additionally, the presence of a universal signature may indicate that microbial community patterns could be used for accurately estimating the PMI. We will discuss this in more detail in Section 10.7.

By reviewing available microbial marker gene studies using human, pig, and mouse models, we identified a number of microbial taxa that are commonly found in gravesoils and on skin and that were observed in all three models (see Table 10.1). The family Xanthomonadaceae increased in both mice (including gravesoil) and human cadavers as decomposition progressed, and although this increase was not observed on pig corpses, Xanthomonadaceae were detected on pig corpses nonetheless. Interestingly, the Xanthomonadaceae genus *Ignatzschineria* was found on human corpses, not on the mice, which decomposed indoors. As described in Section 10.5, *Ignatzschineria* are associated with fly larvae, and this likely explains the absence of this genus in the mouse model. However, the fact that Xanthomonadaceae are present on all three decomposition models indicates that this bacterial family likely has an important decomposition role.

Other common patterns were also observed; for example, the Clostridiaceae family was found at higher abundances at later decomposition stages in all three models and was additionally detected in gravesoils associated with decomposing mice. Additionally, Moraxellaceae increased during human decomposition and for the first 24 hours of swine decomposition, while the Pseudomonadaceae (in the same class as that of Moraxellaceae) increased on decomposing mice and soils associated with those mice.

These observations suggest that some bacterial taxa, at least at the family level, may be universally involved in terrestrial decomposition and may even be similarly involved at different stages of decomposition in all three models. Further studies will better enable us to define whether such a universal decomposer community exists and the precise metabolic role of each member of this community in decomposition. As the availability of human corpses often limits the number and thoroughness of forensic microbiological studies, using animal models will continue to be a key aspect of forensic microbiology research. Therefore, the existence of a universal decomposer community could prove crucial, as such a community would solidify swine and mouse models of decomposition as important proxies for humans in the context of the microbial ecology of decomposition.

10.7 Using microbial signatures to estimate PMI

An ever-present question in medicolegal death investigation is how to accurately estimate the PMI or the time that has passed since an individual or organism died. Entomology has proven to be an effective means to estimate the PMI during the wet stages of decomposition (Nuorteva, 1977; Smith, 1986; Castner, 2001), though estimating the PMI at later time points once the corpse has reached the dry decay stages and during which insect activity is negligible or absent is much more difficult. Additionally, life stages of the same insect can vary with geography, leading to as much as a 10% error in entomology-based PMI estimates (Gallagher *et al.*, 2010). Using the microbial communities (e.g., bacterial, fungal, and eukaryotic) present on a corpse and in the soil during decomposition may represent a good approach for improving PMI estimates. We know that insects can establish phoretic relationships with a number of nematodes (Poinar, 1983; Richter, 1993), which, as described earlier, have been observed

Table 10.1 Presence of several bacterial taxa in three decomposition models: mouse, swine, and human

	Mouse			Swine			Human		
	Present in soil?	Present early decomp?	Present late decomp?	Present in soil?	Present early decomp?	Present late decomp?	Present in soil?	Present early decomp?	Present late decomp?
Phylum									
Proteobacteria	Y	Y	Y	Y	Y, higher	Y, lower	na	Y, higher	Y, lower
Firmicutes	Y	Y	Y	Y	Y, lower	Y, higher	na	Y, lower	Y, higher
Family									
Xanthomonadaceae	Y	Y, lower	Y, higher	Y	Y	Y	na	Y, lower	Y, higher
Clostridiaceae	Y	Y, lower	Y, higher	na	Y, lower	Y, higher	na	Y, lower	Y, higher
Moraxellaceae	N	N	N	Y	Y, higher	Y, lower	na	Y, lower	Y, higher
Pseudomonadaceae	Y	Y, lower	Y, higher	na	na	na	na	na	na
Sphingomonadaceae	Y	Y, lower	Y, higher	na	na	na	na	na	na
Genus									
Corynebacterium	na	na	na	na	na	na	na	Y, lower	Y, higher
Clostridium	na	na	na	na	Y, lower	Y, higher	na	Y, lower	Y, higher
Ignatzschineria	N	N	N	Y	na	na	na	Y, higher	Y, lower
Pseudomonas	na	na	na	na	Y, higher	Y, lower	na	Y, higher	Y, lower

Presence of the microbial taxa in the soil is also indicated for each model.

in a number of vertebrate decomposition settings. Additionally, bacterial volatiles serve to attract insects (Dekeirsschieter *et al.*, 2009; Zheng *et al.*, 2013); therefore, it seems likely that an intimate relationship between microbes and insects exists during decomposition, and if so, utilizing both insect and microbial community patterns may aid PMI estimates (see Chapter 11). Additionally, insect communities may be greatly reduced in abundance or biodiversity, as with burial, or may even be excluded, as with some indoor scenarios; in these cases, forensic scientists may only have microbial communities innate to the corpse on which to base PMI estimates. It is therefore surprising that only recently work has been done to determine the validity of using microbial communities in estimating the PMI.

10.7.1 Estimating PMI in terrestrial systems using gene marker data in nonhuman models of decomposition

By regressing the known PMI on the relative abundances of microbial taxa (based on gene marker datasets) of samples, a model can be developed to predict the PMI for an unknown sample obtained from a specific decomposition event—and may eventually be applied to different decomposition events. A commonly used model is the Random Forests classifier, which is a machine learning technique that creates random decision trees based on subsets of features (such as abundance of taxa) and identifies subsets of these features that classify samples with the highest accuracy.

Pechal *et al.* (2014) used two approaches to identify bacterial community patterns that could estimate the time since placement of swine remains, which served as a proxy for the PMI as remains were placed in the field within 2–3 hours of death. First, they used a multi-response permutation procedure combined with indicator species analysis to identify bacterial taxa that appeared to drive the separation between samples collected at different times during decomposition. They additionally used a Random Forest classifier to identify predictor bacterial taxa, defined as those that contributed most to the variation in physiological time and then used the highest ranking predictor taxa in a generalized additive model or in a generalized linear model to evaluate each taxon as significant variables in decomposition. Using these two approaches, they were able to identify several taxa and construct several models, one of which that could estimate physiological time (i.e., ADDs), explaining an impressive 94.4% of the time since placement of the swine. They found that although a "full" model containing all the five phyla identified by the Random Forest classifier explained 84.1% of the variation in physiological time, the best results were observed using a model built on the four phyla: Bacteroidetes, Proteobacteria, Actinobacteria, and Firmicutes. While bacterial signatures at the phylum level were able to explain a majority of physiological time, indicating that even broad phylogenetic patterns in bacterial communities could be useful PMI predictors, indicator species analysis-based models built on communities resolved to the family level were more successful. The bacterial families that were part of highly successful models included Campylobacteriaceae, Enterococcaceae,

Moraxellaceae, Prevotellaceae, and Pasteurellaceae, indicating that these taxa may be useful for estimating the PMI in terrestrial decomposition scenarios.

Metcalf *et al.* (2013) used a similar approach to that used by Pechal *et al.* (2014), building regression models informed by the Random Forest classifier on bacterial and microbial eukaryotic communities identified on decomposing mouse corpses. They compared models built on 5–10 highly predictive taxa to models built on all taxa in the dataset and found a similar prediction accuracy between the two methods. They were able to estimate the PMI within 3.30 ± 2.52 days and had the highest accuracy before 34 days of decomposition. Skin and soil microbial communities provided the most accurate PMI prediction models, compared to the abdominal cavity. Interestingly, although neither the bacterial nor the microbial eukaryotic community alone provided better PMI estimates, combining these two communities did insignificantly increase the accuracy of PMI estimates. These data were the first to combine information from multiple different microbial communities in estimating the PMI and very strongly suggest that microbial communities, especially when used together, may represent an accurate predictor for the PMI.

In a follow-up study, Metcalf *et al.* (2016a) performed a similar mouse model experiment but with mice decomposing on three contrasting soil types. The soil types included mountain pine forest soils (pH ~ 5), shortgrass prairie soils (pH ~ 7.5), and Moab desert soils (pH ~ 8). Including soil type in the regression model for predicting the PMI did not improve the previously reported error, suggesting that soil type does not affect the accuracy of the "microbial clock." Furthermore, we discovered that the smallest errors resulted from microbial samples of the caecum, soil, and skin over the first 2 weeks of decomposition. We discovered that all sample types provided accurate estimates of PMI (2–3 days over the first 25 days of decomposition) with the mouse cecum and gravesoils providing the most accurate estimates (Metcalf *et al.*, 2016a). This highlights gravesoils as a particularly attractive source of microbial forensic evidence in outdoor death scenarios since it is easy to access. Similarly, skin microbes may also provide an accurate source of microbial evidence. However, only similar studies using human models would indicate the extent to which the results obtained using mouse models could be applied to human decomposition. As described in the following text, Metcalf *et al.* (2016a) have indeed begun to explore the generalized applicability of soil- and corpse-associated microbial communities in estimating the PMI in nonhuman and human models of decomposition.

10.7.2 Estimating PMI in terrestrial systems using gene marker data in human models

Metcalf *et al.* (2016a) estimated the error of time since placement as a proxy for the PMI in a human model experiment in an outdoor scenario and discovered that error was higher than in controlled animal model experiments but still accurate compared to current forensic approaches relying on entomological factors. Importantly, they created their regression model based on data collected from corpses placed in one season and applied their model to corpses placed in a different season. As in animal model experiments, a repeatable succession of microbes on skin and in soil during decomposition was detected. One extremely exciting result in this study was that accurate time since placement estimates were achieved across seasons. Metcalf *et al.* (2016a) used 16S rRNA data from a winter season human decomposition experiment of two bodies and predicted the time since placement for two bodies decomposing in the same location in springtime using a Random Forests regression model that used ADD instead of chronological time. These data suggest that using microbial communities to estimate PMI may be useful for circumventing some of the seasonal limitations that exist with forensic entomology tools. Nevertheless, this data must be seen as preliminary, as sample sizes, geographical locations, and seasons were limited. Further studies using more corpses placed in different geographical locations and during more than two seasons (especially locations with colder winters than those characteristic of Southeast Texas) will solidify the usage of microbial community signatures in predicting the PMI of human corpses.

10.8 Conclusions

The microbial ecology of decomposition is a fast-changing, but predictable, ecosystem. Because of the predictable nature of microbial succession during decomposition, the potential for new microbial forensic technologies is excellent. Microbes may prove indispensable for improving PMI estimates, not only in scenarios that rely heavily on entomological clues but also especially in those scenarios where insect activity is greatly reduced or absent. Additionally, microbes could provide additional "fingerprint"-like patterns that could indicate whether a corpse had been buried at one point or moved geographically. Microbial signatures that remain in the soil for months or years after corpse remaining under the soil have completely decomposed may also prove useful for identifying clandestine graves. In all of these scenarios, microbes can act as an incredible source of forensic information for forensic scientists and law enforcement agents alike and may soon become an important fixture in the forensic science field.

Acknowledgments

The authors of this chapter are funded by the Office of Justice Programs National Institute of Justice Awards NIJ-2011-DN-BX-K533 (JLM, DOC, RK), NIJ-2012-DN-BX-K023 (SRB and AML), and 2014-R2-CX-K011 (EH, JLM, DOC, RK).

References

Allison, P.A., Smith, C.R., Kukert, H. *et al.* (1991) Deep-water taphonomy of vertebrate carcasses: a whale skeleton in the bathyal Santa Catalina Basin. *Paleobiology*, **17** (1), 78–89.

Benbow, M.E., Pechal, J.L., Lang, J.M. *et al.* (2015) The potential of high-throughput metagenomic sequencing of aquatic bacterial communities to estimate the postmortem submersion interval. *Journal of Forensic Sciences*, **60** (6), 1500–1510.

Bjornlund, L. and Christensen, S. (2005) How does litter quality and site heterogeneity interact on decomposer food webs of a semi-natural forest? *Soil Biology and Biochemistry*, **37**, 203–213.

Bowman, J. and Nowak, B. (2004) Salmonid gill bacteria and their relationship to amoebic gill disease. *Journal of Fish Diseases*, **27**, 483–492.

Brickell, D.C. and Goering, J.J. (1970) Chemical effects of salmon decomposition on aquatic ecosystems, in *Proceedings Symposium on Water Pollution Control in Cold Climates* (ed R.S. Murphy), U.S. Government Printing Office, Washington, D.C..

Carter, D.O. and Tibbett, M. (2006) The decomposition of skeletal muscle tissue (*Ovis aries*) in a sandy loam soil incubated at different temperatures. *Soil Biology and Biochemistry*, **38**, 1139–1145.

Carter, D.O. and Tibbett, M. (2008) Does repeated burial of skeletal muscle tissue (Ovis aries) in soil affect subsequent decomposition? *Applied Soil Ecology*, **40** (3), 529–535.

Carter, D.O., Yellowlees, D. and Tibbett, M. (2007) Cadaver decomposition in terrestrial ecosystems. *Naturwissenschaften*, **94** (1), 12–24.

Carter, D., Yellowless, D. and Tibbett, M. (2008) Temperature affects microbial decomposition of cadavers (*Rattus rattus*) in contrasting soils. *Applied Soil Ecology*, **40** (1), 129–137.

Carter, D.O., Yellowlees, D. and Tibbett, M. (2010) Moisture can be the dominant environmental parameter governing cadaver decomposition in soil. *Forensic Science International*, **200** (1–3), 60–66.

Carter, D.O., Metcalf, J.L., Bibat, A. and Knight, R. (2015) Seasonal variation of postmortem microbial communities. *Forensic Science, Medicine, and Pathology*, **11**, 202–207.

Casamatta, D.A. and Verb, R.G. (2000) Algal colonization of submerged carcasses in a mid-order woodland stream. *Journal of Forensic Sciences*, **45** (6), 1280–1285.

Castner, J. (2001) General Entomology and Arthropod Biology, in *Forensic Entomology: The Utility of Arthropods in Legal Investigations* (eds J. Byrd and J. Castner), CRC Press, Boca Raton, FL, pp. 17–41.

Child, A.M. (1995) Towards an understanding of the microbial decomposition of archaeological bone in the burial environment. *Journal of Archaeological Science*, **22** (2), 165–174.

Cobaugh, K.L., Schaeffer, S.M. and DeBruyn, J.M. (2015) Functional and structural succession of soil microbial communities below decomposing human cadavers. *PLoS One*, **10** (6), e0130201.

Coe, M. (1978) The decomposition of elephant carcasses in the Tsavo (East) National Park, Kenya. *Journal of Arid Environments*, **1** (1), 71–86.

Dekeirsschieter, J., Verheggen, F.J., Gohy, M. *et al.* (2009) Cadaveric volatile compounds released by decaying pig carcasses (*Sus domesticus* L.) in different biotopes. *Forensic Science International*, **189**, 46–53.

Dickson, G.C., Poulter, R.T.M., Maas, E.W. *et al.* (2011) Marine bacterial succession as a potential indicator of postmortem submersion interval. *Forensic Science International*, **209** (1–3), 1–10.

Duarte, S., Cássio, F., Ferreira, V. *et al.* (2016) Seasonal variability may affect microbial decomposers and leaf decomposition more than warming in streams. *Microbial Ecology*, **72**, 263–276.

Durbin, A.G., Nixon, S.W. and Oviatt, C.A. (1979) Effects of the spawning migration of the alewife, *Alosa pseudoharengus* on freshwater ecosystems. *Ecology*, **60** (1), 8–17.

Evans, W. (1963) *The Chemistry of Death*, CC Thomas Publishers, Springfield, IL.

Fiedler, S. and Graw, M. (2003) Decomposition of buried corpses, with special reference to the formation of adipocere. *Naturwissenschaften*, **90** (7), 291–300.

Fründ, H.C. and Schoenen, D. (2009) Quantification of adipocere degradation with and without access to oxygen and to the living soil. *Forensic Science International*, **188** (1–3), 18–22.

Gallagher, M.B., Sandhu, S., Kimsey, R. *et al.* (2010) Variation in development time for geographically distinct populations of the common green bottle fly, *Lucilia sericata* (Meigen). *Journal of Forensic Sciences*, **55**, 438–442.

Goffredi, S.K. and Orphan, V.J. (2010) Bacterial community shifts in taxa and diversity in response to localized organic loading in the deep sea. *Environmental Microbiology*, **12** (2), 344–363.

Goffredi, S.K., Wilpiszeski, R., Lee, R., Orphan, V.J. *et al.* (2008) Temporal evolution of methane cycling and phylogenetic diversity of archaea in sediments from a deep-sea whale-fall in Monterey Canyon, California. *The ISME Journal*, **2** (2), 204–220.

Gupta, A.K., Dharne, M.S., Rangrez, A.Y. *et al.* (2011) *Ignatzschineria indica* sp. nov. and *Ignatzschineria ureiclastica* sp. nov., isolated from adult flesh flies (Diptera: Sarcophagidae). *International Journal of Systematic and Evolutionary Microbiology*, **61**, 1360–1369.

Haefner, J.N., Wallace, J.R. and Merritt, R.W. (2004) Pig decomposition in lotic aquatic systems: the potential use of algal growth in establishing a postmortem submersion interval (PMSI). *Journal of Forensic Sciences*, **49** (2), 1–7.

Hobischak, N.R. and Anderson, G.S. (2002) Time of submergence using aquatic invertebrate succession and decompositional changes. *Journal of Forensic Sciences*, **47** (1), 142–151.

Hopkins, D.W., Wiltshire, P.E.J. and Turner, B.D. (2000) Microbial characteristics of soils from graves: an investigation at the interface of soil microbiology and forensic science. *Applied Soil Ecology*, **14** (3), 283–288.

Howard, G.T., Duos, B. and Watson-Horzelski, E.J. (2010) Characterization of the soil microbial community associated with the decomposition of a swine carcass. *International Biodeterioration and Biodegradation*, **64** (4), 300–304.

Hyde, E.R., Haarmann, D.P., Lynne, A.M. *et al.* (2013) The living dead: bacterial community structure of a cadaver at the onset and end of the bloat stage of decomposition. *PLoS One*, **8** (10), e77733.

Hyde, E.R., Haarmann, D.P., Petrosino, J.F. *et al.* (2015) Initial insights into bacterial succession during human decomposition. *International Journal of Legal Medicine*, **129** (3), 661–671.

Janaway, R.C., Percival, S.L. and Wilson, A.S. (2009) Decomposition of Human Remains, in *Microbiology and Aging* (ed S.L. Percival), Springer Science+Business Media, LLC, New York, pp. 313–334.

Johnson, D., Booth, R.E., Whiteley, A.S. *et al.* (2003) Plant community composition affects the biomass, activity and diversity of microorganisms in limestone grassland soil. *European Journal of Soil Science*, **54**, 671–678.

Johnston, N.T., MacIsaac, E.A., Tschaplinski, P.J., Hall, K.J. *et al.* (2004) Effects of the abundance of spawning sockeye salmon (Oncorhynchus nerka) on nutrients and algal biomass in forested streams. *Canadian Journal of Fisheries and Aquatic Sciences*, **403**, 384–403.

Kakizaki, E., Takahama, K., Seo, Y. *et al.* (2008) Marine bacteria comprise a possible indicator of drowning in seawater. *Forensic Science International*, **176** (2–3), 236–247.

Lauber, C.L., Metcalf, J.L., Keepers, K. *et al.* (2014) Vertebrate decomposition is accelerated by soil microbes. *Applied and Environmental Microbiology*, **80** (16), 4920–4929.

Loch, T. and Faisal, M. (2015) Emerging flavobacterial infections in fish: a review. *Journal of Advanced Research*, **6**, 283–300.

Mant, A.K. (1950) A study in exhumation data. MD thesis, London University.

Mathisen, O.A., Parker, P.L., Goering, J.J. *et al.* (1988) Recycling of marine elements trans-ported into freshwater systems by anadromous salmon. *Verhandlungen des Internationalen Verein Limnologie*, **23**, 2249–2258.

Melvin, J.R., Jr, Cronholm, L.S., Simson, L.R., Jr and Isaacs, A.M. (1984) Bacterial transmigration as an indicator of time of death. *Journal of Forensic Sciences*, **29** (2), 412–417.

Metcalf, J.L., Wegener Parfrey, L., Gonzalez, A. *et al.* (2013) A microbial clock provides an accurate estimate of the postmortem interval in a mouse model system. *eLife*, **2**, 1–19.

Metcalf, J.L., Xu, Z.Z., Weiss, S. *et al.* (2016a) Microbial community assembly and metabolic function during mammalian corpse decomposition. *Science*, **351**, 158–162.

Metcalf, J.L., Carter, D.O. and Knight, R. (2016b) Microbiology of death. *Current Biology*, **26**, R561–R563.

Meyers, M.S. and Foran, D.R. (2008) Spatial and temporal influences on bacterial profiling of forensic soil samples. *Journal of Forensic Sciences*, **53** (3), 652–660.

Micozzi, M.S. (1986) Experimental study of postmortem change under field conditions: effects of freezing, thawing and mechanical injury. *Journal of Forensic Sciences*, **31**, 953–961.

Micozzi, M. (ed) (1991) Postmortem Putrefaction and Decay, in *Postmortem Change in Human and Animal Remains: A Systematic Approach*, CC Thomas Publishers, Springfield, pp. 37–47.

Mora-gómez, J., Elosegi, A., Duarte, S. *et al.* (2016) Differences in the sensitivity of fungi and bacteria to season and invertebrates affect leaf litter decomposition in a Mediterranean stream. *FEMS Microbiology Ecology*, **92**, 1–35.

Moreno, L.I., Mills, D., Fetscher, J. *et al.* (2011) The application of amplicon length heterogeneity PCR (LH-PCR) for monitoring the dynamics of soil microbial communities associated with cadaver decomposition. *Journal of Microbiological Methods*, **84** (3), 388–393.

Naganuma, T., Wada, H. and Fujioka, K. (1996) Biological community and sediment fatty acids associated with the deep-sea whale skeleton at the Torishima Seamount. *Journal of Oceanography*, **52** (1), 1–15.

National Research Council, Committee on Identifying the Needs of the Forensic Sciences Community (2009) *Strengthening Forensic Science in the United States: A Path Forward*, National Academies Press, Washington, DC.

Nuorteva, P. (1977) Sarcosaphrophagous Insects as Forensic Indicators, in *Forensic Medicine: A Study in Trauma and Environmental Hazards* (eds C.G. Tedeschi, W.G. Eckert and L.G. Tedeschi), Saunders Company, Toronto, pp. 1072–1095.

Pace, N.R., Stahl, D.A., Lane, D.L. and Olsen, G.J. (1986) The analysis of natural microbial populations by ribosomal RNA sequences. *Advances in Microbial Ecology*, **9**, 1–55.

Parkinson, R.A., Dias, K.-R., Horswell, J. *et al.* (2009) Microbial Community Analysis of Human Decomposition on Soil, in *Criminal and Environmental Soil Forensics* (ed K. Ritz), Springer Science+Business Media, Dordrecht, pp. 379–394.

Parmenter, R.R. and Lamarra, V.A. (1991) Nutrient cycling in a freshwater marsh: the decomposition of fish and waterfowl carrion. *Limnology and Oceanography*, **36**, 976–987.

Paul, E.A. and Clark, F.E. (1996) *Soil Microbiology and Biochemistry*, 2nd edn, Academic Press, New York.

Pechal, J.L. and Benbow, M.E. (2016) Microbial ecology of the salmon necrobiome: evidence salmon carrion decomposition influences aquatic and terrestrial insect microbiomes. *Environmental Microbiology*, **18** (5), 1511–1522.

Pechal, J.L., Crippen, T.L., Tarone, A.M. *et al.* (2013) Microbial community functional change during vertebrate carrion decomposition. *PLoS One*, **8** (11), 1–11.

Pechal, J.L., Crippen, T.L., Benbow, M.E. *et al.* (2014) The potential use of bacterial community succession in forensics as described by high throughput metagenomic sequencing. *International Journal of Legal Medicine*, **128** (1), 193–205.

Pfeiffer, S., Milne, S. and Stevenson, R.M. (1998) The natural decomposition of adipocere. *Journal of Forensic Sciences*, **43** (2), 368–370.

Pinheiro, J. (2006) Decay Process of a Cadaver, in *Forensic Anthropology and Medicien: Complementary Sciences from Recovery to Cause of Death* (eds A. Schmitt, E. Cunha and J. Pinheiro), Humana Press, Totowa, NJ, pp. 100–131.

Poinar, G.O. (1983) *The Natural History of Nematodes*, Prentice Hall, Inc., Englewood Cliffs.

Putman, R.J. (1978) Flow of energy and organic matter from a carcase during decomposition. Decomposition of small mammal carrion in temperate systems 2. *Oikos*, **31**, 58–68.

Richey, J.E., Perkins, M.A. and Goldman, C.R. (1975) Effects of kokanee salmon (*Oncorhynchus nerka*) decomposition on the ecology of a subalpine stream. *Journal of the Fisheries Research Board of Canada*, **32**, 817–820.

Richter, S. (1993) Phoretic association between the dauer juveniles of *Rhabditis stammeri* (Rhabditidae) and life history stages of the burying beetle *Nicrophorus vespilloides* (Coleoptera: Silphidae). *Nematologica*, **39**, 346–355.

Roggenbuck, M., Bærholm Schnell, I., Blom, N. *et al.* (2014) The microbiome of New World vultures. *Nature Communications*, **5**, 5498.

Schoenen, D. and Schoenen, H. (2013) Adipocere formation—the result of insufficient microbial degradation. *Forensic Science International*, **226** (1–3), 301.

Schoenly, K.G., Haskell, N.H., Mills, D.K. *et al.* (2006) Recreating death's acre in the school yard: using pig carcasses as model corpses. *The American Biology Teacher*, **68** (7), 402–410.

Singh, B., Crippen, T.L., Zheng, L. *et al.* (2015) A metagenomic assessment of the bacteria associated with Lucilia sericata and Lucilia cuprina (Diptera: Calliphoridae). *Applied Microbiology and Biotechnology*, **99** (2), 869–883.

Smith, K. (1986) *A Manual of Forensic Entomology*, British Museum of Natural History, Cornell University Press, London/Ithaca.

Smith, C.R. (1992) Whale falls: chemosynthesis on the deep seafloor. *Oceanus*, **35**, 74–78.

Smith, C.R. and Kukert, H. (1989) Vent fauna on whale remains. *Nature*, **341** (6237), 27–28.

Smith, C.R., Glover, A.G., Treude, T. *et al.* (2015) Whale-fall ecosystems: recent insights into ecology, paleoecology, and evolution. *Annual Review of Marine Science*, **7** (1), 571–596.

Swift, M., Heal, O. and Anderson, J. (1979) Decomposition in terrestrial ecosystems, in *Studies in Ecology*, University of California Press, Berkeley/Los Angeles.

Tibbett, M. and Carter, D.O. (2008) *Soil Analysis in Forensic Taphonomy: Chemical and Biological Effects of Buried Human Remains*, CRC Press; Taylor & Francis Group, Boca Raton, FL.

Todd, T.C., Powers, T.O. and Mullin, P.G. (2006) Sentinel nematodes of landuse change and restoration. *Journal of Nematology*, **38** (1), 20–27.

Towne, E.G. (2000) Prairie vegetation and soil nutrient responses to ungulate carcasses. *Oecologia*, **122**, 232–239.

Treude, T., Smith, C.R., Wenzhöfer, F. *et al.* (2009) Biogeochemistry of a deep-sea whale fall: sulfate reduction, sulfide efflux and methanogenesis. *Marine Ecology Progress Series*, **382**, 1–21.

VanLaerhoven, S. and Anderson, G. (1999) Insect succession on buried carrion in two biogeoclimatic zones of British Colombia. *Journal of Forensic Sciences*, **44** (1), 32–43.

Vass, A. (2001) Beyond the grave—understanding human decomposition. *Microbiology Today*, **28** (28), 190–192.

Weiss, S., Carter, D.O., Metcalf, J.L. and Knight, R. (2015) Carcass mass has little influence on the structure of gravesoil microbial communities. *International Journal of Legal Medicine*, **130**, 253–263.

Wold, A.K.F. and Hershey, A.E. (1999) Effects of salmon carcass decomposition on biofilm growth and wood decomposition. *Canadian Journal of Fisheries and Aquatic Sciences*, **56** (5), 767–773.

Young, G.M., Amid, D. and Miller, V.L. (1996) A bifunctional urease enhances survival of pathogenic *Yersinia enterocolitica* and *Morganella morganii* at low pH. *Journal of Bacteriology*, **178** (22), 6487–95.

Zheng, L., Crippen, T.L., Singh, B. *et al.* (2013) A survey of bacterial diversity from successive life stages of black soldier fly (Diptera: Stratiomyidae) by using 16S rDNA pyrosequencing. *Journal of Medical Entomology*, **50** (3), 647–658.

Zimmerman, K.A. and Wallace, J.R. (2008) The potential to determine a postmortem submersion interval based on algal/diatom diversity on decomposing mammalian carcasses in brackish ponds in Delaware. *Journal of Forensic Sciences*, **53** (4), 935–941.

CHAPTER 11

Arthropod–microbe interactions on vertebrate remains: Potential applications in the forensic sciences

Jeffery K. Tomberlin[1], M. Eric Benbow[2], Kate M. Barnes[3] and Heather R. Jordan[4]

[1] *Department of Entomology, Texas A&M University, College Station, TX, USA*

[2] *Departments of Entomology and Department of Osteopathic Medical Specialties, Michigan State University, East Lansing, MI, USA*

[3] *Department of Natural Sciences, University of Derby, Derby, UK*

[4] *Department of Biological Sciences, Mississippi State University, Starkville, MS, USA*

11.1 Introduction

Death and taxes are often posed as the two guarantees of life. However, something that is often overlooked, or possibly avoided, is the promise that Nature is possessive and will not shy away from reclaiming the nutrients loaned to us during life and stored within our bodies. These nutrients represent the currency we all must repay either while we live (i.e., metabolism and discharge into the environment) or eventually when we die (i.e., decomposition). With that said, the death of an organism represents a reverse in entropy, leading from the assimilation of nutrients allowing its growth, formation, and possible reproduction to its breakdown and eventual transition of these nutrients from one state (e.g., muscle and tissue) to another (e.g., arthropod biomass). How this transition occurs and our understanding of the mechanisms regulating decomposition is still not fully described; however, what is known is being applied to benefit humanity in a number of ways, including the forensic sciences.

Forensic Microbiology, First Edition. Edited by David O. Carter, Jeffery K. Tomberlin, M. Eric Benbow and Jessica L. Metcalf.

11.1.1 Decomposition and applications in forensic entomology

Recent papers by Matuszewski (2011) (Figure 11.1) and Tomberlin *et al.* (2011) (Figure 11.2) have proposed frameworks for exploring the mechanisms regulating the phenomena of vertebrate carrion decomposition. These frameworks were proposed primarily with entomology in mind, but can be applied for many other organisms that utilize the carrion patch (e.g., plants, vertebrates, and microbes). In this chapter, we will explore the framework proposed by Tomberlin *et al.* (2011) to explain the process of insect attraction and utilization of vertebrate carrion within the context of microbe–insect interactions, along with the significance of these interactions to forensic entomology in particular and the forensic sciences in general. In order to achieve a complete understanding of the decomposition framework terminology utilized, readers are highly recommended to review Matuszewski *et al.* (2011).

11.1.1.1 Time of death

Entomology is a well-recognized science that is often applied in forensic investigations worldwide. While there are many applications of entomology in forensic science (e.g., movement of human remains from one location to another, abuse and neglect, and entomotoxicology), forensic entomologists are most often asked to assess arthropod fauna associated with deceased individuals to determine the time of colonization. This estimation can then be used to infer a minimum postmortem interval (PMI). In these cases, the predominant methods utilized are based on development data sets for insects that frequent decomposing vertebrates, including human remains.

Recent discussions have begun to explore the temporal variation associated with the initial colonization of human remains by primary colonizers (e.g., blow flies (Diptera:

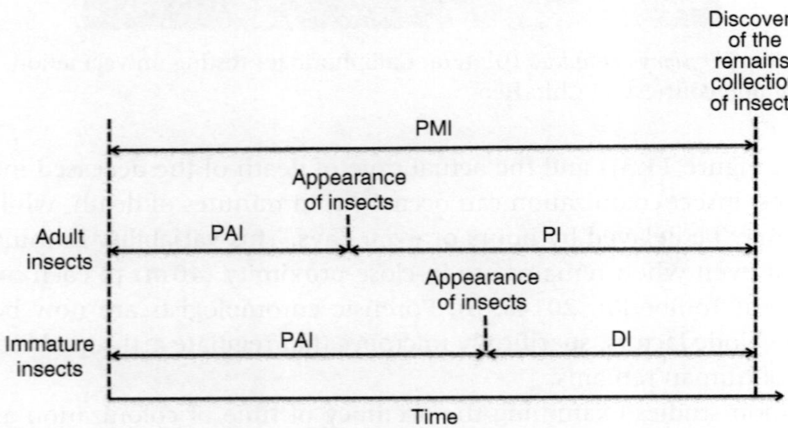

Figure 11.1 A framework of the postmortem interval (PMI) divided into the preappearance interval (PAI), development interval (DI), and presence interval (PI) as proposed by Matuszewski (2011). Source: Matuszewski (2011). Reproduced with permission of Elsevier

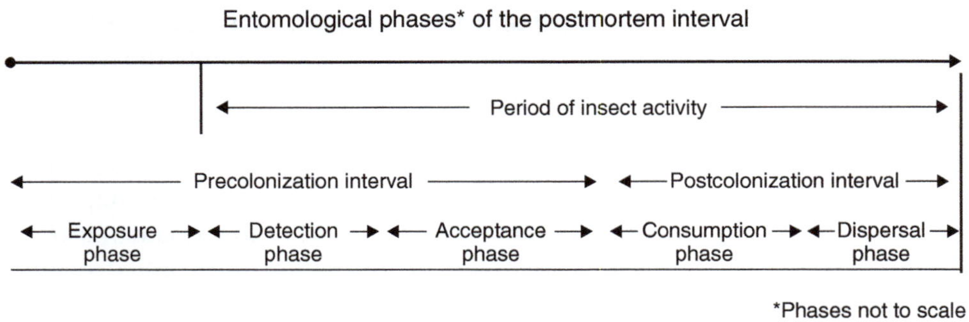

Figure 11.2 Framework proposed by Tomberlin *et al.* (2011) for entomology-based phases of the decomposition process. Source: Tomberlin *et al.* (2011). Reproduced with permission of Annual Reviews

Figure 11.3 Adult *Chrysomya rufifacies* (Diptera: Calliphoridae) resting on vegetation. Source: Photograph courtesy of Chin Heo

Calliphoridae; Figure 11.3)) and the actual time of death of the deceased individual. In some instances, insect colonization can occur within minutes of death, while in others, colonization may be delayed by hours or even days. This variability in initial colonization is evident even when remains are in close proximity (40 m) to each other (Mohr, 2012; Mohr and Tomberlin, 2014a, b). Forensic entomologists are now beginning to appreciate the biotic factors, specifically microbes that regulate arthropod attraction and colonization of human remains.

Few validation studies examining the accuracy of time of colonization as related to time of death of a human (i.e., pig or rat as models) have been completed. VanLaerhoven (2008) examined the accuracy of using accumulated degree-day (ADD) models in combination with blow fly development thresholds and data to estimate the PMI for three mock crime scenes. For each case, estimates were within 1 day of the actual time of death of swine remains. Tarone and Foran (2008) used general additive models to predict the age of blow fly *Lucilia sericata* (Meigen) (Diptera: Calliphoridae) larvae. Estimates based on length and development stage were within 5% of the actual age of the specimen(s) in question; however, predicting the age of postfeeding third instars and pupae was limited. Núñez-Vázquez *et al.* (2013) utilized laboratory-based development data for *Phormia regina* (Meigen) (Diptera: Calliphoridae) to predict the age of specimens reared in glass jars outdoors. As with other validation studies, the age of the early instars (first and second) were most accurately predicted, while third instars and pupae had the greatest variation associated with the estimates.

In contrast to the aforementioned studies, other researchers have examined the use of adult insects arriving at a carrion source to estimate the time elapsed either before or after colonization. Matuszewski (2011, 2012) used coleopteran species (Coleoptera: Silphidae and Staphylinidae) to estimate the precolonization interval, while Mohr and Tomberlin (2014a) examined the ovarian status of dipteran species (Diptera: Calliphoridae) arriving to swine carcasses to estimate a minimum PMI. In all cases, these methods appeared fairly accurate (six of seven estimates of time of placement of swine or human remains were accurate) under the given circumstances of the studies; however, additional validation is needed for different ecoregions and environmental conditions (e.g., season, urban, and rural) to determine the error associated with applying said techniques.

11.1.1.2 Movement of remains

Insect communities can be distinct to a specific geographic location. Such information can be valuable for determining whether human remains have been relocated from one area to another. However, such applications are typically restricted to those sets of remains that have been colonized by the insects prior to their movement to a new location. For example, a review of blow fly species associated with forensic entomology cases in British Columbia determined that some species (e.g., *Calliphora vicina* (Robineau–Desvoidy) *L. sericata*) preferred urban environments, while others (e.g., *Calliphora vomitoria* (Linnaeus), *Ph. regina*, and *Protophormia terraenovae* (Robineau–Desvoidy)) preferred rural settings (Anderson, 1995). Studies in Europe have shown that these blow fly species act as indicators for similar habitats in other parts of the world (Matuszewski and Szafałowicz, 2013; Fremdt and Amendt, 2014). For example, Hwang and Turner (2005) found that in London, UK, an urban habitat was characterized by the presence of blow flies of the species *C. vicina*, *L. sericata*, and *Lucilia illustris* (Meigen) and a rural woodland area by *C. vomitoria*.

Recent efforts have explored molecular analyses to discriminate between different fly populations with degree of relatedness of specimens collected from remains and those from the surrounding environment serving as a means to determine whether the remains were moved from one location to another. Picard and Wells (2012) proposed a method using amplified fragment length polymorphism profiles to determine whether larvae on a corpse were related to one another. These same authors also explored the population genetic structure of *L. sericata* in North America and determined that genetic relatedness could be used to distinguish among populations; assignment tests were 96% accurate with regard to determining the origin of a larval sample (Picard and Wells, 2010).

11.1.2 Microbe–arthropod interactions

Microbe–arthropod interactions have been described in a variety of natural systems (Boissière *et al.*, 2012; Frago *et al.*, 2012; Jones *et al.*, 2013), and recent studies have demonstrated these interactions have deep evolutionary roots for some arthropod species (Russell *et al.*, 2009). Thus, any natural system (e.g., carrion) that is described by the co-occurrence of these two groups of organisms should consider how their possible interactions play a role in the community structure and function of that system. The fundamental understanding of these interactions is thought to have potentially widespread applications ranging from pest management to forensic investigations (Tomberlin *et al.*, 2011; Roy and Muraleedharan, 2014; Jordan *et al.*, 2015). In order to appreciate how these interactions may be important for forensic microbiology, it is first necessary to introduce the broader community context of carrion (or human) decomposition, or what is known as the necrobiome.

The necrobiome is the community of organisms and their genes that interact on or in association with decomposing heterotrophic biomass (Benbow *et al.*, 2013). In other words, the necrobiome consists of the communities of interacting microbes (e.g., bacteria, fungi, and protists), arthropods (e.g., blow flies), and vertebrates (e.g., vultures) that utilize dead animals in some way; whether this use is as a habitat and food source for their offspring or simply as an additional food resource when other resources are scarce or unavailable. Each group within the necrobiome plays a functionally important role in the ultimate recycling of energy and nutrients from carrion into the environment. Recent studies suggest the interactions between the microbes and necrophagous arthropods that colonize decomposing vertebrate carrion affect the rate and timing of decomposition (Pechal *et al.*, 2013; Pechal *et al.*, 2014a; Metcalf *et al.*, 2016). It is within this network of communities where understanding how microbes and arthropods interact could play important roles in forensic science, since in most cases the microbial communities of decomposing human or animal remains do not function without some degree of interaction with other macroorganisms, such as arthropods.

The microbial communities of the necrobiome have been grouped into two primary areas of the body: external and internal. Pechal *et al.* (2014b) first defined the external microbial communities of the necrobiome as epinecrotic communities, or those microbes that are found on the surfaces of carrion including inside the mouth, ears, eyes, or distal reaches of the rectum. The internal microbial communities have been defined by Can *et al.* (2014) as the thanatomicrobiome or the community of microbial organisms found associated with the internal organs, blood, and other fluids of a dead animal.

Figure 11.4 Adult *Nicrophorus marginatus* Fabricius (Coleoptera: Silphidae) present on swine carrion. Source: Photograph courtesy of Chin Heo

The microbial communities most likely coming into contact and interact with necrophagous arthropods will largely depend on the stage of decomposition of the remains (see Section 11.2.1 for factors that affect the initial detection and arrival of arthropods). In general, blow flies and flesh flies (Sarcophagidae) colonize remains early in decomposition, often within the first 1–2 days, while beetles (Figure 11.4) can be found later in decomposition feeding on the fly larvae and using the dry remains for feeding or breeding (e.g., skin, hair, and bones) (Byrd and Castner, 2009). Therefore, during the decomposition process, the microbial communities of carrion and the insects that colonize and use the resource continuously interact. The degree and importance of these interactions on carrion have only recently seen focused research attention (Ma *et al.*, 2012; Tomberlin *et al.*, 2012; Pechal *et al.*, 2013, 2014a).

Understanding the basic ecological types of interactions will help establish a conceptual framework for more thorough investigations into how microbes and arthropods contribute to carrion/human decomposition, and how these relationships influence the use of microbes in forensic investigations. These ecological associations include interactions between individuals of the same species and across species, such as bacteria or worms that live in or on another organism. Here, we briefly discuss six general groups of organismal relationships that have the potential to affect epinecrotic communities and thanatomicrobiomes during carrion decomposition: *intraspecific competition, interspecific competition, predation, parasitism, mutualism,* and *commensalism*. For additional information on these ecological relationships, there are several good texts and reviews available (Lang and Benbow, 2013 and references therein; Benbow *et al.*, 2015; Tomberlin and Benbow, 2015).

11.1.2.1 Intra- and interspecific competition

Two of the most commonly noted ecological interactions are intra- and interspecific competition. The interactions that involve different individuals of the same species are intraspecific, while interspecific is the interactions and competition for resources between individual organisms of different species. Inherently, intraspecific competition is understood to be more intense than interspecific since the organisms are of the same species and share the same niche. For the carrion system, little is understood about the intraspecific competition within microbial species that is likely occurring during decomposition, although several recent studies (Metcalf *et al.*, 2013; Pechal *et al.*, 2013; Hyde *et al.*, 2014; Pechal *et al.*, 2014b) have described microbial successional community patterns that could provide initial insight into such interactions.

Interspecific competition occurs when any organisms of different species, or coarser taxonomic resolution, share a common and limited resource in such a way that when one species consumes or utilizes the resource more than another species, the loss of resource has a direct fitness cost on the species that does not acquire that resource. Janzen (1977) was one of the first to propose an interspecific competitive interaction between microbes and insects in decomposition using rotting fruit as a model system. He suspected that microbes competed with other consumers for resources by producing compounds, such as toxins, that affected resource "appeal" thereby reducing competition. In a more recent study, Burkepile *et al.* (2006) demonstrated that microbes colonizing fish carrion competed with other consumers for the resource (i.e., higher order scavengers ate fresh carrion 2.4 times more frequently than microbe-laden carrion). The authors suspected that microbes released noxious chemicals deterring consumption of carrion by other scavengers (Burkepile *et al.*, 2006). Removing microbes resulted in the resource being attractive to crustaceans for significantly longer periods of time (Burkepile *et al.*, 2006). In support of this hypothesis, Pechal *et al.* (2013) showed that when blow flies were excluded from initial colonization of decomposing vertebrate remains, both the successional trajectory of epinecrotic communities and necrophagous invertebrate communities significantly shifted compared to those same communities that experienced immediate and sustained contact with necrophagous arthropods for 5 days after death. In addition, research by Rozen *et al.* (2008) demonstrated that microbes had a significant negative effect on burying beetle (*Nicrophorus vespilloides* Herbst) reproductive success and larval growth, and this fitness effect was compensated by parental care and other behavior modifications to avoid carrion with dense microbial communities. Although the mechanisms are unknown, these studies collectively indicate that there are probable and substantial competitive interspecific interactions that occur on carrion that could greatly affect the potential of microbial communities being used in forensic investigations. In some instances, interspecies interactions go beyond competing for common resources when one species becomes the resource for another species—predation.

11.1.2.2 Predation

When one organism actively kills and then feeds on that organism, this interaction is defined as predation. Generally, predation is the direct interaction between individuals of different species; however, in some populations, cannibalism is known to occur when individuals of the same species predate upon each other. Predation is typically defined for interactions among animals and some carnivorous plants (e.g., pitcher plants of the Nepenthaceae and Sarraceniaceae), with little research attention focused on predatory interactions among microbes, even though they occur and can be important ecological interactions within microbial communities (Crowley *et al.*, 1980; Bouvy *et al.*, 2011; Kadouri *et al.*, 2013). Microbe–microbe predation has been commonly reported in aquatic ecosystems, often within communities that occupy the open water of lakes and oceans (del Campo *et al.*, 2013), or those that form biofilms on substrates in flowing systems like streams and rivers (Wey *et al.*, 2012). To our knowledge, there are no published studies on bacteria–bacteria predation within the necrobiome, although it is highly likely that such interactions exist based on recent epinecrotic studies of swine and human remains that show complex, interacting microbial communities that change throughout decomposition (Hyde *et al.*, 2014; Pechal *et al.*, 2014b). Here, we will focus our discussion on what is known of arthropod predation on microbes within the carrion system.

Most of the early work on arthropod predation of microbes was in the form of understanding how long and in what way filth flies, such as house flies, *Musca domestica* L. (Diptera: Muscidae), and blow flies, acquired, processed, and eliminated bacterial pathogens of human health interest (Greenberg, 1968; Greenberg *et al.*, 1970; Mumcuoglu *et al.*, 2001; Lerch *et al.*, 2003). For instance, Greenberg and Klowden (1972) fed both larvae and adult house flies different enteric bacterial pathogens within their food medium to study how long they were maintained in the gut and how many cells were evacuated from the gut during defecation. A compilation of such studies for various filth fly species and their feeding-acquired microbial taxa can be found in Greenberg's two volumes on flies and disease (Greenberg, 1971a, b).

Several laboratory studies have evaluated interactions of flies and bacteria often by manipulating how the flies were able to consume the bacterial cells (e.g., inoculated food resource) (Zurek *et al.*, 2000; Lam *et al.*, 2007, 2009). However, we are not aware of studies that have directly tested how fly larvae or adults naturally predate upon bacteria in nature, an activity that has tremendous ecological and evolutionary potential for future research. One of the only field studies that explicitly addressed the interactions of microbial and necrophagous insect communities was by Pechal *et al.* (2013). In the study, they reported that when necrophagous insects were unable to access decomposing swine carrion, the bacterial taxon richness was significantly higher than those microbial communities on carrion where the insects were allowed to naturally colonize and consume the resource. While this study did not directly measure larval fly predation of bacteria on the carrion, the results suggest that necrophagous insects were associated with reduced numbers of microbial taxa, presumably in part by feeding on them. Through the study of predatory interactions between microbes and arthropods, the results suggest that other forms of interactions that do not involve the active killing and consuming of an organism exist in carrion systems. However, these other noncompetitive or predatory relationships between necrobiome arthropods and microbes have not been well studied.

Microbial predation by arthropods has also been studied within the context of identifying new arthropod-mediated methods of pathogen removal from animal manure for food safety concerns. In such studies, manure or some other resource with either estimated levels or known inoculated concentrations of specific bacterial pathogens are fed to flies to determine how well the fly populations reduce the pathogens (Erickson *et al.*, 2004; Liu *et al.*, 2008). Indeed, fly larvae are efficient consumers of bacterial populations of manure, a resource that is often colonized and consumed by necrophagous carrion flies. While it is commonly accepted that necrophagous arthropods regularly consume the microbes of carrion substrates, the ecological importance of such interactions has not received explicit research attention.

11.1.2.3 Symbiosis: parasitism, mutualism, and commensalism

The noncompetitive and nonpredatory interactions among species are considered symbioses, with the main groups of these interactions including parasitism, mutualism, and commensalism. These relationships describe two or more species that interact directly with each other. If the result of the direct interaction is positive to both species, then the relationship is a mutualism, whereas when one species benefits at the expense of the other (the host), the relationship is considered a parasitism of the host. Commensalism defines those interactions that benefit one of the species but with no measurable effect on the other species. Within the necrobiome, there are well-documented cases of all three of these symbioses among arthropod species; however, symbiosis between arthropods and microbial species has received much less research attention beyond the endosymbiotic roles (e.g., from parasitic to mutualistic) of *Wolbachia* for various insects including some necrophagous species (Wei *et al.*, 2014). The research on *Wolbachia* symbioses of arthropods indicates that the nature of the relationship of this bacterial genus depends on the host and environmental contexts (Nikoh *et al.*, 2014; Zug and Hammerstein, 2015), making it difficult to make broad generalizations of the possible nature of this interaction with necrophagous arthropod species. Based on what is known of the microbial communities associated with the carrion resource (e.g., epinecrotic communities) and some necrophagous insect species (Ma *et al.*, 2012; Jordan *et al.*, 2015; Singh *et al.*, 2015), it is likely that all three of these symbioses will be characterized in future research on the interactions between carrion microbes and arthropods.

11.2 Framework for understanding microbe–arthropod interactions on vertebrate remains

11.2.1 Precolonization interval

Traditionally, forensic entomology has relied determining the age of insect larvae collected from human remains using ADD and accumulated degree-hour of insect growth or through describing the suite of insect species that are present and going through succession on the remains (Zurawski *et al.*, 2009; Joseph *et al.*, 2011; Mohr and Tomberlin, 2014b). When using larval growth, scientists use the age of the specimens to provide evidence as to when the female flies first laid their eggs on the human remains (the minimum estimate of the PMI). However, many of the mathematical methods used for this determination fail to take into account mechanisms contributing to insect colonization of human remains, and the time extending from death to the time of invertebrate colonization, also known as the precolonization interval (Mohr and Tomberlin, 2014b; Sharma *et al.*, 2015). This interval, defined by Tomberlin *et al.* (2011), encompasses all of the necessary neurosensory detection and behavioral activation for searching and locating the carcass preceding oviposition and the remaining postcolonization interval and resource utilization.

Little is known of when and under what circumstances adult blow flies appear at a carcass, but information can be gathered from research into invertebrate seeking and attraction to living resources. Studies have shown that volatile organic compounds (VOCs) emitted by living vertebrate animals are widely available to a number of organisms from different trophic levels, potentially leading to nonspecific cues (Pickett *et al.*, 2014). However, many dipteran species have a finely tuned sense of smell and are attracted by specific and microbially mediated host odors (Zhang *et al.*, 2015). Several studies have found that feeding and oviposition are microbially driven (Chaudhury *et al.*, 2010; Chaudhury and Skoda, 2013; Davis *et al.*, 2013). In addition, myiasis (i.e., colonization of living vertebrate tissue by dipteran larvae) is an important problem in both humans and animals, with dipteran attraction and colonization significantly impacted by microbes and/or their products. For instance, extracts from cultures of *Pseudomonas aeruginosa*, *Bacillus subtilis*, *Proteus mirabilis*, and *Enterobacter cloacae* induced oviposition by the Australian blow fly and agent of sheep myiasis, *Lucilia cuprina* (Wiedemann) (Emmens and Murray, 1982).

Wounds and their associated microbes are also attractants that stimulate oviposition for gravid screwworm, *Cochliomyia hominivorax* (Coquerel) (Chaudhury *et al.*, 2014). Hammack (1991) reported that fresh, uninoculated blood stimulated oviposition by *Co. hominivorax*, but the author concluded it was not olfactory based. Interestingly, however, other studies by that group found that steam distillate of cultured *Providencia rettgeri* (formerly *Proteus rettgeri*), a Gram-negative bacterium that is part of a genus known to colonize wounds, led to increased attraction of gravid females (Hammack *et al.*, 1987). Chaudhury *et al.* (2002) exposed *Co. hominivorax* to bovine blood inoculated with eight species of coliform bacteria isolated from screwworm-infested animal wounds. They found that the bacteria-inoculated blood stimulated gravid screwworm flies to land on the same tray and was dependent upon blood incubation time and fly age (Chaudhury *et al.*, 2002). However, the importance of each individual species was not tested in that study. Later work determined attraction of gravid females by inoculation of the individual species compared to composite to determine influence of VOCs emitted from the bacteria-inoculated blood (Chaudhury *et al.*, 2010). They found landing response to bacterial VOCs ranked from highest to lowest response as *Pr. rettgeri* < *Proteus vulgaris* < *Pr. mirabilis* < *Klebsiella oxytoca* < *Pro. stuartii* < *Serratia liquefaciens* < *E. cloacae* < *Enterobacter sakazakii*, and landing response was highest with all species combined. In addition, blood was found in the guts of ovipositing females, suggesting that a blood meal might play a role in stimulating oviposition (Chaudhury *et al.*, 2010). Specific VOCs were not identified in that study; however, more recently, Chaudhury *et al.* (2014) identified dimethyl disulfide, dimethyl trisulfide, phenol *p*-cresol, and indole as VOCs emitted from *Co. hominivorax* waste larval rearing medium (Chaudhury *et al.*, 2014). Waste media and diluted, synthetic blends of the identified compounds were significantly more attractive for gravid *Co. hominivorax* and also to *Cochliomyia macellaria*: (Fabricius) (Diptera: Calliphoridae). Both species also laid significantly more eggs on substrates treated with the waste medium or diluted synthetic blends of the identified VOC compounds (Chaudhury *et al.*, 2014). Researchers investigating colonization cues in insects of forensic importance have also found that *L. sericata* and *C. vicina* are most attracted to the VOCs initially released after death, such as dimethyl disulfide (Frederickx *et al.*, 2012a; Paczkowski *et al.*, 2012), and *Dermestes maculatus* (De Geer) (Coleoptera: Dermestidae) is attracted to indole, which is released later on during the active stage of decay (von Hoermann *et al.*, 2011). It is currently unknown whether these VOCs originate from the corpse itself or the bacteria growing on it, but previous work suggests that certain bacteria associated with animal carcasses release volatiles that attract adult blow flies (Davis *et al.*, 2013).

Indeed, hundreds of VOCs are emitted from the living human body (from blood, breath, skin, sweat, and urine) and usually reflect the metabolic state of the individual (Shirasu and Touhara, 2011; de Lacy Costello *et al.*, 2014). How then does this chemical information shift upon death? In addition, what mechanisms lead to invertebrate olfactory learning to link vertebrate remains as a resource with specific odors released from the resource? As we will show in Sections 11.3.1–11.4.1, many of the mechanisms are likely microbially driven. We will begin this section discussing what is known about VOCs emitted following death and follow with mechanisms influencing invertebrate detection and location of vertebrate remains.

11.2.1.1 Death

Microbial-mediated mechanisms of human decomposition potentially begin immediately after death and are a driving force for conversion of a once living organism to a resource of energy and nutrients. As soft tissues and their components are broken down, VOC by-products are produced and released. While a comprehensive profile of VOCs released immediately after death during human decomposition is unavailable, studies conducted several hours after death have identified oxygenated compounds, aliphatic and aromatic hydrocarbons, organic sulfides, and inorganic compounds (Vass *et al.*, 2004, 2008). Statheropoulos *et al.* (2005) found over 80 volatiles from two human bodies in body bags. Vass *et al.* (2004) reported at least 478 volatile chemicals associated with buried

Figure 11.5 Swine carrion exhibiting (a) fresh, (b) active, and (c) skeletal stages of decomposition. Source: Photograph courtesy of Chin Heo

human remains, with volatile variability association with decomposition succession and biotope (Vass *et al.*, 2004, 2008).

Investigators using nonhuman surrogates at least 4 hours after sacrifice identified 104 specific chemical compounds produced exclusively during the decomposition process (Dekeirsschieter *et al.*, 2009). Some of these compounds have been experimentally found to regulate necrophagous insect behavior (Joseph *et al.*, 2011; Stavert *et al.*, 2014). However, these invertebrates are presumably not present immediately after vertebrate death, but are drawn through olfactory and other sensory mechanisms affected by emitted VOCs. Thus, elucidation of pathophysiological mechanisms underlying microbially derived VOCs at specific time points (i.e., all stages of decomposition (e.g., Figure 11.5)) will likely provide novel insights into understanding ecological interactions and drivers leading to invertebrate attraction and colonization, with potential for allowing application and prediction in forensic sciences.

11.2.1.2 Detection

In order to utilize a resource, arthropods must complete a series of steps. These mechanisms are well described for other systems including parasitoids and herbivorous insects (Dethier, 1954; Vinson, 1976). More recently, Tomberlin *et al.* (2011) described specific processes for carrion-associated arthropods. Specifically, arthropods must (i) detect and become activated by the presence of a resource, (ii) search and locate that resource, (iii) colonize the resource for reproduction or feeding, and (iv) disperse from the resource following complete carrion decomposition or removal of remains (Tomberlin *et al.*, 2011).

Allothetic (i.e., external spatial cues (e.g., visual or tactile) used for orientation) and idiothetic (i.e., utilizing current and past locations for orientation) processes work in tandem for processing of external stimuli triggering neural response and behavior cascades (Tomberlin *et al.*, 2011). Invertebrates use chemical mediators to detect resources for food and refuge, to acquire mates, and for oviposition. Perception of sensory information leads to searching for and locating the resource item. Hierarchical systems for perception and searching for resources have been proposed where arthropods use gross cues indicative of habitat, followed by finer cues such as patch cues, and, finally, individual resource cues (Bell, 1990). Genetic programing and proprioceptively (i.e., responses due to physiological stimuli) derived information help drive these processes (Bell, 1990).

It is no wonder then that adult blow flies can detect chemical cues, such as decomposition-specific VOCs, and have been known to colonize vertebrate remains within minutes following animal death (Catts and Goff, 1992; Anderson, 1999a; VanLaerhoven and Anderson, 1999; Joseph *et al.*, 2011). However, specific physiological mechanisms leading to detection and response are understudied. Variability in blow fly arrival has also been reported where external factors, such as chemical gradients, light and sound intensity, artificial containment, temperature, humidity, or physical barriers, such as clothing, influence blow fly neural stimulation and resulting behavior cascades (Bell, 1990; Tomberlin *et al.*, 2011). Aspects of fly specific genetics and functioning, such as sex, ovarian development, and foraging behavior, also dictate stimuli processing and resulting search behaviors, thus making a reliable prediction of colonization and the PMI range a difficult and complex endeavor (Bell, 1990).

11.2.1.3 Location

Both the temporal aspect of searching and the geographic location of such activities are important for insect ability to locate remains. Mohr and Tomberlin (2014b) conducted a study to characterize dominant Calliphoridae species, sex, population, body size, and arrival patterns at a carcass, and correlated these with temperature, time, season, and PMI. Four trials were conducted where three pig carcasses were placed outdoors in two replicate seasons within 1 hour of sacrifice and hourly measurements of environmental conditions and flies attracted were taken. Both sexes of four blow fly species were collected in significant numbers: *Chrysomya rufifacies* (Macquart), *Co. macellaria*,

C. vicina, and *Ph. regina*. Time of day and wind level for winter collections were significant explanatory variables for population sizes indicating governance by an endogenous circadian rhythm (Mohr and Tomberlin, 2014b). The identification of male flies also indicated the possibility that male flies are attracted to carcasses for mating opportunities (Mohr and Tomberlin, 2014b).

Associated microbes, feeding larvae (Frederickx *et al.*, 2012a, b), and the remains themselves produce VOCs emanating from carrion as it decomposes. Searching is also influenced by circadian rhythm and weather events, locomotive capability, level of chemotaxis, and physiological state of the arthropod (Bell, 1990; Mohr and Tomberlin, 2014b). In addition, insects use topographical features to direct searches when cues from resources are not detected (Bell, 1990). Whether topographical features play a significant role in searching in carrion associated flies is unknown.

Quality assessment of the resource occurs once the arthropod reaches the resource. Assessment may be preconceived, as when genetically specified or by comparison with "what it knows" (Bell, 1990) where acceptance criteria shifts may occur once localized resource assessment actually takes place. Age of the resource and presence of competitors or predators are also factors mitigating acceptance and resource acquisition (Bell, 1990; Schröder and Hilker, 2008). Emission of host or competing organism VOCs may also play a role in influencing resource acceptance (Schröder and Hilker, 2008). Species simply feeding on the carrion resource are under less selection pressure. However, for those insects rearing offspring, the penalty for incorrect quality assessment might lead to failure for eggs to hatch, sterile offspring, or predation (Bell, 1990).

Arthropods are not restricted to olfaction for locating carrion resources as many of the primary colonizers also utilize vision. This dependency in the case of blow flies is not a complete surprise considering the published accounts indicating blow flies typically do not colonize carrion at night (Zurawski *et al.*, 2009; Berg and Benbow, 2013). However, other visual parameters associated with carrion impact its attractiveness to flies. Carcass size along with the presence of conspecific or heterospecific adults can also influence their attraction (Spivak *et al.*, 1991). Also, in some instances, olfaction and visual stimuli are perceived together to stimulate attraction and landing by an insect, such as a blow fly adult (Wall and Fisher, 2001). In the case of *C. vicina*, the orientation of visual cues (vertical vs horizontal), rather than shape, impacted their ability to locate resources (Aak and Knudsen, 2011). In fact, pattern and color could also play roles in regulating the attraction of blow flies to carrion (Gomes *et al.*, 2007).

11.3 Postcolonization interval

Numerous insect species are attracted to a human corpse. Some insects come to feed on the body but do not reproduce there. For example, parasitic Calliphoridae, Sarcophagidae, and Tachinidae visit a corpse solely for a protein source or mating opportunity but will not oviposit on the body itself (Rivers and Dahlem, 2014). Other insects predate or parasitize those already developing in the corpse. For example, Histerid and Staphilinid beetles prey on blow fly larvae (Kočárek, 2002) and the wasp *Nasonia vitripennis* parasitizes blow fly pupae (Grassberger and Frank, 2003). In forensic terms, the most significant insects are those that use the corpse as a larval nutritional resource. These species are found within the orders Diptera (true flies) and Coleoptera (beetles). Generally, the primary colonizers are the blow flies, which are most commonly used in minimum PMI estimations (Anderson, 2004; Oliveira-Costa and Antunes de Mello-Patiu, 2004). Therefore, most research is conducted with this insect family. Necrophagous beetles, such as those in the Silphidae and Dermestidae families, colonize at a later date with significantly less research available on such taxa. This section will concentrate on these two insect families.

11.3.1 Colonization

The species of blow fly that colonize a corpse will differ according to geographical region and habitat (Matuszewski and Szafałowicz, 2013; Fremdt and Amendt, 2014). Necrophagous beetles also differ according to habitat. For example, Dekeirsschieter *et al.* (2011) found differences in the Silphidae distribution between three different habitats with seven species being trapped at a forest site, six at an agricultural site, and none at an urban site. In addition, the season in which death occurred and time of day will influence activity and population numbers of blow flies (Doenier, 1940; Cruickshank and Wall, 2002; Schroeder *et al.*, 2003; Brundage *et al.*, 2011) and beetles (Kočárek, 2002) available for colonization.

Insects such as blow flies and necrophagous beetles are attracted by the odors that emanate from a decomposing corpse (see Section 11.2.1). Once at the corpse, blow fly activity is determined by size, smell, condition, and position of the carcass. Adult blow flies that have just emerged from their puparium use the corpse to obtain a protein meal needed to develop seminal fluid production or egg development. However, gravid female flies will use it as an oviposition site, laying eggs in the dark, moist areas of the body, such as the natural orifices (e.g., eyes, nose, mouth, anus, and genital regions) or wounds (Rodriguez and Bass, 1983; Carvalho *et al.*, 2000; Dix and Graham, 2000; Grassberger and Frank, 2004; Perez *et al.*, 2005). In some flies, such as *L. cuprina* (Wiedemann), tarsal contact with water increases the oviposition rate (Barton Browne, 1962). In addition, the odors released by bacteria, such as *P. aeruginosa*, commonly found in wounds and on human skin facilitate oviposition in this fly species (Watts *et al.*, 1981; Emmens and Murray, 1982). A female blowfly can deposit around 180 eggs in one batch and produces up to 2000–3000 in her lifetime (Erzinçlioğlu, 1996). Unlike Calliphoridae, necrophagous Coleoptera, such as Silphidae and Dermestidae, oviposit in the soil close to the corpse. Once their eggs have hatched, the larvae move onto the corpse to feed on remaining tissues.

The environmental conditions to which a corpse is exposed dictate how and when insect colonization occurs, if at all. Ambient temperature is recognized as one of the most important factors affecting blow fly activity and oviposition (MacLeod, 1949; George *et al.*, 2013b). It is generally accepted that colonization occurs between air temperatures of 12 and 30°C (Gennard, 2007). However, different species of blow fly will become active and oviposit within different temperature ranges (Doenier, 1940), and oviposition has been recorded below this lower threshold (Erzinclioglu, 1986; Faucherre *et al.*, 1999). Research has shown that it is not only the current temperature that is important for colonization, but the previous temperature levels can also influence whether oviposition occurs or not. For example, Berg and Benbow (2013) demonstrated that the abundance of diurnal blow fly oviposition increases when temperatures exceed 20°C the night before. Similarly, there is a significant relationship between temperature and colonization rate in necrophagous beetles, with species such as *Necrodes littoralis* (Linné) (Coleoptera: Silphidae) colonizing more quickly during warmer periods (Matuszewski and Szafałowicz, 2013).

Whereas increasing temperature increases the probability of blow fly oviposition occurring, other environmental conditions, such as increasing humidity, strong wind speeds, and heavy rainfall, have the opposite effect (George *et al.*, 2013a). For example, Mahat *et al.* (2009), working in Malaysia, found that heavy rain delayed colonization of rabbit carcasses by 1–2 days. In general, favorable conditions commonly reported for oviposition to occur are low humidity, light or no rainfall, high light levels, and a wind speed of less than 10 km per hour (Baldridge *et al.*, 2006; Zurawski *et al.*, 2009; Berg and Benbow, 2013; George *et al.*, 2013b). However, even when conditions are favorable, oviposition does not always occur. For example, some field studies have found that oviposition does not occur at night (Tessmer *et al.*, 1995; Anderson, 1999b; Stamper and Derby, 2007; Amendt *et al.*, 2008; Stamper *et al.*, 2009; Zurawski *et al.*, 2009; Berg and Benbow, 2013; Barnes *et al.*, 2015), whereas others have found clear evidence that it is possible (Greenberg, 1990; Singh and Bharti, 2001; Kirkpatrick, 2004; Baldridge *et al.*, 2006; Singh and Bharti, 2008; Pritam and Jayaprakash, 2009; George *et al.*, 2013a). Laboratory work conducted to investigate conditions facilitating nocturnal oviposition has also proved inconclusive. For example, Zurawski *et al.* (2009) demonstrated that *L. sericata* did not oviposit nocturnally and provided evidence that Diptera were unable to fly in the dark. However, Amendt *et al.* (2008) reported that nocturnal oviposition by *L. sericata* took place in two out of six studies.

The manner of disposal of the corpse can affect colonization rates if insect access is restricted. For example, delays are incurred for bodies left in parked vehicles (Voss *et al.*, 2008), inside buildings (Goff, 1991; Pohjoismäki *et al.*, 2010; Reibe and Madea, 2010; Anderson, 2011), and on wrapped bodies (Goff, 1992), but clothed bodies left on the ground surface are colonized simultaneously to unclothed ones (Kelly *et al.*, 2008; Voss *et al.*, 2011). Burial can delay insect colonization and also restrict the number of insect species able to access the body (Erzinçlioğlu, 1985; Turner and Wiltshire, 1999; VanLaerhoven and Anderson, 1999; Bourel *et al.*, 2004; Simmons *et al.*, 2010; Gunn and Bird, 2011; Balme *et al.*, 2012). Generally, Muscidae and Phoridae fly species are more commonly recorded than Calliphoridae under such circumstances (Erzinçlioğlu, 1985; Bourel *et al.*, 2004). However, the ability of insects to colonize buried bodies depends on the depth of the burial. Studies of shallow burials (30–35 cm) have demonstrated that *C. vomitoria, C. vicina*, and *L. illustris* Meigen are able to colonize animal carcasses at this depth (VanLaerhoven and Anderson, 1999; Simmons *et al.*, 2010) along with carrion beetles (VanLaerhoven and Anderson, 1999).

11.3.2 Development

Diptera and Coleoptera are holometabolous, passing through an egg, larval, puparial, and adult stage in their life cycle. The time spent in each stage of the life cycle is species dependent. However, insects are poikilothermic (i.e., dependent on external heat sources to maintain body temperature), and development rate in all species is dependent on the environmental temperature. Development rate increases with temperature within upper and lower thresholds for each species (Grassberger and Reiter, 2001; Richardson and Goff, 2001; Donovan *et al.*, 2006).

When conditions are unfavorable for long periods of time, such as winter in the Northern hemisphere, then blow flies go into diapause (arrested development in response to physiological change). Most blow flies, such as *Lucilia* species, overwinter in the third instar larval stage (Block *et al.*, 1990), but different species may overwinter in different life stages. For example, *Prot. terraenovae* is known to diapause in the adult stage after storing large quantities of food in its fat body (Erzinçlioğlu, 1996). Research has shown that development in different life stages continues at different lower thresholds, and it is the egg stage that is more cold resistant than other life stages in *C. vicina* (Block *et al.*, 1990; Davies and Ratcliffe, 1994). Not all species undergo obligate diapause (Block *et al.*, 1990), and different geographical populations of the same species can behave differently. For example, in Britain, northern populations of *C. vicina* undergo diapause, but southern populations overwinter in a quiescent state of the third instar larvae, pupa or adult, and not in diapause (Block *et al.*, 1990). Knowledge of this condition is important when applying development data to time since death estimations in forensic entomology. Geographical location dictates cold tolerance of a blow fly species. For example, in Europe, northern populations (Edinburgh, UK, and Nallikari, Finland) of blow fly are more cold tolerant than those residing in Barga, Italy (Saunders and Hayward, 1998).

Blow fly larvae go through three molts (instars) while feeding on the body tissues. During the second instar, larvae group together to form large larval masses and feed voraciously, releasing digestive proteases from their salivary glands and use a powerful pharyngeal pump to suck up the liquidized food material. This crowding also increases the temperature to which larvae are exposed and, therefore, their development rate. Studies have observed a 20–26°C difference between ambient and mass temperatures (Turner and Howard, 1992; Grassberger and Frank, 2004), which allows larvae to continue developing even when environmental temperatures become unfavorable (Doenier, 1940) or bodies are stored in a mortuary fridge (Huntington *et al.*, 2007). However, periods of 10 days or more in mortuary refrigeration at 4°C can lead to a disturbance in regular development and therefore time since death estimations (Myskowiak and Doums, 2002). Recently, Charabidze *et al.* (2011) found that heat emission from *L. sericata* depends on a complex mix of larval instar, food availability, weight of the larval mass, number of larvae in the mass, and local temperature, with an optimum ambient temperature between 22 and 25°C producing the maximal heat emission per larva.

The density of the larval mass varies depending on the manner of disposition of the corpse, the time of year, and size of local blow fly populations (Turner and Howard, 1992). Large masses cause overcrowding and increased competition for food resulting in undersized individuals and higher mortality rates (Smith and Wall, 1997). Larger larval densities also increase development rates in blow flies such as *C. vomitoria* (Ireland and Turner, 2006) and *Chrysomya* species (Goodbrod and Goff, 1990). Necrophagous beetles, such as *D. maculatus*, on the other hand, do not aggregate and in fact have higher survival rates when there is a lower density of larvae (Richardson and Goff, 2001).

Temperatures can also differ at different parts of the body (Doenier, 1940), which will influence the development rate of the larvae colonizing each area. However, the type of tissue and, therefore, part of the corpse that blow fly larvae feed on also influences how fast they develop. Studies to date have shown that blow fly larvae of *C. vicina* (Kaneshrajah and Turner, 2004), *L. sericata* (Clark *et al.*, 2006; El-Moaty and Kheirallah, 2013), *L. cuprina*, and *Calliphora augur* Fabricius (Diptera: Calliphoridae) (Day and Wallman, 2006a) that fed on lungs and/or brain developed at a significantly faster rate compared to larvae fed on other organs, such as liver. In addition, the type of animal tissue influences development rates (Flores *et al.*, 2014).

Temperature is the most significant factor affecting development; however, other factors also influence growth. For example, there is evidence to suggest that geographic region influences development rates with significant regional variation having been reported in geographically distinct populations of *L. sericata* (Gallagher *et al.*, 2010) and *Co. macellaria* (Owings *et al.*, 2014).

The condition of the body may also affect development rate. Although, freezing and thawing has no significant effect on development rates in *C. augur* (Day and Wallman, 2006b), Richardson *et al.* (2013) found that the development rate of *C. vicina* reared on decomposed liver was significantly slower than those reared on fresh liver. This observation may be due to the bacterial species present on decomposed compared to fresh liver. Previous work has demonstrated that *Staphylococcus aureus* has no effect on survival of *L. sericata* from egg to adult eclosion (Barnes and Gennard, 2011). However, recent research suggests that ingesting different types of bacteria affects the development of *C. vicina, C. vomitoria*, and *L. sericata* (Crooks *et al.*, 2016).

Necrophagous insects ingest body tissues and, therefore, any controlled substances they may contain. A subdiscipline of forensic entomology, entomotoxicology, researches the effect of different toxins on insect development and their implications for entomologically based time since death estimations (Gosselin *et al.*, 2011). Studies have shown that drugs such as methamphetamine (Mullany *et al.*, 2014) and cocaine accelerate larval growth (De Carvalho *et al.*, 2012), whereas drugs such as morphine slow development (Bourel *et al.*, 1999). However, these results seem to be species specific as results from Bourel *et al.* (1999) using *L. sericata* differed from George *et al.* (2009), which found no effect of morphine on the development rate of *Calliphora stygia* (Fabricius) (Diptera: Calliphoridae). The different methodologies, such as drug level and substrate type, may also contribute to these differences. While some studies have investigated the drug levels detected in necrophagous beetles (Miller *et al.*, 1994; Bourel *et al.*, 2001), no work has been conducted on the effect of controlled substances or toxins on development rates.

11.3.3 Succession

Decomposition is a continuous process, but it is commonly classified into five major stages—fresh, bloat, active decay, advanced decay, and skeletal remains—based on the state of physical decomposition (Matuszewski *et al.*, 2008). Decomposition is driven by temperature and the presence of decomposers such as insects and microbes. There is limited opportunity for research utilizing human corpses, so most forensic studies use swine carcasses as they are considered a good substitute in research investigations (Hewadikaram and Goff, 1991, 1992; Schoenly *et al.*, 2007; Anton *et al.*, 2011) (See also Chapter 3.)

Insects are attracted to a corpse from the moment of death. Different families arrive at different stages of decomposition in response to the odors released as decay progresses (Perez *et al.*, 2005). This insect succession has been investigated in many regions of the world including mainland Europe, Australasia including China, Africa, and North and South America (Bornemissza, 1957; Davis and Goff, 2000; Bharti and Singh, 2003; Arnaldos *et al.*, 2004; Grassberger and Frank, 2004; Chin *et al.*, 2007; Vitta *et al.*, 2007; Eberhardt and Elliot, 2008; Wang *et al.*, 2008; Gomes *et al.*, 2009; Michaud and Moreau, 2009; Segura *et al.*, 2009; Horenstein *et al.*, 2010; Valdes-Perezgasga *et al.*, 2010; Dekeirsschieter *et al.*, 2011; Matuszewski *et al.*, 2011; Zaidi and Chen, 2011; Al-Mesbah *et al.*, 2012; Ortloff *et al.*, 2012; Bygarski and LeBlanc, 2013). In general, these studies have found that fly species dominate the initial stages of decomposition, whereas Coleoptera dominate the later stages. Although insect species colonizing the body vary according to season, geographical location, microclimate, and environmental factors such as temperature, humidity, time of day, and rainfall, the faunal succession of insect families follows a distinct pattern (Walker, 1957; Greenberg, 1990; Campobasso *et al.*, 2001; Koĉárek, 2002; Archer and Elgar, 2003; Arnaldos *et al.*, 2004).

In contrast, limited research has been conducted to assess the microbes colonizing the corpse environment (Howard *et al.*, 2010; Dickson *et al.*, 2011; Moreno, 2011; Hyde *et al.*, 2013; Metcalf *et al.*, 2013; Hyde *et al.*, 2014; Pechal *et al.*, 2014b; Carter *et al.*, 2015). The microbial communities colonizing a corpse vary according to season (Dickson *et al.*, 2011; Carter *et al.*, 2015); environmental conditions, such as temperature and humidity (Carter *et al.*, 2008); and location on the body (Metcalf *et al.*, 2013). However, results to date indicate that, similar to insect succession, the microbial communities change in a predictable order as decomposition progresses. These bacteria originate from the human body, in particular, the intestines, mouth, and skin, but are also introduced from the environment, such as the soil in which the corpse is located, visiting scavengers, and insect colonizers (Moreno, 2011; Hyde *et al.*, 2013). The initial microbiome on each individual at death can vary according to the age of the victim (Sommerville, 1969), their diet, and their state of health (Moreno, 2011). However, Hyde *et al.* (2014) found that despite variation in the initial bacterial community richness and diversity for two human corpses, the microbial communities became more similar as decomposition progressed (see also Chapter 10).

The fresh stage of decomposition begins at the time of death and ends when the body begins to swell (Perez *et al.*, 2005). During the fresh stage, human cells die and are mainly broken down by autolytic enzymes (e.g., lipases, proteases, and amylases) in a process called "autolysis." Organs with high enzyme content, such as the liver and brain, undergo the most rapid decomposition (Vass *et al.*, 2002). Eventually, sufficient cells are broken down into a "soup" of nutrients on which bacteria released from human organs, such as the large intestine, are supported (Gennard, 2007).

The Calliphoridae are the initial colonizers of a human body (Rodriguez and Bass, 1983; Greenberg, 1991) arriving during the beginning stages of decomposition to oviposit in natural orifices and wound sites. Emerging blow fly larvae are thus exposed to the bacteria colonizing these areas. For example, *S. aureus* is commonly found in the nasal cavity (Bexfield *et al.*, 2008) and in wounds (Bowler *et al.*, 1999), and *Pseudomonas* dominates the bacterial communities in the human mouth and on the skin during the initial stages of decomposition (Hyde *et al.*, 2014). *Lucilia sericata* and *C. vicina* larvae produce antimicrobial activity that inhibits growth of bacteria, such as *P. aeruginosa* and *S. aureus* which may be an evolutionary strategy that provides a competitive advantage for nutritional resources in the corpse environment (Barnes *et al.*, 2010).

Sarcophagidae colonize a human corpse during the bloat stage (Mégnin, 1894; Bornemissza, 1957; Reed, 1958; Rodriguez and Bass, 1983; Campobasso *et al.*, 2001; Carvalho and Linhares, 2001) when anaerobic fermentation in the gut results in the release of butanoic and propionic acids (Vass *et al.*, 2002). Members of the Muscidae and Fannidae along with predaceous beetle species, such as Staphylinidae and Histeridae, which prey on fly larvae, are also attracted to the bloat stage of decomposition (Bornemissza, 1957; Reed, 1958; Payne, 1965; Rodriguez and Bass, 1983; Gennard, 2007). During this stage, bacteria associated with flies, such as *Ignatzschineria* and *Wohlfahrtiimonas*, are found on the skin (Hyde *et al.*, 2014), and anaerobic bacteria that are known to be common members of the gut community such as *Lactobacillus* and *Bacteroides* increase in the abdominal cavity (Metcalf *et al.*, 2013). Gut bacteria use the circulatory system as conduit to move around the corpse with peak bacterial cell density occurring 24–30 hours post mortem (Rose and Hockett, 1971). Using a mouse model, Melvin *et al.* (1984) demonstrated that *Staphylococcus* species were the first microorganisms to migrate from the small intestine, followed by coliform bacteria, then anaerobic bacteria. Waste gases from bacterial digestion of body tissues (putrefaction); for example, hydrogen sulfide, sulfur dioxide, carbon dioxide, methane, and ammonia accumulate and cause the body to appear swollen (Vass *et al.*, 2002; Gennard, 2007). Blood begins to break down (hemolyse), and the outer layers of skin start to slip from the body (Dix and Graham, 2000). Putrefaction by-products and bacteria also leach from orifices or wound sites along with electrolyte salts such as sodium and potassium (Sachs, 2001; Vass *et al.*, 2002). Waste gases force the skin open and escape marking the beginning of active decay.

The active stage of decay is characterized by aerobic protein decomposition (Rodriguez and Bass, 1983; Koĉárek, 2002). Proteins are broken down into fatty acids such as pyruvic and butanoic acid and amino acids, which are further reduced to amines such as putrescine and cadaverine (Bonte *et al.*, 1976). Body fat is also digested into fatty acids and glycerols, acetone, phosphates, skatole, and acetic acid (Sachs, 2001). With the rupturing of the body cavity comes a shift from an anaerobic to aerobic environment in the abdominal cavity and the colonization of skin sites by anaerobic gut bacteria. For example, Firmicutes, which are one of the most abundant phyla in feces during fresh and bloat stages, increase in abundance on human skin during the later phases of decomposition (Hyde *et al.*, 2013). After rupture occurs, the bacteria driving bloat stage decrease dramatically, and aerobes, such as *Pseudochrobactrum* and *Ochrobactrum*, dominate. In addition, facultative anaerobes, such as *Serratia*, *Escherichia*, *Klebsiella*, and *Proteus*, become abundant (Metcalf *et al.*, 2013).

Piophilidae, Phoridae, and Sphaeroceridae flies are attracted to a corpse during active decay (Bornemissza, 1957; Campobasso *et al.*, 2001), and most beetle species from the families Histeridae, Silphidae, Staphylinidae, Dermestidae, and Carabidae are associated with this stage (Reed, 1958; Payne, 1965; Rodriguez and Bass, 1983; Grassberger and Frank, 2004; Gennard, 2007). Many of these beetles remain on the corpse throughout advanced decay and into the skeletal stages. Research has shown that *D. maculatus* produce antibacterial activity in their larval stages which is more potent against *Pr. mirabilis* than bacteria such as *Staphylococcus* against which the blow flies are more active (Barnes *et al.*, 2010). These differences may be explained ecologically by the different colonization times of each insect species on the corpse and give them a competitive advantage over the different microbial community present at this stage of decay.

By the onset of advanced decay, the majority of soft tissue has been removed. Therefore, this stage is commonly recognized as the beginning of the drier stages of decomposition. Several bacterial families (Sphingobacteriaceae, Brucellaceae, Phyllobacteriaceae, Hyphomicrobiaceae, and Alcaligenaceae) increase in abundance on the skin and in the soil under a carcass during this stage (Metcalf *et al.*, 2013). The presence of environmental species suggests that the bacteria found on the skin in earlier stages have been replaced by soil bacteria, and the microbial communities in this stage reflect the surrounding environment with which they are in contact rather than the indigenous microbiota of the human body. Dipteran larvae and associated predators migrate from the corpse and beetles from the families Nitidulidae, Cleridae, and Dermestidae begin or continue to colonize the remains (Bornemissza, 1957; Reed, 1958; Payne, 1965).

Remaining body tissue, such as skin and cartilage, is broken down; and once removed completely, decomposition enters the skeletonization stage. During this stage, the skeletal bone decomposes (diagenesis) and is degraded by environmental conditions until it is reduced to minerals such as calcium and magnesium that leach into the surrounding soil (Sachs, 2001; Gennard, 2007). Associated with bones and dried remains are beetles from the family Tenebrionidae, such as *Tenebrio molitor* L. (Coleoptera: Tenebrionidae) (Mégnin, 1894; Gennard, 2007), although members of this family are sometimes found feeding on the corpse and its fauna during the earlier stages of decomposition (Arnaldos *et al.*, 2005). Environmental bacteria, such as *Acinetobacter*, which is associated with soil, are still common at most body sites (Hyde *et al.*, 2014; Damann *et al.*, 2015).

11.3.4 Dispersal

Time of larval insect dispersal from a body is controlled by circadian rhythm (Richards *et al.*, 1986) and appears to coincide with the period of least beetle activity during midnight and 8 a.m., thus minimizing the risk of encountering predators (Kočárek, 2001, 2002). Larvae dispersing at outdoor locations travel a few meters over the surface of the soil and pupate in the soil, whereas larvae dispersing from indoor locations may travel further to find a suitable pupation site, such as under a carpet or skirting board. Therefore, it is recommended that forensic entomologists collecting specimens from crime scenes search a radius of between 2 and 10 m from the corpse for postfeeding larvae and puparia (Amendt *et al.*, 2007). However, Lewis and Benbow (2011) reported entire masses of larvae that moved greater than 20 m away from decomposing remains that they had developed on.

There does not appear to be a preference in directional movement for larval dispersal; however, distance traveled and depth of burial vary according to species and size of individuals (Gomes *et al.*, 2005). For example, smaller and lighter larvae travel a further distance for a suitable pupation site so they do not compete with the larger larvae that tend to pupate closer and deeper to the corpse. In addition, the farther an individual travels in the postfeeding stage, the smaller the resulting adult fly, probably due to increased energy expenditure (Mai and Amendt, 2012). The season in which pupation takes place also affects burial depths with larvae burying deeper during periods of adverse weather conditions, such as winter in the Northern Hemisphere and summer in the Southern Hemisphere. The length of the pupation stage is dependent on species and temperature. Adult flies emerge and leave puparial cases behind. These have some forensic relevance in aiding identification of any controlled substances in the corpse (Miller *et al.*, 1994).

Adult blow flies disperse from pupation sites flying up to 20 km in 24 hours (Bagshaw *et al.*, 2014) and taking with them some of the bacteria they acquired while feeding on a decomposing carcass (Pechal and Benbow, 2016) and from the soil during pupation. In this way, they potentially carry these bacteria a considerable distance in search of food and oviposition sites. Previous work has demonstrated that blow flies carry bacterial species, such as *Aeromonas hydrophila, Edwardsiella tarda, Vibrio cholera, Aeromonas sobria, Citrobacter freundii, Escherichia coli, Providencia alcalifaciens, P. aeruginosa, S. aureus, Streptococcus pyogenes,* and *Streptococcus pneumoniae* (Sukontason *et al.*, 2000; Sukontason *et al.*, 2007; Aigbodion *et al.*, 2013; Wei *et al.*, 2014; Singh *et al.*, 2015). There is also evidence that microbial community structures are dependent on season. For example, Wei *et al.* (2014) found that the predominant genera associated with blow flies were found to be *Staphylococcus* in spring, *Ignatzschineria* in summer, and *Vagococcus, Dysgonomonas,* and an unclassified Acetobacteraceae in autumn.

Flies may carry bacteria on their body surfaces or in their saliva, feces, and internal organs. In laboratory experiments, Fischer *et al.* (2004) isolated *Mycobacterium avium paratuberculosis* from the abdomen, head, thorax, and wings of 15 experimentally infected adult *C. vicina* but not from the legs. However, Sukontason *et al.* (2006) indicated that the ultrastructure of adhesive devices on the feet were potential attachment sites for microorganisms. Dissection of blow flies collected during field work in Japan determined H5 influenza A virus was in the intestinal organs, crop, and gut of *Calliphora nigribarbis,* Vollenhoven, and *Aldrichina graham* (Aldrich) (Sawabe *et al.*, 2006).

Wei *et al.* (2013) found there was a clear change in the bacterial communities associated with three life stages (larvae, pupae, and adult) of the house fly. Overall, there was a change from microorganisms with a lower guanine–cytosine (GC) content to those with a higher GC content in the 16S ribosomal RNA genes. Proteobacteria and Bacteroidetes were detected in samples from all three life stages indicating that some species do survive development, while there were shifts in the other predominant genera through the life cycle; *Koukoulia* and *Schineria* [*Ignatzschineria*] were detected in larvae, *Neisseria* in pupae and *Macrococcus, Lactococcus,* and *Kurthia* were detected in adult flies. Incidence of Gram-positive strains increased as the house fly developed.

There is evidence to suggest that the bacteria carried by flies are the same strains as those found in animal carcasses (Sawabe *et al.*, 2006). However, there is currently limited information available regarding from where the natural bacterial communities associated with adult Calliphoridae originate. Bromel *et al.* (1983) reported a reduction in bacterial abundance from larvae through to adult flies of the New World screwworm fly, *Co. hominivorax*. The normal flora of newly emerged flies contained bacterial species similar to that of the larval stages, including *Proteus* species and coliforms. In the larval and pupal stages, members of the *Proteus–Providencia* group were predominant, while in the adult stage, the enteric bacteria occurred more frequently. In an attempt to understand where the bacteria originated, the larval growth environment represented by wounds and rearing media was tested, and results indicated that the microbiota originate from larval food sources. Pechal and Benbow (2016) have shown that when using high-throughput amplicon-based sequencing, the internal microbial communities of blow fly adults and larvae share microbial taxa with the epinecrotic communities of the decomposing salmon carcasses. These data suggest that both larvae and adults of blow fly may acquire microbes from a decomposing carcass with the potential of dispersing these communities into the surrounding environment.

Singh *et al.* (2015) suggest that the majority of bacteria associated with *L. sericata* and *L. cuprina* are within the phyla Proteobacteria, Firmicutes, and Bacteroidetes and are acquired from the environment. Bacterial communities associated with the egg stage were proportionally lower than other life stages and contained bacteria different from those of the maternal flies, such as high numbers of Acidobacteria and Actinobacteria. These bacteria may provide protection from harmful bacteria and fungi by producing antimicrobial substances.

11.4 Future directions and conclusion

The bridge between microbiology and entomology extends back well over one century. The following discussion will serve to highlight a few applications that have been produced from these cross-discipline interactions in research that are having a major impact on society including the forensic sciences.

11.4.1 Forensic sciences

Over the course of the past 5 years, a number of publications have come out that highlight the diversity of microbes, more specifically bacteria and archaea, associated with decomposing vertebrate remains. These studies have demonstrated that microbial data associated with decomposing remains could be used for determining the time of death of an individual. From an entomological perspective, these studies are demonstrating that bacteria are serving as a key mechanism regulating arthropod attraction, colonization, and decomposition of vertebrate remains—all aspects that are key for determining critical information about a descendent associated with a crime. From the perspective of the chapter, understanding the precolonization interval (Tomberlin *et al.*, 2011), or preappearance interval, PAI (Matuszewski, 2011; Matuszewski, 2012), is critical for understanding the full amount of time to pass from death until discovery of the deceased and applying entomological data to make such estimates. However, like any field in its infancy, more research is needed to understand the variability associated with microbe–insect interactions as related to the true PMI. How does microbiome variability across individuals influence arthropod attraction and colonization? Can this variability be quantified? If so, can methods produced from these data be validated with regard to estimating the time of death of a person? These represent only a few questions related to the forensic sciences and microbe–insect interactions. However, as one can see, this is an exciting time for such research.

11.4.2 Environmental sciences

Much along the same vein as the forensic sciences, microbe–insect interaction research is becoming relevant for fields such as the environmental sciences. This application relates to a number of different venues including, but not limited to, environmental health and pathogen movement from confined animal facilities into urban areas.

11.4.3 Medical research

Data generated from decomposition studies, in the field or laboratory, are now generating exciting pathways of research leading to novel nonantibiotic approaches for suppressing pathogens associated with wounds. This relationship between microbes that are beneficial to arthropod larvae feeding on necrotic tissue and their pathogens in the wound has resulted in unique compounds associated with the insect, and these beneficial microbes, being identified, synthesized, and applied in wound care facilities today.

Acknowledgments

The authors would like to thank the anonymous reviews provided on earlier drafts of the chapter. JKT and MEB would also like to recognize the National Institute of Justice (2010-DN-BX-K243) and HRJ and MEB (2014-DN-BX-K008) for providing financial support leading toward the preparation of the chapter. Points of view in the chapter are those of the authors and do not necessarily represent the official position or policies of the U.S. Department of Justice.

References

Aak, A. and Knudsen, G.K. (2011) Sex differences in olfaction-mediated visual acuity in blowflies and its consequences for gender-specific trapping. *Entomologia Experimentalis et Applicata*, **139**, 25–34.

Aigbodion, F.I., Egbon, I.N. and Obuseli, A.J. (2013) Pathogens of medical importance isolated from *Phaenicia (Lucilia) sericata* (Diptera: Calliphoridae) in Benin City, Nigeria. *Journal of Biological Sciences*, **16**, 1791–1795.

Al-Mesbah, H., Moffatt, C., El-Azazy, O.M.E. and Majeed, Q.A.H. (2012) The decomposition of rabbit carcasses and associated necrophagous Diptera in Kuwait. *Forensic Science International*, **217**, 27–31.

Amendt, J., Campobasso, C.P., Gaudry, E. *et al.* (2007) Best practice in forensic entomology—standards and guidelines. *International Journal of Legal Medicine*, **121**, 90–104.

Amendt, J., Zehner, R. and Reckel, F. (2008) The nocturnal oviposition behavior of blowflies (Diptera: Calliphoridae) in Central Europe and its forensic implications. *Forensic Science International*, **175**, 61–64.

Anderson, G.S. (1995) The use of insects in death investigations: an analysis of cases in British Columbia over a five year period. *Canadian Society of Forensic Science Journal*, **28**, 277–292.

Anderson, G.S. (1999a) Insect succession on carrion and its relationship to determining time of death, in *The Utility of Arthropods in Legal Investigations* (ed J.H. Byrd), CRC Press, Boca Raton, FL.

Anderson, G.S. (1999b) Wildlife forensic entomology: determining time of death in two illegally killed black bear cubs. *Journal of Forensic Sciences*, **44**, 856–859.

Anderson, G.S. (2004) Determining time of death using blow fly eggs in the early postmortem interval. *International Journal of Legal Medicine*, **118**, 240–241.

Anderson, G.S. (2011) Comparison of decomposition rates and faunal colonisation of carrion in indoor and outdoor environments. *Journal of Forensic Sciences*, **56**, 136–142.

Anton, E., Niederegger, S. and Beutel, R.G. (2011) Beetles and flies collected on pig carrion in an experimental setting in Thuringia and their forensic implications. *Medical and Veterinary Entomology*, **25**, 353–364.

Archer, M.S. and Elgar, M.A. (2003) Yearly activity patterns in southern Victoria (Australia) of seasonally active carrion insects. *Forensic Science International*, **132**, 173–176.

Arnaldos, M.I., Romera, E., Presa, J.J. *et al.* (2004) Studies on seasonal arthropod succession on carrion in the southeastern Iberian Peninsula. *International Journal of Legal Medicine*, **118**, 197–205.

Arnaldos, M.I., Garcia, M.D., Romera, E. *et al.* (2005) Estimation of postmortem interval in real cases based on experimentally obtained entomological evidence. *Forensic Science International*, **149**, 57–65.

Bagshaw, C., Isdell, A.E., Thiruvaiyaru, D.S. *et al.* (2014) Molecular detection of canine parvovirus in flies (Diptera) at open and closed canine facilities in the eastern United States. *Preventive Veterinary Medicine*, **114**, 276–284.

Baldridge, R.S., Wallace, S.G. and Kirkpatrick, R. (2006) Investigation of nocturnal oviposition by necrophilous flies in central Texas. *Journal of Forensic Sciences*, **51**, 125–126.

Balme, G.R., Denning, S.S., Cammack, J.A. and Watson, D.W. (2012) Blow flies (Diptera: Calliphoridae) survive burial: evidence of ascending vertical dispersal. *Forensic Science International*, **216**, e1–e4.

Barnes, K.M. and Gennard, D.E. (2011) The effect of bacterially-dense environments on the development and immune defences of the blowfly *Lucilia sericata*. *Physiological Entomology*, **36**, 96–100.

Barnes, K.M., Gennard, D.E. and Dixon, R.A. (2010) An assessment of the antibacterial activity in larval excretion/secretion of four species of insects recorded in association with corpses, using *Lucilia sericata* Meigen as the marker species. *Bulletin of Entomological Research*, **100**, 635–640.

Barnes, K.M., Grace, K.A. and Bulling, M.T. (2015) Nocturnal oviposition behaviour of forensically-important Diptera in Central England. *Journal of Forensic Sciences*, **60** (6), 1601–1604.

Barton Browne, L. (1962) The relationship between oviposition in the blowfly *Lucilia cuprina* and the presence of water. *Journal of Insect Physiology*, **8**, 383–390.

Bell, W.J. (1990) Searching behavior patterns in insects. *Annual Review of Entomology*, **35**, 447–467.

Benbow, M.E., Lewis, A.J., Tomberlin, J.K. and Pechal, J.L. (2013) Seasonal necrophagous insect community assembly during vertebrate carrion decomposition. *Journal of Medical Entomology*, **50**, 440–450.

Benbow, M.E., Tomberlin, J.K. and Tarone, A.M. (eds) (2015) *Carrion Ecology, Evolution, and Their Applications*, CRC Press, Boca Raton, FL.

Berg, M.C. and Benbow, M.E. (2013) Environmental factors associated with *Phormia regina* (Diptera: Calliphoridae) oviposition. *Journal of Medical Entomology*, **50**, 451–457.

Bexfield, A., Bond, A.E., Roberts, E.C. *et al.* (2008) The antibacterial activity against MRSA strains and other bacteria of a <500 Da fraction from maggot excretions/secretions of *Lucilia sericata* (Diptera: Calliphoridae). *Microbes and Infection*, **10**, 325–333.

Bharti, M. and Singh, D. (2003) Insect faunal succession on decaying rabbit carcasses in Punjab, India. *Journal of Forensic Sciences*, **48**, 1–11.

Block, W., Erzinçlioğlu, Z. and Worland, M.R. (1990) Cold resistance in all life stages of two blowfly species (Diptera: Calliphoridae). *Medical and Veterinary Entomology*, **4**, 213–219.

Boissière, A., Tchioffo, M.T., Bachar, D. *et al.* (2012) Midgut microbiota of the malaria mosquito vector *Anopheles gambiae* and interactions with *Plasmodium falciparum* infection. *PLoS Pathogens*, **8**, e1002742.

Bonte, W., Bleifuss, J. and Volck, J. (1976) Experimental investigations in post-mortem protein degradation. *Forensic Science*, **7**, 9–22.

Bornemissza, G.F. (1957) An analysis of arthropod succession in carrion and the effect of its decomposition on the soil fauna. *Australian Journal of Zoology*, **5**, 1–12.

Bourel, B., Hedouin, V., Martin-Bouyer, L. *et al.* (1999) Effects of morphine in decomposing bodies on the development of *Lucilia sericata* (Diptera: Calliphoridae). *Journal of Forensic Sciences*, **44**, 354–358.

Bourel, B., Tournel, G., Hedouin, V. *et al.* (2001) Morphine extraction in necrophagous insects remains for determining ante-mortem opiate intoxication. *Forensic Science International*, **120**, 127–131.

Bourel, B., Tournel, G., Hédouin, V. and Gosset, D. (2004) Entomofauna of buried bodies in northern France. *International Journal of Legal Medicine*, **118**, 215–220.

Bouvy, M., Bettarel, Y., Bouvier, C. *et al.* (2011) Trophic interactions between viruses, bacteria and nanoflagellates under various nutrient conditions and simulated climate change. *Environmental Microbiology*, **13**, 1842–1857.

Bowler, P., Davies, B. and Jones, S. (1999) Microbial involvement in chronic wound malodour. *Journal of Wound Care*, **8**, 216–218.

Bromel, M., Duh, F.M. and Erdmann, G.R. (1983) Bacteria associated with the screwworm fly [*Cochliomyia hominivorax* (Coquerel)] and their metabolites. *Endocytobiology*, **2**, 791–800.

Brundage, A., Bros, S. and Honda, J.Y. (2011) Seasonal and habitat abundance and distribution of some forensically important blow flies (Diptera: Calliphoridae) in Central California. *Forensic Science International*, **212**, 115–120.

Burkepile, D.E., Parker, J.D., Woodson, C.B. *et al.* (2006) Chemically mediated competition between microbes and animals: microbes as consumers in food webs. *Ecology*, **87**, 2821–2831.

Bygarski, K. and LeBlanc, H.N. (2013) Decomposition and arthropod succession in Whitehorse, Yukon Territory, Canada. *Journal of Forensic Sciences*, **58**, 413–418.

Byrd, J. and Castner, J. (eds) (2009) *Forensic Entomology: The Utility of Arthropods in Legal Investigations*, CRC Press, Boca Raton, FL.

del Campo, J., Not, F., Forn, I. *et al.* (2013) Taming the smallest predators of the oceans. *ISME Journal*, **7**, 351–358.

Campobasso, C.P., Di Vella, G. and Introna, F. (2001) Factors affecting decomposition and Diptera colonization. *Forensic Science International*, **120**, 18–27.

Can, I., Javan, G.T., Pozhitkov, A.E. and Noble, P.A. (2014) Distinctive thanatomicrobiome signatures found in the blood and internal organs of humans. *Journal of Microbiological Methods*, **106**, 1–7.

Carter, D.O., Yellowlees, D. and Tibbett, M. (2008) Temperature affects microbial decomposition of cadavers (*Rattus rattus*) in contrasting soils. *Applied Soil Ecology*, **40**, 129–137.

Carter, D.O., Metcalf, J.A., Bibat, A. and Knight, R. (2015) Seasonal variation of postmortem microbial communities. *Forensic Science, Medicine, and Pathology*, **11** (2), 202–207.

Carvalho, L.M.L. and Linhares, A.X. (2001) Seasonality of insect succession and pig carcass decomposition in a natural forest area in southeastern Brazil. *Journal of Forensic Sciences*, **46**, 604–608.

Carvalho, L.M.L., Thyssen, P.J., Linhares, A.X. and Palhares, F.A.B. (2000) A checklist of arthropods associated with pig carrion and human corpses in Southeastern Brazil. *Memórias do Instituto Oswaldo Cruz*, **95**, 135–138.

Catts, E.P. and Goff, M.L. (1992) Forensic entomology in criminal investigations. *Annual Review of Entomology*, **37**, 253–272.

Charabidze, D., Bourel, B. and Gosset, D. (2011) Larval-mass effect: characterisation of heat emission by necrophagous blowflies (Diptera: Calliphoridae) larval aggregates. *Forensic Science International*, **211**, 61–66.

Chaudhury, M.F. and Skoda, S.R. (2013) An artificial diet for rearing *Cochliomyia macellaria* (Diptera: Calliphoridae). *Journal of Economic Entomology*, **106**, 1927–1931.

Chaudhury, M.F., Welch, J.B. and Alvarez, L.A. (2002) Responses of fertile and sterile screwworm (Diptera: Calliphoridae) flies to bovine blood inoculated with bacteria originating from screwworm-infested animal wounds. *Journal of Medical Entomology*, **39**, 130–134.

Chaudhury, M.F., Skoda, S.R., Sagel, A. and Welch, J.B. (2010) Volatiles emitted from eight wound-isolated bacteria differentially attract gravid screwworms (Diptera: Calliphoridae) to oviposit. *Journal of Medical Entomology*, **47**, 349–354.

Chaudhury, M.F., Zhu, J.J., Sagel, A. *et al.* (2014) Volatiles from waste larval rearing media attract gravid screwworm flies (Diptera: Calliphoridae) to oviposit. *Journal of Medical Entomology*, **51**, 591–595.

Chin, H.C., Marwi, M.A., Salleh, A.F.M. *et al.* (2007) A preliminary study of insect succession on a pig carcass in a palm oil plantation in Malaysia. *Tropical Biomedicine*, **24**, 23–27.

Clark, K., Evans, L. and Wall, R. (2006) Growth rates of the blowfly, *Lucilia sericata*, on different body tissues. *Forensic Science International*, **156**, 145–149.

Crooks, E.R., Bulling, M.T. and Barnes, K.M. (2016) Microbial effects on the development of forensically important blow fly species. *Forensic Science International*, **266**, 185–190.

Crowley, P.H., Straley, S.C., Craig, R.J. *et al.* (1980) A model of prey bacteria, predator bacteria, and bacteriophage in continuous culture. *Journal of Theoretical Biology*, **86**, 377–400.

Cruickshank, I. and Wall, R. (2002) Population dynamics of the sheep blowfly *Lucilia sericata*: seasonal patterns and implications for control. *Journal of Applied Ecology*, **39**, 493–501.

Damann, F.E., Williams, D.E. and Layton, A.C. (2015) Potential use of bacterial community succession in decaying human bone for estimating postmortem interval. *Journal of Forensic Sciences*, **60** (4), 844–850.

Davies, L. and Ratcliffe, G.G. (1994) Development rates of some pre-adult stages in blowflies with reference to low temperatures. *Medical and Veterinary Entomology*, **8**, 245–254.

Davis, J.B. and Goff, M.L. (2000) Decomposition patterns in terrestrial and intertidal habitats on Oahu Island and Coconut Island, Hawaii. *Journal of Forensic Sciences*, **45**, 836–842.

Davis, T.S., Crippen, T.L., Hofstetter, R.W. and Tomberlin, J.K. (2013) Microbial volatile emissions as insect semiochemicals. *Journal of Chemical Ecology*, **37** (9), 840–859.

Day, D.M. and Wallman, J.F. (2006a) Influence of substrate tissue type on larval growth in *Calliphora augur* and *Lucilia cuprina* (Diptera: Calliphoridae). *Journal of Forensic Sciences*, **51**, 657–663.

Day, D.M. and Wallman, J.F. (2006b) A comparison of frozen/thawed and fresh food substrates in development of *Calliphora augur* (Diptera: Calliphoridae) larvae. *International Journal of Legal Medicine*, **120**, 391–394.

De Carvalho, L.M.L., Linhares, A.X. and Palhares, F.A.B. (2012) The effect of cocaine on the development rate of immatures and adults of *Chrysomya albiceps* and *Chrysomya putoria* (Diptera: Calliphoridae) and its importance to postmortem interval estimate. *Forensic Science International*, **220**, 27–32.

Dekeirsschieter, J., Verheggen, F.J., Gohy, M. *et al.* (2009) Cadaveric volatile organic compounds released by decaying pig carcasses (*Sus domesticus* L.) in different biotopes. *Forensic Science International*, **189**, 46–53.

Dekeirsschieter, J., Verheggen, F.J., Haubruge, E. and Brostaux, E. (2011) Carrion beetles visiting pig carcasses during early spring in urban, forest and agricultural biotopes of Western Europe. *Journal of Insect Science*, **11**, 73.

Dethier, V.G. (1954) Evolution of feeding preferences in phytophagous insects. *Evolution*, **8**, 33–54.

Dickson, G.C., Poulter, R.T.M., Maas, E.W. *et al.* (2011) Marine bacterial succession as a potential indicator of postmortem submersion interval. *Forensic Science International*, **209**, 1–10.

Dix, J. and Graham, M. (2000) *Time of Death, Decomposition and Identification: An Atlas*, CRC Press, Boca Raton, FL.

Doenier, C.C. (1940) Carcass temperature and their relationship to winter blowfly populations and activity in the Southwest. *Journal of Economic Entomology*, **33**, 166–170.

Donovan, S.E., Hall, M.J., Turner, B.D. and Moncrieff, C.B. (2006) Larval growth rates of the blowfly, *Calliphora vicina*, over a range of temperatures. *Medical and Veterinary Entomology*, **20**, 106–114.

Eberhardt, T.L. and Elliot, D.A. (2008) A preliminary investigation of insect colonisation and succession on remains in New Zealand. *Forensic Science International*, **176**, 217–223.

El-Moaty, Z. and Kheirallah, A.E. (2013) Developmental variation of the blow fly *Lucilia sericata* (Meigen, 1826) (Diptera: Calliphoridae) by different substrate tissue types. *Journal of Asia-Pacific Entomology*, **16**, 297–300.

Emmens, R.L. and Murray, M.D. (1982) The role of bacterial odours in oviposition by *Lucilia cuprina* (Wiedemann) (Diptera: Calliphoridae), the Australian sheep blowfly. *Bulletin of Entomological Research*, **72**, 367–375.

Erickson, M.C., Islam, M., Sheppard, C. *et al.* (2004) Reduction of *Escherichia coli* O157:H7 and *Salmonella enterica* Serovar enteritidis in chicken manure by larvae of the black soldier fly. *Journal of Food Protection*, **67**, 685–690.

Erzinçlioğlu, Z. (1985) The entomological investigation of a concealed corpse. *Medicine, Science, and the Law*, **25**, 228–230.

Erzinclioglu, Y.Z. (1986) Areas of research in forensic entomology. *Medicine, Science, and the Law*, **26**, 273–278.

Erzinçlioğlu, Z. (1996) *Blowflies, Naturalists' Handbooks 23*, The Richmond Publishing Co., Slough, UK.

Faucherre, J., Cherix, D. and Wyss, C. (1999) Behaviour of *Calliphora vicina* (Diptera: Calliphoridae) under extreme conditions. *Journal of Insect Behavior*, **12**, 687–690.

Fischer, O.A., Matlova, L., Dvorska, L. *et al.* (2004) Blowflies *Calliphora vicina* and *Lucilia sericata* as passive vectors of *Mycobacterium avium* subsp. *avium*, *M. a. paratuberculosis* and *M. a. hominissuis*. *Medical and Veterinary Entomology*, **18**, 116–122.

Flores, M., Longnecker, M. and Tomberlin, J.K. (2014) Effects of temperature and tissue type on *Chrysomya rufifacies* (Diptera: Calliphoridae) (Macquart) development. *Forensic Science International*, **245**, 24–29.

Frago, E., Dicke, M. and Godfray, H.C.J. (2012) Insect symbionts as hidden players in insect-plant interactions. *Trends in Ecology & Evolution*, **27**, 705–711.

Frederickx, C., Dekeirsschieter, J., Verheggen, F.J. and Haubruge, E. (2012a) Responses of *Lucilia sericata* Meigen (Diptera: Calliphoridae) to cadaveric volatile organic compounds. *Journal of Forensic Sciences*, **57**, 386–390.

Frederickx, C., Dekeirsschieter, J., Brostaux, Y. *et al.* (2012b) Volatile organic compounds released by blowfly larvae and pupae: new perspectives in forensic entomology. *Forensic Science International*, **219**, 215–220.

Fremdt, H. and Amendt, B. (2014) Species composition of forensically important blow flies (Diptera: Calliphoridae) and flesh flies (Diptera: Sarcophagidae) through space and time. *Forensic Science International*, **236**, 1–9.

Gallagher, M.B., Sandhu, S. and Kimsey, R. (2010) Variation in development time for geographically distinct populations of the common green bottle fly, *Lucilia sericata* (Meigen). *Journal of Forensic Sciences*, **55**, 438–442.

Gennard, D.E. (2007) *Forensic Entomology: An Introduction*, 2nd edn, Wiley-Blackwell, Chichester, UK.

George, K.A., Archer, M.S., Green, L.M. *et al.* (2009) Effect of morphine on the growth rate of *Calliphora stygia* (Fabricius) (Diptera: Calliphoridae) and possible implications for forensic entomology. *Forensic Science International*, **193**, 21–25.

George, K.A., Archer, M.S. and Toop, T. (2013a) Nocturnal colonization behaviour of blowflies (Diptera: Calliphoridae) in Southeastern Australia. *Journal of Forensic Sciences*, **58**, S112–S116.

George, K.A., Archer, M.A. and Toop, T. (2013b) Abiotic environmental factors influencing blowfly colonisation patterns in the field. *Forensic Science International*, **229**, 100–107.

Goff, M.L. (1991) Comparison of insect species associated with decomposing remains recovered inside dwellings and outdoors on the Island of Oahu, Hawaii. *Journal of Forensic Sciences*, **36**, 748–753.

Goff, M.L. (1992) Problems in estimation of postmortem interval resulting from wrapping of the corpse: a case study from Hawaii. *Journal of Agricultural Entomology*, **9**, 237–243.

Gomes, L., Sanches, M.R. and Von Zuben, C.J. (2005) Dispersal behaviour in larvae of *Chrysomya megacephala* and *Chrysomya albiceps* (Diptera: Calliphoridae). *Journal of Insect Behavior*, **18**, 281–292.

Gomes, L., Gomes, G., Casarin, F.E. *et al.* (2007) Interação entre fatores visuais e olfativo em localização de recursos pela mosca-varejeira, *Chrysomya megacephala* (Fabricius) (Diptera: Calliphoridae), em condições naturais. *Neotropical Entomology*, **36**, 633–639.

Gomes, L., Gomes, G. and Desuo, I.C. (2009) A preliminary study of insect fauna on pig carcasses located in sugarcane in winter in southeastern Brazil. *Medical and Veterinary Entomology*, **23**, 155–159.

Goodbrod, J.R. and Goff, M.L. (1990) Effects of larval population density on rates of development and interactions between two species of *Chrysomya* (Diptera: Calliphoridae) in laboratory culture. *Journal of Medical Entomology*, **27**, 338–343.

Gosselin, M., Wille, S.M.R., del Mar Ramirez Fernandez, M. *et al.* (2011) Entomotoxicology, experimental set-up and interpretation for forensic toxicologists. *Forensic Science International*, **208**, 1–9.

Grassberger, M. and Frank, C. (2003) Temperature-related development of the parasitoid wasp *Nasonia vitripennis* as forensic indicator. *Medical and Veterinary Entomology*, **17**, 257–262.

Grassberger, M. and Frank, C. (2004) Initial study of arthropod succession on pig carrion in a central European urban habitat. *Journal of Medical Entomology*, **41**, 511–523.

Grassberger, M. and Reiter, C. (2001) Effect of temperature on *Lucilia sericata* (Diptera: Calliphoridae) development with special reference to the isomegalen- and isomorphen-diagram. *Forensic Science International*, **120**, 32–36.

Greenberg, B. (1968) Model for destruction of bacteria in the midgut of blow fly maggots. *Journal of Medical Entomology*, **5** (1), 31–38.

Greenberg, B. (1971a) *Flies and Disease, Vol. 1, Ecology, Classification and Biotic Associations*, vol. **1**, Princeton University Press, Princeton, NJ.

Greenberg, B. (1971b) *Flies and Disease, Vol. 2, Biology and Disease Transmission*, vol. **2**, Princeton University Press, Princeton, NJ.

Greenberg, B. (1990) Nocturnal oviposition behavior of blow flies (Diptera, Calliphoridae). *Journal of Medical Entomology*, **27**, 807–810.

Greenberg, B. (1991) Flies as forensic indicators. *Journal of Medical Entomology*, **28**, 565–577.

Greenberg, B. and Klowden, M. (1972) Enteric bacterial interactions in insects. *The American Journal of Clinical Nutrition*, **25**, 1459–1466.

Greenberg, B., Kowalski, J.A. and Klowden, M.J. (1970) Factors affecting the transmission of salmonella by flies: natural resistance to colonization and bacterial interference. *Infection and Immunity*, **2**, 800–809.

Gunn, A. and Bird, J. (2011) The ability of blowflies *Calliphora vomitoria* (Linnaeus), *Calliphora vicina* (Rob-Desvoidy) and *Lucilia sericata* (Meigen) (Diptera: Calliphoridae) and the muscid flies *Muscina stabulans* (Fallén) and *Muscina prolapse* (Harris) (Diptera: Muscidae) to colonise buried remains. *Forensic Science International*, **207**, 198–204.

Hammack, L. (1991) Oviposition by screwworm flies (Diptera: Calliphoridae) on contact with host fluids. *Journal of Economic Entomology*, **84**, 185–190.

Hammack, L., Bromel, M., Duh, F.M. and Gassner, G. (1987) Reproductive factors affecting responses of the screwworm fly, *Cochliomyia hominivorax* (Diptera: Calliphoridae) to an attractant of bacterial origin. *Annals of the Entomological Society of America*, **80**, 775–780.

Hewadikaram, K.A. and Goff, M.L. (1991) Effect of carcass size on rate of decomposition and arthropod succession patterns. *The American Journal of Forensic Medicine and Pathology*, **12**, 235–240.

von Hoermann, C., Ruther, J., Reibe, S. *et al.* (2011) The importance of carcass volatiles as attractants for the hide beetle *Dermestes maculatus* (De Geer). *Forensic Science International*, **212**, 173–179.

Horenstein, M.B., Linhares, A.X., De Ferradas, B.R. and García, D. (2010) Decomposition and dipteran succession in pig carrion in central Argentina: ecological aspects and their importance in forensic science. *Medical and Veterinary Entomology*, **24**, 16–25.

Howard, G.T., Duos, B. and Watson-Horzelski, E.J. (2010) Characterization of the soil microbial community associated with the decomposition of a swine carcass. *International Biodeterioration and Biodegradation*, **64**, 300–304.

Huntington, T.E., Higley, L.G. and Baxendale, F.P. (2007) Maggot development during morgue storage and its effect on estimating the post-mortem interval. *Journal of Forensic Sciences*, **52**, 453–458.

Hwang, C. and Turner, B.D. (2005) Spatial and temporal variability of necrophagous Diptera from urban to rural areas. *Medical and Veterinary Entomology*, **19**, 379–391.

Hyde, E.R., Haarmann, D.P., Lynne, A.M. *et al.* (2013) The living dead: bacterial community structure of a cadaver at the onset and end of the bloat stage of decomposition. *PLoS ONE*, **8**, e77733.

Hyde, E.R., Haarmann, D.P., Petrosino, J.F. *et al.* (2014) Initial insights into bacterial succession during human decomposition. *International Journal of Legal Medicine*, **129** (3), 1–11.

Ireland, S. and Turner, B. (2006) The effects of larval crowding and food type on the size and development of the blowfly, *Calliphora vomitoria*. *Forensic Science International*, **159**, 175–181.

Janzen, D.H. (1977) Why fruits rot, seeds mold, and meat spoils. *The American Naturalist*, **111**, 691–713.

Jones, R.T., Sanchez, L.G. and Fierer, N. (2013) A cross-taxon analysis of insect-associated bacterial diversity. *PLoS ONE*, **8**, e61218.

Jordan, H.R., Tomberlin, J.K., Wood, T.K. and Benbow, M.E. (2015) Interkingdom ecological interactions of carrion decomposition, in *Carrion Ecology, Evolution, and Their Applications* (eds M.E. Benbow, J.K. Tomberlin and A.M. Tarone), CRC Press, Boca Raton, FL, pp. 433–460.

Joseph, I., Mathew, D.G., Sathyan, P. and Vargheese, G. (2011) The use of insects in forensic investigations: an overview on the scope of forensic entomology. *Journal of Forensic Dental Sciences*, **3**, 89–91.

Kadouri, D.E., To, K., Shanks, R.M.Q. and Doi, Y. (2013) Predatory bacteria: a potential ally against multidrug-resistant Gram-negative pathogens. *PLoS ONE*, **8**, e63397.

Kaneshrajah, G. and Turner, B. (2004) *Calliphora vicina* larvae grow at different rates on different body tissues. *International Journal of Legal Medicine*, **118**, 242–244.

Kelly, J.A., Van Der Linde, T.C. and Anderson, G.S. (2008) The influence of clothing and wrapping on carcass decomposition and arthropod succession: a winter study in Central South Africa. *Canadian Society of Forensic Science Journal*, **41**, 135–147.

Kirkpatrick, R.S. (2004) Nocturnal light and temperature influences on necrophagous, carrion-associating blow fly species (Diptera: Calliphoridae) of forensic importance in Central Texas. M.S. thesis, Texas A&M University, College Station.

Kočárek, P. (2001) Diurnal patterns of postfeeding dispersal in carrion blowflies (Diptera: Calliphoridae). *European Journal of Entomology*, **98**, 117–119.

Koĉárek, P. (2002) Diel activity patterns of carrion-visiting Coleoptera studied by time-sorting pitfall traps. *Biologia, Bratislava*, **57**, 199–211.

de Lacy Costello, B., Amann, A., Al-Kateb, H. *et al.* (2014) A review of the volatiles from the healthy human body. *Journal of Breath Research*, **8**, 014001.

Lam, K., Babor, D., Duthie, B. *et al.* (2007) Proliferating bacterial symbionts on house fly eggs affect oviposition behaviour of adult flies. *Animal Behaviour*, **74**, 81–92.

Lam, K., Geisreiter, C. and Gries, G. (2009) Ovipositing female house flies provision offspring larvae with bacterial food. *Entomologia Experimentalis et Applicata*, **133**, 292–295.

Lang, J.M. and Benbow, M.E. (2013) Species interactions and competition. *Nature Education Knowledge*, **4**, 8.

Lerch, K., Linde, H.J., Lehn, N. and Grifka, J. (2003) Bacteria ingestion by blowfly larvae: an in vitro study. *Dermatology*, **207**, 362–366.

Lewis, A.J. and Benbow, M.E. (2011) When entomological evidence crawls away: *Phormia regina* en masse larval dispersal. *Journal of Medical Entomology*, **48**, 1112–1119.

Liu, Q., Tomberlin, J., Brady, J. *et al.* (2008) Black soldier fly (Diptera: Stratiomyidae) larvae reduce *Escherichia coli* in dairy manure. *Environmental Entomology*, **37**, 1525–1530.

Ma, Q., Fonseca, A., Liu, W. *et al.* (2012) *Proteus mirabilis* interkingdom swarming signals attract blow flies. *The ISME Journal*, **6**, 1356–1366.

MacLeod, A. (1949) The climatology of blowfly myiasis. II. Oviposition and daily weather indices. *Bulletin of Entomological Research*, **40**, 179–201.

Mahat, N.A., Zafarina, Z. and Jayaprakash, P.T. (2009) Influence of rain and malathion on the oviposition and development of blowflies (Diptera: Calliphoridae) infesting rabbit carcasses in Kelantan, Malaysia. *Forensic Science International*, **192**, 19–28.

Mai, M. and Amendt, J. (2012) Effect of different post-feeding intervals on the total time of development of the blowfly *Lucilia sericata* (Diptera: Calliphoridae). *Forensic Science International*, **221**, 665–669.

Matuszewski, S. (2011) Estimating the pre-appearance interval from temperature in *Necrodes littoralis* L. (Coleoptera: Silphidae). *Forensic Science International*, **212**, 180–188.

Matuszewski, S. (2012) Estimating the preappearance interval from temperature in *Creophilus maxillosus* L. (Coleoptera: Staphylinidae). *Journal of Forensic Sciences*, **57**, 136–145.

Matuszewski, S. and Szafałowicz, M. (2013) Temperature-dependent appearance of forensically useful beetles on carcasses. *Forensic Science International*, **229**, 92–99.

Matuszewski, S., Bajerlein, D., Konwerski, S. and Szpila, K. (2008) An initial study of insect succession and carrion decomposition in various forest habitats of Central Europe. *Forensic Science International*, **180**, 61–69.

Matuszewski, S., Bajerlein, D., Konwerski, S. and Szpila, K. (2011) Insect succession and carrion decomposition in selected forests of Central Europe. Part 3: succession of carrion fauna. *Forensic Science International*, **2007**, 150–163.

Mégnin, P. (1894) *La faune des cadavres, application de l'entomologie a la médecine légale*, Number 101B. Encyclopédie Scientifique des Aide-Mémoire, Paris.

Melvin, J.R., Cronholm, L.S., Simson, L.R. and Isaacs, A.M. (1984) Bacterial transmigration as an indicator of time of death. *Journal of Forensic Sciences*, **29**, 412–417.

Metcalf, J.L., Wegener Parfrey, L., Gonzalez, A. *et al.* (2013) A microbial clock provides an accurate estimate of the postmortem interval in a mouse model system. *eLife*, **2**, e01104.

Metcalf, J.L., Xu, Z.Z., Weiss, S. *et al.* (2016) Microbial community assembly and metabolic function during mammalian corpse decomposition. *Science*, **351**, 158–162.

Michaud, J.-P. and Moreau, G. (2009) Predicting the visitation of carcasses by carrion-related insects under different rates of degree-day accumulation. *Forensic Science International*, **185**, 78–83.

Miller, M.L., Lord, W.D., Goff, M.L. *et al.* (1994) Isolation of amitriptyline and nortriptyline from fly puparia (Phoridae) and beetle exuviae (Dermestidae) associated with mummified human remains. *Journal of Forensic Sciences*, **39**, 1305–1313.

Mohr, R.M. (2012) Female blow fly (Diptera: Calliphoridae) arrival patterns and consequences for larval development on ephemeral resources. Ph.D. dissertation, Texas A&M University, College Station.

Mohr, R.M. and Tomberlin, J.K. (2014a) Development and validation of a new technique for estimating a minimum postmortem interval using adult blow fly (Diptera: Calliphoridae) carcass attendance. *International Journal of Legal Medicine*, **129** (4), 851–859.

Mohr, R.M. and Tomberlin, J.K. (2014b) Environmental factors affecting early carcass attendance by four species of blow flies (Diptera: Calliphoridae) in Texas. *Journal of Medical Entomology*, **51**, 702–708.

Moreno, L. (2011) The application of amplicon length heterogeneity PCR (LH-PCR) for monitoring the dynamics of soils microbial communities associated with cadaver decomposition. *Journal of Microbiological Methods*, **84**, 388–393.

Mullany, C., Keller, P.A., Nugraha, A.S. and Wallman, J.F. (2014) Effects of methamphetamine and its primary human metabolite, p-hydroxymethamphetamine, on the development of the Australian blowfly *Calliphora stygia*. *Forensic Science International*, **241**, 102–111.

Mumcuoglu, K.Y., Miller, J., Mumcuoglu, M. *et al.* (2001) Destruction of bacteria in the digestive tract of the maggot of *Lucilia sericata* (Diptera: Calliphoridae). *Journal of Medical Entomology*, **38**, 161–166.

Myskowiak, J.-B. and Doums, C. (2002) Effects of refrigeration on the biometry and development of *Protophormia terraenovae* (Robineau-Desvoidy) (Diptera: Calliphoridae) and its consequences in estimating post-mortem interval in forensic investigations. *Forensic Science International*, **125**, 254–261.

Nikoh, N., Hosokawa, T., Moriyama, M. *et al.* (2014) Evolutionary origin of insect–*Wolbachia* nutritional mutualism. *Proceedings of the National Academy of Sciences*, **111**, 10257–10262.

Núñez-Vázquez, C., Tomberlin, J.K., Cantú-Sifuentes, M. and García-Martínez, O. (2013) Laboratory development and field validation of *Phormia regina* (Diptera: Calliphoridae). *Journal of Medical Entomology*, **50**, 252–260.

Oliveira-Costa, J. and Antunes de Mello-Patiu, C. (2004) Application of forensic entomology to estimate of the postmortem interval (PMI) in homicide investigations by the Rio de Janeiro Police Department in Brazil. *Anil Aggrawal's Internet Journal of Forensic Medicine and Toxicology*, **5**, 40–44.

Ortloff, A., Peña, P. and Riquelme, M. (2012) Preliminary study of the succession pattern of necrobiont insects, colonising species and larvae on pig carcasses in Temuco (Chile) for forensic applications. *Forensic Science International*, **222**, e36–e41.

Owings, C., Spiegelman, C., Tarone, A. and Tomberlin, J. (2014) Developmental variation among *Cochliomyia macellaria* Fabricius (Diptera: Calliphoridae) populations from three ecoregions of Texas, USA. *International Journal of Legal Medicine*, **128**, 709–717.

Paczkowski, S., Maibaum, F., Paczkowska, M. and Schutz, S. (2012) Decaying mouse volatiles perceived by *Calliphora vicina* Rob.-Desv. *Journal of Forensic Sciences*, **57**, 1497–1506.

Payne, J.A. (1965) A summer carrion study of the baby pig *Sus scrofa* Linnaeus. *Ecology*, **46**, 592–602.

Pechal, J.L. and Benbow, M.E. (2016) Microbial ecology of the salmon necrobiome: evidence salmon carrion decomposition influences aquatic and terrestrial insect microbiomes. *Environmental Microbiology*, **18** (5), 1511–1522.

Pechal, J.L., Crippen, T.L., Tarone, A.M. *et al.* (2013) Microbial community functional change during vertebrate carrion decomposition. *PLoS ONE*, **8** (11), e79035.

Pechal, J., Benbow, M., Crippen, T. *et al.* (2014a) Delayed insect access alters carrion decomposition and necrophagous insect community assembly. *Ecosphere*, **5**, art45.

Pechal, J.L., Crippen, T.L., Benbow, M.E. *et al.* (2014b) The potential use of bacterial community succession in forensics as described by high throughput metagenomic sequencing. *International Journal of Legal Medicine*, **128**, 193–205.

Perez, S.P., Duque, P. and Wolff, M. (2005) Successional behavior and occurrence matrix of carrion-associated arthropods in the urban area of Medellin, Colombia. *Journal of Forensic Sciences*, **50**, 448–454.

Picard, C.J. and Wells, J.D. (2010) The population genetic structure of North American *Lucilia sericata* (Diptera: Calliphoridae), and the utility of genetic assignment methods for reconstruction of postmortem corpse relocation. *Forensic Science International*, **195**, 63–67.

Picard, C.J. and Wells, J.D. (2012) A test for carrion fly full siblings: a tool for detecting postmortem relocation of a corpse. *Journal of Forensic Sciences*, **57**, 535–538.

Pickett, J.A., Barasa, S. and Birkett, M.A. (2014) Vertebrate pheromones and other semiochemicals: the potential for accommodating complexity in signalling by volatile compounds for vertebrate management. *Biochemical Society Transactions*, **42**, 846–850.

Pohjoismäki, J.L.O., Karhunen, P.J., Goebeler, S. *et al.* (2010) Indoors forensic entomology: colonization of human remains in closed environments by specific species of sarcosaprophagous flies. *Forensic Science International*, **199**, 38–42.

Pritam, H.M.H. and Jayaprakash, P.T. (2009) Nocturnal oviposition behavior of necrophagous dipterans in Kelantan, Malaysia. *Journal of Forensic Sciences*, **54**, 1135–1140.

Reed, H.B. (1958) A study of dog carcass communities in Tennessee, with special reference to the insects. *The American Midland Naturalist*, **59**, 213–245.

Reibe, S. and Madea, B. (2010) Use of *Megaselia scalaris* (Diptera: Phoridae) for post-mortem interval estimation indoors. *Parasitology Research*, **106**, 637–640.

Richards, D.S., Saunders, D.S., Egan, V.M. and Thomson, R.C.K. (1986) The timing of larval wandering and puparial formation in the flesh fly *Sarcophaga argyrostoma*. *Physiological Entomology*, **11**, 53–60.

Richardson, M.S. and Goff, M.L. (2001) Effects of temperature and intraspecific interaction on the development of *Dermestes maculatus* (Coleoptera: Dermestidae). *Journal of Medical Entomology*, **38**, 347–351.

Richardson, C.S., Rowlinson, C.C., Cuttiford, L. *et al.* (2013) Decomposed liver has a significantly adverse affect on the development rate of the blow fly *Calliphora vicina*. *International Journal of Legal Medicine*, **127**, 259–262.

Rivers, D.B. and Dahlem, G.A. (2014) *The Science of Forensic Entomology*, Wiley-Blackwell, Chichester, UK.

Rodriguez, W.C. and Bass, W.M. (1983) Insect activity and its relationship to decay rates of human cadavers in East Tennessee. *Journal of Forensic Sciences*, **28**, 423–432.

Rose, G.W. and Hockett, R.N. (1971) The microbiologic evaluation and enumeration of post-mortem specimens from human remains. *Health Laboratory Science*, **8**, 75–78.

Roy, S. and Muraleedharan, N. (2014) Microbial management of arthropod pests of tea: current state and prospects. *Applied Microbiology and Biotechnology*, **98**, 5375–5386.

Rozen, D.E., Engelmoer, D.J.P. and Smiseth, P.T. (2008) Antimicrobial strategies in burying beetles breeding on carrion. *Proceedings of the National Academy of Sciences*, **105**, 17890–17895.

Russell, J.A., Moreau, C.S., Goldman-Huertas, B. *et al.* (2009) Bacterial gut symbionts are tightly linked with the evolution of herbivory in ants. *Proceedings of the National Academy of Sciences*, **106**, 21236–21241.

Sachs, J.S. (2001) *Nature, Forensics and the Struggle to Pinpoint Time of Death*, Perseus Publishing Services, Cambridge, MA.

Saunders, D.S. and Hayward, S.A.L. (1998) Geographical and diapause-related cold tolerance in the blow fly, *Calliphora vicina*. *Journal of Insect Physiology*, **44**, 541–551.

Sawabe, K., Hoshino, K., Isawa, H. *et al.* (2006) Detection and isolation of highly pathogenic H5N1 avian Influenza A virus from blow flies collected in the vicinity of an infected poultry farm in Kyoto, Japan, 2004. *The American Journal of Tropical Medicine and Hygiene*, **75**, 327–332.

Schoenly, K.G., Haskell, N.H., Hall, R.D. and Gbur, J.R. (2007) Comparative performance and complementarity of four sampling methods and arthropod preference tests from human and porcine remains at the Forensic Anthropology Center in Knoxville, Tennessee. *Journal of Medical Entomology*, **44**, 881–894.

Schröder, R. and Hilker, M. (2008) The relevance of background odor in resource location by insects: a behavioral approach. *BioScience*, **58**, 308–316.

Schroeder, H., Klotzbach, H. and Püschel, K. (2003) Insects' colonisation of human corpses in warm and cold season. *Legal Medicine*, **5**, S372–S374.

Segura, N.A., Usaquén, W. and Sánchez, M.C. (2009) Succession pattern of cadaverous entomofauna in a semi-rural area of Bogotá, Colombia. *Forensic Science International*, **187**, 66–72.

Sharma, R., Garg, R.K. and Gaur, J.R. (2015) Various methods for the estimation of the post mortem interval from Calliphoridae: a review. *Egyptian Journal of Forensic Sciences*, **5**, 1–12.

Shirasu, M. and Touhara, K. (2011) The scent of disease: volatile organic compounds of the human body related to disease and disorder. *Journal of Biochemistry*, **150**, 257–266.

Simmons, T., Cross, P.A., Adlam, R.E. and Moffatt, C. (2010) The influence of insects on decomposition rate in buried and surface remains. *Journal of Forensic Sciences*, **55**, 889–892.

Singh, D. and Bharti, M. (2001) Further observations on the nocturnal oviposition behaviour of blow flies (Diptera: Calliphoridae). *Forensic Science International*, **120**, 124–126.

Singh, D. and Bharti, M. (2008) Some notes on the nocturnal larviposition by two species of Sarcophaga (Diptera: Sarcophagidae). *Forensic Science International*, **177**, e19–e20.

Singh, B., Crippen, T.L., Zheng, L. *et al.* (2015) A metagenomic assessment of the bacteria associated with *Lucilia sericata* and *Lucilia cuprina* (Diptera: Calliphoridae). *Applied Microbiology and Biotechnology*, **99**, 869–883.

Smith, K.E. and Wall, R. (1997) Asymmetric competition between larvae of the blowflies *Calliphora vicina* and *Lucilia sericata* in carrion. *Ecological Entomology*, **22**, 468–474.

Sommerville, D.A. (1969) The normal flora of the skin of different age groups. *The British Journal of Dermatology*, **81**, 248–258.

Spivak, M., Conlon, D. and Bell, W.J. (1991) Wind-guided landing and search behavior in fleshflies and blowflies exploiting a resource patch (Diptera: Sarcophagidae, Calliphoridae). *Annals of the Entomological Society of America*, **84**, 447–452.

Stamper, T. and Derby, R.W. (2007) The nocturnal oviposition behavior of carrion flies in rural and urban environments: methodological problems and forensic implications. *Canadian Society of Forensic Science Journal*, **40**, 173–182.

Stamper, T., Davis, P. and Debry, R.W. (2009) The nocturnal ovipositing behaviour of carrion flies in Cincinnati, Ohio. *Journal of Forensic Sciences*, **54**, 1450–1452.

Statheropoulos, M., Spiliopoulou, C. and Agapiou, A. (2005) A study of volatile organic compounds evolved from the decaying human body. *Forensic Science International*, **153**, 147–155.

Stavert, J.R., Drayton, B.A., Beggs, J.R. and Gaskett, A.C. (2014) The volatile organic compounds of introduced and native dung and carrion and their role in dung beetle foraging behaviour. *Ecological Entomology*, **39**, 556–565.

Sukontason, K.L., Bunchoo, M., Khantawa, B. *et al.* (2000) Mechanical carrier of bacterial enteric pathogens by *Chrysomya megacephala* (Diptera: Calliphoridae) in Chiang Mai, Thailand. *Southeast Asian Journal of Tropical Medicine and Public Health*, **31**, 157–161.

Sukontason, K.L., Bunchu, N., Methanitikorn, R. *et al.* (2006) Ultrastructure of adhesive device in fly in families Calliphoridae, muscidae and Sarcophagidae, and their implication as mechanical carriers of pathogens. *Parasitology Research*, **98**, 477–481.

Sukontason, K.L., Bunchoo, M., Khantawa, B. *et al.* (2007) Comparison between *Musca domestica* and *Chrysomya megacephala* as carriers of bacteria in northern Thailand. *Southeast Asian Journal of Tropical Medicine and Public Health*, **38**, 38–44.

Tarone, A.M. and Foran, D.R. (2008) Generalized additive models and *Lucilia sericata* growth: assessing confidence intervals and error rates in forensic entomology. *Journal of Forensic Sciences*, **53**, 942–949.

Tessmer, J.W., Meek, C.L. and Wright, V.L. (1995) Circadian patterns of oviposition by necrophilous flies (Diptera: Calliphoridae) in southern Louisiana. *Southwestern Entomologist*, **20**, 439–445.

Tomberlin, J.K. and Benbow, M.E. (eds) (2015) *Forensic Entomology: International Dimensions and Frontiers*, CRC Press, Boca Raton, FL.

Tomberlin, J.K., Mohr, R., Benbow, M.E. *et al.* (2011) A roadmap for bridging basic and applied research in forensic entomology. *Annual Review of Entomology*, **56**, 401–421.

Tomberlin, J.K., Crippen, T.L., Tarone, A.M. *et al.* (2012) Interkingdom responses of flies to bacteria mediated by fly physiology and bacterial quorum sensing. *Animal Behaviour*, **84**, 1449–1456.

Turner, B. and Howard, T. (1992) Metabolic heat generation in dipteran larval aggregations: a consideration for forensic entomology. *Medical and Veterinary Entomology*, **6**, 179–181.

Turner, B.D. and Wiltshire, P.E.J. (1999) Experimental validation of forensic evidence: a study of the decomposition of buried pigs in a heavy clay soil. *Forensic Science International*, **101**, 113–122.

Valdes-Perezgasga, M.T., Sanchez-Ramos, F.J., Garcia-Martinez, O. and Anderson, G.S. (2010) Arthropods of forensic importance on pig carrion in the Coahuilan semidesert, Mexico. *Journal of Forensic Sciences*, **55**, 1098–1101.

VanLaerhoven, S.L. (2008) Blind validation of postmortem interval estimates using developmental rates of blow flies. *Forensic Science International*, **180**, 76–80.

VanLaerhoven, S.L. and Anderson, G.S. (1999) Insect succession on buried carrion in two biogeoclimatic zones of British Columbia. *Journal of Forensic Sciences*, **44**, 32–43.

Vass, A.A., Barshick, S.A., Sega, G. *et al.* (2002) Decomposition chemistry of human remains: a new methodology for determining the postmortem interval. *Journal of Forensic Sciences*, **47**, 542–553.

Vass, A.A., Smith, R.R., Thompson, C.V. *et al.* (2004) Decompositional odor analysis database. *Journal of Forensic Sciences*, **49**, 760–769.

Vass, A.A., Smith, R.R., Thompson, C.V. *et al.* (2008) Odor analysis of decomposing buried human remains. *Journal of Forensic Sciences*, **53**, 384–391.

Vinson, S.B. (1976) Host selection by insect parasitoids. *Annual Review of Entomology*, **21**, 109–133.

Vitta, A., Pumidonming, W., Tangchaisuriya, U. *et al.* (2007) A preliminary study on insects associated with pig (*Sus scrofa*) carcasses in Phitsanulok, northern Thailand. *Tropical Biomedicine*, **24**, 1–5.

Voss, S.C., Forbes, S.L. and Dadour, I.R. (2008) Decomposition and insect succession on cadavers inside a vehicle environment. *Forensic Science, Medicine, and Pathology*, **4**, 22–32.

Voss, S.C., Cook, D.F. and Dadour, I.R. (2011) Decomposition and insect succession of clothed and unclothed carcasses in Western Australia. *Forensic Science International*, **211**, 67–75.

Walker, T.J. (1957) Ecological studies of the arthropods associated with certain decaying materials in four habitats. *Ecology*, **38**, 262–276.

Wall, R. and Fisher, P. (2001) Visual and olfactory cue interaction in resource-location by the blowfly, *Lucilia sericata*. *Physiological Entomology*, **26**, 212–218.

Wang, J., Li, Z., Chen, Y. *et al.* (2008) The succession and development of insects on pig carcasses and their significances in estimating PMI in south China. *Forensic Science International*, **179**, 11–18.

Watts, J.E., Merritt, G.C. and Goodrich, B.S. (1981) The ovipositional response of the Australian sheep blowfly, *Lucilia cuprina*, to fleece-rot odours. *Australian Veterinary Journal*, **57**, 450–454.

Wei, T., Hu, J., Miyanaga, K. and Tanji, Y. (2013) Comparative analysis of bacterial community and antibiotic-resistant strains in different developmental stages of the house fly (*Musca domestica*). *Applied Microbiology and Biotechnology*, **97**, 1775–1783.

Wei, T., Ishida, R., Miyanaga, K. and Tanji, Y. (2014) Seasonal variations in bacterial communities and antibiotic-resistant strains associated with green bottle flies (Diptera: Calliphoridae). *Applied Microbiology and Biotechnology*, **98** (9), 4197–4208.

Wey, J.K., Jürgens, K. and Weitere, M. (2012) Seasonal and successional influences on bacterial community composition exceed that of protozoan grazing in river biofilms. *Applied and Environmental Microbiology*, **78**, 2013–2024.

Zaidi, F. and Chen, X.X. (2011) A preliminary survey of carrion breeding insects associated with the Eid ul Azha festival in remote Pakistan. *Forensic Science International*, **209**, 186–194.

Zhang, X., Crippen, T.L., Coates, C.J. *et al.* (2015) Effect of quorum sensing by *Staphylococcus epidermidis* on the attraction response of female adult yellow fever mosquitoes, *Aedes aegypti aegypti* (Linnaeus) (Diptera: Culicidae), to a blood-feeding source. *PLoS ONE*, **10**, e0143950.

Zug, R. and Hammerstein, P. (2015) Bad guys turned nice? A critical assessment of *Wolbachia* mutualisms in arthropod hosts. *Biological Reviews*, **90**, 89–111.

Zurawski, K.N., Benbow, M.E., Miller, J.R. and Merritt, R.W. (2009) Examination of nocturnal blow fly (Diptera: Calliphoridae) oviposition on pig carcasses in mid-Michigan. *Journal of Medical Entomology*, **46**, 671–679.

Zurek, L., Schal, C. and Watson, D.W. (2000) Diversity and contribution of the intestinal bacterial community to the development of *Musca domestica* (Diptera: Muscidae) larvae. *Journal of Medical Entomology*, **37**, 924–928.

CHAPTER 12
Microbes, anthropology, and bones

Franklin E. Damann[1] and Miranda M.E. Jans[2]
[1] *Central Identification Laboratory—Offutt Air Force Base, Defense POW/MIA Accounting Agency, Bellevue, NE, USA*
[2] *Central Identification Laboratory—Hickam Air Force Base, Defense POW/MIA Accounting Agency, Honolulu, HI, USA*

12.1 Introduction

During the time between death and discovery, a corpse experiences myriad taphonomic processes to which microbes are significant contributors. Certain postmortem events leave signatures on the skeleton that can be readily identified, while others may be inferred as necessary transitional processes that produced the taphonomic effects observed at recovery. Understanding the depositional environment and associated microbial taphonomic processes are essential for interpreting observations, as microbes may alter the macro-, micro-, and molecular components of bone and affect interpretation (DeNiro, 1985; Piepenbrink, 1986; Grupe and Piepenbrink, 1988; Antoine *et al.*, 1992; Balzer *et al.*, 1997; Fernández-Jalvo *et al.*, 2002; Geigl, 2002).

Human bone is the traditional medium of a physical/forensic anthropologist. With advancements in technology, anthropologists are turning to various laboratory methods to assist with the discovery and interpretation of skeletal evidence on the micro- and molecular scale in medicolegal and archaeological contexts. For example, isotope analyses of bone are conducted to estimate place of origin (Font *et al.*, 2014), year of death (Ubelaker, 2001; Buchholz, 2009), and reconstruct diet (Walker and DeNiro, 1986). DNA analyses are conducted for human identification (Edson *et al.*, 2004) and reconstructing evolutionary relationships (Ovchinnikov *et al.*, 2000; Schmitz *et al.*, 2002). Microbial DNA analyses of decaying soft tissue and bone are being used to better understand taphonomic processes and estimate postmortem intervals (PMIs) (Metcalf *et al.*, 2013; Pechal *et al.*, 2013; Carter *et al.*, 2015; Damann *et al.*, 2015; Hauther *et al.*, 2015). Histological analyses are performed to differentiate human from nonhuman bone

The views expressed are those of the authors and should not be construed to represent those of the U.S. government.

Forensic Microbiology, First Edition. Edited by David O. Carter, Jeffery K. Tomberlin, M. Eric Benbow and Jessica L. Metcalf.

(Hillier and Bell, 2007; Cuijpers, 2009; Crescimanno and Stout, 2012; Mulhern and Ubelaker, 2012), estimate age at death (Kerley, 1965; Stout and Paine, 1992), evaluate disease etiologies (Schultz, 2001; Von Hunnius *et al.*, 2006; Mays and Turner-Walker, 2008), and interpret postmortem histories (Jans *et al.*, 2004; Dominguez-Rodrigo and Barba, 2006; Turner-Walker and Jans, 2008; Fernández-Jalvo *et al.*, 2010; Hollund *et al.*, 2012).

In order to interpret skeletal evidence from medicolegal and archaeological settings at investigative scales currently used, anthropologists must understand microbially mediated processes that affect the living and dead skeletal system. These processes may negatively affect interpretation if not properly accounted while at the same time may permit new and more accurate interpretations of postmortem histories than previously revealed (Turner-Walker and Jans, 2008; Hollund *et al.*, 2012).

To that end, microbial interactions with the human skeleton are being explored through histomorphology and high throughput amplicon-based sequencing. As a result, our understanding about the relationships between microbes and the human body (both living and dead) are rapidly changing.

For example, in life, microbial communities of the gastrointestinal tract produce essential nutrients such as vitamins K and B12 (Wilson, 2008). Anaerobic bacteria of the gut produce menaquinones—a specific type of vitamin K that facilitates bone growth while inhibiting bone resorption (Tsukamoto, 2004). Menaquinones are required for the bacterial biosynthesis of osteocalcin, a noncollagenous protein that connects the primary organic and inorganic components of bone (Conly and Stein, 1992).

And in death, microbes are taphonomic agents guiding specific skeletal decay processes. Microbially mediated bioerosion of bone increases porosity, alters normal microstructure, and increases the likelihood of carbonate exchange of the mineral component with the environment (Nielsen-Marsh and Hedges, 2000).

This chapter discusses the connections among decaying bones, associated microbial communities, and taphonomic effects of bone in the burial environment. By understanding the relationships between microbes and skeletal remains, interpreting postmortem histories, estimating PMIs, and differentiating microbially mediated diagenesis from disease processes and human-induced markers are possible.

12.2 Bone microstructure

Bone is a recalcitrant organic resource, protected by a rigid inorganic substance. As a result, access to bone tissue by microbes is typically achieved through voids in internal microstructures and vasculature networks.

Bone is a composite tissue consisting of a mineral and organic component. The mineralized collagen fibril is the basic structural unit that can be organized "randomly" in woven or fibrous bone and with a high degree of organization in sheets of lamellar bone. Within these sheets, orientation of the fibrils is constant, and their orientation alternates between adjacent sheets and lamellae (Currey, 2002).

The fibrils of collagen are arranged in a "head to tail" pattern with gaps between fibrils, creating a repetitive pattern to the collagen matrix. These gaps are suspected as being the spot for initial mineral deposition (Shipman *et al.*, 1985; Currey, 2002). Bone mineral, most often in the form of hydroxyapatite ($Ca_{10}(PO_4)_6OH_2$) is believed to deposit first into the 64–67 nm gap spaces of the individual collagen fibril framework (Shipman *et al.*, 1985). Bone minerals are small crystals with a high surface area to volume ratio (Currey, 2002) that continue to grow along the long axis of the collagen fibrils.

At the molecular level, bone is formed of collagen microfibrils and mineralized with calcium phosphate crystals, giving bone its elastic and rigid material properties. The association between collagen and mineral is one where only the water molecule can infiltrate the space between the two (Child, 1996). Collagen comprises greater than 90% of the proteins in bone (Leaver, 1979). Glycine is the primary amino acid (Shipman *et al.*, 1985), and the amino acid composition of collagen follows a glycine–X–Y pattern with proline and hydroxyproline substituted most often for X and Y, respectively (Child, 1995; Currey, 2002). Currey (2002) explains that collagen synthesis occurs in the bone-forming cells (osteoblasts), where two polypeptides of identical amino acid sequence are combined with a third polypeptide of a slightly different amino acid sequence, forming a tropocollagen complex, which is a triple helix structure. It is at this molecular level where microbial destruction of bone occurs. Collagenase-positive organisms hydrolyze the tropocollagen bonds and cleave peptides following a specific amino acid sequence pattern (Child, 1996).

As an organ, bone is a dynamic tissue with a micromorphology that adapts according to growth speed, wear (age), and mechanical demands (Francillon-Vieillot *et al.*, 1990). Primary bone is formed during embryogenesis and initial bone development. The speed of growth is a determinant of the micromorphology of primary bone (Francillon-Vieillot *et al.*, 1990). Bone morphology differences between fast-growing and slow-growing mammals allow us to differentiate between nonhuman animal and human (not to the exclusion of other mammals) in fragmented remains. Secondary bone is a highly organized bone that is generated through increased vascularization and mechanical stress to form mature lamellar bone (Francillon-Vieillot *et al.*, 1990; Currey, 2002). The structure of secondary bone is organized into units of Haversian systems or osteons and delineated by a hypermineralized reversal line (Figure 12.1). The process of remodeling is the underlying mechanism by which Haversian systems emerge, fragment, and reappear, all depending on biomechanical need. This change over time allows for the estimation of biological age from bone microstructure. As bone formation responds to mechanical demands, the orientation of collagen fibrils, as visualized using polarized light, can aid in the reconstruction of bone use patterns (Bromage *et al.*, 2003).

Diseased bone, on the other hand, presents alternatives to normal bone remodeling and microarchitecture. General disease classifications that have the ability to change bone microstructure arise from vascular problems, inflammatory responses, nutritional

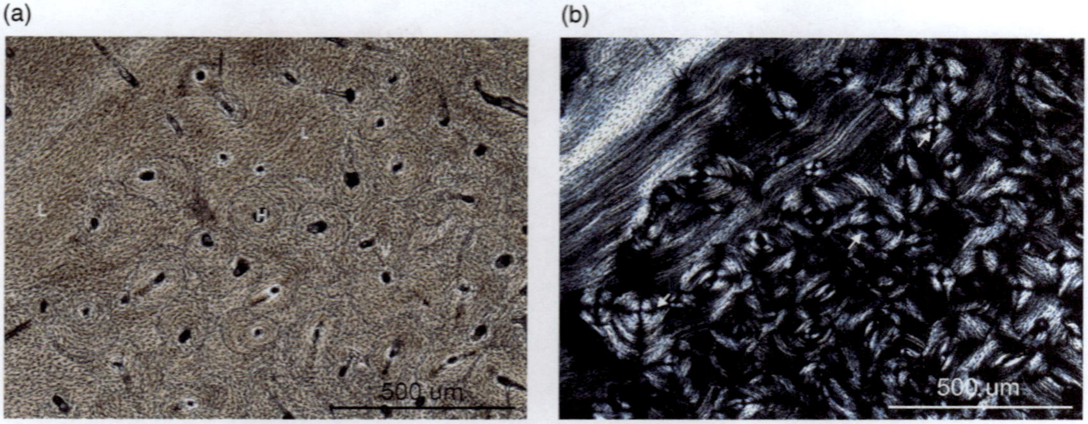

Figure 12.1 Thin section of a human femur. Secondary osteons, Haversian canals (H), a Volkmann's canal (V), and lamellar bone (L) are visualized under transmitted light and marked accordingly (a). Birefringence of mineralized collagen fibrils arranged in concentric lamellae in osteons is observed in polarized light (b). Several osteons display Maltese cross extinction patterns of birefringence (arrows).

stresses, metabolic diseases, or neoplasms (Ragsdale and Lehmer, 2012). In response to disease, bone may be thickened or thinned as lesions are typically proliferative or destructive, and/or specific microarchitectural features consistent with general disease processes may be present (Turner-Walker and Mays, 2008). As infectious agents, microbes may elicit skeletal change that can be observed micro- and macroscopically (Aufderheide and Rodriguez-Martin, 1998; Ortner, 2003). Destructive foci of bone often occur as a result of the host immune response. For example, bacilli of *Mycobacterium tuberculosis*, may spread from the primary focus of the lungs via blood and lymph to the skeleton (Aufderheide and Rodriguez-Martin, 1998), and an aggressive host immunity against the invasive organism often destroys adjacent tissues in the process (Figure 12.2). Further, von Hunnius *et al.* (2006) cite Schultz's earlier works (1994, 2001) suggesting the presence of banded lines separating cortical tissue from new periosteal bone, and villous polyp-like formations of dense lamellae are formed "during bouts of infection" with venereal disease (Von Hunnius *et al.*, 2006). The presence of these features, along with other macroscopic observations, precipitated diagnosis of venereal syphilis in pre-Columbian Europe (Von Hunnius *et al.*, 2006).

12.3 Microbially mediated decomposition

While recognition of altered microstructure due to disease rather than diagenesis is crucial for understanding disease etiology and epidemiology of archaeological populations (Schultz, 2001), the same is true for understanding taphonomic processes that

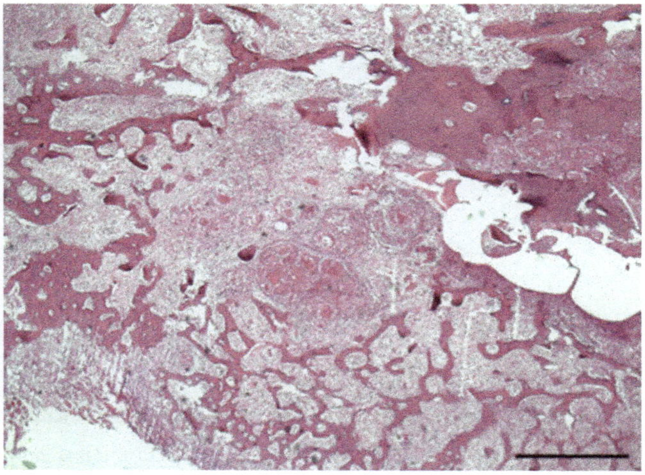

Figure 12.2 Section of a metacarpal from a 64-year-old male with disseminated coccidiodomycosis, a fungal infection caused by *Coccidoides immitis* endemic to the San Joaquin Valley in California and the Rio Grande River Valley of Texas and New Mexico. The section shows chronic inflammation with fibrosis and granuloma formation, causing loss of bone (B). The section is decalcified and stained with haematoxylin and eosin (National Museum of Health and Medicine 2009.0004; AFIP Orthopathology Study Set A, MS78. Source: Image by Michelle L. Davenport, scale bar is 1 mm).

cause microstructural changes to bone after death. Understanding taphonomic effects of microorganisms is crucial for interpreting postmortem histories of recovered skeletal evidence.

Microbially mediated decomposition begins with autolysis. Autolysis is the enzymatic self-destruction of the body's cells. It is initiated with the rapid depletion of oxygen and the cessation of ATP synthesis. While this early stage of decomposition has little to no effect on calcified tissue, various substrata (i.e., mucosa, skin) of the body that are engineered to protect the host from infection in life are no longer viable, permitting the migration of indigenous microbes across various tissue planes and vascular networks, catabolizing organic structures, eventually gaining access to bone through vascular channels.

The early postmortem period is dominated by endogenous obligate and facultative anaerobic organisms of the Bacteroidetes and Firmicutes, which is discussed in greater detail elsewhere (Hyde *et al.*, 2013; Metcalf *et al.*, 2013; Pechal *et al.*, 2013; Can *et al.*, 2014) and also in other chapters of this textbook. Post-rupture flushes the decomposing resource with oxygen and triggers a succession to copiotrophic (fast-growing) organisms on the body and in the associated gravesoil (Metcalf *et al.*, 2013; Cobaugh *et al.*, 2015). The transition is accompanied by a decrease in microbial community function, with members of Bacteroidetes and Firmicutes persisting in soil (Carter *et al.*, 2015; Cobaugh *et al.*, 2015). As the bone decays, it is exposed to a consortium of persistent enteric and soil microbial communities. The transition from Bacteroidetes and Firmicutes to

Proteobacteria and Actinobacteria in bone is accompanied by a reemergence of Acidobacteria in gravesoils and bones (Cobaugh *et al.*, 2015; Damann *et al.*, 2015). These data are not only contributing to a greater understanding of decomposition ecology, nutrient cycling, and microbial foraging behavior, but are also demonstrating utility in medicolegal investigations by being able to estimate a PMI (Metcalf *et al.*, 2013; Pechal *et al.*, 2013; Damann *et al.*, 2015).

12.4 Bone bioerosion

Microbial alteration of bone microstructure likely originates with members of endogenous and local soil bacterial and fungal communities. These alterations can be visualized histologically (Figures 12.3 and 12.4). Over the years, several publications have covered this topic. Observations have coalesced into four basic descriptive types of microscopical focal destructions (MFD) that are characterized by their morphology and hypermineralized cuff (Jans *et al.*, 2002). Linear longitudinal, lamellate, and budded have been ascribed to bacterial processes (Hackett, 1981), while Wedl tunneling is understood to arise from fungal hyphae perforating bone surfaces and extending toward the central cortical region (Wedl, 1864; Hackett, 1981; Trueman and Martill, 2002). More recently, cyanobacteria have been implicated in Wedl-type tunneling of bone in aquatic environments (Turner-Walker and Jans, 2008).

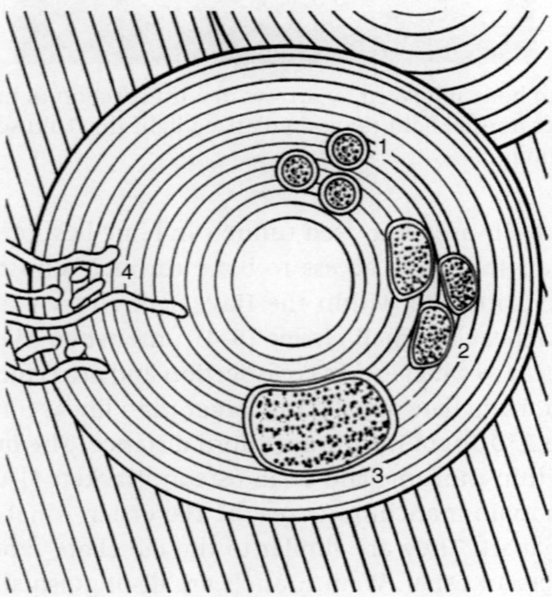

Figure 12.3 Schematic representation of microbial alteration of bone at the microscopic level displaying linear longitudinal (1), budded (2), lamellate (3), and (4) Wedl tunneling.

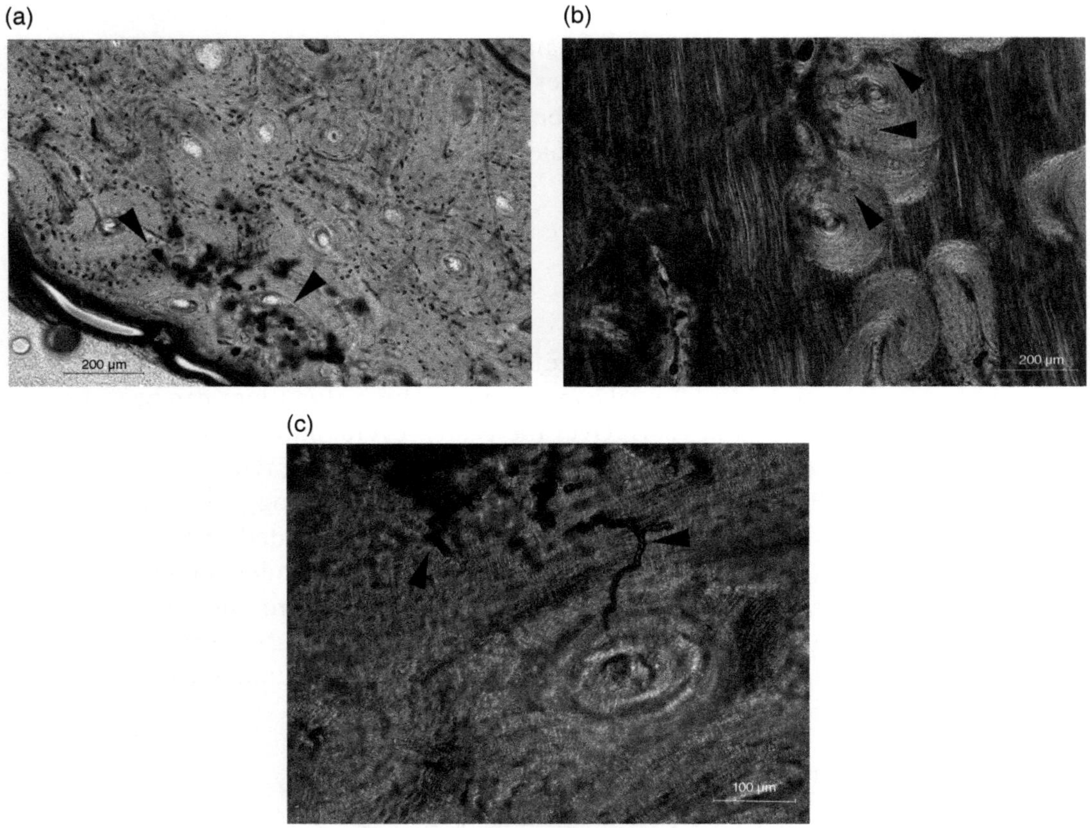

Figure 12.4 Bioerosion of bone indicated by arrows in a thin section of human femur displaying bacterial alteration under normal light (a) and polarized light (b). Thin section of cattle bone displaying fungal alteration (c).

Linear longitudinal MFD are excavated tunnels in bone best observed in longitudinal sections as bacteria generally gain access to bone through sinusoids of the medullary canal and vasculature contained within the Haversian and Volkmann's canals. Linear longitudinal tunnels appear as small circles (5–10 µm) when viewed in cross section. Lamellate tunnels appear as small foci in cross section and are typically within the lamellate rings of an osteon. Lamellate are slightly larger than linear longitudinal and vary in size from 10–20 µm to 250 µm in maximum chord distance. The budded-type of bioerosion is often found within the lamellate of an osteon. Hackett (1981) described budded foci as frond-like, as a main branch traverses the Haversian canal and buds at distances approximately 85 µm apart. They are similar to the lamellate-type, but they are found tangential to the Haversian canal. Wedl tunnels are distinguished from the other three types by having a relatively uniform tunnel diameter (5–10 µm) with a meandering tract that passes away from the cortical surfaces and osteonal canals (Hackett, 1981).

Trueman and Martill (2002) differentiated Wedl tunneling into two subcategories based on location. Type 1 Wedl tunnels appear at the periosteal and endosteal surfaces, while Type 2 tunnels are generally located in osteonal bone (Trueman and Martill, 2002).

12.4.1 Mechanisms, timing, and source of microbial interaction

Bioerosion is believed to occur as an early diagenetic change (Jans, 2014), where microbial access occurs via vascular networks and across decomposing tissue planes. Alternatively, cyanobacteria, fungi, and edaphic organisms gain access to bones by colonizing surfaces, where they may produce the Wedl type of tunneling (Turner-Walker and Jans, 2008).

Bell *et al.* (1996) observed MFD in a bone sample seven months postmortem that passed through the gastrointestinal tract of a predator/scavenger. Fernández-Jalvo *et al.* (2002) demonstrated that predator digestion exposed bone to enzymatic activity and low pH, changed the element composition, and altered the mineral phase. These changes weaken the structural integrity of bone and may promote rapid colonization by environmental microbes.

Yoshino *et al.* (1991) observed alterations in excised and buried humeral sections as early as two and a half years postmortem. Most buried bones had some form of MFD by 5 years postmortem. With time, a pattern of bioerosion emerged in the buried samples. The external and internal surfaces of long bones and internal concentric lamellae of Haversian systems located in the outer cortical zone were eroded first. Subsequent erosion in older samples included the middle zone of compact bone, and hypermineralized boards around erosive lesions appeared. In the same study (Yoshino *et al.*, 1991), bones left on the surface did not show signs of bioerosion, except for one 15-year postmortem sample. This sample included bacteria and fungi in an area of low radiographic density along the outer zone of the external cortex. Clearly this study indicated that buried bones facilitated the destruction of bones more than those bones that were left on the surface. This result was likely achieved due to the greater moisture retention of the surrounding soil than that of the bone exposed on the surface. Moisture may facilitate mineral dissolution, thereby creating more pore spaces for soil microbes to colonize bone.

However, Jans *et al.* (2004) determined that bone elements entering the burial environment disarticulated had less evidence of microbial bioerosion than those that were interred as a complete body, concluding that bioerosion originated with the spread of endogenous microbes. Contrasting this finding with Yoshino *et al.* (1991) would explain why bioerosion in the Yoshino samples followed a pattern from the exposed bone surfaces to the internal structures with advancing time as they were disarticulated prior to deposition and did not go through autolysis and putrefaction.

In another study evaluating the presence of bioerosion between surface and buried bone, White and Booth (2014) decomposed pigs, thereby exposing the bone to endogenous microorganisms. They reported increased presence of amalgamated lacunae in immature pigs buried for 1 year, relative to those decomposing on the surface for the same time period. They suggested that amalgamated osteocyte lacunae are a taphonomic effect caused by prolonged exposure to putrefactive bacteria. Surface samples experienced rapid biomass loss, had a short exposure time to putrefactive bacteria, and had less evidence of bioerosion. Interestingly, MFD sensu stricto were not observed in any samples, and the authors proposed that amalgamated lacunae are precursor features of non-Wedl-type MFD. If amalgamated lacunae are products of bacterial bioerosion, then the results provide further evidence on the role of endogenous organisms in the destruction of bone, as most samples in the study of Yoshino *et al.* (1991) did not show evidence of bioerosion until 5 years postmortem.

12.4.2 Exploration of bioerosion and bacterial community analysis

Despite the amount of work produced thus far on microbial bioerosion of forensic and archaeological bone, there is still very little known about the communities of organisms in postmortem bone samples. Studies isolating microorganisms from archaeological bones and then inoculating the identified organisms on unaltered bones in an effort to confirm a causative agent of bioerosion have been met with mixed results (Marchiafava *et al.*, 1974; Grupe and Piepenbrink, 1988; Child *et al.*, 1993). This finding, coupled with those of Jans *et al.* (2004) and White and Booth (2014), suggests that autolysis and putrefaction are necessary precursors for the colonization of bones by microbes. The close association of collagen with mineral is what typically promotes the survival of bone in the archaeological record (Child, 1996). Since bone is a complex structure evolved to secure nutrients, exposure to these decomposition processes may lessen the intimate association of collagen with the mineral component of bone.

In order to investigate any potential links between microbial communities and histological bone preservation, bone samples (ribs) were taken from 14 bodies at different surface locations at the University of Tennessee Anthropology Research Facility in Knoxville, TN, USA. They were evaluated using thin section histology (Jans, 2014) and clonal amplification and sequencing of the 16S rRNA gene (Damann *et al.*, 2015). The PMI of the remains sampled ranged from less than 1 month to 4 years.

The general histological index (GHI) developed by Hollund *et al.* (2012), which is analogous to the oxford histological index by Hedges *et al.* (1995), was applied to the analyses of these bones. The GHI consists of a scale representing the percentage of well-preserved microstructure in a thin section, ranging from 0 (no original bone structure left) to 5 (pristine bone). The thin sections showed no visible degradation of the bone structure, and no recognizable soft tissue remained in any sample with the exception of adipocere (see the following text), which was found in some samples (Figure 12.5a). The amount and presence of adipocere, which was found in the natural bone pores such as Haversian canals and osteocyte lacunae, was the only microscopical variable that showed change over time (Figure 12.5b).

Adipocere has a whitish-grey appearance and an acicular crystalline structure formation (Evans, 1963). It is visible in thin section appearing as a birefringent

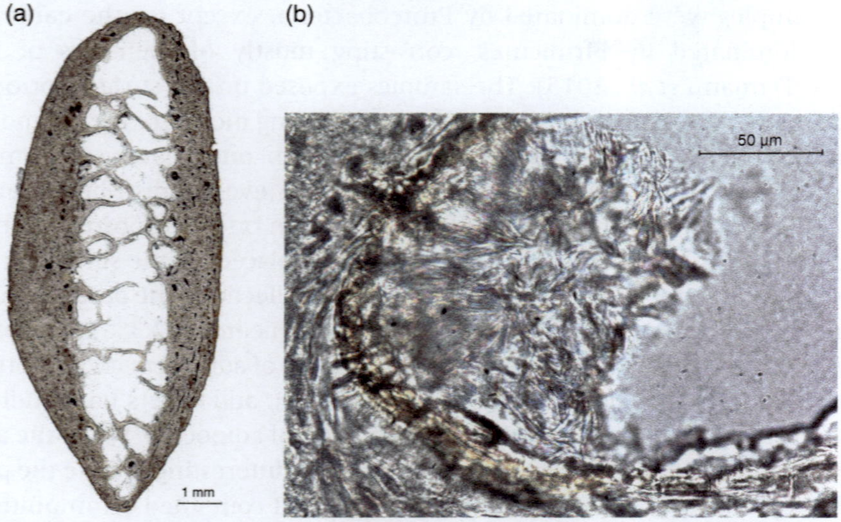

Figure 12.5 Complete transverse thin section of a human rib (a) with a magnified image showing adipocere crystalline structures (b).

amorphous to slightly crystalline yellowish material under polarized light. Adipocere forms from the hydrolysis and hydrogenation of the body's triglycerides in adipose tissue (Forbes *et al.*, 2004). Hydrolysis occurs through internal lipases and "in the presence of bacterial enzymes, especially those of the Clostridia group" (Evans, 1963). Adipocere may slow decomposition, preserving adjacent or underlying soft and skeletal tissues as well as chemical signatures of medicolegal importance (Mant and Furbank, 1957; Evans, 1963; Ubelaker and Zarenko, 2011).

The amount of adipocere present in a bone section clearly decreased with advancing time (Jans, 2014). There was nothing visible in the two oldest samples, while samples exposed between 12 and 24 months showed intermediate amounts of adipocere (adipocere present in scattered Haversian canals or osteocyte lacunae). All but two samples younger than 12 months showed large amounts of adipocere formation. Those that did not show large amounts of adipocere were exposed for less than a month in the late summer and had been completely skeletonized by insect activity. Presumably there had not been sufficient time or substrate for adipocere formation.

Despite the lack of MFD (Hackett, 1981) or presumed precursor MFD (White and Booth, 2014) observed within the bone microstructure, robust communities of bacteria were identified based on molecular analyses. Prior to pulverizing, the bone samples were abraded to remove the external and internal surfaces and cleaned in a series of ultrasonic washes following Loreille *et al.* (2007); thus the communities of organisms identified originate from pulverized cortical bone of the ribs.

All bone samples were dominated by Proteobacteria, except for the earliest sample, which was dominated by Firmicutes, consisting mostly of members of the order Clostridiales (Damann *et al.*, 2015). The samples exposed on the surface for one year or less were dominated by Gammaproteobacteria, consisting mostly of Pseudomonadaceae. Those that decomposed on the surface for more than one year were dominated by Alphaproteobacteria, but the community was more evenly distributed among the bacteria present than compared to the samples less than one year postmortem.

Three samples from three different individuals were placed on the surface in the fall of 2006 and decomposed for 12 months, prior to sample collection. The bacterial community of one sample was represented overwhelmingly by Firmicutes (92.3%), of which 41.8% were *Clostridium*. This specimen contained an abundance of adipocere crystal formations in nearly every Haversian canal. Evans (1963), Vass (2012), and others have indicated previously the suspected role of *Clostridia* in the formation of adipocere. While the association of these two findings is in line with previous reports, it is interesting to note the persistence of *Clostridia* when the two other samples of a similar PMI contained communities that did not include *Clostridia*. Conditions within the rib facilitated adipocere formation by early colonizers (Firmicutes), and the reduced aerobic decay delayed community succession. These observations would suggest that skeletal samples with high concentrations of adipocere formations may not be appropriate for estimating PMI based on bacterial community assemblage, as the estimates would underestimate the actual PMI.

12.5 Reconstructing postmortem histories

Differentiating microbially erosive types and the associated community, coupled with detailed descriptions on inclusions and infiltrations, provides information on the postmortem history of the remains, which may prove useful for understanding time since death and location of decay. Matter included in the pore spaces of decaying bone may include soil particles, fruiting fungal bodies, framboidal pyrites—spherical aggregates of iron sulfide mineral—and various crystal formations of brushite and adipocere (Garland, 1989). The infiltration of humic acids, pigments of microbial biofilm production, and oxidized metal ions and microbial metabolites from the surrounding burial context may also provide insight to the changing ecological footprint of the decaying bone.

Turner-Walker and Jans (2008) demonstrated that the ecological niche of decaying skeletal remains could be recreated. Bacterial and fungal bioerosion increases bone porosity, which occurs predominantly in aerobic environments with well-drained soils (Jans *et al.*, 2002; Turner-Walker and Syversen, 2002). When skeletal elements transit to a burial environment where oxygen permeability and pH are reduced, carbonate exchange of the mineral fraction between bone and soil occurs, and framboidal pyrites may develop (Turner-Walker and Jans, 2008). Anoxic environments promote increased activity of sulfate-reducing bacteria, which are responsible for pyrite synthesis (Turner-Walker and Jans, 2008). Over time, bioturbation and other site formation processes may alter the burial environment, exposing the skeleton to aerated soils, and infiltration of pigments from oxidized heavy metals and crystals introduced by hydrology and fungi (Grupe and Piepenbrink, 1988) or absorbed as a result of increased bone porosity.

Knowing the taphonomic and ecological history of a bone is essential for subsequent interpretation and overcoming problems of equifinality (a concept where in an open system many different avenues may produce a single end state). A perfect example of this concept is the hypothesis championed by paleontologist Charles Brain that bone fractures in animal remains recovered at Makapansgat—a cave site in South Africa—resulted from normal ecological processes (Brain, 1981) rather than Dart's earlier hypothesis where human ancestors (australopithecines) were intentionally smashing bones to salvage a necessary nutrient load (Dart, 1957).

Similarly, Dominguez-Rodrigo and Barba (2006) reinterpreted Blumenschine's (1995) carnivore–hominid–carnivore hypothesis for the FLK Zinj assemblage of animal remains at Olduvai Gorge in East Africa. Dominguez-Rodrigo and Barba (2006) suggested that many of the previously believed felid tooth marks were in fact biochemical taphonomic effects of saprotrophic fungi. By paying attention to the larger ecological context, their findings changed the foraging behavior concept from casual scavengers acquiring resources after felid activity, to one where they were hunting and processing food sources.

To medicolegal contexts, microbial interactions with the recently dead (Hyde *et al.*, 2013; Metcalf *et al.*, 2013; Pechal *et al.*, 2013; Can *et al.*, 2014; Carter *et al.*, 2015; Hauther *et al.*, 2015) and the decaying skeleton (Damann *et al.*, 2015) are being explored in order to improve PMI estimations. Furthermore, Koehn (2013) recently investigated the origin of unknown PCR products generated during forensic STR analysis of bones, concluding that the spurious peaks originated from microbial agents that had taken up residence in the bone. Unaccounted-for spurious peaks that appear within specific STR-call regions may complicate individual identification efforts based on DNA analyses.

At the same time, these forensically oriented data may prove beneficial to areas of investigation that rely on chemical and biological material contained within bone. Grupe and Piepenbrink (1988) identified the presence of elevated heavy metal contaminants in bone from infiltrating microorganisms. Similarly, Balzer *et al.* (1997) demonstrated a significant reduction of carbon through the selective loss of amino acids that constitute Type I collagen, presumably as a result of selective bacterial metabolism. Bulk nitrogen levels, on the other hand, were enriched in the same samples, suggesting a recycling and immobilization of nitrogen within the substrate (Balzer *et al.*, 1997). Stable isotope values for nitrogen would then be artificially elevated, suggesting an interpretation that foraging behavior was reliant on meat (Balzer *et al.*, 1997).

12.6 Conclusions

Microbes are significant taphonomic agents of human decomposition and bone diagenesis. Their postmortem activity alters bone microstructure and selectively catabolizes endogenous compounds, complicating an anthropologist's use of bone as evidence for diet reconstruction and histological age-at-death determinations, and may even create potential problems with individual identifications based on STR analyses of degraded bone. The interactions between microorganisms, the body, and skeleton are more than just a cautionary tale, however. The recent inclusion of a microbe-centric point of view to questions of anthropological and medicolegal importance not only is identifying new methods for estimating PMI based on succession of microbial communities with advancing time, but also is adding to our understanding of taphonomic and ecological processes and even human biological variation, a central theme of the physical/forensic anthropologist.

References

Antoine, S.E., Child, A.M., Nicholson, R.A. and Pollard, A.M. (1992) The biochemistry and microbiology of buried human bone, in relation to dietary reconstruction. *Circaea*, **9**, 65–79.

Aufderheide, A.C. and Rodriguez-Martin, C. (1998) *The Cambridge Encyclopedia of Human Paleopathology*, Cambridge University Press, Cambridge.

Balzer, A., Gleixner, G., Grupe, G. *et al.* (1997) In vitro decomposition of bone collagen by soil bacteria: the implications for stable isotope analysis in archaeometry. *Archaeometry*, **39**, 415–429.

Bell, L.S., Skinner, M.F. and Jones, S.J. (1996) The speed of post mortem change to the human skeleton and its taphonomic significance. *Forensic Science International*, **82**, 129–140.

Blumenschine, R.J. (1995) Percussion marks, tooth marks and the experimental determinations of the timing of hominid and carnivore access to long bones at FLK Zinjanthropus, Olduvai Gorge, Tanzania. *Journal of Human Evolution*, **29**, 21–51.

Brain, C.K. (1981) *Hunters or the Hunted? An Introduction to African Cave Taphonomy*, University of Chicago Press, Chicago.

Bromage, T.G., Goldman, H.M., McFarlin, S.C. *et al.* (2003) Circularly polarized light standards for investigations of collagen fiber orientation in bone. *Anatomical Record. Part B, New Anatomist*, **274B**, 157–168.

Buchholz, B.A. (2009) Bomb-pulse dating, in *Wiley Encyclopedia of Forensic Science* (eds A. Jamieson and A. Moenssens), John Wiley & Sons, Inc., New York, pp. 418–422.

Can, I., Javan, G.T., Pozhitkov, A.E. and Noble, P.A. (2014) Distinctive thanatomicrobiome signatures found in the blood and internal organs of humans. *Journal of Microbiological Methods*, **106**, 1–7.

Carter, D.O., Metcalf, J.L., Bibat, A. and Knight, R. (2015) Seasonal variation of post-mortem microbial communities. *Forensic Science, Medicine and Pathology*, **11**, 202–207.

Child, A.M., Gillard, R.D. and Pollard, A.M. (1993) Microbially-induced promotion of amino acid racemization in bone: isolation of the organisms and detection of their enzymes. *Journal of Archaeological Science*, **20**, 159–168.

Child, A.M. (1995) Towards an understanding of the microbial decomposition of archaeological bone in the burial environment. *Journal of Archaeological Science*, **22**, 165–174.

Child, A.M. (1996) Amino acid racemization and the effects of microbial diagenesis, in *Archaeological Chemistry: Organic, Inorganic, and Biochemical Analysis*, ACS Symposium Series 625 (ed M.V. Orna), American Chemical Society, Washington, DC, pp. 366–377.

Cobaugh, K.L., Schaeffer, S.M. and DeBruyn, J.M. (2015) Functional and structural succession of soil microbial communities below decomposing human cadavers. *PLoS ONE*, **10**, e0130201.

Conly, J.M. and Stein, K. (1992) The production of menaquinones (vitamin K2) by intestinal bacteria and their role in maintaining coagulation homeostasis. *Progress in Food and Nutrition Science*, **16**, 307–343.

Crescimanno, A. and Stout, S. (2012) Differentiating fragmented human and nonhuman long bone using osteon circularity. *Journal of Forensic Sciences*, **57**, 287–294.

Cuijpers, A.F.G.M. (2009) *The Application of Bone Histology for Species Identification in Archaeological; with a Photo Catalogue*, Geoarchaeological and Bioarchaeological Studies 12, Institute for Geo and Bioarchaeology, Vrije Universiteit, Amsterdam.

Currey, J.D. (2002) *Bones: Structure and Mechanics*, Princeton University Press, Princeton.

Damann, F.E., Williams, D.E. and Layton, A.L. (2015) Potential use of bacterial community succession in decaying human bone for estimating postmortem interval. *Journal of Forensic Sciences*, **60**, 844–850.

Dart, R.A. (1957) *The Osteodontokeratic Culture of Australopithecus Prometheus*, Transvaal Museum Memoir, No 10, Transvaal Museum, Pretoria.

DeNiro, M.J. (1985) Postmortem preservation and alteration of in vivo bone collagen isotope ratios in relation to palaeodietary reconstruction. *Nature*, **317**, 806–809.

Dominguez-Rodrigo, M. and Barba, R. (2006) New estimates of tooth mark and percussion mark frequencies at the FLK Zinj site: the carnivore-hominid-carnivore hypothesis falsified. *Journal of Human Evolution*, **50**, 170–194.

Edson, S.M., Ross, J.P., Coble, M.D. *et al.* (2004) Naming the dead, confronting the realities of rapid identification of degraded skeletal remains. *Forensic Science Review*, **16**, 63–88.

Evans, W.E.D. (1963) *The Chemistry of Death*, Charles C Thomas, Springfield, IL.

Fernández-Jalvo, Y., Sánchez-Chillón, B., Andrews, P. *et al.* (2002) Morphological taphonomic transformations of fossil bones in continental environments, and repercussions on their chemical composition. *Archaeometry*, **44**, 353–362.

Fernández-Jalvo, Y., Andrews, P., Pesquero, D. *et al.* (2010) Early bone diagenesis in temperate environments part I: surface features and histology. *Paleogeography, Palaeoclimatology, Palaeoecology*, **288**, 62–81.

Font, L., Van der Peijl, G., Van Leuwen, C. *et al.* (2014) Identification of the geographical origin of an unidentified individual by multi-isotope analysis. *Science and Justice*, **55**, 34–42.

Forbes, S.L., Stuart, B.H., Dadour, I.R. and Dent, B.B. (2004) A preliminary investigation of the stages of adipocere formation. *Journal of Forensic Sciences*, **49**, 566–574.

Francillon-Vieillot, H., de Buffrenil, V., Castanet, J. *et al.* (1990) Microstructure and mineralization of vertebrate skeletal tissues, in *Skeletal Biomineralization Patterns, Processes, and Evolutionary Trends* (ed J.G. Carter), Van Nostrand Reinhild, New York, pp. 471–548.

Garland, A.N. (1989) Microscopical analysis of fossil bone. *Applied Geochemistry*, **4**, 215–229.

Geigl, E.M. (2002) On the circumstances surrounding the preservation and analysis of very old DNA. *Archaeometry*, **44**, 337–342.

Grupe, G. and Piepenbrink, H. (1988) Trace element contaminations in excavated bones by microorganisms, in *Trace Elements in Environmental History* (eds G. Grupe and B. Herrmann), Springer-Verlag, Berlin, pp. 103–112.

Hackett, C.L. (1981) Microscopical focal destruction (tunnels) in exhumed human bones. *Medicine, Science and the Law*, **21**, 243–265.

Hauther, K.A., Cobaugh, K.L., Jantz, L.M. *et al.* (2015) Estimating time since death from postmortem human gut microbial communities. *Journal of Forensic Sciences*, **60** (5), 1234–1240.

Hedges, R.E.M., Millard, A.R. and Pike, A.W.G. (1995) Measurement s and relationships of diagenetic alteration of bone from three archaeological sites. *Journal of Archaeological Science*, **22**, 201–209.

Hillier, M.L. and Bell, L.S. (2007) Differentiating human bone from animal bones: a review of histological methods. *Journal of Forensic Sciences*, **52**, 249–263.

Hollund, H.I., Jans, M.M.E., Collins, M.J. *et al.* (2012) What happened here? Bone histology as a tool in decoding the postmortem histories of archaeological bone from Castricum, the Netherlands. *International Journal of Osteoarchaeology*, **22**, 537–548.

Hyde, E.R., Haarmann, D.P., Lynne, A.M. *et al.* (2013) The living dead: bacterial community structure of a cadaver at the onset and end of the bloat stage of decomposition. *PLoS ONE*, **8**, e77733.

Jans, M.M.E. (2014) Microscopic destruction of bone, in *Manual of Forensic Taphonomy* (eds J.T. Pokines and S.A. Symes), CRC Press, Boca Raton, pp. 19–35.

Jans, M.M., Kars, H., Nielsen-Marsh, C.M. *et al.* (2002) In situ preservation of archaeological bone: a histological study within multidisciplinary approach. *Archaeometry*, **44**, 343–352.

Jans, M.M.E., Nielsen-Marsh, C.M., Smith, C.I. *et al.* (2004) Characterisation of microbial attack on archaeological bone. *Journal of Archaeological Science*, **31**, 87–95.

Kerley, E.R. (1965) The microscopic determination of age in human bone. *American Journal of Physical Anthropology*, **23**, 149–164.

Koehn, A. (2013) Identification of unknown PCR products generated during STR analysis of DNA from bone samples. Unpublished MS thesis, University of North Texas Health Science Center, Fort Worth.

Leaver, A.G. (1979) Noncollagenous proteins, in *Skeletal Research: An Experimental Approach* (eds D.J. Simmons and A.S. Kunin), Academic Press, New York, pp. 194–224.

Loreille, O.M., Diegoli, T.M., Irwin, J.A. *et al.* (2007) High efficiency DNA extraction from bone by total demineralization. *Forensic Science International: Genetics*, **1**, 191–195.

Marchiafava, V., Bonucci, E. and Ascenzi, A. (1974) Fungal osteoclasia: a model of dead bone resorption. *Calcified Tissue Research*, **14**, 195–210.

Mant, A.K. and Furbank, R. (1957) Adipocere—a review. *Journal of Forensic Medicine*, **4**, 18–35.

Mays, S. and Turner-Walker, G. (2008) A possible case of renal osteodystrophy in a skeleton from medieval Wharram Percy, England. *International Journal of Osteoarchaeology*, **18**, 307–316.

Metcalf, J.L., Wegener Parfrey, L., Gonzalez, A. *et al.* (2013) A microbial clock provides an accurate estimate of the postmortem interval in a mouse model system. *eLife*, **2**, e01104.

Mulhern, D.M. and Ubelaker, D.H. (2012) Differentiating human from nonhuman bone microstructure, in *Bone Histology. An Anthropological Perspective* (eds C. Crowder and S. Stout), CRC Press, Boca Raton, pp. 109–134.

Nielsen-Marsh, C.M. and Hedges, R.E.M. (2000) Patterns of diagenesis in bone II; effects of acetic acid treatment and the removal of diagenetic CO_3^{3-}. *Journal of Archaeological Science*, **27**, 1151–1159.

Ortner, D. (2003) *Identification of Pathological Conditions in Human Skeletal Remains*, Elsevier Science (USA), San Diego.

Ovchinnikov, I.V., Götherström, A., Romanoval, G.P. *et al.* (2000) Molecular analysis of Neanderthal DNA from the northern Caucasus. *Nature*, **404**, 490–493.

Pechal, J.L., Crippen, T.L., Benbow, M.E. *et al.* (2013) The potential use of bacterial community succession in forensics as described by high throughput metagenomic sequencing. *International Journal of Legal Medicine*, **128**, 193–205.

Piepenbrink, H. (1986) Two examples of biogenous dead bone decomposition and their consequences for taphonomic interpretation. *Journal of Archaeological Science*, **13**, 417–430.

Ragsdale, B.D. and Lehmer, L.M. (2012) A knowledge of bone at the cellular (histological) level is essential to paleopathology, in *A Companion to Paleopathology* (ed A.L. Grauer), Blackwell Publishing, Malden, pp. 227–249.

Schmitz, R., Serre, D., Bobani, G. *et al.* (2002) The Neandertal type site revisited: interdisciplinary investigations of skeletal remains from the Neander Valley, Germany. *Proceedings of the National Academy of Sciences*, **99**, 13342–13347.

Schultz, M. (1994) Comparative histopathology of syphilitic lesions in prehistoric and historic human bones, in *The Origin of Syphilis in Europe: Before or After 1493?* (eds O. Dutour, G. Pa´lfi, J. Be´rato and J.-P. Brun), Centre Archeologique du Var, Toulon, pp. 63–67.

Schultz, M. (2001) Paleohistopathology of bone: a new approach to the study of ancient diseases. *Yearbook of Physical Anthropology*, **44**, 106–147.

Shipman, P., Walker, A. and Bichell, D. (1985) *The Human Skeleton*, Harvard University Press, Cambridge, MA.

Stout, S.D. and Paine, R.R. (1992) Brief communication: histological age estimation using rib and clavicle. *American Journal of Physical Anthropology*, **87**, 111–115.

Trueman, C.N. and Martill, D.M. (2002) The long-term survival of bone: the role of bioerosion. *Archaeometry*, **44**, 371–382.

Tsukamoto, Y. (2004) Studies on action of menaquinone-7 in regulation of bone metabolism and its preventative role of osteoporosis. *Biofactors*, **22**, 5–19.

Turner-Walker, G. and Jans, M. (2008) Reconstructing taphonomic histories using histological analysis. *Paleogeography, Palaeoclimatology, Palaeoecology*, **266**, 227–235.

Turner-Walker, G.M. and Mays, S. (2008) Histological studies of ancient bone, in *Advances in Human Paleopathology* (eds R. Pinhasi and S. Mays), John Wiley & Sons, West Sussex, pp. 121–146.

Turner-Walker, G. and Syversen, U. (2002) Quantifying histological changes in archaeological bones using BSE-SEM image analysis. *Archaeometry*, **44**, 461–468.

Ubelaker, D.H. (2001) Artificial radiocarbon as an indicator of recent origin of organic remains in forensic cases. *Journal of Forensic Sciences*, **46**, 1285–1287.

Ubelaker, D.H. and Zarenko, K. (2011) Adipocere: what is known after over two centuries of research. *Forensic Science International*, **208**, 167–172.

Vass, A.A. (2012) Odor mortis. *Forensic Science International*, **222**, 234–241.

Von Hunnius, T.E., Roberts, C.A., Boylston, A. and Saunders, S.R. (2006) Histological identification of syphilis in Pre-Colombian England. *American Journal of Physical Anthropology*, **129**, 559–566.

Walker, P.L. and DeNiro, M.J. (1986) Stable nitrogen and carbon isotope ratios as indices of prehistoric dietary dependence on marine and terrestrial resources in Southern California. *American Journal of Physical Anthropology*, **71**, 51–61.

Wedl, C. (1864) Ueber einen Zahnbein und Knochen keimenden Pilz. *Sitzungsberict der Mathematische Naturwissenschaftliche Classe der Kaiserliche Akademie der Wissenschaften Wein Band L.* Kl 50, 171–193.

White, L. and Booth, T.J. (2014) The origin of bacteria responsible for bioerosion to the internal bone microstructure: results from experimentally-deposited pig carcasses. *Journal of Archaeological Sciences*, **239**, 92–102.

Wilson, M. (2008) *Bacteriology of Humans: An Ecological Perspective*, Blackwell Publishing, Malden.

Yoshino, M., Kimijima, T., Miyasaka, S. *et al.* (1991) Microscopical study on time since death in skeletal remains. *Forensic Science International*, **49**, 143–158.

CHAPTER 13

Forensic microbiology in built environments

Simon Lax[1,2] and Jack A. Gilbert[1,2,3,4]

[1] *Department of Ecology and Evolution, University of Chicago, Chicago, IL, USA*
[2] *Institute for Genomics and Systems Biology, Argonne National Laboratory, Lemont, IL, USA*
[3] *Department of Surgery, University of Chicago, Chicago, IL, USA*
[4] *Marine Biological Laboratory, Woods Hole, MA, USA*

13.1 Introduction

Humans are host to a vast array of microbial diversity on our skin as well as in our gastrointestinal and respiratory tracts. These microbes are readily dislodged to our surrounding environments, even as we pick up new microbes with each interaction we share with our environments or with other people, creating a constant microbial flux with our surroundings. In the built environment (e.g., homes, offices) where people in the developed world spend as much as 90% of their time (Custovic *et al.*, 1994), the microbes we deposit account for the vast majority of observed microbial diversity, making it possible to link individuals to objects and spaces through microbial similarity (Lax *et al.*, 2015). In this chapter, we review our current understanding of the factors shaping microbial diversity in built environments and discuss how microbial interaction between these environments and their occupants can be used as trace evidence in forensic science.

13.2 The human skin microbiome

We are born mostly sterile (Aagaard *et al.*, 2014) and acquire our foundational microbiota from our mothers' vaginal canal or their skin, depending on the delivery method (Dominguez-Bello *et al.*, 2010). In early life, our microbiota is highly plastic as our limited neonatal microbial community increases in diversity and may be radically altered by infection, exposure to antibiotics, exposure to healthcare environments, or changes in

Forensic Microbiology, First Edition. Edited by David O. Carter, Jeffery K. Tomberlin, M. Eric Benbow and Jessica L. Metcalf.

diet. By 3–5 years of age, our intestinal microbiota is mostly stabilized (La Rosa *et al.*, 2014), with major differences observed between modern urban settings and rural, less-industrialized societies (De Filippo *et al.*, 2010; Tyakht *et al.*, 2013; Clemente *et al.*, 2015).

Human skin is a complex ecosystem comprised of numerous folds, invaginations, and specialized niches that are colonized by a wide array of microorganisms including bacteria, fungi, viruses, and mites (Grice and Segre, 2011). We have numerous symbiotic relationships with these microorganisms, which may protect against invasion by pathogens or educate the billions of T cells found in our skin (Grice and Segre, 2011).

The majority of our symbiotic bacteria, both on our skin and in inner mucosal surfaces, fall into four phyla: the Actinobacteria, Bacteroidetes, Firmicutes, and Proteobacteria. Although both skin and gut communities are of low diversity at the phylum level, they are characterized by very high diversity at the species level, which we are only beginning to delineate through culture-independent molecular approaches (Grice *et al.*, 2009). Skin microbiome studies have found that the bacteria which colonize a skin site are dependent on the physiology of that site, with different microbes adapted to moist, dry, and sebaceous regions. *Propionibacterium* spp., for example, are especially abundant on sebaceous skin sites such as the forehead and back, while staphylococci and corynebacteria are most abundant on moist sites such as the axilla and foot sole. Dry areas such as the forearm and hand are the most diverse skin sites and are not usually dominated by a single genus (Grice *et al.*, 2009).

As our primary interface with our surroundings, the skin ecosystem is subject to constant perturbations, both endogenous (such as selection driven by host genotype) and exogenous (e.g., showering). Temporal variability of the skin microbiome varies by site, and molecular analyses have revealed that community membership and structure are most consistent in the ear, nose, and inguinal crease (Grice *et al.*, 2009). There are numerous factors that influence an individual's skin microbiome, including host physiology (e.g., sex, age), immune responses (e.g., previous exposure to infective pathogens, inflammation), host genotype, host lifestyle (e.g., occupation and hygiene), and the climate in which the individual lives (reviewed in Grice and Segre, 2011). Although skin microbial communities are volatile in their membership and structure, the interpersonal variation of symmetric skin sites that results from these myriad factors is generally greater than interpersonal variation over time (Gao *et al.*, 2007; Costello *et al.*, 2009; Bouslimani *et al.*, 2015), making possible the forensic identification of individuals from their skin microbiota.

13.3 The microbiota of the built environment

Humans occupy an ecological niche unprecedented in evolutionary history: we have become an indoor species. In the developed world, we inhabit spaces that consist of surfaces and environmental systems that are specifically designed to reduce our exposure to microbial life. This biome of indoor spaces is expansive and constantly growing, with estimates of the extent of residential and commercial buildings ranging from 1.3% (Kitzes *et al.*, 2007) to 6% (Hooke *et al.*, 2012) of ice-free land area. As urbanization has accelerated over the last century, we have become increasingly bound to this ecosystem: from the hospital we are born in to the house we grow up in, the microbial diversity of these environments plays a profound role in shaping our own microbiota and in developing our immune system.

Recent studies have uncovered the extent to which humans influence the microbiota of the buildings they inhabit. From studies encompassing homes (Lax *et al.*, 2014), classrooms (Meadow *et al.*, 2014), residential kitchens (Flores *et al.*, 2013), gyms (Wood *et al.*, 2015), public restrooms (Flores *et al.*, 2011; Gibbons *et al.*, 2015), and hospitals (Hewitt *et al.*, 2013), human skin and respiratory cavities are the primary source of microorganisms to built environments.

13.3.1 Human–home microbial dynamics

The strength of the microbial interaction between a home and its occupants was recently demonstrated by the Home Microbiome Project (Lax *et al.*, 2014), which surveyed the microbial communities associated with six home surfaces and three human skin sites in seven homes over the course of 4–6 weeks. These longitudinal surveys found a very strong correlation between the abundance of microbial taxa on home surfaces and on the skin of home occupants, with an average of two-thirds of major operational taxonomic units (OTUs) (those with greater than 100 total observations) not significantly differentiated in abundance between the two surface types. Using a Bayesian source tracking algorithm, the relative contributions of different occupants, including pets, to the microbial communities of home surfaces were modeled. When an occupant was traveling away from their home for multiple days, his/her microbial signatures either vanished or were substantially diminished, exemplifying how rapidly occupants can affect the microbiota of their surroundings (Figure 13.1). The study also used dynamic Bayesian networks to track microbial taxa as they moved between different surfaces and found that transfer from human skin to a home surface was many times more common than the transfer in the other direction, with transfer from one occupant's skin to another also shown to be widespread.

The strong association between human skin and the built environment microbiome allows for the forensic matching of spaces to their occupants. For example, samples taken from different surfaces in the same home have been shown to be significantly more similar than samples taken from the same surface in different homes, suggesting that occupants play a larger role in shaping home microbial communities than building materials or surface functions (Lax *et al.*, 2014). Models trained to predict which house a sample was taken from performed very well, correctly classifying which home a sample originated from 20 times better than expected by chance.

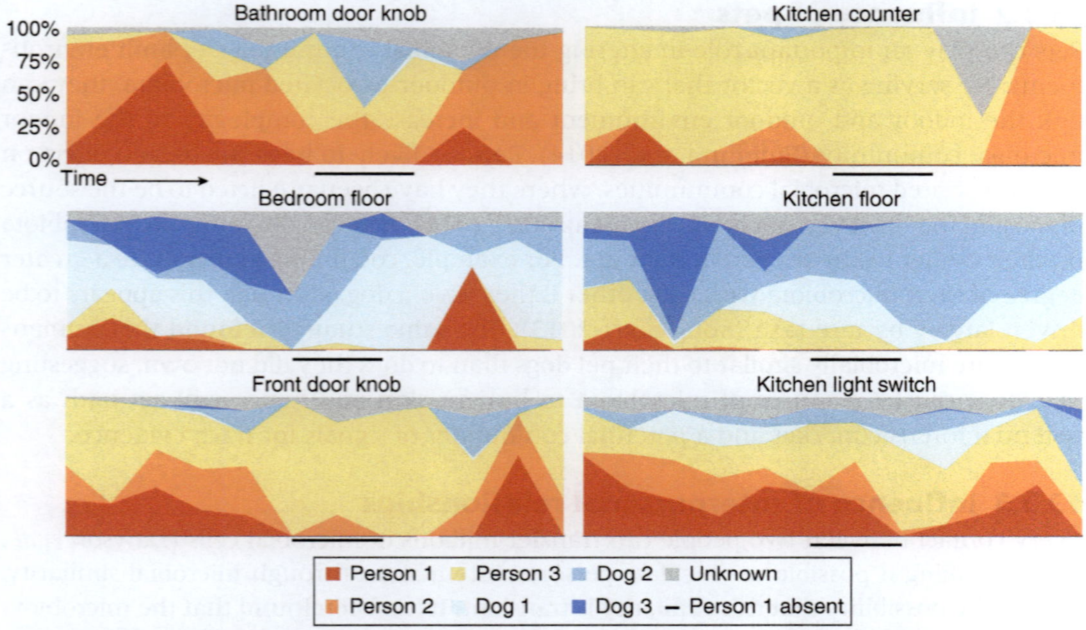

Figure 13.1 Predicted relative contributions of different occupants to the microbial communities associated with six home surfaces over the course of a 4-week sampling window. Microbial sources were inferred through the Source Tracker algorithm. Note that when Person 1 leaves the residence for travel, as indicated by the black bars along the *x*-axis, their predicted microbial contribution to the home surfaces either vanishes or diminishes, and increases after their return to the home

When only floor samples were considered, the classifiers performed approximately 50 times better than expected by chance. By contrast, classifiers trained to determine which home surface a sample was taken from were only three times better than expected by chance. This is an important consideration for forensic studies, as trace evidence samples collected from different parts of a home environment could all be equally effective for identification of the people active in that space. However, this also belies potential differences in building architecture and how people interact with different surfaces. Someone coming into a room for the first time may not interact with surfaces in the same way as someone who resides in that space. This difference in behavior could shape where effective trace evidence samples could be found or identified as a priority location for sample collection. Additionally, the design and layout of a space can affect how people interact with it, and as such these components need to be considered if trace evidence standards were to be developed. Source Tracker models have shown that the surface that is most easily matched to the skin of a home's occupants varies between houses (Lax *et al.*, 2014). This difference must be considered when developing appropriate forensic tools.

13.3.2 Influence of pets

Pets can play an important role in altering the microbial communities of built environments. By serving as a vector that can bring in outdoor-associated microbiota, they can link the indoor and outdoor environment and increase the complexity of the indoor microbial community (Fujimura *et al.*, 2014). Pets are likely to have the biggest effect on floor-associated microbial communities, where they have been predicted to be the source of most of the observed microbial taxa (Lax *et al.*, 2014). Pets can also alter the microbiota on their owner's skin or even in their gut. For example, cohabiting adults share a greater degree of their microbiota with each other if they own a dog, although this appears to be driven largely by rare taxa (Song *et al.*, 2013). The same study also found that humans were more microbially similar to their pet dogs than to dogs they did not own, suggesting that dogs can be a source of microbiota to human skin and hence will act both as a potential forensic marker and a potential confounder of signals for trace evidence.

13.3.3 Influence of interpersonal relationships

Every contact between two people can transfer millions of microbial cells (Dawson *et al.*, 2007), making it possible to elucidate personal interactions through microbial similarity, despite the possibility that this similarity is transient. It has been found that the microbiota of hands, feet, and noses are each significantly more similar between individuals inhabiting the same home than between individuals living in different homes, with the effect especially pronounced in nasal microbial communities (Lax *et al.*, 2014). Even within the same home, the microbial communities at those three sites were found to be significantly more similar between two individuals living as a physically interacting couple than between those individuals and a roommate living in a different room (Lax *et al.*, 2014).

Significant microbial transfer between individuals can occur even with only short-term contact. A study of roller derby players found that all players' skin microbiomes were more similar to one another after competition, suggesting a homogenization of skin-associated bacteria through contact sports (Meadow *et al.*, 2013). Even personal effects such as shoes and cell phones can become microbially more similar when people temporarily inhabit the same space. A recent study in which attendees at three different scientific conferences sampled their shoes and personal cell phones revealed that samples taken from the same conference were significantly more similar than samples taken from different conferences (Lax *et al.*, 2015).

13.4 Tools for the forensic classification of the built environment microbiome

13.4.1 Sampling and sequencing considerations

Culture-independent surveys have revolutionized the way in which researchers characterize microbial diversity. Rather than growing microbes in clonal cultures, these methods allow us to characterize the community holistically, providing a more comprehensive assessment compared to a small subset of taxa in isolation. Microbial communities are influenced not only by environmental factors but also by the myriad interactions between different taxa, making it imperative that we view diversity in an ecological context. We can now gain deep insights into community membership and function by directly sequencing a microbial sample, which in the case of human skin or the built environment is often collected on a sterile, moistened cotton swab rubbed against the surface of interest.

There are a number of culture-independent molecular tools that can be used to characterize a microbial community, but they can generally be classified in one of two categories: amplicon sequencing and shotgun sequencing (Chapter 5). Amplicon sequencing relies on the polymerase chain reaction (PCR) amplification of a small region of DNA, which in the case of microbial surveys is often the gene encoding the small subunit of the ribosomal RNA, known as 16S rRNA. Ribosomal genes are universal to all microbial life and are sufficiently conserved in some sequence regions to allow for the use of primers with the potential to capture a broad phylogenetic diversity; although recent work demonstrates that even "state-of-the-art" assessments are likely considerably biased against uncultured taxa with highly divergent ribosomal sequences (Brown *et al.*, 2015; Luef *et al.*, 2015). However, especially in human and built environment contexts, amplicon sequencing of the 16S rRNA gene with only a single primer set enables the rapid and inexpensive characterization of the vast majority of the microbial communities. Although this method can scale to a very high resolution of community membership, inconsistencies can result from different DNA extraction methods or from primer bias that favors the amplification of certain taxa (Feinstein *et al.*, 2009; Wesolowska-Andersen *et al.*, 2014). Additionally, because genomic content is so variable in bacteria, even between closely related phylotypes, 16S rRNA is a poor window into the functional capacity of the community. Finally, because 16S rRNA is so highly conserved, even distantly related phylotypes may have sufficiently similar sequences that they are indistinguishable in amplicon-based surveys. Although this can confound our ability to match microbial taxa to individuals in forensic studies, the ease and depth of ribosomal amplicon sequencing has made it an incredibly useful tool in forensic microbiology and it is the method employed by many studies discussed in this chapter. 16S sequencing, in its benefits and shortcomings, is analogous to 18S amplicon sequencing in eukaryotes and internal transcribed spacer (ITS) sequencing in fungi.

An alternative method is to employ shotgun metagenomic analysis, which randomly sheers DNA extracted from a sample into small fragments that can be directly sequenced. These raw sequences encompass all genomic DNA extracted from the sample, so they can be used to assess the relative abundance of gene ontologies and infer the functional potential of a community. When the sample is sequenced deeply enough, it is often possible to assemble these short reads into full metagenomes that provide much greater insight into inter-sample differentiation than amplicon sequencing. As these methods will sequence DNA indiscriminately, human DNA will often account for a large percentage of reads in surveys of human skin or built environments with shed skin cells (Lax *et al.*, 2014; Gibbons *et al.*, 2015).

13.4.2 Machine learning and statistical classification

The last decade has seen an incredible rise in the throughput and capacity of DNA sequencing technologies, which has precipitated a vast decrease in the financial cost. Next-generation sequencers like the Illumina HiSeq and MiSeq have made it possible to simultaneously sequence thousands of samples to a depth of many thousands of reads per sample. This has led to an unprecedented increase in the size and complexity of microbial datasets, which have required the development of new bioinformatic and statistical tools to enable appropriate interpretation.

Most studies store their data in an ecological data matrix, which has sample names on one axis and taxa names on the other, with the observation counts of each taxon in each sample populating the matrix, often measured as 16S amplicon read counts. In general, these matrices are incredibly sparse, with the vast majority of entries being zero because many taxa are found only in a small subset of samples. This is fundamentally different from other large biological datasets such as regulatory networks and microarray data, which are much denser.

Machine learning methods are particularly useful for recognizing patterns in these complex datasets (Knights *et al.*, 2011a). Supervised classification is a form of machine learning that develops models based on a subset of the data that includes input features, such as person identity. The goal of these classification methods is to train the models to determine taxa that are associated with a metadata criterion in a subset of the data and to then test their predictive power on the data not used for training. For example, 16S gene sequences derived from human skin samples can be used to find which taxa discriminate best between the individuals within a study. These models attempt to optimize the expected prediction error of the model on future data and provide importance scores to individual taxa that can be used to determine the predictive value of individual species.

Numerous supervised learning models have been tested on microbial communities (Knights *et al.*, 2011a), though random forest (RF) classifiers have emerged as a particularly well-performing method. RFs are an ensemble learning method that operate by constructing a multitude of decision tress in training and then output the mode of the classes of the individual trees. The ratio of model error to the error expected from random classification is used as a metric for the predictive power of the microbial community in discriminating between sample categories. The decrease in decision tree accuracy when a taxon is not included in training is used as a metric for the feature importance (discriminatory power) of that taxon.

13.4.3 Sequence clustering

Determining which sorts of data optimize the success of machine learning algorithms remains an area of active research. Overcoming the high variability of microbial community composition between individuals is difficult, and it is important to collapse or cluster observed taxa or genes in order to overcome this complexity (Knights *et al.*, 2011b).

Canonically, amplicon sequences are clustered into OTUs based on a predetermined threshold of similarity (commonly 97% for species-level taxonomic resolution). The problem lies in making sure these clusters are not too specific or too broad so that they lose their discriminative signal. A study of six human-associated microbial communities found that optimal clustering thresholds ranged from 76 to 99% depending on the factor being classified, implying that predictive models are likely to benefit from a flexible approach to OTU clustering (Knights *et al.*, 2011b). This study further suggested that information about nucleotide similarity and phylogenetic relationships between OTUs be supplied directly to machine learning classifiers in order to improve their accuracy. Novel sequence clustering methods, such as oligotyping (Eren *et al.*, 2013) and minimum entropy decomposition (Eren *et al.*, 2015) that use information theory approaches such as the Shannon entropy at different nucleotide positions to cluster sequence, may further enable us to fine-tune how we process raw sequence data before analysis.

These tools provide us with the ability to refine bacterial communities associated with humans and human trace evidence down to the strain level. The gold standard for this analysis is the genotype, as bacterial organisms show strain-level association with each person (Gilbert, 2015). However, oligotyping can be used to identify strain level associations for 16S rRNA data that could be highly effective in creating a cost-effective and accurate mechanism to identify individuals with a subspecies-level microbial fingerprint.

13.5 Forensic microbiology of the built environment

13.5.1 Tracking disease in hospital environments

The ability to track disease-causing organisms in hospital environments represents a forensic epidemiology goal that, if realized, could significantly influence healthcare-associated infection rates in hospital settings (Westwood *et al.*, 2014). Using genomic sequencing approaches for cultured pathogenic strains, researchers have been able to forensically track microbial pathogens retrospectively to determine the risk factors in disease spread (Sebaihia *et al.*, 2006; He *et al.*, 2013). However, these retrospective studies do not facilitate rapid tracking of the spread of microbial infection. This is changing with the genotype detection and analysis in near real time for cultures detected in disease outbreaks (Quick *et al.*, 2015), but there is also the potential of being able to model the microbial ecosystem of a building in a predictive way. By monitoring the longitudinal dynamics of microbial communities in a hospital and by tracking as many potentially influential variables about the building and its occupants as possible, it is likely that a model could be constructed to determine how pathogens may survive and spread in that ecosystem. Using such a model it may be possible to test interventions *in silico*, to determine the most effective strategy for combating these infections. This is the current focus of studies such as the Hospital Microbiome Project (www.hospitalmicrobiome.com) that aims to build such a model by exploring the microbial communities that patients shed into hospital environments, and by routinely monitoring how these strains move around the hospital, through patients, equipment, air, water, and staffs. In the future, such models could be highly effective in helping to design new hospital buildings or in augmenting how these buildings are managed. New sequencing assays, such as propidium monoazide (PMA)–quantitative PCR (PMA–qPCR) that allows for differentiation between live and dead bacteria in culture-independent sequencing, will further aid in the development of these models.

13.5.2 Tracking occupancy and activity in a built environment

Developing new forms of trace evidence is going to be extremely important for the next generation of forensic scientists. Genotype and amplicon sequencing technology could assist the investigator immediately when the case is most relevant. It is almost a scientific law, as well as an investigative law, that the most beneficial time to assess a case is soon after it occurs. The human skin and mucosal surface microbiome have a unique signature that represents a composite of all the physical and environmental interactions a human has throughout their life (Flores *et al.*, 2014). This microbial signature is preserved when it is transferred to a new environment for a defined period of time (Lax *et al.*, 2014). The unique character of the human skin microbiome can provide clues as to that person's lifestyle, diet, ethnicity, type of work, and even whom they regularly interact with (Lax *et al.*, 2014; Gibbons *et al.*, 2015). That this signature left behind in a space provides a unique forensic potential to provide trace evidence that can be used in investigations. Once the microbiome stabilizes around ages 3–5 years, a unique microbial composition and structure forms, which can fluctuate, but retains a core identity (Bäckhed *et al.*, 2015). Human microbial signatures associated with the indoor and outdoor built environment demonstrate the integrity of the signature we leave behind and even show, hour by hour, the way in which that signature changes on different surfaces. Using this information it will likely be possible to track time since the last occupancy for a space and even capture information about the lifestyles, geographic origin, and personal relationships of a previous or current occupant.

13.6 Conclusion

Forensic microbiology is a nascent field, but one with an extraordinary potential to transform not only criminal investigations but also personalized medicine and disease prevention. Investigations of homes, hospitals, restrooms, and other urban environments are fundamentally altering our ability to track the microbial interactions we have with the environment we have built for ourselves.

References

Aagaard, K., Ma, J., Antony, K.M. *et al.* (2014) The placenta harbors a unique microbiome. *Science Translational Medicine*, **6** (237), 237ra65.

Bäckhed, F., Roswall, J., Peng, Y. *et al.* (2015) Dynamics and stabilization of the human gut microbiome during the first year of life. *Cell Host & Microbe*, **17** (5), 690–703.

Bouslimani, A., Porto, C., Rath, C.M. *et al.* (2015) Molecular cartography of the human skin surface in 3D. *Proceedings of the National Academy of Sciences*, **112** (17), E2120–E2129.

Brown, C.T., Hug, L.A., Thomas, B.C. *et al.* (2015) Unusual biology across a group comprising more than 15% of domain Bacteria. *Nature*, **523** (7559), 208–211.

Clemente, J.C., Pehrsson, E.C., Blaser, M.J. *et al.* (2015) The microbiome of uncontacted Amerindians. *Science Advances*, **1** (3), e1500183.

Costello, E.K., Lauber, C.L., Hamady, M. *et al.* (2009) Bacterial community variation in human body habitats across space and time. *Science*, **326** (5960), 1694–1697.

Custovic, A., Taggart, S.C.O. and Woodcock, A. (1994) House dust mite and cat allergen in different indoor environments. *Clinical and Experimental Allergy*, **24** (12), 1164–1168.

Dawson, P., Han, I., Cox, M. *et al.* (2007) Residence time and food contact time effects on transfer of *Salmonella typhimurium* from tile, wood and carpet: testing the five-second rule. *Journal of Applied Microbiology*, **102** (4), 945–953.

De Filippo, C., Cavalieri, D., Di Paola, M. *et al.* (2010) Impact of diet in shaping gut microbiota revealed by a comparative study in children from Europe and rural Africa. *Proceedings of the National Academy of Sciences*, **107** (33), 14691–14696.

Dominguez-Bello, M.G., Costello, E.K., Contreras, M. *et al.* (2010) Delivery mode shapes the acquisition and structure of the initial microbiota across multiple body habitats in newborns. *Proceedings of the National Academy of Sciences*, **107** (26), 11971–11975.

Eren, A.M., Maignien, L., Sul, W.J. *et al.* (2013) Oligotyping: differentiating between closely related microbial taxa using 16S rRNA gene data. *Methods in Ecology and Evolution*, **4** (12), 1111–1119.

Eren, A.M., Morrison, H.G., Lescault, P.J. *et al.* (2015) Minimum entropy decomposition: unsupervised oligotyping for sensitive partitioning of high-throughput marker gene sequences. *The ISME Journal*, **9** (4), 968–979.

Feinstein, L.M., Sul, W.J. and Blackwood, C.B. (2009) Assessment of bias associated with incomplete extraction of microbial DNA from soil. *Applied and Environmental Microbiology*, **75** (16), 5428–5433.

Flores, G.E., Bates, S.T., Knights, D. *et al.* (2011) Microbial biogeography of public restroom surfaces. *PLoS One*, **6** (11), e28132.

Flores, G.E., Bates, S.T., Caporaso, J.G. *et al.* (2013) Diversity, distribution and sources of bacteria in residential kitchens. *Environmental Microbiology*, **15** (2), 588–596.

Flores, G.E., Caporaso, J.G., Henley, J.B. *et al.* (2014) Temporal variability is a personalized feature of the human microbiome. *Genome Biology*, **15** (12), 1.

Fujimura, K.E., Demoor, T., Rauch, M. *et al.* (2014) House dust exposure mediates gut microbiome *Lactobacillus* enrichment and airway immune defense against allergens and virus infection. *Proceedings of the National Academy of Sciences*, **111** (2), 805–810.

Gao, Z., Tseng, C.H., Pei, Z. and Blaser, M.J. (2007) Molecular analysis of human forearm superficial skin bacterial biota. *Proceedings of the National Academy of Sciences*, **104** (8), 2927–2932.

Gibbons, S.M., Schwartz, T., Fouquier, J. *et al.* (2015) Ecological succession and viability of human-associated microbiota on restroom surfaces. *Applied and Environmental Microbiology*, **81** (2), 765–773.

Gilbert, J.A. (2015) Our unique microbial identity. *Genome Biology*, **16** (1), 97.

Grice, E.A. and Segre, J.A. (2011) The skin microbiome. *Nature Reviews Microbiology*, **9** (4), 244–253.

Grice, E.A., Kong, H.H., Conlan, S. *et al.* (2009) Topographical and temporal diversity of the human skin microbiome. *Science*, **324** (5931), 1190–1192.

He, M., Miyajima, F., Roberts, P. *et al.* (2013) Emergence and global spread of epidemic healthcare-associated Clostridium difficile. *Nature Genetics*, **45** (1), 109–113.

Hewitt, K.M., Mannino, F.L., Gonzalez, A. *et al.* (2013) Bacterial diversity in two neonatal intensive care units (NICUs). *PLoS One*, **8** (1), e54703.

Hooke, R.L.B., Martín-Duque, J.F. and Pedraza, J. (2012) Land transformation by humans: a review. *GSA Today*, **22** (12), 4–10.

Kitzes, J., Peller, A., Goldfinger, S. *et al.* (2007) Current methods for calculating national ecological footprint accounts. *Science for Environment & Sustainable Society*, **4** (1), 1–9.

Knights, D., Costello, E.K. and Knight, R. (2011a) Supervised classification of human microbiota. *FEMS Microbiology Reviews*, **35** (2), 343–359.

Knights, D., Parfrey, L.W., Zaneveld, J. *et al.* (2011b) Human-associated microbial signatures: examining their predictive value. *Cell Host & Microbe*, **10** (4), 292–296.

La Rosa, P.S., Warner, B.B., Zhou, Y. *et al.* (2014) Patterned progression of bacterial populations in the premature infant gut. *Proceedings of the National Academy of Sciences*, **111** (34), 12522–12527.

Lax, S., Smith, D.P., Hampton-Marcell, J. *et al.* (2014) Longitudinal analysis of microbial interaction between humans and the indoor environment. *Science*, **345** (6200), 1048–1052.

Lax, S., Hampton-Marcell, J.T., Gibbons, S.M. *et al.* (2015) Forensic analysis of the microbiome of phones and shoes. *Microbiome*, **3** (1), 1.

Luef, B., Frischkorn, K.R., Wrighton, K.C. *et al.* (2015) Diverse uncultivated ultra-small bacterial cells in groundwater. *Nature Communications*, **6**, 6372.

Meadow, J.F., Bateman, A.C., Herkert, K.M. *et al.* (2013) Significant changes in the skin microbiome mediated by the sport of roller derby. *PeerJ*, **1**, e53.

Meadow, J.F., Altrichter, A.E., Kembel, S.W. *et al.* (2014) Bacterial communities on classroom surfaces vary with human contact. *Microbiome*, **2** (1), 1.

Quick, J., Ashton, P., Calus, S. *et al.* (2015) Rapid draft sequencing and real-time nanopore sequencing in a hospital outbreak of Salmonella. *Genome Biology*, **16** (1), 1.

Sebaihia, M., Wren, B.W., Mullany, P. *et al.* (2006) The multidrug-resistant human pathogen Clostridium difficile has a highly mobile, mosaic genome. *Nature Genetics*, **38** (7), 779–786.

Song, S.J., Lauber, C., Costello, E.K. *et al.* (2013) Cohabiting family members share microbiota with one another and with their dogs. *eLife*, **2**, e00458.

Tyakht, A.V., Kostryukova, E.S., Popenko, A.S. *et al.* (2013) Human gut microbiota community structures in urban and rural populations in Russia. *Nature Communications*, **4**, 2469.

Wesolowska-Andersen, A., Bahl, M., Carvalho, V. *et al.* (2014) Choice of bacterial DNA extraction method from fecal material influences community structure as evaluated by metagenomic analysis. *Microbiome*, **2** (1), 1.

Westwood, J., Burnett, M., Spratt, D. *et al.* (2014) The hospital microbiome project: meeting report for the UK science and innovation network UK-USA workshop 'beating the superbugs: hospital microbiome studies for tackling antimicrobial resistance', October 14th 2013. *Standards in Genomic Sciences*, **9** (1), 1.

Wood, M., Gibbons, S.M., Lax, S. *et al.* (2015) Athletic equipment microbiota are shaped by interactions with human skin. *Microbiome*, **3** (1), 1.

CHAPTER 14

Soil bacteria as trace evidence

David R. Foran[1,2], Ellen M. Jesmok[1,*] and James M. Hopkins[1,**]

[1] *Forensic Science Program, School of Criminal Justice, Michigan State University, East Lansing, MI, USA*

[2] *Department of Integrative Biology, Michigan State University, East Lansing, MI, USA*

14.1 The forensic analysis of soil

Soil is a common category of forensic evidence, given its ubiquity. The most frequent evidential encounter with soil is in the form of impressions, such as footwear or tire tracks left at a crime scene. In other instances, soil may adhere to items like a shovel, shoe, or clothing, which could end up at a considerable distance from where the soil originated. The forensic examiner's goal, then, is to either show that evidentiary soil is consistent with a specific location, or eliminate the location as the origin of the evidence. Traditionally, soil evidence has been characterized based on physical and chemical properties, including color, moisture content, particle size distribution, pH, and organic and inorganic compounds present (Saferstein, 2002), which can vary widely among soils. Indeed, the US Department of Agriculture recognizes tens of thousands of soil types in that country alone (Soil Survey Staff, 1999). Soil's vast diversity was not lost on Sir Arthur Conan Doyle (1890), whose famous Sherlock Holmes noted soil on shoes and clothing and from its appearance deduced where the wearers had been within the city of London. However, Doyle was ahead of his time, and it took several more decades for the analysis of soil evidence to become an established forensic science.

The earliest documented use of soil evidence in a criminal case dates to April of 1856 (Scientific American, 1856). A barrel of silver coins traveling on the Prussian railroad was found to have been emptied and replaced with sand. Professor Ehrenberg, a Berlin microscopist and geologist, was asked to assist in the case. Ehrenberg requested sand from stations along the train route and compared them to the barrel

* Current address: Minnesota Bureau of Criminal Apprehension Laboratory, St. Paul, MN, USA

** Current address: St. Jude Children's Research Hospital, Memphis, TN, USA

Forensic Microbiology, First Edition. Edited by David O. Carter, Jeffery K. Tomberlin, M. Eric Benbow and Jessica L. Metcalf.

sand using light microscopy. He was able to identify from which station the sand had originated, which quickly led to the guilty party.

It is widely held that Georg Popp—a chemist in Frankfurt, Germany—was the first person to analyze soil in a death investigation (Murray and Tedrow, 1975). Popp became involved in the murder investigation of Eva Disch after her remains were identified in 1908. Disch was discovered in a bean field, her cause of death being strangulation. A soiled handkerchief recovered from the scene contained particles of coal, snuff, and the mineral hornblende. Popp was asked to analyze sediment on the clothing that suspect Karl Laubach had worn the night Disch died and compare it to soil found on the victim's handkerchief and at the crime scene. Popp identified two layers of soil in Laubach's pant cuffs, one of which was consistent with the soil at the scene, while the other contained mica that was consistent with the path between the bean field and Laubach's home. Investigators challenged Laubach with the evidence and he confessed to the murder.

These cases, and the countless others that have followed, illustrate the potential that soil analysis has in a forensic setting. However, the traditional methods for analyzing soil focus mostly on class characteristics, or those that place a specimen in a group but do not individualize it. Despite the huge level of variability that can exist among soils, unless there is some truly unique characteristic in an evidentiary soil sample (e.g., a trace man-made substance existing only at the crime scene), the soil evidence is likely to be designated "consistent with" a location, not reliably assigned to it, and there is no method for placing some level of confidence on such a conclusion. In that regard, assaying features of soils that are more likely to approach individualization, and to be able to test them statistically, remains a key goal of forensic soil analysis, as is true for many of the forensic sciences.

This chapter details a new, powerful, and potentially individualizing strategy for analyzing forensic soil samples through the assessment and comparison of microorganisms within. Although a host of taxa exist in soil, many of which have been assayed for their soil comparison utility, the most studied, and the focus of the chapter, are bacteria due to their ubiquity and diversity. Statistical methods for comparing the microbial makeup of soil samples are also presented, and their potential for meeting the demands of forensic analysis is assessed. Finally, the monetary and legal considerations for crime laboratory implementation of bacterial profiling of soil are examined.

14.2 Assessing the biological components of soil

A single gram of soil has been estimated to harbor up to 10 billion microorganisms (Törsvik and Øvreås, 2002), representing a multitude of taxa—many of which are currently unidentified. These microorganisms are affected by various factors including pH, nutrient availability, moisture content, and plant life, (Kowalchuk *et al.*, 2002; Lauber *et al.*, 2008; Fierer *et al.*, 2012; Ranjard *et al.*, 2013), which result in not only diverse microbial populations among soil types but also diversity across short distances (e.g., Ramette and Tiedje, 2007) and times (e.g., Lauber *et al.*, 2013) within a location. Early methods for measuring this diversity relied on counting individual cells, through either microscopy or microbial plating (e.g., counting bacterial colonies). These methods are known to produce a gross underestimation of microbial diversity (Amann *et al.*, 1995) as many microorganisms are not easily visualized or cultured. Today, diversity is measured largely through DNA-based methods, allowing for the identification of unculturable microorganisms (Riesenfeld *et al.*, 2004) and more accurate estimates of diversity at taxonomic levels ranging from phylum to species or strain.

The vast eukaryotic, fungal, plant, and bacterial diversity that exists among soils holds great potential for forensic analysis. However, all of these do not necessarily provide equal utility for soil identification. Young *et al.* (2014) assayed a different genetic marker for each of the four major soil taxa and found that all conferred some level of discriminatory information that could be utilized for forensic purposes. The last group, bacteria, are the most numerous and diverse soil microorganisms (Sylvia *et al.*, 1997), which may be advantageous in forensic scenarios, particularly when only a very small amount of soil is available. Additionally, many microbiological studies comparing soil types and conditions have been performed by assaying their bacterial makeup (reviewed by van Elsas and Boersma, 2011), producing substantial diversity and complexity information, including the construction of large bacterial reference databases (e.g., Cole *et al.*, 2004; Quast *et al.*, 2013). The extensive knowledge of soil bacterial populations and how they differ across environments makes the genomic assessment of bacteria an obvious target for the forensic differentiation of soils.

14.3 Bacteria in soil

There are an estimated 10,000–50,000 species of bacteria per gram of soil (Roesch *et al.*, 2007). Soil bacterial diversity at higher taxonomic levels is harder to discern; however, Janssen (2006) stated that at least 32 phyla are represented. The most prevalent bacterial phylum across North and South American soils is Proteobacteria, with β-Proteobacteria being dominant in North American soil and γ-Proteobacteria in South American soil (Roesch *et al.*, 2007). Other prevalent phyla include Actinobacteria, Firmicutes, Acidobacteria, Cyanobacteria, and Verrucomicrobia (e.g., Ramette and Tiedje, 2007; Fierer *et al.*, 2013). At more refined levels, bacterial taxa are increasingly varied across soil types, with estimates for shared species as low as 1.5% in ecologically diverse soils (Fulthorpe *et al.*, 2008).

Bacterial populations are impacted by various factors, both natural and anthropogenic. Extreme environments, such as deserts and tundras, harbor bacteria very different from those in more temperate conditions (e.g., Fierer *et al.*, 2012). Additionally, differences in bacterial species or, more likely, species abundance can exist among similar habitat types in close proximity (Jesmok *et al.*, 2016), potentially due to microenvironmental factors such as pH, foliage, and nutrient supply (e.g., Ettema and Wardle, 2002; Eichorst *et al.*, 2007). Along with natural phenomena, human activity can result in highly diverse "anthropogenic soils," including those exposed to chemicals or intermixed in industrialized processes like mining, irrigation, and construction, which result in soils that exhibit strong spatial heterogeneity (IUSS Working Group WRB, 2006). Farming, logging, and other human disturbances can further affect the microbial makeup of soil. For example, Jesmok *et al.* (2016) compared bacteria from several diverse habitats within a few miles of each other. Soils from all habitats shared the most common bacterial types, with the exception of a dirt road, which showed extreme differences in bacterial makeup when compared to all other habitats. This likely resulted from summer treatment of the road with calcium chloride to reduce dust levels. Human actions may also contribute to bacterial heterogeneity at short distances. Ramirez *et al.* (2014) assayed several soil samples from Central Park in New York City, a highly trafficked area, and found that bacterial diversity levels were akin to a global soil sample set, emphasizing the variable effects of urbanization on soil. Less extreme human impact could also lead to spatial bacterial diversity. For example, a residential or commercial lawn treated with pesticides and fertilizers may contain bacteria very different from those in an adjacent untreated lawn. Similarly, certain bacteria may or may not thrive near the downspout of a building drainage system, resulting in differences from soil a short distance away.

The fact that both environmental factors and human activity can affect the bacteria in soils may make their distinguishability possible for the forensic scientist. The ability to compare and differentiate the bacterial communities from multiple locations could be invaluable in linking a suspect, victim, or object to a crime scene. Microbiologists currently employ various genomic techniques to generate a snapshot, or genetic profile, of the bacteria within a soil sample, all of which can have utility in forensic investigations.

14.4 Molecular techniques for the forensic analysis of soil

14.4.1 Analysis of soil bacteria

Multiple molecular methods have been used to assay bacterial variability in soil. The goal of any of these is to generate as much data as possible, and then to detect whatever variation exists within that dataset. Given the massive number and types of bacteria that exist in soil, this can be challenging, both at the detection level and at the data analysis level. A second consideration is which genetic marker to assay. The marker requires three specific attributes to be viable; it must (i) be present in all bacteria (or other microbial taxa), (ii) have regions that are highly similar among all bacteria so that it can be universally amplified, and (iii) have variable regions that differ among (ideally) all species or strains. While several such markers have been utilized, most scientists studying bacteria assay the 16S rRNA gene. First used by Woese and Fox (1977) for phylogenetic analysis, the 16S rRNA gene contains ten conserved regions, between which lie nine variable regions; each variable region has been utilized for bacterial differentiation. Mizrahi-Man *et al.* (2013) examined the variable regions individually and found that region 3 was the most useful for differentiating bacteria, and recommended that it and nearby region 4 be assayed in combination for the most accurate classification.

14.4.2 Denaturing gradient gel electrophoresis

One of the earliest methods for detecting the variability in bacterial populations within a sample was denaturing gradient gel electrophoresis (DGGE), reviewed by Muyzer and Smalla (1998). The DNA marker of interest, often 16S, is amplified and the resultant pool of DNAs is denatured and allowed to randomly reanneal. The reassociated DNA pool is then electrophoresed on a gel containing a gradient of a DNA denaturing chemical such as urea or formamide. Less similar DNAs denature early in the gel, while those that are more similar denature later. A complicated set of bands is created that can be compared among soils. The technique can be somewhat challenging technically, and it is possible for different DNAs to comigrate through a gel by chance, decreasing resolution. Lerner *et al.* (2006) used DGGE to investigate soils from a murder scene, the suspect's home, and the suspect's alibi location. Bacterial profiles generated from soils collected at the alibi site were similar, and differed from the scene, the suspect's home, and unrelated sites. Likewise, profiles generated from soil collected at and near the crime scene grouped together; however, so did the profile from the suspect's home that was not in close proximity, accentuating the somewhat low resolving power of soil DGGE analysis and the problems that can result from it.

14.4.3 Assaying DNA size variability

A second molecular method for assaying bacterial variability is amplicon length heterogeneity-polymerase chain reaction (ALH-PCR). A region of DNA that tends to vary in length among bacterial species is amplified using a fluorescently tagged primer. The DNA pool is then separated using capillary electrophoresis, resulting in a series of peaks in an electropherogram (e.g., Moreno *et al.*, 2006). A third and more commonly used method, terminal restriction fragment length polymorphism (T-RFLP; e.g., Heath and Saunders, 2006), is based on a similar principle, except that the amplified DNA (often 16S) is cut with a restriction enzyme, followed by capillary electrophoresis. Based on restriction site differences in the variable regions, different-sized DNA fragments are produced, whose sizes are reflected in the electropherogram.

Several groups have explored the utility of bacterial T-RFLP analysis for forensic soil comparisons. Horswell *et al.* (2002) simulated two forensic scenarios involving the deposition of soil onto an evidentiary item. In the first scenario, a shoe impression was made and soil was immediately collected from the ground and the shoe, and 8 months later from the same spot on the ground. Soil was also collected from four other locations for comparison purposes. The second scenario involved kneeling in soil wearing clean denim pants and comparing soil from the ground and each knee of the pants. In both scenarios, soil from the clothing and ground were highly similar based on shared peaks in T-RFLP profiles, while soils from the other locations were dissimilar. However, the soil collected 8 months later in the first scenario was equally dissimilar; thus, time can act as an important variable in bacterial profiling of soils for forensic purposes.

Temporal and spatial considerations of soil bacteria profiling were analyzed in greater detail by Meyers and Foran (2008) and Lenz and Foran (2010), who assayed the 16S locus and the *recA* gene specific to nitrogen-fixing rhizobia, respectively, using T-RFLP. Five habitat types were sampled at a single point monthly for a year, and every 3 months approximately 3 m in the cardinal directions from those main collection points. Overall, soils from within a habitat were more similar than among habitats; however, there were often substantial differences both spatially and temporally within habitats. This may have resulted in part from the less than ideal resolution of the technique, in that DNAs from unrelated bacteria can migrate similarly by chance during electrophoresis. Like DGGE, the low resolution of T-RFLP bacterial profiles presents a forensic challenge, potentially resulting in a false inclusion when soils are actually from different locations.

14.4.4 Next-generation DNA sequencing

An increasingly popular and highly informative method for assaying bacterial diversity is next-generation DNA sequencing, given its unparalleled data production capabilities and the rapid decrease in cost over the past few years. As with the other techniques, variable regions between conserved DNA segments are assayed; however, the method does not detect surrogates of bacterial diversity such as migration of DNA through a gel, but instead assays the actual DNA sequences of bacteria present in a sample. These sequences can then be compared to the constantly expanding databases of known bacterial sequences to determine exactly which species (or other taxonomic groups) are present in a soil sample. The fact that next-generation sequencing can result in tens to hundreds of thousands of well-defined bacterial sequences per sample makes it far more discriminating than the techniques described earlier, and today it has become *de rigueur* for bacterial profiling, the methodologies of which are discussed in greater detail in Chapter 5.

The number of studies by microbiologists assessing soil microbes via next-generation sequencing is large, although from a forensic context little research exists. As noted before, Young *et al.* (2014) conducted a preliminary next-generation survey of different microbe types in soil, which was based on the bacterial 16S rRNA gene, the eukaryotic 18S rRNA gene, the plant chloroplast leucine tRNA gene, and the fungal spacer region between rRNA genes. The authors expanded on this work (Young *et al.*, 2015), applying next-generation sequencing of the 18S rRNA gene to soils in a fictional case study. Shoes and a shovel harboring soil from a single location were left in a car for 6 weeks. Known soil samples were later collected from the site and at various distances away ranging from 5 m to 180 km. For the most part, the shoe and shovel soils were most similar to the site itself, followed by 5 m away and then more distant soils. Young *et al.* (2016) also assessed how soil mass affects fungal profile development using next-generation sequencing. The results showed that soil quantity did not significantly affect profile development and even trace amounts of soil, which will often be encountered in forensic casework, provided valuable genetic information. Hopkins (2014) and Jesmok (2015) obtained equally good results when assaying the bacterial 16S rRNA gene via next-generation sequencing, as they were able to differentiate dissimilar and highly similar habitat types over space and time, as well as soil on evidentiary items. These forensic studies also explored statistical methods to analyze the massive datasets produced through next-generation sequencing, with the goal of definitively associating or discriminating soil samples.

14.5 Soil microbial profile data analysis methods

14.5.1 Qualities of ideal forensic data analysis techniques

There are two main qualities that ideal forensic data analysis techniques possess. The first quality, pointed out in the 2009 National Academy of Sciences report (National Research Council, 2009), is objectivity in associating a piece of evidence with a suspect, victim, or scene. Ideally, subjective human determination of associations is completely avoided. The objective method should also provide a definitive statistical value that can be placed on the association, reflecting the probability that it is incorrect (i.e., the method's intrinsic error rate). Further, more than one expert may analyze materials collected in a criminal investigation, and the results obtained and conclusions drawn should be consistent among them.

The second quality of an ideal forensic data analysis technique is ease of explanation to laypeople. Forensic scientists are often summoned to court to present methods, results, and overall conclusions about any associations made and delineate how those conclusions were reached. In many instances, the scientific methodologies used, such as those described earlier, are highly technical, as are the results produced. A data analysis strategy that allows for straightforward presentation of the scientific findings to the judge, jury, and attorneys increases both the ease of the task for the expert witness and the understanding by the trier of fact. If possible, demonstrative presentation of the findings is produced in the form of charts or graphs, augmenting the comprehension of the court.

Of course, there is the possibility that no single data analysis technique can meet all the needs of forensic scientists, and many analysis methods are themselves highly technical. Regarding microbial analyses, the datasets produced can be extremely complicated and, as is true with next-generation sequencing results, massive. An objective method for analyzing these data may be difficult to explain to a jury, while methods that lend themselves to visual interpretation may be less objective than desired. However, by combining more than one data analysis method, it might be possible to develop a strategy that satisfies all desired criteria.

14.5.2 Objective microbial profiling analysis methods

The most straightforward approach for associating a questioned and known soil sample is to directly compare them to one another. This is not dissimilar to what is undertaken in the visual or chemical comparison of soils, except pairwise comparisons introduce a statistical component, resulting in a *p*-value that indicates the likelihood that the two specimens have a common origin. Most of these methods are based on repetitive (Monte Carlo) sampling of the data, the distribution of which is then compared to the original data (e.g., the bacterial makeup of a questioned and known soil sample). A *p*-value can be placed on the likelihood that the two soils are similar or dissimilar by chance, based on where each falls in the generated distribution. Although pairwise techniques have been used in many microbiological studies (e.g., Upchurch *et al.*, 2008; Lauber *et al.*, 2009), few have used them to compare forensic soil samples or help assign soils to a location of origin. Additionally, research evaluating pairwise statistics for forensic science indicates such methods might be overly sensitive to normal fluctuations in the microbial makeup of soil that exist over time and space (Hopkins, 2014), limiting their overall forensic utility.

A second method to compare microbial profiles directly involves calculating a similarity (or dissimilarity) value between each pair of samples in a dataset. These are a measure of how alike two profiles are, and can be calculated based on the presence/absence of shared sequences and/or the abundance differences of shared sequences (Choi *et al.*, 2010). Data matrices are produced by calculating pairwise (dis)similarity values between all samples in a given dataset, which can then be analyzed using multiple methods, including analysis of similarities (ANOSIM; Clarke, 1993) and the similarity profile routine (SIMPROF; Clarke *et al.*, 2008). These methods are akin to standard analysis of variance statistics (e.g., ANOVA); however, soil bacteria sequencing data rarely follow a normal distribution; thus, parametric methods are often not appropriate. ANOSIM is used when microbial profiles can be divided into known groups based on their location of origin (e.g., crime scene, alibi location, or evidentiary item) before analysis. It statistically compares the means of ranked similarity values among groups to the means of ranked similarity values within groups to determine if they are statistically different. Young *et al.* (2015, 2016) employed ANOSIM to compare soil samples collected from various items and diverse locations, illustrating how it might establish baseline information on location and evidentiary similarities in the investigative stage of a criminal case. In realistic forensic scenarios, evidentiary soil samples may be too small to generate multiple bacterial profiles, and ANOSIM cannot analyze a single-sample group. In contrast, SIMPROF can be used to statistically compare a single evidentiary sample to known location profiles. This technique is effective when the general groupings of samples are unknown, with a null hypothesis that all profiles are homogenous. SIMPROF detects structure within a dataset, sorting samples into statistically similar groups. Wakelin *et al.* (2008) used SIMPROF to analyze microbial profiles generated from several diverse locations to determine if terrestrial activity influences soil microbes.

They found that microbial communities largely joined into two groups statistically, primarily based on the pH of the soil. SIMPROF could be useful in a forensic context because single evidentiary samples can be grouped with a set of known samples, potentially associating the sample with one location.

A third statistical approach that holds strong potential for forensic soil analysis involves supervised learning. These techniques are based on a collection of known soil samples that make up a training set (Beebe *et al.*, 1998), which itself can be cross-validated to assess the amount of variation present within each soil location. Questioned samples are then classified as belonging to one of the sites (or to none of them, depending on the method), and a level of confidence in the assignment is determined. Multiple supervised classifiers exist that produce statistical measures of classification strength (e.g., support vector machines, random forest, and decision trees); however, few studies have been conducted examining their utility regarding soil microbial data analysis. Yang *et al.* (2006), based on ALH-PCR data, utilized two supervised classifiers, *k*-nearest neighbor and support vector machines, to categorize soil samples from two habitats, finding an overall 91.5% location of origin assignment success. Jesmok (2015) examined 16S next-generation sequencing data via *k*-nearest neighbor analysis, which resulted in correct assignment to a location of origin for 97.6% of the bacterial profiles generated from soil samples.

14.5.3 Demonstrative microbial profiling analysis methods

The statistical approaches described thus far can result in objective microbial profile analysis and association; however, they do not provide an easy interpretation for an audience of nonscientists/statisticians. Fortunately, several analysis techniques utilized by microbiologists display profile information and relative similarities among profiles in readily interpretable, visual representations, which could have great utility in court.

Abundance charts are the simplest of these visualization techniques. Microbial data are first classified using an appropriate sequence database and quantified. A stacked bar graph is then generated (exemplified in Figure 14.1) with each bar representing a single soil sample consisting of several different microbial taxa at a user-defined taxonomic level. Microbiologists usually generate abundance charts at the phylum level (e.g., Jansson and Tas, 2014), where major changes brought on by factors such as environmental stressors or chemical contamination are evident. Soil samples examined for forensic purposes likely require a deeper taxonomic level (e.g., class or family), where more subtle differences can be identified, given that determining if two soil samples are highly similar is a forensic goal.

Two additional statistical methods that frequently appear in the microbiological literature and offer visualization of the results are multidimensional scaling (MDS; Kruskal and Wish, 1978) and hierarchical cluster analysis (HCA; Ward, 1963). Both techniques compare multiple profiles at once based on the same (dis)similarity values that ANOSIM and SIMPROF operate on, clustering similar samples closer to one another and dissimilar ones farther apart.

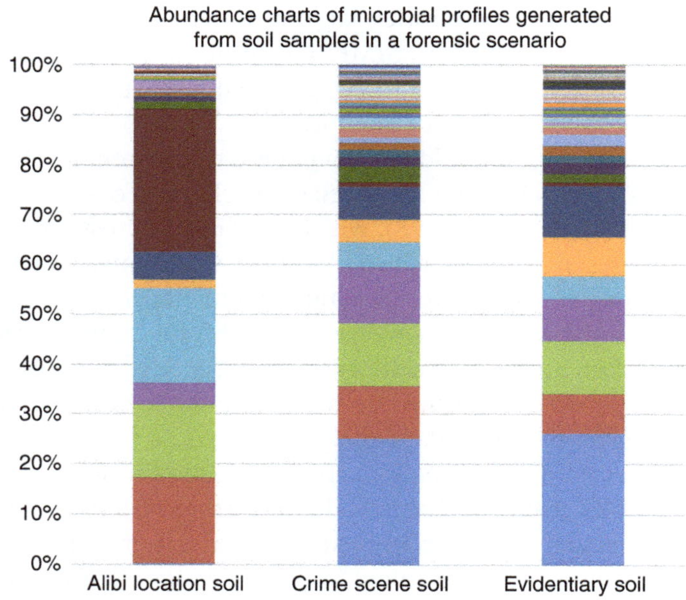

Figure 14.1 Hypothetical abundance chart displaying soil microbial profiles from three soil samples. The evidentiary soil profile is visually more similar to the profile produced from the crime scene soil than the alibi location, indicating the latter can be excluded as the possible source of the evidentiary soil

MDS (or nonmetric multidimensional scaling, NMDS, for nonparametric data, as are commonly found in soil microbial profiles) is an ordination technique that plots microbial profiles, each represented by a single point, in two or more dimensions based on a goodness-of-fit measurement, or stress, that measures how well profile relationships are being represented in a plot (Ramette, 2007) (exemplified in two dimensions in Figure 14.2). MDS plots provide information on which soil microbial profiles are more similar based on their proximity to other samples/clusters. The straightforward interpretability of MDS plots could make them valuable for expert testimony, as they can reveal relationships within complex data and help visualize results produced in other data analysis methods. In a forensic scenario, the location of an evidentiary sample in the plot can provide information on what location(s) it is most closely related to. However, there is neither a single definition of what constitutes a cluster, nor is there a statistical measure of how closely microbial profiles should plot in MDS to be considered associated.

HCA is another clustering method in which microbial profiles are grouped in a branching tree structure or dendrogram (exemplified in Figure 14.3) based on their relative (dis) similarity. Forensic soil samples can be clustered in this manner, where, for instance, samples with higher similarity cluster together first and are then clustered with slightly more dissimilar samples, and so on, until a hierarchy of overall sample similarity is produced. The location of an unknown within this dendrogram allows the analyst to identify the

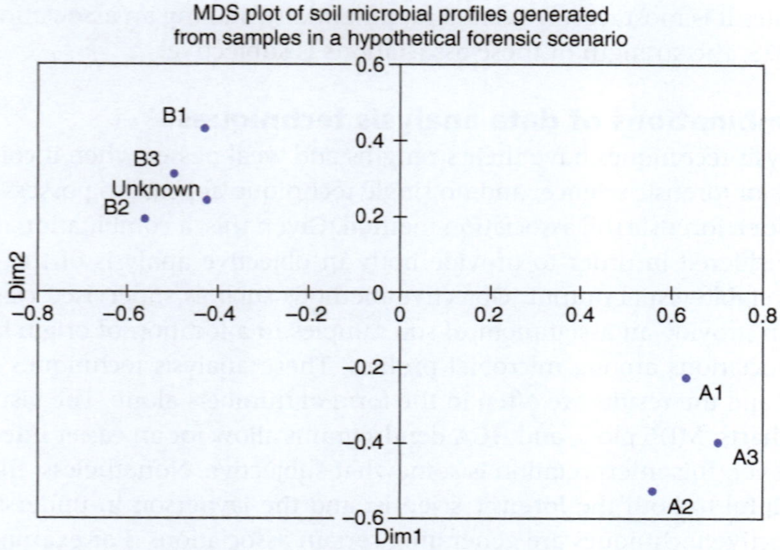

Figure 14.2 Exemplary MDS plot ordinating replicate bacterial profiles from an alibi location (A) and a crime scene (B). An evidentiary sample profile (unknown) is clustering with the profiles from the crime scene; thus, it is more similar to this location than the alibi location

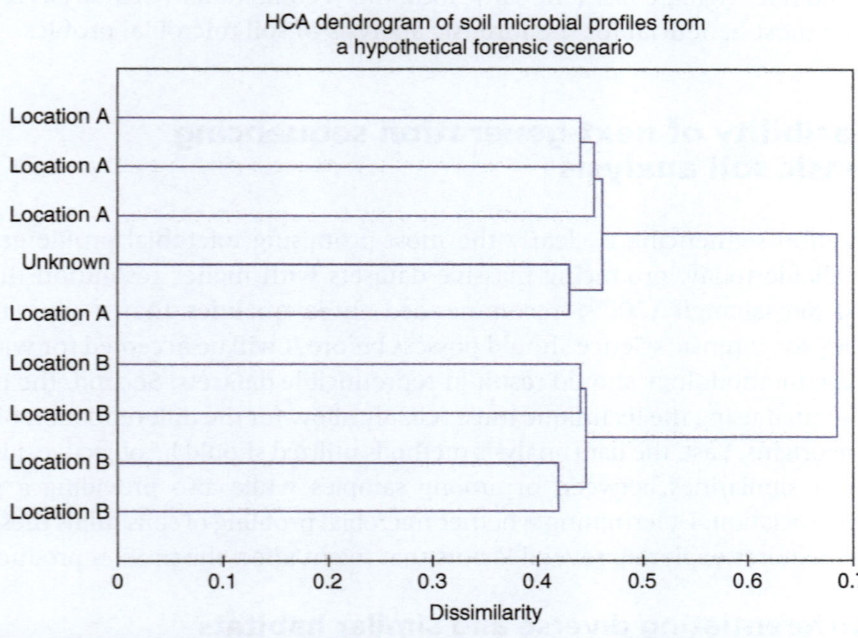

Figure 14.3 Exemplary HCA dendrogram displaying clusters of hypothetical crime scene (A), alibi location (B), and evidentiary (unknown) microbial profiles. The evidentiary profile is clustering with the soil samples collected at the crime scene, indicating it is more similar to those samples than alibi location soils

known samples it is most closely related to, potentially making an association. However, similar to MDS, the strength of these associations is subjective.

14.5.4 Combinations of data analysis techniques

All data analysis techniques have their strengths and weaknesses when it comes to meeting the needs of forensic science, and no single technique appears to possess all the qualities desired for a forensic soil association method. Given this, a combination of techniques might be considered in order to provide both an objective analysis of the data and an easily interpretable visual output. Objective methods such as supervised classification or SIMPROF can provide an assignment of soil samples to a location of origin by producing statistical associations among microbial profiles. These analysis techniques are typically complicated, and the results are often in the form of numbers alone. The visual aspects of abundance charts, MDS plots, and HCA dendrograms allow for an easier interpretation of results; however, this interpretation is somewhat subjective. Nonetheless, their graphical quality is helpful to both the forensic scientist and the layperson in understanding why complex objective techniques are generating certain associations. For example, an expert witness can easily explain an unknown's classification by showing which group of known samples it clusters nearest in an MDS plot. Likewise, evidentiary profiles that are not classified may appear visually different from known profiles in an abundance chart, signifying they likely did not originate from the same location. A combination of analysis methods is probably the most beneficial for the forensic analysis of soil microbial profiles.

14.6 Feasibility of next-generation sequencing for forensic soil analysis

Next-generation sequencing is clearly the most promising microbial profile generating method available today, producing massive datasets with higher resolution than older techniques. Sensabaugh (2009) recommended three qualities that a new microbial methodology for forensic science should possess before it will be accepted for widespread use. First, the methodology should result in reproducible datasets. Second, the microbial profiles generated using the technique must reliably allow for the differentiation of samples with unique origins. Last, the data analysis methods utilized should be objective, identifying differences or similarities between or among samples while also providing a statistical measure of association. Determining whether microbial profiling of soils fulfils these recommendations requires exploring several factors that might affect the profiles produced.

14.6.1 Differentiating diverse and similar habitats

The first step in examining the potential of microbial profiling for forensic soil analysis is to determine whether profiles generated from separate locations can be differentiated. This includes profiles from different habitat types as well as ones that are quite similar.

Forensic soil samples may be contested as being from one location or another, and these two locations might be environmentally similar or in close proximity. Many soil-based forensic studies have been conducted with the goal of examining if a technique can produce differentiating profiles (e.g., Lerner *et al.*, 2006; Meyers and Foran, 2008; Jesmok *et al.*, 2016). T-RFLP analysis has allowed diverse habitat soil differentiation; however, the quantity and diversity of microorganisms in a soil sample often crowd electropherograms, limiting the technique's resolution. Similarly, DGGE analysis of soil microbes can lead to indistinguishable bands on the gel. Next-generation sequencing of soil microbes has shown much stronger profile generation potential, as resolution and objectivity are higher, and profiles from both ecologically diverse and similar locations can be distinguished (Hopkins, 2014; Young *et al.*, 2014, 2015; Jesmok *et al.*, 2016).

14.6.2 Temporal changes in soil microbial profiles

Known forensic soil samples are virtually never collected at the time a crime occurred; therefore, the change in microbial profiles over time could affect the association of known and evidentiary soil samples. Research has shown that microbial profiles do change over time (Meyers and Foran, 2008; Lenz and Foran, 2010; Lauber *et al.*, 2013; Hopkins, 2014), sometimes enough to affect classification and/or association. Seasonal effects and day-to-day weather phenomena may also have an impact on soil profile association. Ideally, a known forensic soil sample will be collected as early as possible; however, if the environment has changed, it may be better to collect them over a number of days or even longer to account for microbial variation occurring temporally. This is clearly an area of research that will require more attention in the future.

14.6.3 Spatial differences in soil microbial profiles

Just as known soil samples will be collected at a time different from when a crime occurs, they also may not be collected at the exact spot where an evidentiary item was exposed to the soil, potentially affecting the association accuracy of profile analysis. Microbial profiles can be impacted by several microenvironmental factors including pH, foliage density, and nutrient supply (e.g., Ettema and Wardle, 2002; Eichorst *et al.*, 2007; King *et al.*, 2010), all of which might vary across a single habitat. Similarly, an evidentiary soil sample stemming from a burial may be a mixture of soils from various depths. Microbial populations are known to differ with depth (Eilers *et al.*, 2012; Hopkins, 2014; Jesmok, 2015), likely due to oxygen availability and the chemical composition of deeper soils; thus, associations between known and evidentiary soils could easily be affected if depth is not considered when collecting known samples from a crime. Spatial variation may be circumvented through the collection of multiple soil samples across a habitat, either on the surface or from several different depths as appropriate. As with time, such spatial considerations are a prime target for future research.

14.6.4 Soil sample collection strategies

Strategies for collecting known soil samples are exceedingly important to ensure a representative microbial profile can be generated for comparison to evidentiary soil. A single profile may not be an accurate representation of an entire location, and it may not provide enough information to confidently classify an evidentiary soil sample to an origin. Further, both ordination and supervised classification techniques require multiple samples to produce associations between soil profiles. Given this, multiple known samples should be collected from each location in question, as this allows for the formation of representative clusters and training sets during analysis. Further, *ad hoc* sampling strategies may be needed depending on the specifics of a case. For example, the collection of multiple samples branching outward from a main site might be beneficial; however, some locations could have obstructions preventing soil collection in certain directions (e.g., a body of water or road). In these cases, crime scene investigators will have to decide on a set of soils to collect based on practicality and common sense.

14.6.5 Evidence storage and changes in bacterial abundance over time

Unknown forensic soil samples will most often be collected from evidentiary items that have been stored for some period of time prior to discovery. Given this, it is important to understand how soil microbial profiles may change upon removal from a habitat, as well as what effects the evidentiary material itself and environmental factors (e.g., storage temperature) have on the microbes. Research has shown that bacterial and eukaryotic populations in the soil change when stored; however, soil association is largely unaffected (Jesmok, 2015; Young *et al.*, 2015). Jesmok (2015) found that similar taxonomic class changes occurred in soil bacterial profiles across various stored evidentiary items. Predictable profile divergence on stored items suggests that evidentiary soil may have the potential to be used as a biological clock indicating how long the soil has been removed from a given location. Similar to how insects can be used to estimate a postmortem interval range for a body, soil could become even more probative if such a biological clock can be developed based on microbial abundance changes.

14.6.6 Costs of next-generation sequencing of forensic soil samples

The cost of next-generation sequencing has fallen rapidly since 2008; faster than Moore's law (1965) would predict. The US National Human Genome Research Institute (https://www.genome.gov/sequencingcosts/) reported that the price per megabase has dropped from US$10,000 to under $0.10 in the last decade. Overall, producing a microbial profile is extremely inexpensive given the amount of data developed. However, implementation of next-generation sequencing into crime laboratories could be difficult given the sheer amount of data generated. A single Illumina MiSeq run can produce over 15 gigabytes of sequence data, equaling 25 million sequences or more. In a standard 96 well format, this equates to an average of over 260,000 sequences per soil sample. That much data can allow for the identification of rare microbia that might lead to associations or exclusions; however, it can also be quite unwieldy. Analysis of such massive datasets requires computers with extensive processing power and immense storage capacity, although these drawbacks are likely to be reduced in the future and thus may not represent a limitation for forensic soil analysis going forward.

On the other hand, asking each crime laboratory to obtain the equipment and validate the methodologies for forensic microbial soil profiling is probably unrealistic. A more viable strategy for effectively profiling soil samples based on next-generation sequencing would be the establishment of a centralized laboratory or laboratories to perform this function, similar to the Federal Bureau of Investigation (FBI) mitochondrial DNA analysis laboratories. A local laboratory could submit soil samples to the specialized laboratory for processing, mitigating the cost per sample, while also requiring fewer scientists be trained in the techniques. Additionally, such a laboratory could be designed to store the massive datasets produced, potentially building a database of forensic soil microbial profiles.

14.6.7 Legal considerations for the implementation of microbial profiling

The microbial profile generation and analysis methodologies developed by microbiologists for soil samples and tailored to forensic science satisfy Sensabaugh's three recommendations for such techniques: reproducible data production, high discriminatory power, and objective data analysis. Further, adopting already well-established methods should make acceptance in the courts a simpler process, as they meet most or all tenants of Daubert (1993), including prior testing, general acceptance, peer-reviewed publication, and known error rates. Given this, and the broad legal acceptance of DNA-based techniques in general, microbial profiling of soil should encounter minimal resistance from the courts.

14.7 Consensus on methodologies for soil collection and analysis

Microbial profiling of forensic soil samples must be standardized before these methods can be effectively implemented. Ideally, a consensus will be reached on how soil is collected, what organisms and which regions of their genomes are assayed, the optimal profile generation technique, and the data analysis methods that will be used. This will allow experts to acquire consistent and comparable results when analyzing soil samples. Free, open-source software packages and databases are currently available on the World Wide Web (e.g., mothur (http://www.mothur.org/), R (https://www.r-project.org/), Qiime (http://qiime.org/), silva (http://www.arb-silva.de/), and greengenes (http://greengenes.lbl.gov/cgi-bin/nph-index.cgi)) to analyze and classify large microbial datasets; however, too many options might lead to conflicting results across crime laboratories if different programs or databases are used. Large standardization bodies already exist within the microbiology community (e.g., the Earth Microbiome Project (http://www.earthmicrobiome.org/)) with the goal of amassing microbial data from around the world for shared use among researchers. Forensic soil analysis could utilize such a repository of data; however, it is still important that standardized procedures be defined so that multiple forensic laboratories will reach the same conclusion when analyzing soil samples. Most likely, an entity such as the Scientific Working Group on Microbial Genetics and Forensics will consider and make such recommendations (https://archives.fbi.gov/archives/about-us/lab/forensic-science-communications/fsc/oct2003/standards-and-guidelines/2003_10_guide01.htm).

Reaching a consensus in all of these areas should increase the probative value of soil evidence far beyond that of traditional forensic analysis techniques. The advancements in next-generation sequencing technology will allow forensic scientists to generate powerful datasets from small amounts of soil. In combination with robust statistical techniques, this technology holds the potential to individualize soil samples, allowing forensic experts to link a suspect, victim, or item to a crime scene.

Acknowledgments

The authors thank past researchers in the Michigan State University's Forensic Biology Laboratory who helped develop bacterial profiling of soil, including Melissa Meyers, Erin Lenz, and Ethan Smith. They also thank Dr. Jeff Landgraf and other members of the Research Technology Support Facility (RTSF) Genomics Core at Michigan State University.

References

Amann, R.I., Ludwig, W. and Schleifer, K.H. (1995) Phylogenetic identification and in situ detection of individual microbial cells without cultivation. *Microbiology and Molecular Biology Reviews*, **59**, 143–169.

Beebe, K.R., Pell, R.J. and Seasholtz, M.B. (1998) Pattern recognition, in *Chemometrics: A Practical Guide* (eds K.R. Beebe, R.J. Pell and M.B. Seasholtz), John Wiley & Sons, Inc., New York, pp. 56–182.

Choi, S.S., Sung-Hyuk, C. and Tappert, C.C. (2010) A survey of binary similarity and distance measures. *Journal of Systemics, Cybernetics and Informatics*, **8** (1), 43–48.

Clarke, K.R. (1993) Non-parametric multivariate analyses of changes in community structure. *Australian Journal of Ecology*, **18**, 117–143.

Clarke, K.R., Somerfield, P.J. and Gorley, R.N. (2008) Testing of null hypotheses in exploratory community analyses: similarity profiles and biota-environment linkage. *Journal of Experimental Marine Biology and Ecology*, **366**, 56–69.

Cole, J.R., Chai, B., Farris, R.J. *et al.* (2004) The Ribosomal Database Project (RDP-II): sequences and tools for high-throughput rRNA analysis. *Nucleic Acids Research*, **33**, D294–D296.

Daubert et al. v. Merrell Dow Pharmaceuticals (1993) 509 U.S. 579, 92–102.

Doyle, A.C. (1890) *The Sign of Four*, Spencer Blackett, London.

Eichorst, S.A., Breznak, J.A. and Schmidt, T.M. (2007) Isolation and characterization of soil bacteria that define *Terriglobus* gen. nov., in the Phylum Acidobacteria. *Applied Environmental Microbiology*, **73**, 2708–2717.

Eilers, K.G., Debenport, S., Anderson, S. and Fierer, N. (2012) Digging deeper to find unique microbial communities: the strong effect of depth on the structure of bacterial and archaeal communities in soil. *Soil Biology and Biochemistry*, **50**, 58–65.

van Elsas, J.D. and Boersma, F.G.H. (2011) A review of molecular methods to study the microbiota of soil and the mycosphere. *European Journal of Soil Biology*, **47**, 77–87.

Ettema, C.H. and Wardle, D.A. (2002) Spatial soil ecology. *Trends in Ecological Evolution*, **17**, 177–183.

Fierer, N., Leff, J.W., Adams, B.J. *et al.* (2012) Cross-biome metagenomic analyses of soil microbial communities and their functional attributes. *Proceedings of the National Academy of Sciences of the United States of America*, **109**, 21390–21395.

Fierer, N., Ladau, J., Clemente, J.C. *et al.* (2013) Reconstructing the microbial diversity and function of pre-agricultural tallgrass prairie soils in the United States. *Science*, **342**, 621–624.

Fulthorpe, R.R., Roesch, L.F.W., Riva, A. and Triplett, E.W. (2008) Distantly sampled soils carry few species in common. *The ISME Journal*, **2**, 301–910.

Heath, L.E. and Saunders, V.A. (2006) Assessing the potential of bacterial DNA profiling for forensic soil comparisons. *Journal of Forensic Sciences*, **51**, 1062–1068.

Hopkins, J.M. (2014) Forensic soil bacterial profiling using 16S rRNA gene sequencing and diverse statistics. Published Master's thesis, Michigan State University, East Lansing.

Horswell, J., Cordiner, S.J., Maas, E.W. *et al.* (2002) Forensic comparison of soils by bacterial community DNA profiling. *Journal of Forensic Sciences*, **47**, 350–353.

IUSS Working Group WRB (2006) *World reference base for soil resources 2006*. World Soil Resources Reports No. 103. FAO, Rome.

Janssen, P.H. (2006) Identifying the dominant soil bacterial taxa in libraries of 16S rRNA and 16S rRNA genes. *Applied and Environmental Microbiology*, **72** (3), 1719–1728.

Jansson, J.K. and Tas, N. (2014) The microbial ecology of permafrost. *Nature Reviews Microbiology*, **12**, 414–425.

Jesmok, E.M. (2015) Adoption of next-generation 16S bacterial sequencing practices for the forensic analysis of soil. Published Master's thesis, Michigan State University, East Lansing.

Jesmok, E.J., Hopkins, J.M. and Foran, D.R. (2016) Next-generation sequencing of the bacterial 16S rRNA gene for forensic soil comparison: a feasibility study. *Journal of Forensic Sciences*, **61**, 607–617.

King, A.J., Freeman, K.R., McCormick, K.F. *et al.* (2010) Biogeography and habitat modelling of high-alpine bacteria. *Nature Communications*, **1**, 53.

Kowalchuk, G.A., De Souza, F.A. and Van Veen, J.A. (2002) Community analysis of arbuscular mycorrhizal fungi associated with *Ammophila arenaria* in Dutch coastal sand dunes. *Molecular Ecology*, **11**, 571–581.

Kruskal, J.B. and Wish, M. (1978) *Multidimensional Scaling*. Sage University Paper Series on Quantitative Application in the Social Sciences, 07-011. Sage Publications, Beverly Hills/London.

Lauber, C.L., Strickland, M.S., Bradford, M.A. and Fierer, N. (2008) The influence of soil properties on the structure of bacterial and fungal communities across land-use types. *Soil Biology and Biochemistry*, **40**, 2407–2415.

Lauber, C.L., Hamady, M., Knight, R. and Fierer, N. (2009) Pyrosequencing-based assessment of soil pH as a predictor of soil bacterial community structure at the continental scale. *Applied Environmental Microbiology*, **75**, 5111–5120.

Lauber, C.L., Ramirez, K.S., Aanderud, Z. *et al.* (2013) Temporal variability in soil microbial communities across land-use types. *ISME Journal*, **7**, 1641–1650.

Lenz, E.J. and Foran, D.R. (2010) Bacterial profiling of soil using genus–specific markers and multidimensional scaling. *Journal of Forensic Sciences*, **55**, 1437–1442.

Lerner, A., Shor, Y., Vinokurov, A. *et al.* (2006) Can denaturing gradient electrophoresis (DGGE) analysis of amplified 16s rDNA of soil bacterial populations be used in forensic investigations? *Soil Biology and Biochemistry*, **38**, 1188–1192.

Meyers, M.S. and Foran, D.R. (2008) Spatial and temporal influences on bacterial profiling of forensic soil samples. *Journal of Forensic Sciences*, **53**, 652–660.

Mizrahi-Man, O., Davenport, E.R. and Gilad, Y. (2013) Taxonomic classification of bacterial 16S rRNA genes using short reads: evaluation of effective study designs. *PLoS One*, **8**, e53608.

Moore, G.E. (1965) Cramming more components onto integrated circuits. *Electronics*, **38**, 114–117.

Moreno, L.I., Mills, D.K., Entry, J. *et al.* (2006) Microbial metagenome profiling using amplicon length heterogeneity-polymerase chain reaction proves more effective than elemental analysis in discriminating soil specimens. *Journal of Forensic Sciences*, **51**, 1–8.

Murray, R.C. and Tedrow, J.C. (1975) Forensic geology—earth sciences and criminal investigation. National Criminal Justice Reference Service NCJ 017479. Rutgers University Press.

Muyzer, G. and Smalla, K. (1998) Application of denaturing gradient gel electrophoresis (DGGE) and temperature gradient gel electrophoresis (TGGE) in microbial ecology. *Antonie van Leeuwenhoek*, **73**, 127–141.

National Research Council (2009) *Strengthening Forensic Science in the United States: A Path Forward*, National Academies Press, Washington, DC.

Quast, C., Pruesse, E., Yilmaz, P. *et al.* (2013) The SILVA ribosomal RNA gene database project: improved data processing and web-based tools. *Nucleic Acids Research*, **41**, D590–D596.

Ramette, A. (2007) Multivariate analyses in microbial ecology. *FEMS Microbiology Ecology*, **62**, 142–160.

Ramette, A. and Tiedje, J.M. (2007) Multiscale responses of microbial life to spatial distance and environmental heterogeneity in a patchy ecosystem. *Proceedings of the National Academy of Sciences of the United States of America*, **104**, 2761–2766.

Ramirez, K.S., Leff, J.W., Barberan, A. *et al.* (2014) Biogeographic patterns in below-ground diversity in New York City's Central Park are similar to those observed globally. *Proceedings of the Royal Society*, **281**, http://rspb.royalsocietypublishing.org/faq (accessed December 7, 2016).

Ranjard, L., Dequiedt, S., Chemidlin Prévost-Bouré, N. *et al.* (2013) Turnover of soil bacterial diversity driven by wide-scale environmental heterogeneity. *Nature Communications*, **4**, 1434.

Riesenfeld, C.S., Schloss, P.D. and Handelsman, J. (2004) Metagenomics: genomic analysis of microbial communities. *Annual Review of Genetics*, **38**, 525–552.

Roesch, L.F.W., Fulthorpe, R.R., Riva, A. *et al.* (2007) Pyrosequencing enumerates and contrasts soil microbial diversity. *The ISME Journal*, **1**, 283–290.

Saferstein, R.E. (2002) *Forensic Science Handbook*, Prentice Hall, Englewood Cliffs, 784 pp.

Scientific American (1856) Science and art: curious use of the microscope. *Scientific American*, **11**, 240.

Sensabaugh, G.F. (2009) Microbial community profiling for the characterization of soil evidence: forensic considerations, in *Criminal and Environmental Soil Forensics* (eds K. Ritz, L. Dawson and D. Miller), Springer, Dordrecht, pp. 49–60.

Soil Survey Staff (1999) *Soil Taxonomy: A Basic System of Soil Classification for Making and Interpreting Soil Surveys*, 2nd edn, Natural Resources Conservation Service, U.S. Department of Agriculture Handbook 436, Washington, DC.

Sylvia, D., Fuhrmann, J., Hartel, P. and Zuberer, D. (1997) *Principles and Applications of Soil Microbiology*, Prentice Hall, Upper Saddle River.

Törsvik, V. and Øvreås, L. (2002) Microbial diversity and function in soil: from genes to ecosystems. *Current Opinion in Microbiology*, **5**, 240–245.

Upchurch, R., Chiu, C.Y., Everett, K. *et al.* (2008) Differences in the composition and diversity of bacterial communities from agricultural and forest soils. *Soil Biology and Biochemistry*, **40**, 1294–1305.

Wakelin, S.A., Macdonald, L.M., Rogers, S.L. *et al.* (2008) Habitat selective factors influencing the structural composition and the functional capacity of microbial communities in agricultural soils. *Soil Biology and Biochemistry*, **40**, 803–813.

Ward, J.H. (1963) Hierarchical grouping to optimize an objective function. *Journal of the American Statistical Association*, **58**, 236–244.

Woese, C.R. and Fox, G.E. (1977) Phylogenetic structure of the prokaryotic domain: the primary kingdoms. *Proceedings of the National Academy of Sciences of the United States of America*, **74**, 5088–5090.

Yang, C., Mills, D., Mathee, K. *et al.* (2006) An ecoinformatics tool for microbial community studies: supervised classification of amplicon length heterogeneity (ALH) profiles of 16S rRNA. *Journal of Microbiological Methods*, **65**, 49–62.

Young, J.M., Weyrich, L.S. and Cooper, A. (2014) Forensic soil DNA analysis using high-throughput sequencing: a comparison of four molecular markers. *Forensic Science International: Genetics*, **13**, 176–184.

Young, J.M., Weyrich, L.S., Breen, J. *et al.* (2015) Predicting the origin of soil evidence: high throughput eukaryote sequencing and MIR spectroscopy applied to a crime scene scenario. *Forensic Science International*, **251**, 22–31.

Young, J.M., Weyrich, L.S. and Cooper, A. (2016) High-throughput sequencing of trace quantities of soil provides reproducible and discriminative fungal DNA profiles. *Journal of Forensic Sciences*, **61**, 478–484.

CHAPTER 15

DNA profiling of bacteria from human hair: Potential and pitfalls

Silvana R. Tridico[1], Dáithí C. Murray[2], Michael Bunce[2] and K. Paul Kirkbride[3]

[1] *Forensic Science & Wildlife Matters, Adelaide, South Australia, Australia*

[2] *Trace and Environmental DNA (TrEnD) Laboratory, Department of Environment and Agriculture, Curtin University, Perth, Western Australia, Australia*

[3] *School of Chemical and Physical Sciences, Flinders University, Adelaide, South Australia, Australia*

15.1 An introduction to human hair as a forensic substrate

15.1.1 Relevance of hair in forensic science

The relevance and value of hair as forensic evidence first gained prominence in the nineteenth century with the first publication of "Hairs of Mammalia from the Medico-Legal Aspect" (Glaister, 1931). Mammalian hairs, unlike ephemeral biological structures such as tissue or body fluids, are remarkably resistant to the effects of degradation as exemplified by their ability to survive, intact, for millennia in disparate environments such as permafrost (Rasmussen *et al.*, 2010), caves (Clack *et al.*, 2012), or on mummified remains (Frei *et al.*, 2015). This resilience is exemplified thus by Lord Byron who observed "I have seen a thousand graves opened, and always perceived that whatever was gone, the teeth and the hair remained of those who had died with them. Is that not odd? They go the very first things in youth and yet the longest to survive in the dust" (Byron, 1820).

Durability and stability are the key qualities that render mammalian hair as relevant and important biological structures in forensic investigations. The extreme hardiness of these filamentous biological structures is due to a protein group found in all mammalian hair, the keratins. Keratins contain a variety of amino acids, of which the sulphur-rich cysteine (Cys) amino acid is present in far greater abundance than the other amino acids. Cysteine is critical to the keratinization (hardening) of the hair shaft, which occurs in the dermis, prior to its emergence through the skin (Robbins, 2012). The ease with which hair is readily shed and transferred during physical contact renders hair as one of the most prevalent types of biological trace evidence[1] encountered in forensic investigations.

---

[1] Materials that are transferred during the commission of a crime.

Forensic Microbiology, First Edition. Edited by David O. Carter, Jeffery K. Tomberlin, M. Eric Benbow and Jessica L. Metcalf.

15.1.2 Historical and current forensic perspectives of hair examination and analysis

The invention of the transmitted light microscope in the seventeenth century (Gest, 2004) played a pivotal role in early forensic investigations in which mammalian hair was central to the investigation. The earliest recorded case in which the microscopy of hairs associated the offender to the crime involved the untimely death of an elderly woman who died as a result of severe and extensive head and facial trauma. The inspection of the alleged murder weapon, a bloodstained hammer, revealed the presence of "short, stiff, white hairs corresponding to the victim's eyebrows" embedded in the blood. This observation was pivotal in refuting the suspect's assertion that the fatal injuries were caused by a kick to the head by a horse (*Regina v Teague*, 1851) (Gerber and Saferstein, 1997). The value of hair in forensic investigations gained traction in the 1890s, in which fledging medico-legal texts emphasized that "any hairs found on the hands or about the body should be preserved, mounted, and given careful microscopic (sic) consideration" (Gerber and Saferstein, 1997). Microscopical examination of mammalian hair features using transmitted visible light microscopy enables a systematic audit of genetically determined characteristics such as color, diameter, or profile (e.g., curly or straight) as well as those characters that are "acquired," such as pathologies, damage, and cosmetic treatments.

The introduction of the comparison microscope in the 1920s was seminal in consolidating the fledgling discipline of forensic microscopy. Prior to its invention, hair examiners relied on memory as hair comparisons were performed on hairs that were on separate microscopes. The comparison microscope allowed, for the first time, questioned and known hairs to be examined simultaneously. Comparative analyses of genetically determined characteristics (e.g., pigment type) and acquired characteristics (i.e., artificially colored hairs) enabled the hair examiner to determine whether an individual may or may not be excluded as the source of the questioned hair. The practice of comparing "known" and "questioned" hairs was the mainstay for forensic investigations for decades and still remains as a useful process today. The morphological examination of microscopic attributes of hairs, both genetic and acquired, can assist law enforcement officers by providing avenues to explore in the early stages of the investigation. Furthermore, these initial examinations provide a basis upon which hairs are selected for additional detailed analyses, such as DNA profiling.

The advent of forensic DNA profiling irrevocably changed the manner in which hairs were processed in forensic investigations. In 1996, postconviction DNA testing exonerated many defendants who were incarcerated solely based on comparative microscopy. The main criticism was leveled at examiners who overstated the significance of "matching" hair characteristics exhibited by the defendant's hair and those from the crime scene hair (Neufeld and Scheck, 1992). The National Academy of Sciences of the United States of America (Committee on Identifying the Needs of the Forensic Sciences Community, National Research Council, 2009) recommended that comparison microscopy of hairs should solely occur in order to determine whether known and questioned hairs "are sufficiently similar to merit comparisons with DNA analysis and for excluding suspects." It is of note that hair examiners (some three centuries ago) demonstrated an appreciation of the limitations in the weight of evidence that may be placed on the term "match" as derived from human hair microscopical comparisons. In 1894, it was first observed that "...this kind of evidence may be of importance in showing that the hair was similar to or different from the assailant or deceased ... identity based on similarity of hairs found on a weapon or on a person ... should be expressed with caution ... there are many persons who have hair similar in colour, size, and length; hence a witness can rarely be in position to swear that there is absolute identity" (Taylor, 1910). This cautious approach to the interpretation of comparative hair evidence appears to have become neglected over the centuries.

The inception of human DNA profiling using short tandem repeats (STRs) resulted in a paradigm shift in the way in which forensic hair samples were processed. Unlike qualitative comparison microscopy of hairs, DNA profiles are amenable to statistical evaluation, which enables a quantitative value to be placed on the significance of hair evidence.

The modern approach, therefore, with regard to hair examination is that microscopical examination is employed early in an investigation to:

1. Determine whether questioned and known hairs do not "match", that is, suspect excluded as the source of the questioned hair (in which case DNA examination is not warranted and police should seek a new suspect);
2. Determine whether the hairs "match," that is, suspect is not excluded as the source of the questioned hair (in which case DNA profiling is warranted);
3. Establish the growth phase of the hair (and therefore indicate exactly which type of DNA profiling is likely to be most successful);
4. Identify whether there are characteristics (e.g., associated materials, dyes, pathologies, parasites, damage, or evidence of forcible removal) that can provide circumstantial evidence or intelligence prior to destructive sampling for genetic analyses.

Human hairs contain variable amounts of nuclear (nu) DNA, and the ability to obtain a complete STR DNA profile from hairs is wholly dependent on the growth phase of the hair root. Mammalian hairs undergo a three-stage, asynchronous growth cycle, each of which may be characterized based on root morphology (Figure 15.1):

1. *Anagen*—actively growing hairs firmly embedded in the follicle and are difficult to dislodge without the application of force. Anagen hair roots and attached tissue are rich and abundant sources of nuDNA (1–750 ng DNA/root) (Butler, 2010) and mtDNA (on average 3,500,000 copies) (Andreasson *et al.*, 2006).
2. *Catagen*—these roots represent an ephemeral growth phase of the hair in which hair growth enters a quiescent phase (duration of a few weeks); during this time, the root gradually atrophies, which results in the concomitant loss of nuclear DNA.
3. *Telogen*—these roots represent the final growth stage in which the hair has ceased to grow; the root is bulbous and lacks a sheath. Telogen hair roots are a poor source of nuDNA (1–10 ng DNA/root) (Butler, 2010) but remain rich in mtDNA (20,000–45,700 copies) (Andreasson *et al.*, 2006).

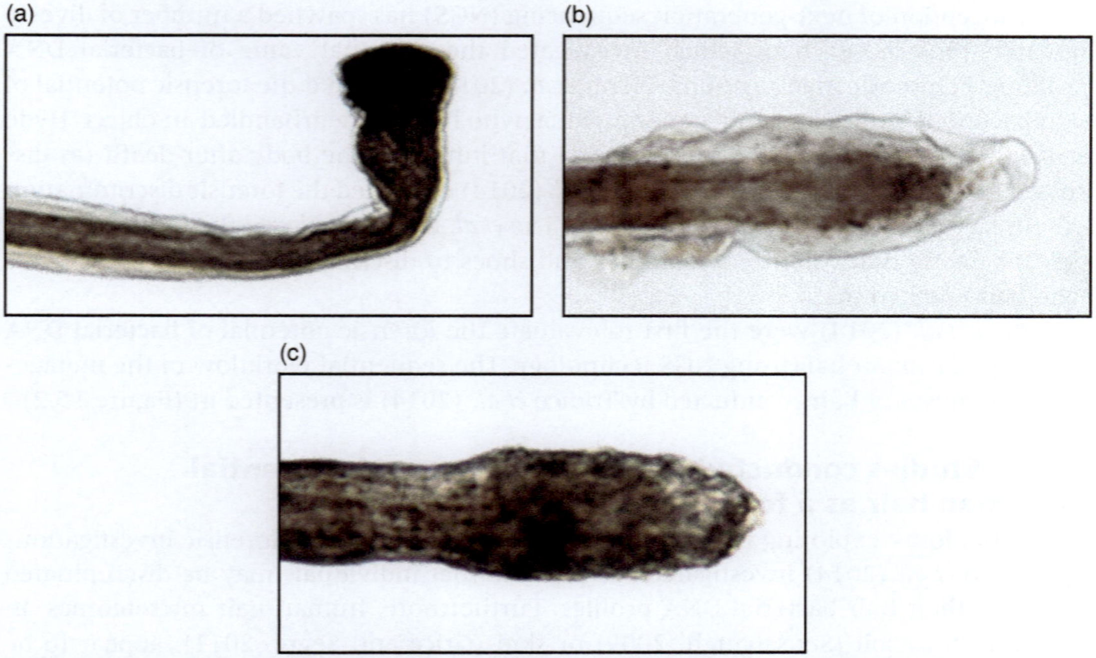

Figure 15.1 Morphology of the three growth stages of human hair roots: anagen (a: 100× magnification), catagen (b: 200×), and telogen (c: 200×)

Hairs identified as being in anagen or early catagen growth phases are routinely and successfully processed for nuDNA STR profiling in many forensic laboratories. However, in the absence of violence, telogen hairs (naturally shed) are the most frequently encountered in forensic investigations. In fact, more than 90% of the hairs retrieved from crime scenes are in the telogen growth phase (Nilsson *et al.*, 2012).

The provision of associative hair evidence through human DNA profiling of the shaft and/or root is a well-established and successful procedure. However, in instances in which human DNA extracted from hairs is too compromised in quality or quantity, examination of DNA profiles of the community of bacteria present on the hairs (i.e., hair microbiomes) may fill this void. Furthermore, bacterial profiles derived from hairs may be used as adjuncts to modest statistical results commonly derived from human mtDNA or Y-STR profiles.

15.2 Current research into hair microbiomes

The genetic analyses of microbial species that inhabit humans may evolve into an additional powerful tool with which to carry out source attribution or gather investigative information in relation to criminal acts.

The inception of next-generation sequencing (NGS) has spawned a number of diverse research projects, each of which investigated the potential value of bacterial DNA profiling in forensic investigations. Fierer *et al.* (2010) evaluated the forensic potential of skin bacterial profiles to identify an individual who has touched/handled an object. Hyde *et al.* (2013) attempted to identify bacteria that inhabited the body after death (as discussed in Chapters 8 and 9). Khodakova *et al.* (2014) evaluated the forensic discrimination of soils based on their bacterial inhabitants. Lax *et al.* (2015) explored the forensic potential of bacteria that colonize telephones and shoes to discriminate between individuals who had touched them.

Tridico *et al.* (2014) were the first to evaluate the forensic potential of bacterial DNA profiling of human hair using NGS technology. The sequential workflow of the metagenomic analyses of hairs conducted by Tridico *et al.* (2014) is presented in (Figure 15.2).

15.2.1 Studies conducted into the metagenomic potential of human hair as a forensic substrate

The initial foray exploring the potential value of hair bacteria in forensic investigations by Tridico *et al.* (2014) investigated the premise that individuals may be discriminated based on their hair bacterial DNA profiles. Furthermore, human hair microbiomes, in comparison to soil (Sensabaugh, 2009) or skin (Grice and Segre, 2011), appear to be more tractable (based on NGS data). Tridico *et al.* (2014) processed 42 bacterial DNA extracts obtained from human scalp and pubic hair (taken from four females and three males) and generated 39,814 Roche 454 reads (post control and abundance filtering).

The results suggested that pubic hair bacteria may play a more prominent role in forensic investigations than those derived from scalp hair. Furthermore, the data also revealed the presence of unique combinations of bacterial taxa that enable discrimination between individuals and signature bacterial taxa prevalent on female pubic hair. The presence of *Lactobacillus* spp. was the defining, niche-specific bacterial taxa that enabled differentiation not only between male and female pubic hair but also between female individuals. Of all the data generated in the study, NGS data generated from pubic hair held the most potential for forensic applications in revealing a general difference between bacterial taxa harbored on male and female pubic hair shafts, which was most probably due to niche-specific bacteria harbored on female pubic hairs. The prevalence of *Lactobacillus* spp. in the human vagina and vaginal secretions is well established (Wilson, 2005; Witkin *et al.*, 2007), and the presence of that species represents one such general difference between males and females. However, Tridico *et al.* observed a bacterial profile from pubic hair taken from a male participant with a confounding result; the pubic hair bacterial profile contained a community of lactobacilli species resembling those present from one of the female participants (at the same sample time interval), which indicates the opposite of the generality discussed earlier. Furthermore, profiling of the microbial taxa on hairs taken from each of these individuals at other times showed the usual, general differences (i.e., only hairs from the female showed the presence of lactobacilli species).

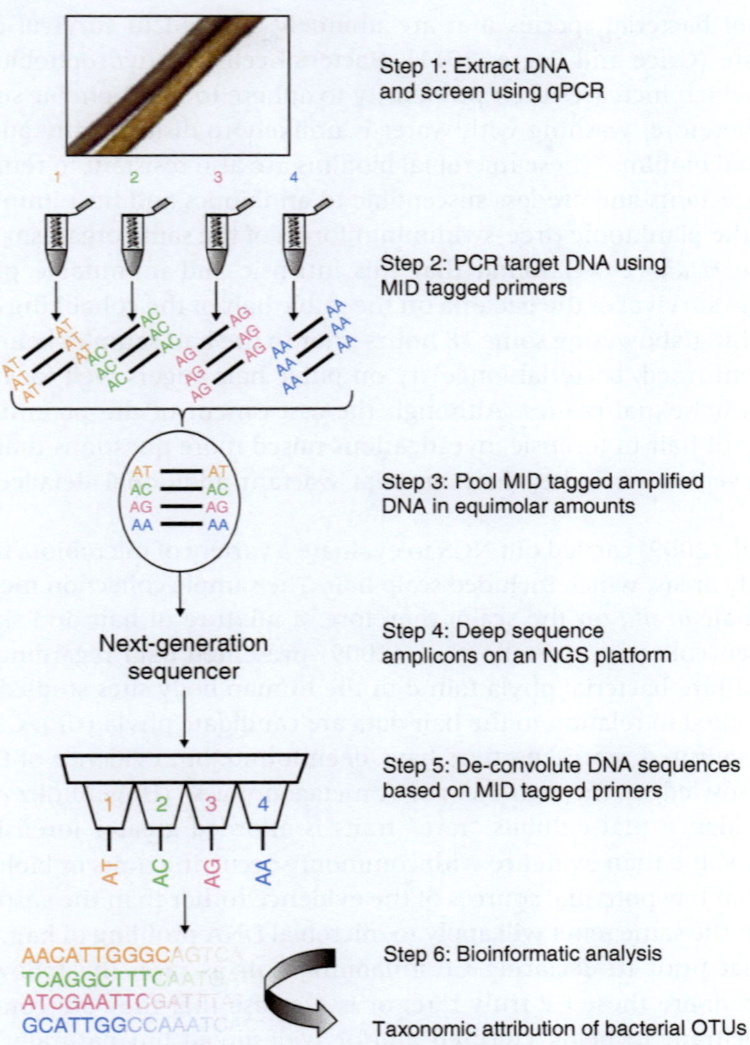

Figure 15.2 Example of an amplicon-based next-generation workflow diagram

Inquiries revealed that the two apparently anomalous hair samples originated from a cohabiting couple that had engaged in sexual intercourse prior to the collection of the hairs; no such activity had taken place prior to the collection of hairs from these two individuals on the two other occasions. Significantly, the mixed bacterial profile originated from pubic hairs that were sampled post showering and approximately 18 hours post coitus. The study of Tridico *et al.* (2014) is the first to suggest cross-transference and persistence of pubic/genital bacteria as a result of sexual intercourse.

Colonization of human body areas by bacteria is largely determined by the ecology of the skin surface. For example, the physical and chemical composition of the skin attracts colonization of bacterial species that are uniquely adapted to survival on a cool and acidic substrate (Grice and Segre, 2011). Bacterial cells are hydrophobic (Shirtliff and Leid, 2009), which increases their propensity to adhere to hydrophobic surfaces such as hair shafts; therefore, washing with water is unlikely to dislodge any adhering hydrophobic bacterial biofilms. These microbial biofilms are also resistant to removal by detergents or disinfectants and are less susceptible to antibiotics and host immune responses compared to the planktonic (free-swimming) forms of the same organism (Vickery *et al.*, 2013). Tridico *et al.* (2014) found that this intrinsic and immutable property might account for the survival of the bacteria on the pubic hair of the cohabiting couple despite postcoital bathing/showering some 18 hours prior to the hair samples being taken. If this property is confirmed, bacterial longevity on pubic hair augers well for future forensic investigations in sexual crimes. Although the assessment of the potential of bacterial DNA profiling of hair in forensic investigations raised more questions than it answered, the results revealed some key findings that warrant additional detailed and focused analyses.

Costello *et al.* (2009) carried out NGS to evaluate a variety of microbiota from a number of human body areas, which included scalp hair. The sample collection method involved swabbing of hair *in situ* on the scalp; therefore, a mixture of hair and skin microbiota may have been collected. Costello *et al.* (2009) presented data regarding "the relative abundances of rare bacterial phyla found in the human body sites studied." Each of the bacterial taxa cited in relation to the hair data are candidate phyla (CP); CP are "bacteria for which no cultured representatives have been found, but evidence of their existence has been acknowledged through 16S rRNA metagenomics" (Hugenholtz *et al.*, 1998).

Forensic evidence that exhibits "rare" traits is afforded greater forensic significance and probative value than evidence with commonly occurring traits or biological profiles, as it implies that few potential sources of the evidence (other than the suspect) exist. It is envisaged that the same tenet will apply to microbial DNA profiling of hair. It is therefore imperative that prior to declaring CP inhabiting hair as rare, the following question should be asked: are these CP truly rare, or is it a case that they are common bacteria that are recalcitrant to being cultured and/or widespread but naturally occur in low abundance? For example, the following CP isolated from scalp hair have been reported as "rare bacterial phyla" (Costello *et al.*, 2009) but in reality appear to be common as indicated in the following text:

- **TM7:** This is a "cosmopolitan and highly divergent division of bacteria" that appear to be a common, and perhaps permanent, member of the oral flora (Brinig *et al.*, 2003) and also in human stools and upper digestive tract (Dewhirst *et al.*, 2010; Segata *et al.*, 2012).
- **Gemmatimonadetes:** Environmental sequence data indicate that this phylum is "widespread in nature," which may also be implicated in clinical niches (Zhang *et al.*, 2003).
- **SR1:** Members of this phylum appears to be prevalent in saliva (Lazarevic *et al.*, 2009; Segata *et al.*, 2012).

- **Chloroflexi:** Members of this phylum include lineages that are adapted to a wide range of aquatic and terrestrial habitats (Bjornsson *et al.*, 2002) and also colonize human skin and oral microbiota (Campbell *et al.*, 2014).

Although these organisms are "new" to science in that they are recently detected, it does not necessarily follow that they are rare or narrowly distributed.

15.3 Importance of hair sample collection, storage, and isolation of microbial DNA

No matter how "state-of-the-art" or effective downstream bioinformatics analyses are in processing microbial data, they will not be able to compensate for the negative impacts of poorly collected or compromised crime scene or reference hair samples. Following the advent of the polymerase chain reaction (PCR), in forensic investigations, the forensic community developed stringent and robust protocols to ensure that the integrity of the biological evidence is not compromised during collection, processing, or storage.

15.3.1 Hair sample collection, storage, and analysis

The collection of forensic biological evidence, including hairs, often starts at the crime scene, which may be outdoors (where environmental conditions may affect biological evidence) or in a relatively protected environment such as a building.

Logically, the microbial community present on crime items arises from three sources: colonization of the items prior to the crime; transferal of bacteria between the assailant and victim during the crime; and transferal of bacteria to the items during crime scene examination or in the laboratory. Clearly, the former two are out of the control of the crime scene examiner and laboratory personnel. The latter however amounts to contamination or cross-contamination, which clearly are under the control of the examiner or scientist. The most common routes through which forensic biological evidence may become compromised are presented in Figure 15.3. If the integrity of the evidence becomes compromised at the crime scene or in the laboratory, or at any time in between, it will taint all subsequent results that the use of state-of-the-art downstream processes and equipment cannot compensate for (Murray *et al.*, 2015). The requirements for microbial DNA profiling are therefore not less stringent than those currently prevailing for profiling of human DNA. However, the current standards for robust crime scene examination practices, item packaging and storage procedures, and laboratory protocols and controls can be expected to be appropriate for metagenomic analyses.

In addition to the collection of crime scene samples, obtaining appropriate control (reference) hair samples from the victim, suspect, and other relevant individuals is important. Results from comparative microscopy of morphological structures and characteristics are only valid if it is conducted between hairs that originate from the same

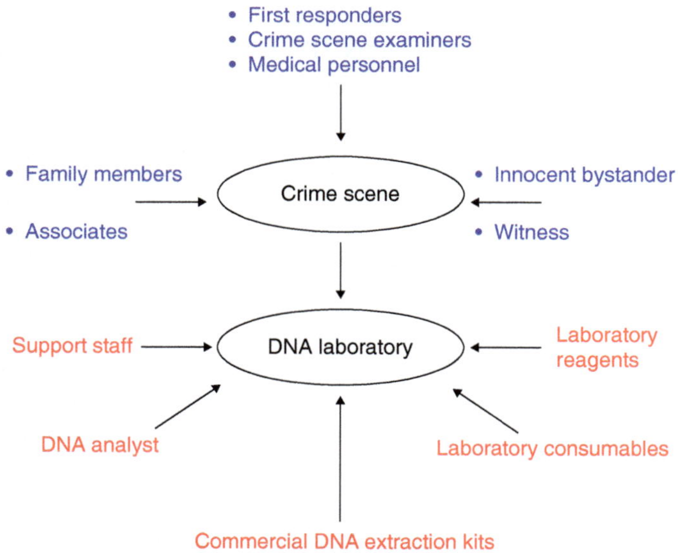

Figure 15.3 Avenues in which contamination of hair evidence may occur prior to sample collection (blue) and post sample collection (red)

somatic (or body) origin; morphological data derived from a control scalp hair sample cannot be used for comparative purposes with morphological data derived from crime scene pubic hairs or body/limb hairs. The findings of Tridico *et al.* (2014) indicate this tenet is also applicable to metagenomic analyses of hairs; for example, the comparison of pubic hair bacterial profile from the victim with the bacterial profile derived from the scalp hair of the suspect will almost certainly result in the false exclusion (type II error) of the suspect as being the offender. However, comparison of the suspect's pubic hair microbial profile with the victim's pubic hair bacterial profile will result in a more accurate and valid basis to determine the exclusion or nonexclusion status of the suspect as the offender.

In contrast to scalp hairs, pubic hairs are less frequently encountered in forensic investigations; Mann (1990) demonstrated that the transference of pubic hairs seldom occurred during heterosexual sexual assaults. Male pubic hair was transferred to the victim in 4% of the cases, and the study found no hairs were transferred from the victim to the male's pubic area. The increased use of prophylactics during the commission of unlawful sexual intercourse (Bradshaw *et al.*, 2013; Vickery *et al.*, 2013, p. 6328) markedly decreases the opportunity to transfer seminal fluid. In instances where there is an absence of pubic hair transfer, it may still be possible to obtain microbial profiles from victim and offender. Tridico *et al.* (2014) show that the victim's and perpetrator's bodies may be regarded as the "crime scenes" in that their pubic hairs may harbor valuable evidence of intimate contact. Meadow *et al.* (2013) found that skin bacteria were transferred during participation in the contact sport of "roller derby." Based on these findings, it is reasonable to propose that sampling the pubic areas of victim and offender be carried out so that bacterial DNA profiles can be generated. This would involve cutting a sample of hairs (preferably) or swabbing the pubic hairs or pubic area. Forensic protocols in relation to the collection of hairs or swabs in forensic investigations are globally well established, differing only in fine detail, an example of which may be found in the 'Best Practice Manual for the microscopic examination and comparison of Human and Animal Hairs' enfsi.org (2015).

All evidence, irrespective of source, requires either short- or long-term storage post collection. Storage conditions are a critical component of human DNA-based analyses; wet or damp hairs stored in sealed plastic bags encourage bacterial growth. Current biological evidence collection protocols mandate that dry biological evidence, such as hair, is packaged in paper or self-seal plastic bags or envelopes to ensure that the integrity of the evidence is maintained. Although studies have yet to be conducted to assess the effects storage conditions may exert on bacteria colonizing hair, Lauber *et al.* (2010) studied the effects of a variety of storage conditions on bacteria harvested from human skin, human gut (feces), and soil. The results suggested that microbial samples collected under "field conditions," which included storage for up to 2 weeks at room temperature, remained relatively unaffected. Bai *et al.* (2012) found that neither storage time nor storage temperature led to "significant" differences between vaginal microbiomes collected on vaginal swabs. The preliminary findings of Lauber *et al.* (2010) and Bai *et al.* (2012) are encouraging regarding storage of crime scene and control hair samples that may be destined for metagenomic analyses.

Providing an exhaustive description of procedural minutiae regarding the collection and storage of hairs, beyond the aforementioned general discussion, is outside the scope of the chapter. The reader interested in procedural detail is referred to the 'Best Practice Manual for the microscopic examination and comparison of Human and Animal Hairs' enfsi.org (2015), which are representative of forensic protocols used in many jurisdictions regarding the collection and storage of human hairs.

15.4 DNA sequencing of hair microbiomes

The study of bacterial populations generally falls into two categories: metagenomic analyses that aim to identify "what (microorganism) is there"; and metatranscriptomics that aims to identify not only "what is there" but also "what are they doing." The forensic analyses of hair microbiomes will, most likely, focus on determining "what is there" rather than "what are they doing." Bacterial profiles originating from hair, like human DNA profiles, will form the basis for comparative analyses of bacterial profiles of victim, suspect, or crime scene in order to achieve source attribution. Amplicon sequencing of hair has the potential to identify bacterial taxa through conducting a comprehensive genetic audit that can be expressed in taxonomic affiliation, known as operational taxonomic units (OTUs). The inception of NGS coupled with metabarcoding offers a forensic potential to profile bacterial DNA composition in any given substrate in a way not previously possible (see Chapter 5).

Research conducted by Mason *et al.* (2013) on bacteria that colonize human dental plaque and saliva microbiome demonstrated a "significant ethnic affiliation and composition of the oral microbiome; to the extent that these microbial signatures appear to be capable of discriminating between ethnicities." Furthermore, these authors suggest that the selection of this ethnicity is a result of genetic, not environmental, influences.

Although the relevance and applicability of oral microbiota to human hair microbiota may not be apparent *prima facie*, Tridico *et al.* (2014) isolated bacterial taxa that colonize oral and naso-pharynx areas of the human body in samples taken from scalp and pubic hair. Taken together, the findings of Mason *et al.* (2013) and Tridico *et al.* (2014) raise the possibility that adventitiously deposited oral bacteria on human hair may allow ethnicity deductions as to the donor of the bacteria to be made.

The presence and prevalence of *Lactobacillus* spp. as diverse and ubiquitous microbes of the female urogenital tract is well documented (Wilson, 2005). Lactobacilli are critical vaginal commensals that play a central and vital role in maintaining a healthy vaginal microbiome (Redondo-Lopez *et al.*, 1990). A number of research endeavors focus on vaginal lactobacilli as potential forensic markers in the identification of vaginal fluids. Fleming and Harbison (2010) suggest that the presence of *L. crispatus* and *L. gasseri* may be considered as potential genetic markers for the identification of vaginal secretions in forensic investigations. However, although Akutsu (Akutsu *et al.*, 2012) agree with Fleming and Harbison (2010) in relation to the potential of *L. crispatus*, they disagree with the suitability of other bacterial species. Benschop *et al.* (2012) found these bacteria on the hand, groin, and penis, which casts doubt on the reliability of *L. crispatus* and *L. gasseri* species as potential markers for the identification of vaginal fluids in forensic investigations.

The initial evaluation of the potential metagenomic analyses of human bacteria in forensic investigations is encouraging. However, further refinements to the selection of primers may be necessary in order to maximize microbial data that may be extracted from hair samples. Likewise, workflows for data generation and the methods of analysis postsequencing microbial data will vary between labs and will all influence the efficacy of the approach (Murray *et al.*, 2015).

15.4.1 Bioinformatics considerations for analyzing microbial hair data

The analysis of the massive amount of data afforded by NGS is arguably one of the greatest challenges in microbial forensic investigations. The obstacles, however, do not arise from the sheer amount of genetic output but surface in the manner in which the preprocessing of sequence data should occur prior to downstream analyses (see Chapters 4 and 5). Despite these challenges, bacterial profiling of hairs could evolve into a valuable adjunct to other forensic analyses of hair, for example mtDNA, or provide valuable investigative data in the initial stages of an investigation.

15.5 Conclusions and future directions

Despite the plethora of 16S DNA metagenomic audits of microbiomes that have been conducted since the inception of NGS technology and platforms, the potential application of metagenomic analysis of human hair in forensic investigations has barely begun. Human DNA profiling has played a central and vital role in forensic science over the past 20 years and will continue to do so. However, in regard to hairs that are unsuitable for nuDNA analyses, microbial DNA profiling may be of value as an adjunct to results garnered through less discriminatory molecular typing, for example, mtDNA. Whereas nuDNA is highly discriminatory for the identification of suspect or victim during criminal proceedings or investigations, identity might be of lesser importance than the circumstances surrounding the crime, which might be revealed by metagenomics analyses. Furthermore, metagenomic analyses of hairs may be able to provide investigative leads in the critical early stages of an investigation.

The introduction of microbial DNA profiling, as a novel technique in the forensic molecular toolkit, will require a measured and considered approach. Following the initial foray exploring the potential metagenomic approaches to profiling hair bacteria, there are many criteria that need to be satisfied prior to this technique being adopted by the forensic community and the courts. As we discuss in further detail later, we advocate that a good starting point will be to sample larger cohorts of individuals, evaluate the efficacy of suites of primer pairs to maximize the microbial data recovered from hair, and develop robust processing methods for the data.

15.5.1 Major challenges and future directions of metagenomic analyses of hairs in forensic science

Based on the preliminary work conducted to date on human hairs, the adoption of current forensic techniques and protocols of existing "front end" procedures pertaining to hair evidence, such as collection of hairs and DNA extraction, may remain applicable for metagenomic analysis of hair. However, the evaluation of transportation and storage effects upon bacterial communities adhering to hair shafts post collection or sampling requires more detailed and comprehensive assessment than currently exists.

The future of microbial DNA profiling of hair is likely to involve inter- or multidisciplinary approaches to the interpretation of the metagenomic data and presentation of evidence in court. The potential collaborators will most likely include the forensic practitioner, microbiologists, and bioinformaticians. The most challenging aspect of this alliance may be in the coordination of the presentation of evidence to the court in the most comprehensible, coherent, and logical manner.

The evolution of metagenomic assessments of scalp and pubic hairs as a novel forensic technique will, to some extent, benefit from "standing on the shoulders of giants" in as much as 20 or so years of human DNA profiling. Associated legal challenges to methodologies, laboratory contamination issues, statistical analyses, and value of reference databases together with forensic laboratory accreditation guidelines have resulted in a robust and mature molecular discipline that can serve a model. Further, metagenomic analysis of hairs will benefit from the legacy of 20 years of development of crime scene examination, contamination mitigation, and sample collection from victim and suspect in a "DNA era."

15.5.2 Future metagenomic assessments of hair samples

The study conducted by Tridico *et al.* (2014) might be regarded as the preface to the potential of metagenomic analyses of hair in forensic investigations; however, confirmation of these initial results will require interrogation of hair bacteria from markedly larger groups of individuals. The potential cross-transference of bacteria during sexual intercourse may be the most appropriate initial focus of future metagenomic evaluations. The rationale behind this recommendation is based on the prevalence of unlawful heterosexual intercourse (rape) cases and limited transference of human biological trace evidence (hair and semen) post assault; furthermore, in contrast to scalp hair, pubic hair microbiomes appear to reflect "indigenous" bacterial species, rather than those derived from the environment.

15.5.3 Development of more focused approaches to detect bacterial population level differences between bacteria inhabiting human hairs

The modest research conducted to date (Tridico *et al.*, 2014) in relation to human scalp and pubic microbiomes shows the presence of a substantial number of common microbes that inhabit these two body niches. Future metagenomic evaluations of scalp and pubic hair, as a novel forensic technique, should focus not only on larger study groups but also on moving beyond the exclusive use of universal primers to conduct bacterial audits of hair.

Universal primers have been shown to lack sequence homology to many of the CP (Winsley *et al.*, 2012); therefore, a more nuanced approach may not only reveal optimal primer pairs that target specific species (e.g., *Lactobacillus* spp.) but also reveal optimal amplicon lengths to encourage a more accurate assessment of scalp and pubic hair microbiomes. Future evaluations may reveal different optimal primer pairs that will enable a more comprehensive and targeted microbial audit to be conducted on scalp and pubic hairs; this approach could ameliorate the bias of universal bacterial primers in evaluating microbial communities (Kawamura and Kamiya, 2012).

There is no doubt that amplicon sequencing technology has the capability to provide powerful insights into the bacterial composition of species-rich environments. The coverage and judicious selection of primers for a more refined approach to audit the bacterial 16S rRNA gene will almost certainly be the linchpin in obtaining more accurate audits of bacterial taxa that colonize human hair and provide the basis for source attribution. As genome sequencing technologies become more rapid, robust, and cost effective, it is likely to result in increased interest in its application to forensic investigations involving the metagenomic evaluation of human hair. One potential challenge will be in relation to improving the probative value associated with the detection and identification of common human commensals. Obvious future work is to develop methodology based upon detection of bacterial SNPs. These are random nucleotide variations at single, particular points within the DNA molecule. The power of bacterial DNA profiling, as described earlier, relies upon the detection and identification of rare species, while the presence of common species is less important. The identification of SNPs confers greater power to microbial profiling because variants can be detected not only within the rare species present but also in the more common species (Holt *et al.*, 2009). For example, the detection of a particular Lactobacillus species (e.g., *L. crispatus*, a common inhabitant of the human vagina) on a suspect represents "weaker" evidence of contact than detecting *L. crispatus* exhibiting a set of SNPs that "match" the SNP set from the victim.

The forensic potential of microorganisms as novel molecular markers is gaining traction within the forensic community. Preliminary results into the applicability of human hair bacterial profiling in forensic investigations are sufficiently provocative as to warrant further, more detailed analyses. If this potential is realized, there will be a need to standardize the manner in which this novel forensic evidence is evaluated (bioinformatics) and the results reported.

15.5.4 General requirements for quality management

The introduction of any novel forensic technique requires scrutiny in order to ensure the production of robust and reliable results and that analyses and processes are fit for purpose.

15.5.4.1 Method validation and evidence evaluation

The International Organization for Standardization's (ISO) general requirements for the competence of testing and calibration laboratories standard (ISO/IEC 17025) is becoming the norm for forensic laboratories seeking accreditation of their quality management system (ISO Organisation, 2015). An integral part of this standard is that regular audits of technical competence of staff and suitability of procedures and processes are required in order to retain accreditation. The uptake of novel techniques in forensic investigations, such as metagenomic analyses of hair, will be contingent upon successful compliance with the ISO/IEC 17025 standard. The most relevant technical requirements to be fulfilled are quality control (the development of appropriate positive and negative controls) and validation (demonstrating the accuracy and reliability of the methodology and sources of error or limitations). Layne Desportes (2009) defines validations as a process that "is a documented program that provides a high degree of assurance that a specific technique will consistently produce a result within the defined specifications and quality parameters."

In relation to conducting bacterial DNA profiling of hair samples, the process will most likely involve validating the efficacy of a suite of primer pairs that are used simultaneously during amplification in order to promote a balanced audit of bacterial taxa. Validation exercises are also a critical step in regard to facilitating the acceptance of metagenomics evidence in courts in those jurisdictions where Daubert or other reliability standards for the admissibility of expert evidence are in place. After all, a novel forensic procedure that may ultimately result in the incarceration of an individual demands a level of scrutiny that requires its strengths, weaknesses, and benefits to be rigorously evaluated prior to use in forensic investigations.

An additional avenue to explore in relation to microbial hair evidence may involve statistical processes that provide a quantitative or probabilistic weighting to the evidentiary significance of microbial profiles from crime samples. Statistical analyses most likely will be an absolute requirement. The forensic community and the courts are accustomed to statistical weighting being placed on human DNA profiles and an equivalent approach in regard to microbial profiles will be expected. As an example, the European Network of Forensic Science Institutes in their "Guideline for Evaluative Reporting in Forensic Science" (ENFSI, 2015, p. 7626) recommends that forensic scientists make use of an evaluative framework for their evidence that produces a likelihood ratio for expression of the strength of their evidence. Microbial profiles may be amenable to hypothesis-based testing in which the relative probabilities of a microbial profile originating from individual A or a person at random can be weighed.

The work conducted to date regarding the potential role of metagenomic analyses of human hair in forensic science is sufficiently encouraging for the forensic community to ponder—"Is human hair DNA profiling enough?"

Acknowledgments

We thank members of the TrEnD laboratory for helpful discussions. MB acknowledges the support of Australian Research Council Future Fellowship (FT0991741).

References

Akutsu, T., Motani, H., Watanabe, K. *et al.* (2012) Detection of bacterial 16S ribosomal RNA genes for forensic identification of vaginal fluid. *Legal medicine (Tokyo, Japan)*, **14**, 160–162.

Andreasson, H., Nilsson, M., Budowle, B. *et al.* (2006) Nuclear and mitochondrial DNA quantification of various forensic materials. *Forensic Science International*, **164**, 56–64.

Bai, G., Gajer, P., Nandy, M. *et al.* (2012) Comparison of storage conditions for human vaginal microbiome studies. *PLoS ONE*, **7**, e36934.

Benschop, C.C., Quaak, F.C., Boon, M.E. *et al.* (2012) Vaginal microbial flora analysis by next generation sequencing and microarrays; can microbes indicate vaginal origin in a forensic context? *International Journal of Legal Medicine*, **126**, 303–310.

Bjornsson, L., Hugenholtz, P., Tyson, G.W. and Blackall, L.L. (2002) Filamentous Chloroflexi (green non-sulfur bacteria) are abundant in wastewater treatment processes with biological nutrient removal. *Microbiology*, **148**, 2309–2318.

Bradshaw, R., Wolstenholme, R., Ferguson, L.S. *et al.* (2013) Spectroscopic imaging based approach for condom identification in condom contaminated fingermarks. *Analyst*, **138**, 2546–2557.

Brinig, M.M., Lepp, P.W., Ouverney, C.C. *et al.* (2003) Prevalence of bacteria of division TM7 in human subgingival plaque and their association with disease. *Applied and Environmental Microbiology*, **69**, 1687–1694.

Butler, J.M. (2010) *Fundamentals of Forensic DNA Typing*, Academic Press, Amsterdam/Boston.

Byron, G.G. (1820) in *Byron's Letters and Journal VII: 'Between two worlds'* (ed L. Marchand), Harvard University Press, Cambridge, MA.

Campbell, A.G., Schwientek, P., Vishnivetskaya, T. *et al.* (2014) Diversity and genomic insights into the uncultured Chloroflexi from the human microbiota. *Environmental Microbiology*, **16**, 2635–2643.

Clack, A.A., Macphee, R.D.E. and Poinar, H.N. (2012) Mylodon darwinii DNA sequences from ancient fecal hair shafts. *Annals of Anatomy—Anatomischer Anzeiger*, **194**, 26–30.

Committee on Identifying the Needs of the Forensic Sciences Community, National Research Council (2009) Front matter, in *Strengthening Forensic Science in the United States: A Path Forward*, The National Academies Press, Washington, DC.

Costello, E.K., Lauber, C.L., Hamady, M. *et al.* (2009) Bacterial community variation in human body habitats across space and time. *Science*, **326**, 1694–1697.

Dewhirst, F.E., Chen, T., Izard, J. *et al.* (2010) The human oral microbiome. *Journal of Bacteriology*, **192**, 5002–5017.

ENFSI, Best Practice Manual for the Microscopic Examination and Comparison of Human and Animal Hairs' enfsi.org (2015)

ENFSI, Strengthening the Evaluation of Forensic Results across Europe (2015) ENFSI Guideline for Evaluative Reporting in Forensic Science, http://www.enfsi.eu/news/enfsi-guideline-evaluative-reporting-forensic-science (accessed December 2, 2016).

Fierer, N., Lauber, C.L., Zhou, N. *et al.* (2010) Forensic identification using skin bacterial communities. *Proceedings of the National Academy of Sciences of the United States of America*, **107**, 6477–6481.

Fleming, R.I. and Harbison, S. (2010) The use of bacteria for the identification of vaginal secretions. *Forensic Science International. Genetics*, **4**, 311–315.

Frei, K., Mannering, U., Kristiansen, K. *et al.* (2015) Tracing the dynamic life story of a Bronze Age Female. *Scientific Reports*, **5**, 10431.

Gerber, S.M. and Saferstein, R. (1997) *More Chemistry and Crime: From Marsh Arsenic Test to DNA Profile*, American Chemical Society, Washington, DC.

Gest, H. (2004) The discovery of microorganisms by Robert Hooke and Antoni Van Leeuwenhoek, fellows of the Royal Society. *Notes and Records of the Royal Society of London*, **58**, 187–201.

Glaister, J. (1931) *A Study of Hairs and Wools Belonging to the Mammalian Group of Animals, Including a Special Study of Human Hair, Considered from the Medico-Legal Aspect*, Misr Press, Cairo.

Grice, E.A. and Segre, J.A. (2011) The skin microbiome. *Nature Reviews. Microbiology*, **9**, 244–253.

Holt, K.E., Teo, Y.Y., Li, H. *et al.* (2009) Detecting SNPs and estimating allele frequencies in clonal bacterial populations by sequencing pooled DNA. *Bioinformatics*, **25**, 2074–2075.

Hugenholtz, P., Goebel, B.M. and Pace, N.R. (1998) Impact of culture-independent studies on the emerging phylogenetic view of bacterial diversity. *Journal of Bacteriology*, **180**, 4765–4774.

Hyde, E.R., Haarmann, D.P., Lynne, A.M. *et al.* (2013) The living dead: bacterial community structure of a cadaver at the onset and end of the bloat stage of decomposition. *PLoS One*, **8**, e77733.

International Standards Organisation (2015) Forensic Science ISO/IEC 17025 Application Document. In: (Australia), NATA (ed.).

Kawamura, Y. and Kamiya, Y. (2012) Metagenomic analysis permitting identification of the minority bacterial populations in the oral microbiota. *Journal of Oral Biosciences*, **54**, 132–137.

Khodakova, A.S., Smith, R.J., Burgoyne, L. *et al.* (2014) Random whole metagenomic sequencing for forensic discrimination of soils. *PLoS ONE*, **9**, e104996.

Lauber, C.L., Zhou, N., Gordon, J.I. *et al.* (2010) Effect of storage conditions on the assessment of bacterial community structure in soil and human-associated samples. *FEMS Microbiology Letters*, **307**, 6.

Lax, S., Hampton-Marcell, J.T., Gibbons, S.M. *et al.* (2015) Forensic analysis of the microbiome of phones and shoes. *Microbiome*, **3**, 21.

Layne Desportes, B. (2009) *Validation of Forensic Science Techniques: Principles and Procedures. Wiley Encyclopedia of Forensic Science*, John Wiley & Sons, Ltd, Chichester.

Lazarevic, V., Whiteson, K., Huse, S. *et al.* (2009) Metagenomic study of the oral microbiota by Illumina high-throughput sequencing. *Journal of Microbiological Methods*, **79**, 266–271.

Mann, M.J. (1990) Hair transfers in sexual assault: a six-year case study. *Journal of Forensic Sciences*, **35**, 951–955.

Mason, M.R., Nagaraja, H.N., Camerlengo, T. *et al.* (2013) Deep sequencing identifies ethnicity-specific bacterial signatures in the oral microbiome. *PLoS ONE*, **8**, e77287.

Meadow, J.F., Bateman, A.C., Herkert, K.M. *et al.* (2013) Significant changes in the skin microbiome mediated by the sport of roller derby. *PeerJ*, **1**, e53.

Murray, D.C., Coghlan, M.L. and Bunce, M. (2015) From benchtop to desktop: important considerations when designing amplicon sequencing workflows. *PLoS One*, **10**, e0124671.

Neufeld, P.J. and Scheck, B.C. (1992) *Wrongful convictions involving unvalidated or improper forensic science that were later overturned through DNA testing* [Online], www.innocenceproject.org (accessed December 2, 2016).

Nilsson, M., Norlin, S. and Allen, M. (2012) Sequencing of mtDNA in shed hairs: a retrospective analysis of material from forensic cases and a pre-screening method. *The Open Forensic Science Journal*, **5**, 13–22.

Rasmussen, M., Li, Y., Lindgreen, S. *et al.* (2010) Ancient human genome sequence of an extinct Palaeo-Eskimo. *Nature*, **463**, 757–762.

Redondo-Lopez, V., Cook, R.L. and Sobel, J.D. (1990) Emerging role of lactobacilli in the control and maintenance of the vaginal bacterial microflora. *Reviews of Infectious Diseases*, **12**, 856–872.

Robbins, C.R. (2012) *Chemical and Physical Behaviour of Human Hairs*, Springer-Verlag, Heidelberg.

Segata, N., Haake, S.K., Mannon, P. *et al.* (2012) Composition of the adult digestive tract bacterial microbiome based on seven mouth surfaces, tonsils, throat and stool samples. *Genome Biology*, **13**, R42.

Sensabaugh, G.F. (2009) Microbial community profiling for the characterisation of soil evidence: forensic considerations, in *Criminal and Environmental Soil Forensics* (eds K. Ritz, L. Dawson and D. Miller), Springer, Dordrecht/London.

Shirtliff, M. and Leid, J.G. (2009) *The Role of Biofilms in Device-Related Infections*, Springer, Berlin.

Taylor, A.S. (1910) *The Principles and Practice of Medical Jurisprudence*, Churchill, London.

Tridico, S.R., Murray, D.C., Addison, J. *et al.* (2014) Metagenomic analyses of bacteria on human hairs: a qualitative assessment for applications in forensic science. *Investigative Genetics*, **5**, 1–13.

Vickery, K., Hu, H., Jacombs, A.S. *et al.* (2013) A review of bacterial biofilms and their role in device-associated infection. *Healthcare Infection*, **18**, 61–66.

Wilson, M. (2005) *Microbial Inhabitants of Humans*, Cambridge University Press, Cambridge.

Winsley, T., Van Dorst, J.M., Brown, M.V. and Ferrari, B.C. (2012) Capturing greater 16S rRNA gene sequence diversity within the domain Bacteria. *Applied and Environmental Microbiology*, **78**, 5938–5941.

Witkin, S.S., Linhares, I.M. and Giraldo, P. (2007) Bacterial flora of the female genital tract: function and immune regulation. *Best Practice & Research. Clinical Obstetrics & Gynaecology*, **21**, 347–354.

Zhang, H., Sekiguchi, Y., Hanada, S. *et al.* (2003) *Gemmatimonas aurantiaca* gen. nov., sp. nov., a gram-negative, aerobic, polyphosphate-accumulating micro-organism, the first cultured representative of the new bacterial phylum Gemmatimonadetes phyl. nov. *International Journal of Systematic and Evolutionary Microbiology*, **53**, 1155–1163.

Perspectives on the future of forensic microbiology

David O. Carter[1], Jeffery K. Tomberlin[2], M. Eric Benbow[3] and Jessica L. Metcalf[4]

[1] *Laboratory of Forensic Taphonomy, Forensic Sciences Unit, Division of Natural Sciences and Mathematics, Chaminade University of Honolulu, Honolulu, HI, USA*

[2] *Department of Entomology, Texas A&M University, College Station, TX, USA*

[3] *Departments of Entomology and Osteopathic Medical Specialties, Michigan State University, East Lansing, MI, USA*

[4] *Department of Animal Sciences, Colorado State University, Fort Collins, CO, USA*

Microorganisms have the potential to be relevant in every forensic investigation, simply because they are ubiquitous. It is crucial for us to develop reliable methods to analyze microorganisms and their ecology because they can provide both spatial and temporal evidence relevant to a broad range of investigative contexts. Recent advances in microbiology have significant implications for criminal investigation and medicolegal death investigation because they represent new forms of evidence or provide new insight into the interpretation of well-established forms of evidence. Also, many of the applications discussed in this book are in the early stages of development, which provides scientists with an excellent opportunity to establish quality assurance protocols that are sufficient for the legal system.

The development of reliable protocols to use microbial evidence needs to be an emphasis over the next decade to ensure that microbial evidence is of the highest quality and results are repeatable across laboratories. The importance of experiment design, data analysis, and sample preservation is discussed in Chapters 3–6. Forensic microbiology finds itself in a relatively unique position where many best practices have yet to be established because the capabilities are so new. As basic research establishes the use and limitations of these methods for forensic science, we will work toward standards and best practices. Identifying robust protocols for sample, collection, analysis, and preservation is critical for the use of microorganisms as physical evidence, particularly in this era where the development of standards and best practices has been designated high priority by the Organization of Scientific Area Committees

Forensic Microbiology, First Edition. Edited by David O. Carter, Jeffery K. Tomberlin, M. Eric Benbow and Jessica L. Metcalf.

(OSAC)—a joint initiative between the US Department of Commerce and Department of Justice (http://www.nist.gov/forensics/osac/).

A significant portion of the book, Chapters 7–12, considers the role of microorganisms in medicolegal death investigation. Important topics include the role of microbiology as an adjunct to autopsy and the microbial communities associated with decomposing remains. Understanding the influence of microorganisms on results from autopsy, toxicology, and histology is absolutely critical because microbes can significantly affect the results of these analyses (Chapters 7–9). Postmortem microbiology has been an adjunct to autopsy for over a century, but how the availability of genomic and metabolomic techniques can enhance forensic pathology remains to be seen. Similarly, the investigation of postmortem bacterial translocation using these modern techniques is of great importance. Postmortem bacterial translocation has confounded forensic pathology for over a century, so it is important that we understand the process and, ideally, use it to our advantage (i.e., develop into a form of evidence in its own right).

Chapters 10–15 cover topics that represent two of the most rapidly developing areas of forensic microbiology: decomposition ecology and trace evidence. The current understanding of postmortem microbial communities is a burgeoning field. Recent and current research focuses on surveying the structure and function of microbial communities in a range of habitats and environmental conditions. This approach is probably similar to the early days of forensic entomology and anthropology. Thus, we recommend that forensic microbiologists collaborate regularly with forensic entomologists and anthropologists as these multitrophic interactions serve as part of the "equations" resulting in forensically relevant information. Surely, as knowledge continues to grow within the field of forensic microbiology, the foundational sciences with which it is bridged will continue to diversify as well. At present, postmortem microbial communities have great potential to complement other estimates of postmortem interval, and it will be exciting to see this application develop as these datasets become more robust.

The use of microorganisms as trace evidence (Chapters 13–15) is not a new idea; even during Edmond Locard's time, microbes were recognized as spatial evidence. However, modern techniques of microbial community analysis have provided the potential to use skin microbial communities as a means to associate individuals with objects and locations. Great potential also exists to use these microbes as temporal evidence so that an estimate of when a person was associated with an object or location can be established. This would be a significant development for forensic sciences because ubiquitous forms of physical evidence that provide spatial and temporal insights are very rare.

As shown in this book, forensic microbiology has undergone significant advances in the past 5 years, and this pace of development is expected for the foreseeable future. Forensic microbiology is truly a vibrant field of scientific endeavor, and it will be exciting to watch its development over the next few decades. Much to our delight the field is diversifying and evolving quickly; however, this is a double-edged sword as new ideas and applications most likely have developed since the publication of this text. With that said, we believe this publication will serve as a foundational, and hopefully historical, source of information for students, instructors, and practitioners alike—hopefully you feel the same.

Index

Forensic Microbiology, First Edition. Edited by David O. Carter, Jeffery K. Tomberlin, M. Eric Benbow and Jessica L. Metcalf.

www.ingramcontent.com/pod-product-compliance
Ingram Content Group UK Ltd.
Pitfield, Milton Keynes, MK11 3LW, UK
UKHW051039091026
15624UKWH00011B/574